IBM Mainframe Developer

Training and Reference Guide

JCL, MVS Utilities, COBOL, VSAM, IMS, DB2, CICS

Robert Wingate

ISBN 13:   9781734584738

Disclaimer

The contents of this book are based upon the author's understanding of and experience with the following IBM products: JCL, MVS Utilities, COBOL, VSAM, IMS, DB2 and CICS. Every attempt has been made to provide correct information. However, the author and publisher do not guarantee the accuracy of every detail, nor do they assume responsibility for information included in or omitted from it. All of the information in this book should be used at your own risk.

Copyright

CONTENTS

Introduction

Welcome

Congratulations on your purchase of **IBM Mainframe Developer Training and Reference Guide.** This book will teach you the basic information and skills you need to develop applications on IBM mainframes running z/OS. The instruction, examples and sample programs in this book are a fast track to becoming productive as quickly as possible using JCL, MVS Utilities, COBOL, VSAM, IMS, DB2 and CICS. The content is easy to read and digest, well organized and focused on honing real job skills. IBM Mainframe Developer Training and Reference Guide is a key step in the direction of mastering IBM application development so you'll be ready to join a technical team.

This is not an "everything you will ever need to know" book. Rather, this text will teach you what you need to know to become **productive quickly** with JCL, MVS utilities, COBOL, VSAM, IMS, DB2 and CICS. For additional detail, you can download and reference the IBM manuals and Redbooks associated with these products.

Assumptions:

While I do not assume that you know a great deal about IBM mainframe programming, I do assume that you've logged into an IBM mainframe and know your way around. Also I assume that you have a working knowledge of computer programming in some language (it can be a language other than the ones we use in this text which is COBOL). All in all, I assume you have:

1. A working knowledge of ISPF navigation and basic operations such as creating data sets.
2. A basic understanding of structured programming concepts.
3. A basic understanding of SQL.
4. Access to a mainframe computer running z/OS, IMS, CICS and DB2 (and having a COBOL compiler available).

Approach to Learning

I suggest you follow along and do the examples yourself in your own test environment. There's nothing like hands-on experience. Going through the motions will help you learn faster.

If you do not have access to a mainframe system through your job, I can recommend Mathru Technologies. You can rent a mainframe account from them at a very affordable rate, and this includes access to (at this writing they offer **DB2 version 10**). Their environment supports COBOL, IMS and CICS as well. The URL to the Mathru web site is:

Besides the instruction and examples, I've included questions at the end of each chapter. I recommend that you answer these and then check yourself against the answers in the back of the book.

Knowledge, experience and practice questions. Will that guarantee that you'll succeed as an IBM z/OS application developer? Of course, nothing is guaranteed in life. But if you put sufficient effort into this well-rounded training plan that includes all three of the above, I believe you have a very good chance of becoming productive as an IBM Application Developer as soon as possible. This is your chance to get a quick start!

Best of luck!

Robert Wingate

IBM Certified Application Developer – DB2 11 for z/OS

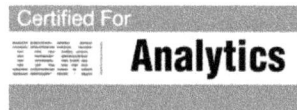

Certified For
IBM | **Analytics**

Chapter 1 : Job Control Language (JCL)

Introduction

Job Control Language (JCL) is an IBM mainframe scripting language that invokes batch programs and other processes to run on a mainframe computer. A batch "job" is an instance of JCL that consists of one or more steps which specify which program to run, what files are to be processed, and the conditions for executing subsequent steps. JCL is submitted to the Job Entry System (JES) via a submit statement or by means of a scheduling tool such as CA7 or Control-M.[1]

JCL Structure and Syntax

JCL Statements

JCL has three statement types: JOB, EXEC and DD.

JOB

A JOB statement is like a header that identifies the beginning of a job, some information about the overall job such as accounting, notification and default output routing, and job class designation (which has implicit time limits). A job can have one or multiple steps that execute programs or procedures.

EXEC

An EXEC statement specifies a particular program or JCL procedure to run. The EXEC statement can also include optional parameters such as condition codes which are explained later in this chapter.

DD

A DD (data definition) statement identifies a data file to be used in the job step. A DD statement includes detailed information about the file if it is a new file. In all cases (new or existing files) the DD statement specifies the disposition of the file when the job step concludes. DD statements link a program file identifier to the actual physical file name.

JCL Coding Rules

Positions 1 and 2 of each line that contains a JCL statement must contain the forward slash characters //. The // indicates to the JCL interpreter that this is a JCL statement. JCL statements begin in position 3 immediately after the //.

---

1 We won't cover CA7 or Control-M in this text. For purposes of application development, you submit a job by typing SUB on the command line while editing your JCL.

EXEC Statement Rules

For an EXEC statement, a step name immediately follows the // in position 3. The step name can be 1-8 alphanumeric and special characters (($, #, @)), but the first character must be alphabetic.

```
//STEP1

//JS010

//STEP001$
```

Most shops have a convention for naming consecutive job steps, such as JS10, JS20, JS30. It is wise to leave some room for adding new steps between the existing steps. For example you may in the future need to add a JS15 between JS10 and JS20.

The job step identifier must be followed by at least one space and the EXEC statement. The EXEC statement must be followed by at least one space and then either the PGM= and the program name, or else simply a proc name. Here are some examples:

```
//JS10      EXEC PGM=IEFBR14

//JS20      EXEC PGM=EMPLOY57

//JS20      EXEC PAYPROC
```

You can optionally add condition codes and other parameters to the EXEC statement. We'll discuss these shortly.

DD Statement Rules

A DD name identifier follows the // in position 3. The DD name identifier can be 1-8 alphanumeric and special characters (($, #, @)), but the first character must be alphabetic.

```
//FILE01

//EMPFILE

//DD4878
```

The file identifier must be followed by a space and then the characters DD. The DD characters should be followed by DSN= and an actual file name. [2]

```
//FILE01  DD DSN=filename

//EMPFILE DD DSN=filename

//DD4878  DD DSN=filename
```

The file name should be followed by additional detailed file information such as the file disposition (status at the beginning and end of the job step), storage unit type (such as tape or disc), space allocation, and data control block (DCB) which specifies the record format, logical record length and block size.

Here are the main keywords you need to know:

DSN
DSN specifies the actual name of the file in the z/OS file system. Here are some examples:

```
TEST.EMPLOYEE.PAYFILE

PROD.HR.RECRUITS
```

DISP
Disposition (DISP) specifies the initial and final status of the file at the end of the job step. It consists of three parts: status at the beginning, action to take if the job step has a normal ending, action to take if the step has an abnormal ending (ABEND).

Beginning Disposition
The status at the beginning is either NEW, SHR, OLD or MOD. The values NEW, MOD and OLD give the job exclusive control over the dataset. The value SHR (share) means that multiple jobs can read data from the file at the same time.

Disposition NEW is used to create a new dataset. If nothing is coded for this parameter, the default is NEW.

---

2 You could also follow the DD with the keyword SYSOUT= and specify a system output class that you want to direct to. In this case you could also specify the default SYSOUT class on the job statement in which case you write SYSOUT=*

Disposition `MOD` means to either extend an existing data set, or to create the dataset if it does not already exist.

Disposition `OLD` means an existing dataset. However for partitioned datasets, `OLD` means if the specified member does not already exist in the PDS file, add it. If it does exist, replace it.

Disposition `SHR` means to share the dataset and is used only for files being used as input, as well as PDS files being used as input or having a member updated.

Ending Disposition

Ending disposition is the action to be taken on the file at the end of the job step, and can be `CATLG`, `UNCATLG`, `KEEP`, `DELETE` or `PASS`. If there is to be no change to the status in the course of the job, then this parameter can be omitted. So for existing input files you are retaining, you could either code:

```
DISP=(SHR,KEEP,KEEP)
```

Or simply

```
DISP=SHR
```

`CATLG` means to catalog the file name into the system so it can be referenced later in this job or by other processes.

`UNCATLG` means to remove the catalog entry (this does not delete the file). `KEEP` means to retain the file.

`KEEP` means to retain the file (assuming it is cataloged).

`DELETE` means to uncatalog and delete the file.

`PASS` means to retain a temporary file for subsequent job steps, but not to catalog it. The job retains the dataset location in the file system so that the system catalog doesn't have to be queried again. Temporary datasets must be passed to any subsequent steps that use them because they are not cataloged.

UNIT

`UNIT` refers to an I/O unit such as a disk or tape machine. It can be an actual hardware device address, a type of device or a group (pool) name. A typical pool name is `SYSDA` or `PERMDA`. One unit name that is not assigned but rather built into the o/s is `SYSALLDA` which means any available direct access volume.

```
UNIT=SYSDA
UNIT=TAPE
UNIT=3390
UNIT=CART
```

SPACE

`SPACE` specifies the amount of primary and secondary space to request for DASD (direct access storage device) file. Unlike other operating systems, z/OS requires you to estimate how much space new disk files will use in order to allocate that space up front. Besides primary space, you can specify secondary space to allow for growth. Up to 16 extents (multiples) of the secondary allocation will be attempted before the job errors with an space unavailable ABEND.

The following example specifies 25 primary tracks and 10 secondary tracks. It also releases any unused space at the end of the step. Coding `RLSE` prevents wasted space.

```
SPACE=(TRK,(25,10),RLSE)
```

When you create a partitioned data set, a third number is included in the allocation, which is the number of directory blocks (which is a directory of each member in the PDS). The following example specifies space for a PDS as 100 primary tracks, 35 secondary tracks and 25 directory blocks.

```
SPACE=(TRK,(100,35,25),RLSE)
```

DCB, RECFM, LRECL, BLKSIZE

`DCB` stands for Data Control Block. The `DCB` parameter can be used with `RECFM`, `LRECL`, and `BLKSIZE`. `DCB` can also be omitted and `RECFM`, `LRECL`, and `BLKSIZE` can simply be specified as standalone parameters.

`RECFM` means record format. You can specify fixed or variable length records. The most common formats are fixed blocked (FB) or variable blocked (VB).

LRECL means logical record length. For FB files, this is the actual length of each record. So for an 80 byte record you would specify LRECL=80. For variable length records, the LRECL is the number of bytes in the longest record plus 4. The additional 4 bytes is required by the system to store the actual record length.

BLKSIZE determines how the data is physically stored on the storage device. Blocking simply means grouping some records together for efficient storage. For FB files, the blocksize must be a multiple of the record length. Here are two examples:

```
RECFM=FB,LRECL=80,BLKSIZE=8000

RECFM=FB,LRECL=80,BLKSIZE=27920
```

DASD devices have certain optimal block sizes. For example, 27,920 is the optimal block size for a file with 80 byte FB records on a 3390 device. A CLIST or table is sometimes provided by the system support group to determine optimum block sizes based on LRECL.

Block size for variable length records is always a minimum of the longest record size plus 8 bytes. Four of the extra bytes store the record length, and the other four bytes store the block size.

DCB statement example:

```
DCB=(RECFM=FB,LRECL=80,BLKSIZE=27920)
```

You could also omit the DCB= and parentheses, and just specify the sub parameters like the following example (which means the same thing):

```
RECFM=FB,LRECL=80,BLKSIZE=27920
```

Additional DCB parameters include the DSORG which means data set organization. By default this is physical sequential (PS). If you are creating a partitioned data set, you must specify DSORG=PO. There is also the direct access type DA (direct access) which allows a read-write function, but in my experience it is somewhat rarely used.

One other parameter is the BUFNO parameter. Buffers are main memory that is used to hold data after it has been read from disc or tape. BUFNO specifies how many buffers are to be used in reading or writing data files. The default is 5 buffers which is usually enough. In some cases of heavy I/O situations, you can improve performance by increasing the number of buffers on the input file. The default value is 5 and the maximum is 255. However, each buffer requires main memory to be allocated for one block of data, and you may quickly reach a memory limit long before the maximum number of buffers can be loaded.

Generation Data Groups

A generation data group is a collection of related datasets that use the same base name and can be referenced by generation number. The most recent version of the file is the zero (0) generation. The first previous generation is (-1), the next previous is (-2), etc.

The absolute file name of each generation is the base GDG file name plus a G00Vxxxx where the xxxx is a sequential number. So the first few generations of a GDG named XXX.FILEABC would be:

```
XXX.FILEABC.G00V0001

XXX.FILEABC.G00V0002

XXX.FILEABC.G00V0003
```

To create a new generation of a file, you specify the (+1) generation of the file.

```
//FILEOUT  DD DSN= XXX.FILEABC(+1)
```

If you are creating more than one generation of a file *in the same job*, then you must increment the relative generation number for each new file. Once you've specified +1 for the first new files, you must specify +2, then +3, etc. for additional files. Once the job ends, the most recent version is +0, the next most recent is -1, etc.

Instream Data

While your DD names normally refer to external files, it is possible to include data "instream", meaning the data is included in the JCL itself. You do this by specifying DDNAME DD * instead of a file name, and then including your data on the next line or lines. For example:

```
//TESTNAME DD *
SAMPLE DATA ONE
SAMPLE DATA TWO
/*
```

The /* value on the last line is a terminator or delimiter for the data.

Instream data is rarely used in production, although I have seen it done. Usually it is used for testing where you can conveniently change the values in the test data by simply modifying your test JCL. It's a handy feature for that.

Concatenating Input Files

If you have multiple similar files that should be processed as a combined file, JCL allows you to concatenate them in one DD statement. Note that the DD name appears only once in the concatenated sequence – it is omitted on each succeeding line. Concatenation is only allowed when the files have the same record length (LRECL), although they can have different block sizes.

```
//EMPFILE DD  DSN=TEST.EMPFILE1,DISP=SHR
//          DD  DSN=TEST.EMPFILE2,DISP=SHR
//          DD  DSN=TEST.EMPFILE3,DISP=SHR
```

STEPLIB

One DD name that has special meaning is STEPLIB. The STEPLIB DD tells the system which library to find the executable program in. STEPLIB applies only to that particular job step, which means that other steps may use a different program library. In practice an application typically uses the same library (or concatenation of libraries) for an entire job. See the note on JOBLIB at the end of this chapter.

JCL Comments

Columns 1, 2, and 3 of a JCL comment statement contain //*

```
//* This is a JCL comment
```

JCL Condition Code Processing

This is one of the more confusing topics to JCL newcomers, partly because it's a bit counter intuitive. Once you get the hang of it, it's not really that hard. Let's jump right in.

First, we have to understand that z/OS traps return codes from the various job steps that indicate whether or not the step was successful. This is referred to as a return code or "condition code". The JCL keyword for a condition code is COND and you can evaluate it to determine whether to continue running further steps in a job.
Some common return codes are as follows:

- 0 = Normal - all OK

- 4 = Warning - minor errors or problems

- 8 = Error - significant errors or problems

- 12 = Severe error - major errors or problems, the results (e.g. files or reports produced) should not be trusted.

- 16 = Terminal error - very serious problems, do not use the results!

A condition code specifies a particular value and then a Boolean condition which will evaluate to either true or false. For example, COND=(4,LT) establishes a condition that the value 4 is less than the highest return code in previously executed job steps. Other Boolean values are:

```
EQ - equal to
NE - not equal to
GT - greater than
GE - greater than or equal to
LT - less than
LE - less than or equal to
```

You can specify the condition for the entire job (highest condition code encountered from all previous steps), or the condition code for a specific prior step. The counter-intuitive part of it is that if the condition you specify is TRUE, then the current step will be **bypassed**. Let's do an example.

The following says if the highest condition code encountered by previous steps in the job is not equal to zero, then bypass this current step. Literally it is saying if zero is not equal to the highest condition code encountered so far in the job, then do not execute this step.

```
//JS040   EXEC PGM=USERPGM,COND=(0,NE)
```

Another way of saying the above is to say if the condition code is greater than zero, do not execute the step. You could also code it this way. This says if zero is less than the actual condition code, then bypass the job step.

```
//JS040   EXEC PGM=USERPGM,COND=(0,LT)
```

Now let's put it in context of an entire job. The following JCL has two steps. STEP01 executes program USERPGM1, and then STEP02 executes USERPGM2. However if STEP01 has a return code greater than zero, STEP02 will NOT be executed.

```
//USER01D  JOB (123),'USER01',CLASS=A,MSGCLASS=A,MSGLEVEL=(1,1),
//             NOTIFY=&SYSUID
//*
//*  RUN TWO PROGRAMS SEQUENTIALLY
//*
//STEP01   EXEC PGM=USERPGM1
//STEPLIB  DD DSN=MYLIB.LOADLIB,DISP=SHR
//DATAIN   DD DSN=MYFILEA.TEST,DISP=OLD
//DATAOUT  DD DSN=MYFILEB.TEST,
//            DISP=(NEW,CATLG,DELETE),
//            SPACE=(TRK,(15,15),RLSE),
//            UNIT=SYSDA,
//            DCB=(DSORG=PS,RECFM=FB,LRECL=80,BLKSIZE=800)
//SYSPRINT DD  SYSOUT=*
//SYSUDUMP DD  SYSOUT=*
//*
//STEP02   EXEC PGM=USERPGM2,COND=(0,NE)
//STEPLIB  DD  DSN=MYLIB.LOADLIB,DISP=SHR
//DATAIN   DD  DSN=MYFILEB.TEST,DISP=SHR
//SYSPRINT DD  SYSOUT=*
//SYSUDUMP DD  SYSOUT=*
//
```

In the above case, there is only one prior step to STEP02, so you needn't code the specific prior step name on your condition code. However you can code one or more prior steps if you specify them by name. Simply add the step name as a third comma separated value on the COND parameter.

```
//*
//STEP02   EXEC PGM=USERPGM2,COND=(0,NE,STEP01)
//STEPLIB  DD  DSN=MYLIB.LOADLIB,DISP=SHR
//DATAIN   DD  DSN=MYFILEB.TEST,DISP=SHR
//SYSPRINT DD  SYSOUT=*
//SYSUDUMP DD  SYSOUT=*
//
```

Finally there are some special values that enable you to run or not run a step depending on whether or not a prior step abnormally terminated regardless of the specific return code. These values are EVEN and ONLY. COND=EVEN means this step will be executed even if a prior step caused an abend. `COND=ONLY` means this step will only be executed if a prior step caused an abend. This gives you the means of ensuring that certain steps execute no matter what (`EVEN`), and of running certain steps only when an abend condition has occurred (`ONLY`). So in the following example, `STEP02` will be executed **even if** `STEP01` abends:

```
//STEP02   EXEC PGM=USERPGM2,COND=EVEN
//STEPLIB   DD   DSN=MYLIB.LOADLIB,DISP=SHR
//DATAIN    DD   DSN=MYFILEB.TEST,DISP=SHR
//SYSPRINT DD   SYSOUT=*
//SYSUDUMP DD   SYSOUT=*
//
```

And in the next example, `STEP02` will be executed **only if** `STEP01` abended.

```
//STEP02   EXEC PGM=USERPGM2,COND=ONLY
//STEPLIB   DD   DSN=MYLIB.LOADLIB,DISP=SHR
//DATAIN    DD   DSN=MYFILEB.TEST,DISP=SHR
//SYSPRINT DD   SYSOUT=*
//SYSUDUMP DD   SYSOUT=*
//
```

You can also use `EVEN` and `ONLY` referring to specific steps, such as the following which says to execute `STEP03` even if `STEP01` abended.

```
//STEP03   EXEC PGM=USERPGM3,COND=(EVEN,STEP01)
//STEPLIB   DD   DSN=MYLIB.LOADLIB,DISP=SHR
//DATAIN    DD   DSN=MYFILEB.TEST,DISP=SHR
//SYSPRINT DD   SYSOUT=*
//SYSUDUMP DD   SYSOUT=*
```

You can also do multiple condition tests on the same `EXEC` statement. Suppose we want to bypass `STEP03` if `STEP01` has a condition code greater than 0, **or if** `STEP02` has a condition code greater than 4. Again, to read it literally:

- If zero is less than the condition code from `STEP01`, then do not execute `STEP03`
- If four is less than the condition code from `STEP02`, then do not execute `STEP03`

```
//STEP03   EXEC PGM=USERPGM3,COND=((0,LT,STEP01),(4,LT,STEP02))
//STEPLIB   DD   DSN=MYLIB.LOADLIB,DISP=SHR
//DATAIN    DD   DSN=MYFILEB.TEST,DISP=SHR
//SYSPRINT DD   SYSOUT=*
//SYSUDUMP DD   SYSOUT=*
```

To repeat something that I hope is making sense by now, remember that a COND parameter that evaluates as TRUE means **do not execute the step** (unless the COND statement is EVEN or ONLY which have their own meanings).

Now having said all this about condition code processing, I must also say that quite some time ago IBM added IF/THEN logic to the JCL product. The IF/THEN is more intuitive and works more like a programming construct. We'll consider that topic next. But just be forewarned that there is still a lot of existing JCL out there that is coded with classic COND parameters. So you need to know how to read and interpret JCL using both methods.

JCL IF/THEN Logic

Let's build a composite job from some previous examples and then see how we could convert it to use IF/THEN logic instead of classic condition codes. Here is a JCL using condition codes on the EXEC statement. We'll follow that with a version that uses IF/THEN.

```
//USER01D  JOB (123),'USER01',CLASS=A,MSGCLASS=A,MSGLEVEL=(1,1),
//              NOTIFY=&SYSUID
//*
//*   RUN TWO PROGRAMS SEQUENTIALLY
//*
//STEP01   EXEC PGM=USERPGM1
//STEPLIB  DD  DSN=MYLIB.LOADLIB,DISP=SHR
//DATAIN   DD  DSN=MYFILEA.TEST,DISP=OLD
//DATAOUT  DD  DSN=MYFILEB.TEST,DISP=(NEW,CATLG,DELETE),
//             SPACE=(TRK,(15,15),RLSE),UNIT=SYSDA,
//             DCB=(DSORG=PS,RECFM=FB,LRECL=80,BLKSIZE=800)
//SYSPRINT DD  SYSOUT=*
//SYSUDUMP DD  SYSOUT=*
//*
//STEP02   EXEC PGM=USERPGM2,COND=(0,NE)
//STEPLIB  DD  DSN=MYLIB.LOADLIB,DISP=SHR
//DATAIN   DD  DSN=MYFILEB.TEST,DISP=SHR
//SYSPRINT DD  SYSOUT=*
//SYSUDUMP DD  SYSOUT=*
//*
//STEP03   EXEC PGM=USERPGM3,COND=((0,LT,STEP01),(4,LT,STEP02))
//STEPLIB  DD  DSN=MYLIB.LOADLIB,DISP=SHR
//DATAIN   DD  DSN=MYFILEB.TEST,DISP=SHR
//SYSPRINT DD  SYSOUT=*
//SYSUDUMP DD  SYSOUT=*
//
```

Here is the same JCL using IF/THEN statements:

```
//USER01D  JOB (123),'USER01',CLASS=A,MSGCLASS=A,MSGLEVEL=(1,1),
//              NOTIFY=&SYSUID
//*
//*   RUN TWO PROGRAMS SEQUENTIALLY
//*
//STEP01  EXEC PGM=USERPGM1
//STEPLIB  DD DSN=MYLIB.LOADLIB,DISP=SHR
//DATAIN   DD DSN=MYFILEA.TEST,DISP=OLD
//DATAOUT  DD DSN=MYFILEB.TEST,
//            DISP=(NEW,CATLG,DELETE),
//            SPACE=(TRK,(15,15),RLSE),
//            UNIT=SYSDA,
//            DCB=(DSORG=PS,RECFM=FB,LRECL=80,BLKSIZE=800)
//SYSPRINT DD  SYSOUT=*
//SYSUDUMP DD  SYSOUT=*
//*
//*
//  IF (STEP01.RC = 0) THEN
//*
//STEP02  EXEC PGM=USERPGM2
//STEPLIB  DD  DSN=MYLIB.LOADLIB,DISP=SHR
//DATAIN   DD  DSN=MYFILEB.TEST,DISP=SHR
//SYSPRINT DD  SYSOUT=*
//SYSUDUMP DD  SYSOUT=*
//*
//  ENDIF
//*
//  IF (STEP01.RC = 0 AND STEP02.RC <= 4) THEN
//*
//STEP03  EXEC PGM=USERPGM3
//STEPLIB  DD  DSN=MYLIB.LOADLIB,DISP=SHR
//DATAIN   DD  DSN=MYFILEB.TEST,DISP=SHR
//SYSPRINT DD  SYSOUT=*
//SYSUDUMP DD  SYSOUT=*
//*
//  ENDIF
```

You may find the **IF/THEN/ELSE/ENDIF** coding much easier and more intuitive to understand than the COND= parameter, and I recommend that you use it for new development. Again be advised that you are likely to encounter both!

Now let's look at some more examples of JCLs, some of which execute IBM utility programs, and some of which execute application programs.

Sample JCLs

Delete a File

```
//USER01A  JOB (123),'USER01',CLASS=A,MSGCLASS=A,MSGLEVEL=(1,1),
//            NOTIFY=&SYSUID
//*************************************************************
//* DELETE A DATA SET USING IEFBR14 UTILITY
//*************************************************************
//STEP20   EXEC PGM=IEFBR14
//SYSPRINT DD SYSOUT=*
//SYSOUT   DD SYSOUT=*
//SYSUDUMP DD SYSOUT=*
//DD1      DD DSN=USER01.TEST.FILE,
//            DISP=(OLD,DELETE,DELETE)
//*
```

Delete a File That May or May Not Exist

```
//USER01B  JOB (123),'USER01',CLASS=A,MSGCLASS=A,MSGLEVEL=(1,1),
//            NOTIFY=&SYSUID
//*************************************************************
//* DELETE A DATA SET USING IEFBR14 UTILITY
//*************************************************************
//STEP30   EXEC PGM=IEFBR14
//SYSPRINT DD SYSOUT=*
//SYSOUT   DD SYSOUT=*
//SYSUDUMP DD SYSOUT=*
//DD1      DD DSN=USER01.TEST.FILE,
//            DISP=(MOD,DELETE,DELETE),
//            UNIT=SYSDA,
//            SPACE=(TRK,(1,1),RLSE)
```

Read a file (SYSUT1) and write output to new file (SYSUT2).

Here we are using an IBM utility to copy a file with DD name SYSUT1 to a file with DD name SYSUT2.

```
//USER01C  JOB (123),'USER01',CLASS=A,MSGCLASS=A,MSGLEVEL=(1,1),
//            NOTIFY=&SYSUID
//*************************************************
//* TO COPY PS DATA SET TO ANOTHER PS DATA SET
//*************************************************
//STEP10   EXEC PGM=IEBGENER
//SYSPRINT DD SYSOUT=*
//SYSOUT   DD SYSOUT=*
//SYSUDUMP  DD SYSOUT=*
//SYSUT1   DD DSN=USER01.SPUFI.WORK,
//            DISP=SHR
//SYSUT2   DD DSN=USER01.SPUFI.LAST,
//            DISP=(NEW,CATLG,DELETE),
//            SPACE=(TRK,(15,15),RLSE),
//            UNIT=SYSDA,
//            DCB=(DSORG=PS,RECFM=FB,LRECL=80,BLKSIZE=800)
//SYSIN    DD DUMMY
//
```

Run an Application Program

This is a plain vanilla execution of an application program.

```
//USER01C  JOB (123),'USER01',CLASS=A,MSGCLASS=A,MSGLEVEL=(1,1),
//            NOTIFY=&SYSUID
//*
//*  RUN A PROGRAM
//*
//STEP01  EXEC PGM=USERPGM1
//STEPLIB  DD  DSN=MYLIB.LOADLIB,DISP=SHR
//SYSPRINT DD  SYSOUT=*
//SYSUDUMP DD  SYSOUT=*
```

Run a job that executes two application programs

In this case we will only run program USERPGM2 if the previous step was successful.

```
//USER01D  JOB (123),'USER01',CLASS=A,MSGCLASS=A,MSGLEVEL=(1,1),
//               NOTIFY=&SYSUID
//*
//*  RUN TWO PROGRAMS SEQUENTIALLY
//*
//STEP01   EXEC PGM=USERPGM1
//STEPLIB  DD DSN=MYLIB.LOADLIB,DISP=SHR
//DATAIN   DD DSN=MYFILEA.TEST,DISP=OLD
//DATAOUT  DD DSN=MYFILEB.TEST,
//            DISP=(NEW,CATLG,DELETE),
//            SPACE=(TRK,(15,15),RLSE),
//            UNIT=SYSDA,
//            DCB=(DSORG=PS,RECFM=FB,LRECL=80,BLKSIZE=800)
//SYSPRINT DD  SYSOUT=*
//SYSUDUMP DD  SYSOUT=*
//*
//STEP02   EXEC PGM=USERPGM2,COND=(0,NE)
//STEPLIB  DD  DSN=MYLIB.LOADLIB,DISP=SHR
//DATAIN   DD  DSN=MYFILEB.TEST,DISP=SHR
//SYSPRINT DD  SYSOUT=*
//SYSUDUMP DD  SYSOUT=*
//
```

JCL Procedures
A JCL procedure (PROC) is a collection of one or more JCL statements that can be invoked by another JCL with a single EXEC statement. PROCs are a way of creating reusable JCL steps, similar to the way we write subprograms to reuse program code.

Unlike a JCL, a procedure does not include a JOB statement. Typically procedures are stored in a library other than the JCL library. We call this a PROC lib. You can also code procedures inline within a JCL. Inline coding is not usually done for production, but it can streamline development and testing. We'll look at that later in this section.

Why use Procedures?
One benefit of creating procedures is that you can call them from more than one JCL, so you've centralized the job steps by storing them in a procedure. Another benefit is when you have to make changes to those job steps, you only need to make the changes in one place instead of many.

Let's do an example where we convert the three-step JCL we were looking at earlier into a three-step procedure. We'll use the same content except there is no job statement. Instead a procedure begins with a proc name identifier and the keyword PROC. We'll also change our step names from STEP** to PS** (the PS meaning proc step). The identifiers names are arbitrary but you'll want to come up with a system and stick with it as a standard.

```
//EMPPAY PROC
//PS01   EXEC PGM=USERPGM1
//STEPLIB  DD DSN=MYLIB.LOADLIB,DISP=SHR
//DATAIN   DD DSN=MYFILEA.TEST,DISP=OLD
//DATAOUT  DD DSN=MYFILEB.TEST,
//            DISP=(NEW,CATLG,DELETE),
//            SPACE=(TRK,(15,15),RLSE),
//            UNIT=SYSDA,
//            DCB=(DSORG=PS,RECFM=FB,LRECL=80,BLKSIZE=800)
//SYSOUT   DD  SYSOUT=*
//SYSPRINT DD  SYSOUT=*
//SYSUDUMP DD  SYSOUT=*
//*
//  IF (PS01.RC = 0) THEN
//*
//PS02   EXEC PGM=USERPGM2
//STEPLIB  DD  DSN=MYLIB.LOADLIB,DISP=SHR
//DATAIN   DD  DSN=MYFILEB.TEST,DISP=SHR
//SYSOUT   DD  SYSOUT=*
//SYSPRINT DD  SYSOUT=*
//SYSUDUMP DD  SYSOUT=*
//*
//  ENDIF
//*
//  IF (PS01.RC = 0 AND PS02.RC <= 4) THEN
//*
//PS03   EXEC PGM=USERPGM3
//STEPLIB  DD  DSN=MYLIB.LOADLIB,DISP=SHR
//DATAIN   DD  DSN=MYFILEB.TEST,DISP=SHR
//SYSOUT   DD  SYSOUT=*
//SYSPRINT DD  SYSOUT=*
//SYSUDUMP DD  SYSOUT=*
//*
//  ENDIF
//
```

Now let's say we have saved this procedure code into a procedure library named `MYPROC.LIB`, and the member name is `EMPPAY`. You can invoke it by running this JCL:

```
//USER01D  JOB (123),'USER01',CLASS=A,MSGCLASS=A,MSGLEVEL=(1,1),
//             NOTIFY=&SYSUID
//PROCLIB  DD DSN=MYPROC.LIB,DISP=SHR
//*
//STEP01  EXEC EMPPAY
//SYSPRINT DD  SYSOUT=*
//SYSUDUMP DD  SYSOUT=*
```

Inline Procedures

You can code a procedure to be inline with a JCL. To do so, you include the PROC definition right after the job card. Then you code a PEND statement which marks the end of the inline proc. Next you code the job step to execute the proc. So all of the procedure steps and job steps are self-contained inside one JCL. This is handy for testing before you move the proc code to its own file.

Here is how it would work using the previous example.

```
//USER01D  JOB (123),'USER01',CLASS=A,MSGCLASS=A,MSGLEVEL=(1,1),
//             NOTIFY=&SYSUID
//*
//EMPPAY PROC
//PS01  EXEC PGM=USERPGM1
//STEPLIB  DD DSN=MYLIB.LOADLIB,DISP=SHR
//DATAIN   DD DSN=MYFILEA.TEST,DISP=OLD
//DATAOUT  DD DSN=MYFILEB.TEST,
//            DISP=(NEW,CATLG,DELETE),
//            SPACE=(TRK,(15,15),RLSE),
//            UNIT=SYSDA,
//            DCB=(DSORG=PS,RECFM=FB,LRECL=80,BLKSIZE=800)
//SYSOUT   DD  SYSOUT=*
//SYSPRINT DD  SYSOUT=*
//SYSUDUMP DD  SYSOUT=*
//*
//  IF (PS01.RC = 0) THEN
//*
//PS02  EXEC PGM=USERPGM2
//STEPLIB  DD  DSN=MYLIB.LOADLIB,DISP=SHR
//DATAIN   DD  DSN=MYFILEB.TEST,DISP=SHR
//SYSOUT   DD  SYSOUT=*
//SYSPRINT DD  SYSOUT=*
//SYSUDUMP DD  SYSOUT=*
//*
//  ENDIF
//*
//  IF (PS01.RC = 0 AND PS02.RC <= 4) THEN
//*
```

```
//PS03   EXEC PGM=USERPGM3
//STEPLIB  DD   DSN=MYLIB.LOADLIB,DISP=SHR
//DATAIN   DD   DSN=MYFILEB.TEST,DISP=SHR
//SYSOUT   DD   SYSOUT=*
//SYSPRINT DD   SYSOUT=*
//SYSUDUMP DD   SYSOUT=*
//*
//   ENDIF
//*
//   PEND
//*
//STEP01   EXEC EMPPAY
//SYSPRINT DD   SYSOUT=*
//SYSUDUMP DD   SYSOUT=*
//
```

Symbolics in Procedures

Another benefit of procedures is that you can code symbolic parameters that are later passed by the job that executes the procedure. A good example is when you have a procedure that can be used both for test and production. At a minimum you would be using different data set names, and let's suppose that the second level node for production is PROD and the second level node for test is TEST. So you could have MYFILEA.PROD and MYFILEA.TEST, MYFILEB.PROD and MYFILEB.TEST, etc.

In the above scenario, the utility of a procedure is that you can code a symbolic for the second level node and then use the same procedure for both environments by passing the correct value in the JCL at runtime. To do this, you must create a symbolic identifier and add it to the proc identifier. Let's create a symbolic to represent the environment called &ENV and this is the first line of our proc now:

```
//EMPPAY PROC ENV=
```

 If you want to put a default value, you can code that too. Suppose we want the default value to be PROD.

```
//EMPPAY PROC ENV=PROD
```

Now we must change our DD statements to use this new symbolic. To do this we replace the node named TEST with an ampersand (&) followed by the symbolic name ENV. Here's our revised first proc step:

```
//PS01   EXEC PGM=USERPGM1
//STEPLIB   DD DSN=MYLIB.LOADLIB,DISP=SHR
//DATAIN    DD DSN=MYFILEA.&ENV,DISP=OLD
//DATAOUT   DD DSN=MYFILEB.&ENV,
//             DISP=(NEW,CATLG,DELETE),
//             SPACE=(TRK,(15,15),RLSE),
//             UNIT=SYSDA,
//             DCB=(DSORG=PS,RECFM=FB,LRECL=80,BLKSIZE=800)
```

Once we've completed all the changes to our file names to use the ENV symbolic, we can call the proc with the execute statement and the symbolic name with the value we want to pass to the proc. If we are running for test purposes, we would code:

```
//STEP01   EXEC EMPPAY,ENV=TEST
```

Or if running in production use:

```
//JS01   EXEC EMPPAY,ENV=PROD
```

Since we've coded a default value in the proc, if we omit the ENV symbolic on our JCL EXEC statement, then the proc will use the default PROD value of ENV.

You can use symbolic parameters to override file names, programs and pretty much any other value on a job or proc step. If you have more than one symbolic, you simply separate them with commas both when you code them at the top of the proc, and in the JCL when you call the proc. Symbolic parameters need not be overridden in any order since they are based upon parameter name (not position).

Non-Symbolic Overrides in Procedures

Besides passing symbolic parameter values to a proc, JCL can also override DD parameters such as the file name and DCB values without using symbolic. You simply refer to the proc step name and DD name, and then supply the correct parameter and value.

Let's look back at the first step of our proc EMPPAY.

```
//PS01   EXEC PGM=USERPGM1
//STEPLIB   DD DSN=MYLIB.LOADLIB,DISP=SHR
//DATAIN    DD DSN=MYFILEA.TEST,DISP=OLD
//DATAOUT   DD DSN=MYFILEB.TEST,
//             DISP=(NEW,CATLG,DELETE),
//             SPACE=(TRK,(15,15),RLSE),
//             UNIT=SYSDA,
//             DCB=(DSORG=PS,RECFM=FB,LRECL=80,BLKSIZE=800)
```

Suppose that we want to override the file name on the DATAIN DD from MYFILEA.TEST to MYFILEA.MODEL. If we only need to do this on a temporary basis, we need not change the PROC. We can simply override it after the EXEC statement. We code the PROC step, followed by a period, followed by the DD name, followed by the desired override values. In this case we are overriding the file name to be MYFILE.MODEL.

```
//USER01D  JOB (123),'USER01',CLASS=A,MSGCLASS=A,MSGLEVEL=(1,1),
//            NOTIFY=&SYSUID
//*
//JS10      EXEC EMPPAY
//PS01.DATAIN DD DSN=MYFILEA.MODEL
//SYSPRINT DD  SYSOUT=*
//SYSUDUMP DD  SYSOUT=*
```

That's all there is to it. Similarly, you can change output datasets and their sub parameters. Be aware that for these non-symbolic overrides, you must code them in the order that the DD names occur in the proc. So if you have two DDs you want to override, you must override them in the order they appear in the proc.

Other JCL Statements

In many cases your system will have certain libraries available for you by default, meaning you do not need to specify them. Often JCL and proc libs, as well as loadlibs for executables are available by default. If they are not included, you may need to specify them. Here are two examples.

JOBLIB

JOBLIB takes the place of STEPLIB and it applies to the entire job, not just a single step. This statement is often used for testing but in some shops you are required to include it to point to your application load libraries. Place the JOBLIB statement after the JOB statement and before any job steps. Note: if you code both JOBLIB and STEPLIB in the same job, the STEPLIB will be used and the JOBLIB ignored.

PROCLIB

PROCLIB defines a procedure library from which PROCs can be called. It is also often used for testing. Place the PROCLIB statement after the JOB statement and before any job steps.

This chapter has given you the basic of JCL. We'll look at many more examples of JCL as we move on to MVS utilities in the next chapter. You can also obtain a PDF version of the MVS JCL Users Guide by using the link here: [3]

---

3 https://www-05.ibm.com/e-business/linkweb/publications/servlet/pbi.wss?CTY=US&FNC=SRX&PBL=SA22-7598-07

Chapter 1 Review Questions

1.   What statement do you use to delimit an instream procedure?

2.   What does COND=(0,NE) mean?

3.   What keyword on a SPACE allocation statement returns unused space to the system when the dataset is closed?

4.   How do you reference the latest version of a GDG-based file?

5.   Explain JOBLIB and STEPLIB in JCL.

6.   How is a temporary dataset coded?

7.   How do you concatenate datasets in JCL?

8.   How do you designate a comment in JCL?

9.   What does a disposition of `(NEW,CATLG,DELETE)` mean?

10.   What are the required attributes for an output DD statement?

11.   What does COND=EVEN mean in JCL?

12.   What does the following statement mean?
        ```
        SYSIN  DD  *
        ```

Chapter 2 : MVS JCL UTILITIES

You can accomplish many routine file operations using the MVS utility programs. This chapter of the text book provides ready reference material for the most common utility programs, and I encourage their use. Use of these utilities will contribute to programmer productivity by minimizing or eliminating custom programming for simple, routine operations.

IDCAMS
Copying a Sequential Dataset

IDCAMS is a very versatile utility. You can use it to copy files, but be forewarned that **if you get an error (such as a B37 ABEND), the new file will still get written** – regardless of the (NEW, CATLG, DELETE) file disposition. Since this is usually not what you want to happen, you should, always check the return codes from IDCAMS to ensure everything worked as intended.

```
//JOB00101 JOB (ACCTCODE),'NAME',
//  CLASS=A,MSGCLASS=A,NOTIFY=USERID
//*
//*************************************************
//* IDCAMS TO COPY A DATA SET
//*************************************************
//*
//STEP1     EXEC PGM=IDCAMS
//SYSPRINT  DD SYSOUT=*
//SYSUDUMP  DD SYSOUT=*
//FILEIN    DD DSN=DSNAME.TEST.FILE,DISP=SHR
//FILEOUT   DD DSN=DSNAME.TEST.FILE2,
//             DISP=(NEW,CATLG,DELETE),
//             UNIT=SYSDA,
//             SPACE=(TRK,(5,5),RLSE),
//             RECFM=FB,LRECL=4096,BLKSIZE=4096
//SYSIN     DD *
 REPRO INFILE (FILEIN) OUTFILE (FILEOUT)
/*
//
```

Copying a Specified Number of Records

Use the **COUNT** parameter to specify a limited number of records to copy from a source file. This can be useful when you want to create test data from a production file, or when you simply want to determine the contents of a file. The following example writes the first 500 records of the input file to an output file.

```
//JOB00101 JOB (ACCTCODE),'NAME',
//   CLASS=A,NOTIFY=USERID,MSGCLASS=A
//*
//**************************************************
//*  COPY A FILE, BUT STOP AFTER XX RECORDS.
//**************************************************
//*
//STEP1     EXEC PGM=IDCAMS
//FILEIN    DD DSN=DSNAME.TEST.FILE1,
//             DISP=SHR
//FILEOUT   DD DSN=DSNAME.TEST.FILE2,
//             DISP=(NEW,CATLG,DELETE),
//             UNIT=SYSDA,
//             SPACE=(TRK,(5,5),RLSE),
//             RECFM=FB,LRECL=80,BLKSIZE=0,DSORG=PS
//SYSIN     DD *
     REPRO   INFILE(FILEIN)   -
             OUTFILE(FILEOUT) -
             COUNT  (500)
//SYSPRINT  DD SYSOUT=*
//SYSUDUMP  DD SYSOUT=*
//
```

Skipping a Specified Number of Records

Use the **SKIP** parameter when you want to omit a number of records from a file copy operation. You will find this useful if you know precisely how many records you want to omit. For example, you may need to strip some records from a file that is too big to edit online. You can create a smaller output file by skipping over some older records at the front of the file. NOTE: You can use the **SKIP** and **COUNT** parameters together. For example, you can skip the first 1,000 records and then write the next 500 by also including a **COUNT (500)** clause.

```
//JOB00101 JOB (ACCTCODE),'NAME',
//   CLASS=A,NOTIFY=USERID,MSGCLASS=A
//*
//***************************************************
//* COPY A FILE, BUT SKIP XX RECORDS.
//***************************************************
//*
//STEP1     EXEC PGM=IDCAMS
//FILEIN    DD DSN=DSNAME.TEST.FILE1,DISP=SHR
//FILEOUT   DD DSN=DSNAME.TEST.FILE2,
//             DISP=(NEW,CATLG,DELETE),
//             UNIT=SYSDA,
//             SPACE=(TRK,(5,5),RLSE),
//             RECFM=FB,LRECL=80,BLKSIZE=0,DSORG=PS
//SYSIN     DD *
    REPRO   INFILE(FILEIN)   -
            OUTFILE(FILEOUT) -
            SKIP  (1000)
            COUNT (500)
//SYSPRINT  DD SYSOUT=*
//SYSUDUMP  DD SYSOUT=*
//
```

Initializing a File to Empty

You can create an empty file by using the IDCAMS REPRO function and providing a dummy input file specification. The BLKSIZE on input and output must match or you will get an error.

```
//JOB00101 JOB (ACCTCODE),'NAME',
//    CLASS=A,NOTIFY=USERID,MSGCLASS=A
//*
//STEP1     EXEC PGM=IDCAMS
//INFILE    DD DUMMY,
//             BLKSIZE=23250
//OUTFILE   DD DSN=DSNAME.FILE.TEST,
//             DISP=(NEW,CATLG,DELETE),
//             UNIT=TAPE,
//             RECFM=FB,LRECL=250,BLKSIZE=23250
//SYSPRINT  DD SYSOUT=*
//SYSIN     DD *
  REPRO INFILE (INFILE) OUTFILE (OUTFILE)
/*
//
```

Verifying Data Exists in a File

Sometimes you will need to verify whether there is any data in a file. For example, you may have a job that should conditionally process a file depending on whether there is data in it. One way is to use an IDCAMS read, code a dummy output file, and specify a COUNT of 1. If the file is empty, you will get a 04 return code from IDCAMS. Subsequent job steps can test the return code, and only execute if there is data – for example COND=(0,NE) or COND= (0,LT).

```
//JOB00101 JOB (ACCTCODE),'NAME',
//   CLASS=A,NOTIFY=USERID,MSGCLASS=A
//*
//STEP1     EXEC PGM=IDCAMS
//SYSUDUMP  DD SYSOUT=*
//SYSPRINT  DD SYSOUT=*
//FILEIN    DD DSN=DSNAME.TEST.FILE1,DISP=SHR
//FILEOUT   DD DUMMY,RECFM=FB,LRECL=80,BLKSIZE=8000
//SYSIN     DD *
      REPRO   INFILE(FILEIN) -
              OUTFILE(FILEOUT) -
  COUNT(1)
/*
```

Renaming a File

Since you can generally rename a file online, you might not have many occasions when you need a batch job for this purpose. One possible use is if you have several files to rename as part of a standardization project. In this case you use **ALTER** with **RENAME** to get this result.

```
//JOB00101 JOB (ACCTCODE),'NAME',
// CLASS=A,NOTIFY=USERID,MSGCLASS=A
//*
//*********************************************
//* IDCAMS TO RENAME A DATA SET
//*********************************************
//*
//STEP1     EXEC PGM=IDCAMS
//SYSPRINT  DD SYSOUT=*
//SYSUDUMP  DD SYSOUT=*
//SYSIN     DD *
  ALTER DSNAME.TEST.FILE1 -
  NEWNAME(DSNAME.TEST.FILE2)
/*
//
```

Deleting a File

There are many utilities that allow you to delete a file. Here is the IDCAMS function to perform a delete.

```
//JOB00101 JOB (ACCTCODE),'NAME',
// CLASS=A,MSGCLASS=A,NOTIFY=USERID
//*
//*********************************************
//*   IDCAMS TO DELETE A DATA SET
//*********************************************
//*
//STEP1     EXEC PGM=IDCAMS
//SYSPRINT  DD SYSOUT=*
//SYSIN     DD *
 DELETE DSNAME.TEST.FILE
/*
//
```

Building a GDG

You can create a GDG base using the **DEFINE GDG** command. NAME is the dataset name for the base. LIMIT specifies how many generations are to be created before one or more generations has to "roll off", meaning it gets automatically deleted by the o/s. Normally you include the SCRATCH parameter so that the oldest generation gets deleted when the limit is exceeded. If you instead specify NOSCRATCH then the oldest generation will get uncataloged but not deleted. I can't think of any use for this option.

There is also an optional EMPTY and NOEMPTY mutually exclusive clause. If you specify EMPTY then when the limit is reached, all generations of the GDG are deleted. If you specify NOEMPTY, then only the oldest generation is deleted. The default is NOEMPTY.

```
//JOB00101 JOB (ACCTCODE),'NAME',
//    CLASS=A,NOTIFY=YOURID,MSGCLASS=A
//*
//**************************************************
//*     DEFINE GDG
//**************************************************
//*
//STEP1    EXEC PGM=IDCAMS
//SYSPRINT  DD SYSOUT=*
//SYSIN     DD *
  DEFINE GDG -
        (NAME(DSNAME.FILEGDG.FILENAME) -
        LIMIT(2) SCRATCH)
/*
//
```

Adjusting the Number of Gens on a GDG

You can change the maximum number of generations on a GDG by using the **ALTER** command and the **LIMIT** parameter. You can increase or decrease the number of gens. This JCL sets the number of gens to 5. Note that if more than 5 generations are already cataloged, the oldest generation will be deleted leaving the most recent 5 (provided that the SCRATCH clause was used when the GDG was set up).

```
//JOB001I  JOB (ACCTCODE), 'NAME',
//  CLASS=A,NOTIFY=USERID,MSGCLASS=A
//*
//JS001    EXEC PGM=IDCAMS
//SYSPRINT  DD SYSOUT=*
//SYSIN     DD *
 ALTER    DATASET.GDG.BANAME   SCR LIMIT(5)
//
```

42

Deleting all Generations of a GDG

You may have occasion to delete all gens of a GDG (such as accumulating several gens, processing them all, and then deleting them all). The following JCL will delete all generations of a GDG by using a wild card (*) as the last node of the file name to delete (the G00 part). The GDG base will remain. When you create the next generation, it will use the G0001V00 generation number.

```
//JOB001ID JOB (ACCTCODE), 'NAME',
//  CLASS=A,MSGCLASS=A,NOTIFY=USERID
//*
//****************************************************
//*   IDCAMS TO DELETE ALL GENS OF A GDG
//****************************************************
//*
//STEP1    EXEC PGM=IDCAMS
//SYSPRINT DD SYSOUT=*
//SYSIN    DD *
 DELETE DSNAME.FILEGDG.FILENAME.*
/*
//
```

Deleting an Entire GDG and Index

You may have occasion to completely delete and "un-define" a GDG, such as if you are cleaning up or standardizing file names. The following JCL will delete all generations of a GDG and then remove the base definition from the system. Note the use of the GDG and FORCE parameters. FORCE will cause existing generations of the GDG to be deleted.

```
//JOB001GP JOB (ACCTCODE),'NAME',
//     CLASS=A,MSGCLASS=A,NOTIFY=USERID
//*
//****************************************************
//*   DELETE GDG INDEX AND ALL ENTRIES           *
//****************************************************
//*
//STEP1    EXEC PGM=IDCAMS
//SYSPRINT DD SYSOUT=*
//SYSIN    DD *
   DELETE DSNAME.TEST.FILEX GDG FORCE
/*
//
```

Listing Catalog Entries

When you want a listing of catalog information about a dataset, use the LISTCAT command. The following provides catalog information about a GDG whose base name is DSNAME. GDGFILE.TEST1.

```
//JOB001LC  JOB (ACCTCODE),'NAME',
//  CLASS=B,NOTIFY=USERID,MSGCLASS=A
//*
//STEP1     EXEC PGM=IDCAMS
//SYSPRINT  DD SYSOUT=X
//SYSIN     DD *
 LISTCAT GDG ENT('DSNAME.GDGFILE.TEST1') ALL
/*
//
```

Printing a File

Several IBM utilities allow you to print a file. Here is the IDCAMS print command. In this case, we are requesting character format.

```
//JOB001IP JOB (ACCTCODE),'NAME',
//  CLASS=A,MSGCLASS=A,NOTIFY=USERID
//*
//************************************************
//*    IDCAMS TO PRINT A DATA SET
//************************************************
//*
//STEP1     EXEC PGM=IDCAMS
//SYSPRINT  DD SYSOUT=*
//FILEIN    DD DSN=DSNAME.TEST.FILE,
//             DISP=SHR
//SYSIN     DD *
 PRINT INFILE (FILEIN) CHAR
/*
//
```

44

Printing a File Dump

Use this JCL if you need a hex dump of a file. It may be useful when you are diagnosing a problem, such as a data related abend where a simple character print does not show you the exact value.

```
//JOB001IP JOB (ACCTCODE),'NAME',
//  CLASS=A,MSGCLASS=A,NOTIFY=USERID
//*
//**************************************************
//*    IDCAMS TO PRINT A HEX DUMP OF A FILE
//**************************************************
//*
//STEP1    EXEC PGM=IDCAMS
//SYSPRINT  DD SYSOUT=*
//FILEIN    DD DSN=DSNAME.TEST.FILE,
//             DISP=SHR
//SYSIN     DD *
 PRINT INFILE (FILEIN) -
      DUMP
/*
//
```

IEBCOMPR

Compare Two Sequential Datasets

IEBCOMPR is used to compare two files for difference. An example is if you're performing a regression test, and you want to make sure your baseline and test files are identical. The following provides the compare function. A zero return code means the files are identical. A non-zero RC means the files are not identical.

```
//JOB001IC JOB (ACCTCODE),'NAME',
//  CLASS=A,MSGCLASS=A,NOTIFY=USERID
//**************************************************
//*    COMPARE TWO DATASETS //****************************************************
//*
//STEP1    EXEC PGM=IEBCOMPR
//SYSUT1    DD DSN=DSNAME.TEST.FILE1,DISP=SHR
//SYSUT2    DD DSN=DSNAME.TEST.FILE2,DISP=SHR
//SYSIN     DD DUMMY
//SYSPRINT DD SYSOUT=*
//SYSUDUMP DD SYSOUT=*
//
```

If there are differences between the files, you will receive a report like the following which includes a hex dump of the content of unequal records. In this case only the first record was unequal out of six total records compared.

```
                                    COMPARE UTILITY                                    PAGE 0001
 IEB221I    RECORDS ARE NOT EQUAL
     DDNAME = SYSUT1
     PHYSICAL RECORD NUMBER = 00000001  LOGICAL RECORD NUMBER WITHIN PHYSICAL RECORD = 00000001
 F1F1F2F2D1C5D5D2C9D5E24040404040404040404040404040404040404040404040404040C4C5C2D6D9C1C840404040404040404040404040F0F5F2F0F1
 F660F0F960F0F14040404040404040404040404040
     DDNAME = SYSUT2
     PHYSICAL RECORD NUMBER = 00000001  LOGICAL RECORD NUMBER WITHIN PHYSICAL RECORD = 00000001
 F1F1F2F2D1C5D5D2C9D54040404040404040404040404040404040404040404040404040404040C4C5C2C2C9C54040404040404040404040404040F0F5F2F0F1
 F660F0F960F0F14040404040404040404040404040
     END OF JOB-TOTAL NUMBER OF RECORDS COMPARED = 00000006
```

Compare Two Partitioned Datasets

This example works the same as the above – the only difference is that we are comparing partitioned data sets.

```
//JOB001IC JOB (ACCTCODE),'NAME',
//   CLASS=A,MSGCLASS=A,NOTIFY=USERID
//************************************************
//*    COMPARE TWO DATASETS //*****************************************************
//*
//STEP1    EXEC PGM=IEBCOMPR
//SYSPRINT DD SYSOUT=A
//SYSUT1   DD DSN=DSNAME.TEST.PDSFILE1,DISP=SHR
//SYSUT2   DD DSN=DSNAME.TEST.PDSFILE2,DISP=SHR
//SYSIN    DD  *
   COMPARE  TYPORG=PO
/*
//
```

IEBGENER

Copying a File

IEBGENER is one of the most used MVS utilities. Typically it is used to copy or print a file, although it can also be used to edit a file. Here is the JCL for a plain vanilla copy. You must use SYSUT1 and SYSUT2 as the original and new files respectively. Specifying SYSIN as a dummy dataset means that IEBGENER will perform it's default action, which is to copy SYSUT1 to SYSUT2.

```
//JOB001GN JOB (ACCTCODE),'NAME',
//  CLASS=A,NOTIFY=USERID,MSGCLASS=A
//*
//STEP1    EXEC PGM=IEBGENER
//SYSUT1   DD DSN=DSNAME.TEST.FILE1,
//            DISP=SHR
//SYSUT2   DD DSN=DSNAME.TEST.FILE2,
//            DISP=(NEW,CATLG,DELETE),
//            UNIT=SYSDA,
//            SPACE=(TRK,(5,5),RLSE),
//            RECFM=FB,LRECL=80,BLKSIZE=0,DSORG=PS
//SYSPRINT DD SYSOUT=*
//SYSUDUMP DD DUMMY
//SYSIN    DD DUMMY
//
```

Print a File

To print a file, just use IEBGENER and make sure SYSUT2 specifies a valid print class.

```
//JOB001GN JOB (ACCTCODE),'NAME',
//  CLASS=A,NOTIFY=USERID,MSGCLASS=A
//*
//STEP1    EXEC PGM=IEBGENER
//SYSUT1   DD DSN=DSNAME.TEST.FILE1,
//            DISP=SHR
//SYSUT2   DD SYSOUT=A
//SYSPRINT DD SYSOUT=*
//SYSUDUMP DD DUMMY
//SYSIN    DD DUMMY
//
```

Editing a File - Reformatting

IEBGENER provides a facility for editing a file, although the commands are a bit cryptic. This example reformats a file. That's basically what this "editing" feature is good for: it can save you from having to write and compile a separate program to do a fairly simple edit.

The **GENERATE** and **RECORD FIELD** parameters are mandatory. **MAXFLDS** defines the total number of fields you will specify. **MAXLITS** specifies the total field length of fields for which you will use literals.

IMPORTANT: Remember that on each FIELD parameter, the first numeric value is the **length** of the field you want to manipulate. The second value is either a literal or else the **starting position** of the data field you want to reference. The last number is the **target displacement** (position) to which you want to put the referenced data. Once you have these three values down, this is a very easy utility to use.

In this case, we are reformatting an 60-byte record by adding a literal "ABCD" as the first 4 bytes of the new file, and then taking the first 44 bytes of the original file and applying them to positions 5 through 48 of the new file. Finally, we pad the last 12 bytes of each record with blanks.

```
//JOB001GN JOB (ACCTCODE),'NAME',
//   CLASS=A,NOTIFY=USERID,MSGCLASS=A
//*
//STEP1     EXEC PGM=IEBGENER
//SYSUT1    DD DSN=DSNAME.TEST.FILE1,DISP=SHR
//SYSUT2    DD DSN=DSNAME.TEST.FILE2,
//             DISP=(NEW,CATLG,DELETE),
//             UNIT=SYSDA,
//             SPACE=(TRK,(5,5),RLSE),
//             RECFM=FB,LRECL=60,BLKSIZE=0,DSORG=PS
//SYSPRINT  DD SYSOUT=*
//SYSUDUMP  DD DUMMY
//SYSIN     DD *
         GENERATE MAXFLDS=4,MAXLITS=40
      RECORD FIELD=(4,'ABCD',,1),
             FIELD=(44,1,,5),
             FIELD=(12,'"              ',,49)
/*
//
```

IEBCOPY

Copying a Partitioned Dataset

You will typically use **IEBCOPY** to make a copy of a partitioned dataset. It is most useful for backing up or restoring a PDS. You can specify the input and output DDNAMEs in the COPY statement.

```
//JOB001C2 JOB (ACCTCODE),'NAME',
//   CLASS=A,MSGCLASS=A,NOTIFY=USERID
//**************************************************
//*   COPY A PARTITIONED DS
//**************************************************
//*
//STEP1     EXEC PGM=IEBCOPY
//SYSIN     DD *
  COPY INDD=SYSUT1,OUTDD=SYSUT2
/*
//SYSUT1     DD DSN=DSNAME.PDS.FILE,DISP=SHR
//SYSUT2     DD DSN=DSNAME.PDS.FILE.BACKUP,
//              DISP=(NEW,CATLG,DELETE),
//              UNIT=TAPE,RECFM=FB,LRECL=80,
//              DSORG=PO,BLKSIZE=27920
//SYSUT3     DD UNIT=SYSDA,
//              SPACE=(TRK,(100,100),RLSE)
//SYSUT4     DD UNIT=SYSDA,
//              SPACE=(TRK,(100,100),RLSE)
//SYSPRINT  DD SYSOUT=*
//SYSUDUMP  DD SYSOUT=*
//
```

Copying Specific Members of a PDS

You can also use IEBCOPY to copy a subset of an original PDS. To do this, use the SELECT MEMBER option. The following JCL copies four members of a PDS.

```
//JOB001C2 JOB (ACCTCODE),'NAME',
//      CLASS=A,MSGCLASS=A,NOTIFY=USERID
//*************************************************
//*   COPY MEMBERS OF A PDS TO ANOTHER PDS
//*************************************************
//*
//STEP1    EXEC PGM=IEBCOPY
//SYSIN    DD *
  COPY INDD=SYSUT1,OUTDD=SYSUT2
  SELECT MEMBER=(MEMBER1,MEMBER2,MEMBER3,MEMBER4)
//*
//SYSUT1   DD DSN=DSNAME.PDS.FILE,DISP=SHR
//SYSUT2   DD DSN=DSNAME.PDS.NEWFILE,
//            DISP=(NEW,CATLG,DELETE),
//            UNIT=SYSDA,
//            SPACE=(TRK,(20,20,10),RLSE),
//          RECFM=FB,LRECL=80,DSORG=PO,BLKSIZE=27920
//SYSUT3   DD UNIT=SYSDA,
//            SPACE=(TRK,(60,90),RLSE)
//SYSUT4   DD UNIT=SYSDA,
//            SPACE=(TRK,(60,90),RLSE)
//SYSPRINT DD SYSOUT=*
//SYSUDUMP DD SYSOUT=*
//
```

Compressing a PDS

You can use IEBCOPY to compress the space used by a PDS. To do this, simply specify the input and output dataset as the same name. The following JCL shows this technique.

```
//JOB001C2 JOB (ACCTCODE),'NAME',
//      CLASS=A,MSGCLASS=A,NOTIFY=USERID
//*************************************************
//*   COPY MEMBERS OF A PDS TO ANOTHER PDS
//*************************************************
//*
//STEP1    EXEC PGM=IEBCOPY
//SYSIN    DD *
  COPY INDD=SYSUT1,OUTDD=SYSUT2
//*
//SYSUT1   DD DSN=DSNAME.PDS.FILE,DISP=OLD
//SYSUT2   DD DSN=DSNAME.PDS.FILE,DISP=OLD
//SYSPRINT DD SYSOUT=*
//SYSUDUMP DD SYSOUT=*
//
```

IEFBR14

`IEFBR14` has a variety of uses. Most often it is used to delete a file.

The following JCL deletes the referenced data sets (the second example catalogs a data set). Also, by specifying (MOD, DELETE, DELETE) as the disposition, IEFBR14 first creates the catalog entry for the files if they do not exist. Why do it this way? Because it is flexible, allowing for the possibility that the files to be deleted do not actually exist every time the job runs. If the files do exist, they get deleted. If they do not exist, you prevent a runtime error (trying to delete a non-existent file) by specifying MOD as the first sub-parameter of the disposition.

Deleting a File

```
//JOB001IF JOB (ACCTCODE),'NAME',
//   CLASS=A,NOTIFY=USERID,MSGCLASS=A
//*
//STEP1     EXEC PGM=IEFBR14
//DD1       DD DSN=DSNAME.TEST.FILE1,
//               DISP=(MOD,DELETE,DELETE),
//               UNIT=SYSDA,SPACE=(TRK,(0,0),RLSE)
//SYSPRINT  DD SYSOUT=*
//SYSUDUMP  DD SYSOUT=*
//
```

Cataloging a File

```
//JOB001IF JOB (ACCTCODE),'NAME',
//   CLASS=A,NOTIFY=USERID,MSGCLASS=A
//*
//               NOTIFY=&SYSUID
//*************************************************************
//* CATALOG PS DATA SET USING IEFBR14 UTILITY
//*************************************************************
//STEP1     EXEC PGM=IEFBR14
//SYSPRINT DD SYSOUT=*
//SYSOUT   DD SYSOUT=*
//SYSDUMP  DD SYSOUT=*
//DD1      DD DSN= DSNAME.TEST.FILE1,
//               DISP=(NEW,CATLG,DELETE),
//               SPACE=(TRK,(1,1),RLSE),
//               UNIT=SYSDA,
//               DCB=(DSORG=PS,RECFM=FB,LRECL=23,BLKSIZE=230)
```

IEHLIST

Listing Dataset Information for a Volume

IEHLIST is used to obtain dataset information about a particular DASD volume. Use IEHLIST with the LISTVTOC command to obtain this information. **VOLNAME1** is not a keyword — **it** refers to the actual DASD volume name (you must use a valid volume name for your environment).

```
//JOB001I  JOB (ACCTCODE),'NAME',
//  CLASS=A,NOTIFY=USERID,MSGCLASS=A
//*
//STEP1    EXEC PGM=IEHLIST
//SYSPRINT  DD SYSOUT=*
//SYSUDUMP  DD SYSOUT=*
//DD1       DD UNIT=SYSDA,VOL=SER=VOLNAME1,DISP=SHR
//SYSIN     DD *
  LISTVTOC VOL=SYSDA=VOLNAME1,FORMAT
```

Listing Dataset Information for a PDS

In addition to volume information, you can get detailed information about a partitioned dataset by using IEHLIST with the LISTPDS option. This JCL shows how to do this.

```
//JOB001LP JOB (ACCTCODE),'NAME',
//  CLASS=A,NOTIFY=USERID,MSGCLASS=A
//*
//STEP1    EXEC PGM=IEHLIST
//SYSPRINT  DD SYSOUT=*
//SYSUDUMP  DD SYSOUT=*
//DD1       DD UNIT=SYSALLDA,VOL=SER=VOLNAME,DISP=SHR
//SYSIN     DD *
  LISTPDS    DSNAME=PDS.FILE.NAME,VOL=SYSALLDA=VOLNAME,FORMAT
//
```

SYNCSORT

Strictly speaking, SORT is a utility program, but it is not an MVS JCL utility per se. It does not come with z/OS automatically.[4] As of late 2017 SYNCSORT was owned by Centerbridge Partners. Even though it is not built into z/OS, it is such a commonly used program (I've never worked in a shop that didn't have it) that I've included samples of the operations you can do with it.

SORT is a very common and efficient utility for copying large files. It also has other features that make it more flexible for file manipulation than IEBGENER or IDCAMS. SORT is often the tool of choice for many file operations.

Copying a File

The following example does a simple file copy. SORTIN and SORTOUT are always the DD names for the input and output files respectively.

```
//JOB001SS JOB (ACCTCODE),'NAME',
//   CLASS=A,NOTIFY=USERID,MSGCLASS=A
//*
//***************************************************
//*      DO A SIMPLE COPY OF A FILE USING SORT
//***************************************************
//*
//STEP1     EXEC PGM=SORT
//SORTIN    DD DSN=DSNAME.TEST.FILE1,DISP=SHR
//SORTOUT   DD DSN=DSNAME.TEST.FILE2,
//             DISP=(NEW,CATLG,DELETE),
//             UNIT=SYSDA,
//             SPACE=(TRK,(5,5),RLSE),
//             RECFM=FB,LRECL=80,BLKSIZE=0,DSORG=PS
//SORTWK01  DD UNIT=SYSDA,
//             SPACE=(TRK,(5,5),RLSE)
//SORTWK02  DD UNIT=SYSDA,
//             SPACE=(TRK,(5,5),RLSE)
//SYSIN     DD *
  SORT FIELDS=COPY
//SYSOUT    DD SYSOUT=*
//SYSPRINT  DD SYSOUT=*
//
```

---

4 I wrote the examples for SYNCSORT but your installation may instead use DFSORT which is an IBM product. The examples in this book should work for both SYNCSORT and DFSORT. Shops typically use one or the other but not both since they don't want to pay licensing costs for two similar products.

Sorting a File

Naturally, the SORT utility can be used for sorting a file. The following example shows the syntax for the sort action, which uses four parameters:

```
(Starting position, length, data type, ascending or descending sequence)
```

You can use more than one sort parameter. In the case below, it sorts a file into ascending sequence based on positions 1 through 4 of the file using character data. Then **within** that major order, it further sorts the records into descending sequence using the packed decimal (PD) values in positions 12 through 14.

```
//JOB001SS JOB (ACCTCODE),'NAME',CLASS=A,
//  NOTIFY=USERID,MSGCLASS=A
//**************************************************
//* SORT - USE POSITIONS 1-4 ASCENDING AS PRIMARY //* SORTKEY, AND 12-14 DE-
SCENDING AS SECONDARY KEY
//**************************************************
//STEP1     EXEC PGM=SORT
//SYSOUT    DD SYSOUT=X
//SORTIN    DD DSN=DSNAME.FILE.TOSORT,DISP=SHR
//SORTOUT   DD DSN=DSNAME.FILE.SORTED,
//             DISP=(NEW,CATLG,DELETE),
//             UNIT=TAPE,
//             RECFM=FB,BLKSIZE=8160,LRECL=20
//SORTWK01  DD UNIT=SYSDA,
//             SPACE=(TRK,(90,9),RLSE)
//SORTWK02  DD UNIT=SYSDA,
//             SPACE=(TRK,(90,9),RLSE)
//SYSIN     DD *
  SORT FIELDS=(1,4,CH,A,12,3,PD,D)
 END
 /*
//
```

Sorting Out Selected Records

Sometimes you may need to select certain records from a file. Here are two samples of what you can do with SORT. In the first example, only records with a "Z" value in position one will be written to the SORTOUT file. The second example shows an example of multiple select conditions.

Single Select Condition

```
//JOB001SS JOB (ACCTCODE),'NAME',
//  CLASS=A,NOTIFY=USERID,MSGCLASS=A
//************************************************
//*  SORT OUT RECORDS BASED UPON CRITERIA
//************************************************
//*
//STEP1     EXEC PGM=SORT
//SORTIN    DD DSN=DSNAME.FILE.TESTIN,DISP=SHR
//SORTOUT   DD DSN=DSNAME.FILE.TESTOUT,
//             DISP=(NEW,CATLG,DELETE),
//             UNIT=SYSDA,
//             SPACE=(TRK,(60,10),RLSE),
//             LRECL=250,BLKSIZE=0,RECFM=FB,DSORG=PS
//SORTWK01  DD UNIT=SYSDA,
//             SPACE=(TRK,(50,50),RLSE)
//SORTWK02  DD UNIT=SYSDA,
//             SPACE=(TRK,(50,50),RLSE)
//SYSOUT    DD SYSOUT=*
//SYSPRINT  DD SYSOUT=*
   SORT FIELDS=COPY
   INCLUDE COND=(1,1,CH,EQ,C'Z')
   OUTREC FIELDS=(1,250)
 END
//*
//
```

Multiple Select Condition

Here is what a control card would look like to select all records for which the first character is "Z" and characters 40-41 do not contain "AA".

```
        SORT FIELDS=COPY
        INCLUDE COND=(1,1,CH,EQ,C'Z',AND,
                    40,2,CH,NE,C'AA')
```

As you probably guessed, the Booleans for comparison are "EQ" for equal, "NE" for not equal, etc. Check the references if you need full documentation.

Omitting Selected Records

Sometimes you may need to remove certain records from a file. Here are two samples of how you can do that, and it is exactly the opposite of the INCLUDE examples. In the first example, records with a "Z" value in position one will be omitted from the SORTOUT file. The second example shows an example of multiple OMIT conditions.

Single OMIT Condition

```
//JOB001SS JOB (ACCTCODE),'NAME',
// CLASS=A,NOTIFY=USERID,MSGCLASS=A
//***************************************************
//*   SORT OUT RECORDS BASED UPON CRITERIA
//***************************************************
//*
//STEP1      EXEC PGM=SORT
//SORTIN     DD DSN=DSNAME.FILE.TESTIN,DISP=SHR
//SORTOUT    DD DSN=DSNAME.FILE.TESTOUT,
//              DISP=(NEW,CATLG,DELETE),
//              UNIT=SYSDA,SPACE=(TRK,(60,10),RLSE),
//              LRECL=250,BLKSIZE=0,RECFM=FB,DSORG=PS
//SORTWK01   DD UNIT=SYSDA,SPACE=(TRK,(50,50),RLSE)
//SORTWK02   DD UNIT=SYSDA,SPACE=(TRK,(50,50),RLSE)
//SYSOUT     DD SYSOUT=*
//SYSPRINT   DD SYSOUT=*
   SORT FIELDS=COPY
   OMIT COND=(1,1,CH,EQ,C'Z')
   OUTREC FIELDS=(1,250)
 END
//*
```

Multiple OMIT Condition

Here is what a control card would look like to OMIT all records for which the first character is "Z" and characters 40-41 are NOT "AA".

```
   SORT FIELDS=COPY
   OMIT COND=(1,1,CH,EQ,C'Z',AND,
              40,2,CH,NE,C'AA')
```

Note: Besides **AND**, you can use the Boolean **OR** condition in establishing multiple conditions for both **INCLUDE** and **OMIT**. Just remember what your requirement is and what you intend to do. Oddly, one of the most frequent logic errors I've seen in programming is using one of these Boolean operators where the other was intended.

Sorting Into a Smaller Sized Record

Sometimes you may need to create a new file with a smaller record length than the original. SORT allows you to do this by specifying **OUTREC FIELDS**. The length you specify must match the file parameters on the SORTOUT DD.

In the following example, we have an input file for which the records are 80 bytes long. We extract only positions 1-4, and 40-56 by specifying these as the OUTREC fields. This results in an output file for which the records are 20 positions in length.

```
//JOB001SS JOB (ACCTCODE),'NAME',
// CLASS=A,NOTIFY=USERID,MSGCLASS=A
//************************************************
//*  SORT OUT RECORDS BASED UPON CRITERIA
//************************************************
//*
//STEP1     EXEC PGM=SORT
//SORTIN    DD DSN=DSNAME.FILEIN,DISP=SHR
//SORTOUT   DD DSN=DSNAME.FILEOUT,
//             DISP=(NEW,CATLG,DELETE),
//             UNIT=SYSDA,
//             SPACE=(TRK,(10,5),RLSE),
//             RECFM=FB,LRECL=20,BLKSIZE=0,DSORG=PS
//SORTWK01  DD UNIT=SYSDA,SPACE=(TRK,(16,16),RLSE)
//SORTWK02  DD UNIT=SYSDA,SPACE=(TRK,(16,16),RLSE)
//SYSOUT    DD SYSOUT=*
//SYSIN     DD *
  SORT FIELDS=COPY
  OUTREC FIELDS=(1,4,40,56)
 END
/*
```

Removing Duplicate Records

You can remove duplicate records from a file by using the SUM clause. Whatever you specified as the sort fields will be used to determine whether a record is a duplicate of another record. The example removes duplicate records where the "key" is defined as positions 1-4, and 40-49.

```
//JOB001SR  JOB (ACCTCODE),'NAME',CLASS=A,
//    NOTIFY=USERID,MSGCLASS=A
//************************************************
//*  SORT
//************************************************
//STEP1     EXEC PGM=SORT
//SYSOUT    DD SYSOUT=*
//SYSPRINT  DD SYSOUT=*
//SORTIN    DD DSN=DSNAME.FILEIN,DISP=SHR
//SORTOUT   DD DSN=DSNAME.FILEOUT,
//             DISP=(NEW,CATLG,DELETE),
//             UNIT=SYSDA,
//             SPACE=(TRK,(50,10),RLSE),
//             RECFM=FB,LRECL=250,BLKSIZE=0,DSORG=PS
//SORTWK01  DD UNIT=SYSDA,
//             SPACE=(TRK,(50,10),RLSE)
//SORTWK02  DD UNIT=SYSDA,
//             SPACE=(TRK,(50,10),RLSE)
//SYSIN     DD *
  SORT FIELDS=(1,4,CH,A,40,49,CH,A)
  SUM FIELDS=NONE
 END
 /*
//
```

Removing Unmatched Records

You can do quite a lot with SORT such as creating a file with all matched records from two files, or removing unmatched records from a file. The example I provide here will write a file of records from file 2 that do not exist in file 1. In this case the records with values 234 and 678 in file 2 are not matched to any values in File 1. **Note:** we are using instream data.

The JOINKEYS statements tell us which fields in each file we are comparing. The JOIN UNPAIRED, F2, ONLY means we are only interested in the unpaired records in the second file. REFORMAT tells us which fields we are preserving for output.

```
//JOB001SR  JOB (ACCTCODE),'NAME',CLASS=A,
//   NOTIFY=USERID,MSGCLASS=A
//*************************************************
//*  SORT
//*************************************************
//STEP1     EXEC PGM=SORT
//SYSOUT    DD SYSOUT=*
//SYSPRINT  DD SYSOUT=*
//SORTIN1   DD *
123
456
789
//SORTIN2   DD *
123
234
456
678
789
//SORTOUT   DD DSN=DSNAME.SORTOUT,
//             DISP=(NEW,CATLG,DELETE),
//             UNIT=SYSDA,
//             SPACE=(TRK,(50,10),RLSE),
//             RECFM=FB,LRECL=250,BLKSIZE=0,DSORG=PS
//SYSIN     DD *
//SYSIN     DD *
  JOINKEYS F1=SORTIN1,FIELDS=(1,3,A)
  JOINKEYS F2=SORTIN2,FIELDS=(1,3,A)
  JOIN UNPAIRED,F2,ONLY
  REFORMAT FIELDS=(F2:1,3)
  SORT FIELDS=(1,3,CH,A)
/*
```

Summing Detail Records into Summary Records

You can use SORT to provide a summation of detail records based on a certain field. To do this you specify the field to be summed as follows:

```
SUM FIELDS=(start position, length, data type).
```

For example, suppose we have a file of pay data for some employees that covers several pay periods. We want a summary that shows the total pay by employee number for each employee. Here is a sample file of employee numbers and pay amounts. The employee id is in positions 1-4. Pay date is in positions 7-16. Pay amount is in positions 24-30 (assume an implied decimal point before the last two positions, e.g., 270833 is 2708.33).

```
    ----+----1----+----2----+----3----+----4
    *****************************************
    3217  2017-01-15      0270833
    3217  2017-01-31      0270833
    3217  2017-02-15      0270833
    3217  2017-02-28      0270833
    4720  2017-01-15      0333333
    4720  2017-01-31      0333333
    4720  2017-02-15      0333333
    4720  2017-02-28      0333333
    6288  2017-01-15      0291666
    6288  2017-01-31      0291666
    6288  2017-02-15      0291666
    6288  2017-02-28      0291666
```

To summarize this data, we should sort on the employee id, and sum on the pay amount.

Here is the JCL we can run to do the totals.

```
//USER01A JOB MSGLEVEL=(1,1),NOTIFY=&SYSUID
//*
//*************************************************
//*   SORT WITH SUMMATION
//*************************************************
//STEP1    EXEC PGM=SORT
//SYSOUT   DD SYSOUT=*
//SYSPRINT DD SYSOUT=*
//SORTIN   DD DSN=USER01.PAYDETL,DISP=SHR
//SORTOUT  DD DSN=USER01.PAYTOTL,
//            DISP=(OLD,KEEP,KEEP)
//SORTWK01 DD UNIT=SYSDA,
//            SPACE=(TRK,(1,1),RLSE)
//SORTWK02 DD UNIT=SYSDA,
//            SPACE=(TRK,(1,1),RLSE)
//SYSIN    DD *
  SORT FIELDS=(1,4,CH,A)
  SUM FIELDS=(24,7,ZD)
  END
```

```
/*
//
```

And here is our resulting output file:

```
----+----1----+----2----+----3----+----4-
***************************** Top of
3217   2017-01-15        1083332
4720   2017-01-15        1333332
6288   2017-01-15        1166664
```

We can check the first employee id to make sure our total is correct. And in fact 4 paychecks times 270833 is 1083332 (or 10833.32 if we added the decimal point).

Notice that the pay date is included but only for the January 1st paycheck date. This is because pay date is not part of the sort key, so only a single date made it to the summation records. SORT took the first record for each employee and included the pay date on the summary record. At this point the pay date field is irrelevant because it played no part in summing the records. We could remove it completely by using the OUTREC option to include only the employee id and pay amount fields in the output file. The next JCL example shows this.

```
//USER01A JOB MSGLEVEL=(1,1),NOTIFY=&SYSUID
//*
//****************************************************
//*   SORT WITH SUMMATION
//****************************************************
//STEP1     EXEC PGM=SORT
//SYSOUT    DD SYSOUT=*
//SYSPRINT  DD SYSOUT=*
//SORTIN    DD DSN=USER01.PAYDETL,DISP=SHR
//SORTOUT   DD DSN=USER01.PAYTOTL,
//             DISP=(OLD,KEEP,KEEP)
//SORTWK01  DD UNIT=SYSDA,
//             SPACE=(TRK,(1,1),RLSE)
//SORTWK02  DD UNIT=SYSDA,
//             SPACE=(TRK,(1,1),RLSE)
//SYSIN     DD *
  SORT FIELDS=(1,4,CH,A)
  SUM FIELDS=(24,7,ZD)
  OUTREC FIELDS=(1,4,1C' ',24,7,68C' ')
 END
/*
```

The above specifies we are to include in the output the first four bytes from the input, plus a single space, then bytes 24-30 from the input file, and finally 68 spaces. This fills up an 80 byte record. Of course you could also use a 12 byte output record and not need the last 68 bytes. We're just using 80 byte records for convenience here.

Here is the result:

```
----+----1----+----2----+----3----+----4----+----5----+----6----+----7----+----8
******************************* Top of Data ***********************************
3217 1083332
4720 1333332
6288 1166664
```

To conclude, SORT is a very powerful tool, especially when you put together combinations of the various options. Besides data selection and formatting, you can use it to aggregate totals for data fields, thus avoiding having to write a program for simple operations. I recommend that you use SORT wherever possible.

SUPERC

SUPERC is a handy file comparison utility that is often invoked online but has a batch option as well. If you specify batch mode using the online utility you can capture the JCL and subsequently run it yourself. The JCL looks like the following. OLDDD and NEWDD are the DD-NAMEs for the datasets to be compared.

Comparing the Contents of Two Files

You compare the contents of two files as follows:

```
//JOB001B  JOB  (ACCTCODE),'PROGRAMMER',
//    CLASS=A,MSGCLASS=A,NOTIFY=USERID
//*
//STEP1 EXEC PGM=ISRSUPC,
//           PARM=(CHNGL,LINECMP,
//              '',
//              '')
//NEWDD  DD DSN=PDSFILE1(MEMBER1),
//         DISP=SHR
//OLDDD  DD DSN=PDSFILE2(MEMBER),
//         DISP=SHR
//OUTDD  DD SYSOUT=*
```

Searching a Dataset for a String Value

You can search a file for a particular string s follows:

```
//JOB001W  JOB ACCTCODE),'PROGRAMMER',CLASS=A,MSGCLASS=A,
//         NOTIFY=USERID
//*
//SEARCH  EXEC PGM=ISRSUPC,
//             PARM=(SRCHCMP,
//               '')
//NEWDD  DD DSN=FILE.TO.BE.SEARCHD,
//         DISP=SHR
//OUTDD  DD SYSOUT=(X)
//SYSIN  DD *
SRCHFOR   'STRING VALUE I AM LOOKING FOR'
/*
//
```

Chapter 2 Review Questions

1.   What is the purpose of the `IEFBR14` utility?

2.   What `IDCAMS` keyword do you specify to copy a file?

3.   Which IBM JCL utility is used to duplicate a partitioned dataset?

4.   How do you catalog an uncataloged dataset with a JCL?

5.   What are some capabilities of the `IDCAMS` utility in JCL?

Chapter 3 : COBOL Programming Language

Introduction

COBOL is an acronym for Common Business-Oriented Language. It's a third generation procedural language that has been around since 1959. COBOL was developed primarily for business use. It generally focuses on record or database input/out, as well as calculations and reports.

You may hear that COBOL is gone or on the way out, but it's demise is probably exaggerated. While lots of mainframe COBOL programming has been rewritten in other languages (or replaced by commercial packages), COBOL is still heavily used in legacy applications that run on IBM mainframe computers. This is true especially for banking and finance applications. [5]

If you are used to programming in languages such as C or Java, you'll find that COBOL is a bit more English-like. This can be an advantage or disadvantage depending on how you look at it. For example, a data assignment to a variable in most programming languages can be as simple as:

```
X = 27;
```

In COBOL you would code this same operation as:

```
MOVE 27 TO X;
```

The latter is somewhat more verbose. It also has the receiving variable on the right side of the equation which is different.

Also, in COBOL you can spell out Booleans such as > or < using English phrases, such as:

```
IF X IS GREATER THAN 100
```

Whatever your view of the verbosity, COBOL is a very usable language and worth learning. As legacy COBOL programmers retire, there is still much code to be maintained or converted. This chapter will help make you productive for these tasks quickly.

---

[5] https://thenewstack.io/cobol-everywhere-will-maintain/

Language Basics

Programming Format

Unlike more freeform languages, COBOL is very particular about exactly where you can put executable code. Here is a summary of the formatting rules you must follow in COBOL.

Area Name	Column(s)	Usage
Sequence number area	1-6	Originally used for card/line numbers, this area is ignored by the compiler
Indicator area	7	The following characters are allowed here: • * – Comment line • / – Comment line that will be printed on a new page of a source listing • - – Continuation line, where words or literals from the previous line are continued • D – Line enabled in debugging mode, which is otherwise ignored
Area A	8-11	This contains: DIVISION, SECTION and procedure headers; 01 and 77 level numbers and file/report descriptors
Area B	12-72	Any other code not allowed in Area A
Program name area	73-	Historically up to column 80 for punched cards, it is used to identify the program or sequence the card belongs to

If you do not follow these rules, you will receive compiler errors which tend to cascade (one error causes multiple others). The model programs in this book will help keep you out of trouble.

Four Divisions

The statements, entries, paragraphs, and sections of a COBOL source program are grouped into the following four divisions:

1. Identification Division

2. Environment Division

3. Data Division

4.   Procedure Division

Only the Identification division is required, but you can't do much without the others.

IDENTIFICATION DIVISION

The identification division can specify the program name, author, data written and a few other pieces of information. The only mandatory element is the program name.

ENVIRONMENT DIVISION

The environment division consists of a configuration section and an input-output section.

Configuration Section.

The Configuration Section is optional. If you include it you can specify the source and object computer upon which the program was written and run, respectively. This is simply documentation and has no affect when the program is compiled.

Input-Output Section.

The Input-Output Section is used to define external input and output files, including the record description that the program requires to read and write to the files.

File Control.

We'll get to this in the file I/O section of this text.

I/O Control.

The I/O-CONTROL paragraph is optional. It specifies the storage areas to be shared by different files. I've never actually seen this used.

DATA DIVISION

The Data Division is divided into four sections:

File Section

Describes externally stored data (including sort-merge files).

Working-Storage Section
Describes internal data such as variables and structures used for computation, strings manipulation, reports, etc.

Local-Storage Section
Describes internal data that is allocated on a per-invocation basis.

Linkage Section
Describes data made available by another program. It appears in the called program and describes data items that are provided by the calling program and are referred to by the called program.

PROCEDURE DIVISION
The Procedure Division contains executable statements and sentences organized into sections or paragraphs. Statements start with a verb such as MOVE, COMPUTE or PERFORM. Sentences consist of one or more statements. Sentences always end with a period.

The program execution starts with the first statement. The Procedure Division ends at the physical end of the source code (i.e., when no more statements appear).

Optional Entries and Sections

The following Identification Division entries are optional and have no affect on the compiler.

AUTHOR
Name of the author of the program. It is syntax checked only.

INSTALLATION
Name of the company or location. It is syntax checked only.

DATE-WRITTEN
Date the program was written. It is syntax checked only.

DATE-COMPILED
Date the program was compiled.

SECURITY
Level of confidentiality of the program.

Variables, Data Types and Assignment

Like most programming languages, COBOL requires you to declare the data elements (variables and structures) you will use in your program. You put your variable declarations in the Data Division of the program, typically either the Working Storage or Linkage sections.

A variable declaration must include a level number, variable name and a variable type. It can also include an initial value by using the VALUE clause. Variable names are always preceded by a level number. If you use level 01 or 77 for variable declarations, your declaration must begin in Area A. All other level numbers must begin in Area B. Variable names must use alphanumeric characters and they can include a dash character for readability (see examples below).

For top level or standalone variables, the number is 1 (or 01 by convention), or else 77. The difference between a 01 level and a 77 level is that the 77 level can only define a single variable. In contrast, the 01 level can either be a single variable or it can define a structure with multiple variables (sometimes referred to as a record).

Here are some examples of declaring individual variables.

```
01 EMPLOYEE-NAME     PIC X(30).

77 EMPLOYEE-COUNT    PIC S9(4) USAGE IS BINARY.
```

Data Types

You'll notice the PIC clause in the above variable declarations. A PICTURE (or PIC) clause is a sequence of characters, each of which represents a portion of the data item and what it can contain. Alphabetic-only data can use an A picture. More commonly, we use an alphanumeric picture X to define data that is not exclusively numeric. You also specify the length of the picture variable. In the case of EMPLOYEE-NAME above we have specified 30 bytes – PIC X(30). Notice also the declaration must end with a period.

For numeric variables you specify PIC 9 and the number of digits, and optionally a sign character which is S. The EMPLOYEE-COUNT variable is 4 digits plus a sign character S (meaning the positive or negative sign is stored with the numeric value). Our declaration also specifies USAGE IS BINARY which indicates it is to be stored in binary format as opposed to display format. You can always leave off the USAGE IS part of the phrase and simply specify the usage. Variables which are to be used for computation or indexing usually are defined as `BINARY` or `COMP`.

Note finally that for `EMPLOYEE-COUNT` you can specify either `9(4)` or `9999` for `EMPLOY-EE-COUNT`. The meaning is the same.

```
77 EMPLOYEE-COUNT    PIC S9999 BINARY.
77 EMPLOYEE-COUNT    PIC S9(4) BINARY.
```

You can also specify an **implied** decimal point in your numeric data by including a V in the picture definition. For example:

```
77 EMPLOYEE-PAY    PIC 9(5)V9(2).
```

The above specifies 7 digits, two of which are on the right hand side of an implied decimal point. This actually does not change the storage of the variable, but tells the compiler how to align the results of computations.

Finally, a bit more on the `USAGE IS` clause which specifies the data format that the data is stored in. The default is `DISPLAY` format. The formats are:

`USAGE DISPLAY`, the default format, where data is stored as a string.

`USAGE BINARY` where a minimum size is either specified by the `PICTURE` clause or by a `USAGE` clause such as `BINARY-LONG`.

`USAGE COMPUTATIONAL` (or `USAGE COMP`) where data may be stored in whatever format the implementation provides; often equivalent to `USAGE BINARY`.

`USAGE PACKED-DECIMAL`, where data is stored in the smallest possible decimal format (typically packed binary-coded decimal).

COMP Variables

`COMP` variables are binary numeric types that vary in the precision and type of storage as follows:

COMP (Computational) – the compiler decides the data type which is usually the most efficient (typically `BINARY` format).

70

COMP-1 – typically a single precision floating point value, stored in 4 bytes.

COMP-2 – typically a double precision floating point value, stored in 8 bytes.

COMP-3 - stores data in a binary coded decimal format – the sign is placed after the least significant digit. This is often referred to as **packed decimal**.

When declaring variables you can also assign an initial value using the VALUE clause. To continue the examples of the variables we declared earlier:

```
01 EMPLOYEE-NAME      PIC X(30) VALUE SPACES.
77 EMPLOYEE-COUNT     PIC S9(4) BINARY VALUE 0.
```

Data Structures

Data items in COBOL are declared hierarchically through the use of level-numbers which indicate if a data item is a grouping of items, a part of a grouping or a standalone data item. An item with a higher level-number is subordinate to an item with a lower one. Top-level data items, with a level-number of 01, are sometimes called records. Items that have subordinate aggregate data items are called group items; those that do not are called elementary items. Level-numbers used to describe standard data items are between 1 and 49. Here is a sample COBOL data structure.

```
01 EMPLOYEE-INFO,
    05 EMP-NAME,
        10 EMP-LAST-NAME      PIC X(25).
        10 EMP-FIRST-NAME     PIC X(15).
        10 EMP-MI             PIC X(01).
    05 EMP-ADDRESS,
        10 EMP-ADDR-1         PIC X(30).
        10 EMP-ADDR-2         PIC X(30).
        10 EMP-CITY           PIC X(25).
        10 EMP-STATE          PIC X(15).
        10 EMP-ZIP            PIC 9(05).
```

Sample Program

Okay, so let's do the obligatory "Hello World" program. We only need the Identification division and the Procedure division for this program, so that's all we'll include. We'll name the program COBHELO. We'll use the DISPLAY verb and the literal value "HELLO WORLD" to implement the program. Here is our code.

```
IDENTIFICATION DIVISION.
PROGRAM-ID.    COBHELO.
********************************************************
```

```
*       SIMPLE HELLO WORLD PROGRAM                        *
*****************************************************
      PROCEDURE DIVISION.

* THIS IS A COMMENT
      MAIN-PARA.

            DISPLAY 'HELLO WORLD'
            GOBACK
```

The GOBACK statement returns control to the operating system.

Now we must compile/link the program according to the procedures in our shop. I'll show you the procedure I use in my shop, but you'll need to get the correct procedure from your technical leader or supervisor. I run the following JCL to compile the program:

```
//USER01D JOB MSGLEVEL=(1,1),NOTIFY=&SYSUID
//*
//*   COMPILE A COBOL PROGRAM
//*
//CL      EXEC COBOLCL,
//              COPYLIB=USER01.COPYLIB,          <= COPYBOOK LIBRARY
//              LOADLIB=USER01.LOADLIB,          <= LOAD LIBRARY
//              SRCLIB=USER01.COBOL.SRCLIB,      <= SOURCE LIBRARY
//              MEMBER=COBHELO                   <= SOURCE MEMBER
```

The output appears in SDSF and you can browse it there. If there are any errors, they will be flagged here. In this case, we do have an error. We forgot to put a period at the end of the last sentence in the program, so our source code was not ended correctly.

```
PP 5655-S71 IBM Enterprise COBOL for z/OS  4.2.0          COBHELO   Date 01/31/2018
 Defined    Cross-reference of programs      References

       2   COBHELO
PP 5655-S71 IBM Enterprise COBOL for z/OS  4.2.0          COBHELO   Date 01/31/2018
LineID  Message code  Message text

     15  IGYSC1082-E   A period was required.  A period was assumed before "END OF PROGRAM"
Messages    Total    Informational    Warning    Error    Severe    Terminating
Printed:      1                                     1
* Statistics for COBOL program COBHELO:
*     Source records = 15
*     Data Division statements = 0
*     Procedure Division statements = 2
End of compilation 1,  program COBHELO,  highest severity 8.
Return code 8
z/OS V1 R13 BINDER     03:20:23 WEDNESDAY JANUARY 31, 2018
BATCH EMULATOR  JOB(USER01D ) STEP(CL      ) PGM= IEWBLINK  PROCEDURE(LKED    )
```

So let's fix the program. You can see that the revised version has the period after the last sentence.

```
    IDENTIFICATION DIVISION.
    PROGRAM-ID.    COBHELO.

   *************************************************
   *       SIMPLE HELLO WORLD PROGRAM             *
   *************************************************

    PROCEDURE DIVISION.

   * THIS IS A COMMENT

    MAIN-PARA.

       DISPLAY 'HELLO WORLD'
       GOBACK.
```

Once we correct the error we can recompile. This time we have a good compile because it says no statements flagged (and we received a zero return code on both the compile and link edit steps).

```
    Defined    Cross-reference of procedures    References

        12    MAIN-PARA
   PP 5655-S71 IBM Enterprise COBOL for z/OS  4.2.0             COBHELO   Date 01/31/2018
    Defined    Cross-reference of programs      References

         2    COBHELO
   * Statistics for COBOL program COBHELO:
   *    Source records = 15
   *    Data Division statements = 0
   *    Procedure Division statements = 2
   End of compilation 1,  program COBHELO,  no statements flagged.
   Return code 0
   z/OS V1 R13 BINDER      03:20:23 WEDNESDAY JANUARY 31, 2018
   BATCH EMULATOR  JOB(USER01D ) STEP(CL      ) PGM= IEWBLINK  PROCEDURE(LKED    )
   IEW2008I 0F03 PROCESSING COMPLETED.  RETURN CODE = 0.
```

Now we can run the program using this JCL:

```
//USER01D JOB MSGLEVEL=(1,1),NOTIFY=&SYSUID
//*
//*  RUN A COBOL PROGRAM
//*
//STEP01  EXEC PGM=COBHELO
//STEPLIB  DD   DSN=USER01.LOADLIB,DISP=SHR
//SYSOUT   DD   SYSOUT=*
```

And we can review output on SDSF. As you can see, the "HELLO WORLD" was printed.

```
SDSF OUTPUT DISPLAY USER01D  JOB08443 DSID   101 LINE 1      COLUMNS 02- 81
 COMMAND INPUT ===>                                          SCROLL ===> CSR

HELLO WORLD
******************************* BOTTOM OF DATA ********************************
```

Ok for our next program, let's include all four divisions and we'll also include some optional entries in the Identification division. Instead of displaying the literal "HELLO WORLD", let's declare a variable and we'll load the literal value into the variable at run time. Finally, we'll display the content of the variable. This program will produce exactly the same result as the COBHELO program.

Our program name is COBTRN1 and here is the listing. In the Working Storage section we have declared a variable named WS-MESSAGE as a 12 byte container for an alphanumeric value. In the procedure MAIN-PARA we copy the literal "HELLO WORLD" into WS-MESSAGE. Finally we display the value of WS-MESSAGE.

```
        IDENTIFICATION DIVISION.
        PROGRAM-ID.    COBTRN1.
        AUTHOR.        ROBERT WINGATE.
        INSTALLATION.  SUNSET SERVICES.
        DATE-WRITTEN.  JANUARY 15, 2018.
        DATE-COMPILED. JANUARY 15, 2018.
        SECURITY.      NON-CONFIDENTIAL.
        *****************************************************
        *        SIMPLE HELLO WORLD PROGRAM                *
        *****************************************************

        ENVIRONMENT DIVISION.
        CONFIGURATION SECTION.
        SOURCE-COMPUTER. IBM-ZOS.
        OBJECT-COMPUTER. IBM-ZOS.

        DATA DIVISION.
        WORKING-STORAGE SECTION.

        01 WS-MESSAGE    PIC X(12).

        PROCEDURE DIVISION.

        * THIS IS A COMMENT

        MAIN-PARA.
```

```
MOVE "HELLO WORLD" TO WS-MESSAGE
DISPLAY WS-MESSAGE
GOBACK.
```

This program is just to demonstrate some of the optional entries, and to introduce the use of variables. Our next programs will not include most of the optional Identification division entries. Just be aware that you can use these if you want to.

Sequence, Selection, Iteration

Structured programming involves writing code that controls the execution of the program. This includes the primary concepts sequence, selection and iteration.

Sequence

Sequence means that program statements are executed sequentially according to the order in which they occur either in the main procedure of a program, or within sub-procedures which are called either paragraphs or sections.[6] For example, assume we have two variables VARI-ABLE-A and VARIABLE-B already declared in a program. The following code will execute sequentially.

```
ADD +1 to VARIABLE-A
MULTIPLY VARIABLE-A BY 2 GIVING VARIABLE-B
DISPLAY VARIABLE-B
```

The above is an example of the **sequence** control structure and it will occur throughout the program unless one of the other two control structures intervenes. Sequence also involves invoking other paragraphs or sections of the program by using the PERFORM verb (this is the same as the CALL verb in other languages).

Selection

Selection means the program will execute (or not execute) statements based on a condition. For example, given a variable RECORD-COUNTER, we could display the number of records if the value is greater than zero, or display a literal if it is zero.

```
IF RECORD-COUNTER > 0 THEN
    DISPLAY 'NUMBER OF RECORDS IS ' RECORD-COUNTER
ELSE
    DISPLAY 'NO RECORDS WERE PROCESSED'.
```

Another example: You may have several paragraphs that you may call depending on your data

---

6 The only practical difference between paragraphs and sections is that sections can contain multiple paragraphs. If you perform a section then all paragraphs in the section will be executed. If you perform a paragraph, only the code in that one paragraph is executed.

values. You can use IF/THEN logic to control which paragraph gets called.

```
IF COUNTRY EQUAL 'USA' THEN
    PERFORM P100-PROCESS-DOMESTIC
ELSE
    PERFORM P200-PROCESS-INTERNATIONAL
END-IF.
```

Iteration

Iteration means repeating an action until some condition is met. The condition can use a counter to ensure the action is executed for a specified number of times, or it can be a switch whose value indicates a condition is true.

COBTRN2

The COBTRN2 example program will include sequence, selection and iteration. We will implement the following:

- The flow of a sequential set of instructions

- Branching both with IF/THEN logic and with EVALUATE (case logic)

- Iteration using PERFORM X TIMES, and PERFORM UNTIL

Here is our program code. We use a counter variable CNTR. We follow a sequence of statements. We use both IF/THEN and EVALUATE logic for branching. We perform a procedure a specific number of times. Finally we perform a procedure until the value of the counter reaches a certain value.

```
IDENTIFICATION DIVISION.
PROGRAM-ID. COBTRN2.

****************************************************
*     PROGRAM WITH SEQUENCE, SELECTION AND         *
*     ITERATION.                                   *
****************************************************

ENVIRONMENT DIVISION.
DATA DIVISION.
WORKING-STORAGE SECTION.

77 CNTR          PIC S9(9) USAGE COMP VALUE +0.
```

76

```
PROCEDURE DIVISION.

MAIN-PARA.
    DISPLAY 'COBOL WITH SEQUENCE, SELECTION AND ITERATION'.

    DISPLAY '** PROCESSING IF/THEN SELECTION'

    IF CNTR = 0 THEN
       PERFORM P100-ROUTINE-A

    ADD +1 TO CNTR

    IF CNTR GREATER THAN 0 THEN
       PERFORM P200-ROUTINE-B
     END-IF

    DISPLAY '** PROCESSING CASE TYPE SELECTION'

    MOVE ZERO TO CNTR

    EVALUATE CNTR
       WHEN 0          PERFORM P100-ROUTINE-A
       WHEN 1          PERFORM P200-ROUTINE-B
       WHEN OTHER      DISPLAY 'NO ROUTINE TO PERFORM'
    END-EVALUATE

    ADD +1 TO CNTR

    EVALUATE CNTR
       WHEN 0          PERFORM P100-ROUTINE-A
       WHEN 1          PERFORM P200-ROUTINE-B
       WHEN OTHER      DISPLAY 'NO ROUTINE TO PERFORM'
    END-EVALUATE

    ADD +1 TO CNTR

    EVALUATE CNTR
       WHEN 0          PERFORM P100-ROUTINE-A
       WHEN 1          PERFORM P200-ROUTINE-B
       WHEN OTHER      DISPLAY 'NO ROUTINE TO PERFORM'
    END-EVALUATE.

    PERFORM P100-ROUTINE-A 3 TIMES

    MOVE 0 TO CNTR

    PERFORM P300-ROUTINE-C VARYING CNTR FROM +1 BY +1
       UNTIL CNTR = 5

    GOBACK.

P100-ROUTINE-A.
```

```
        DISPLAY 'PROCESSING IN P100-ROUTINE-A'.
        DISPLAY 'LEAVING P100-ROUTINE-A'.

   P200-ROUTINE-B.

        DISPLAY 'PROCESSING IN P200-ROUTINE-B'.
        DISPLAY 'LEAVING P200-ROUTINE-B'.

   P300-ROUTINE-C.

        DISPLAY 'PROCESSING IN P300-ROUTINE-C'.
        DISPLAY 'ITERATOR VALUE IS '  CNTR
        DISPLAY 'LEAVING P300-ROUTINE-C'.
```

When we compile and execute this program, the results are as follows:

```
COBOL WITH SEQUENCE, SELECTION AND ITERATION
** PROCESSING IF/THEN SELECTION
PROCESSING IN P100-ROUTINE-A
LEAVING P100-ROUTINE-A
PROCESSING IN P200-ROUTINE-B
LEAVING P200-ROUTINE-B
** PROCESSING CASE TYPE SELECTION
PROCESSING IN P100-ROUTINE-A
LEAVING P100-ROUTINE-A
PROCESSING IN P200-ROUTINE-B
LEAVING P200-ROUTINE-B

NO ROUTINE TO PERFORM
PROCESSING IN P100-ROUTINE-A
LEAVING P100-ROUTINE-A
PROCESSING IN P100-ROUTINE-A
LEAVING P100-ROUTINE-A
PROCESSING IN P100-ROUTINE-A
LEAVING P100-ROUTINE-A
PROCESSING IN P300-ROUTINE-C
ITERATOR VALUE IS 000000001
LEAVING P300-ROUTINE-C
PROCESSING IN P300-ROUTINE-C
ITERATOR VALUE IS 000000002
LEAVING P300-ROUTINE-C
PROCESSING IN P300-ROUTINE-C
ITERATOR VALUE IS 000000003
LEAVING P300-ROUTINE-C
PROCESSING IN P300-ROUTINE-C
ITERATOR VALUE IS 000000004
LEAVING P300-ROUTINE-C
```

File I/O

Starting with program COBTRN3 we will be working with a fictitious Human Resource application. Program COBTRN3 will read a file of employee pay information, reformat it and then write it to an output file. This program will include the following:

- File definition
- Read file input
- Write file output
- Use 88 level variables for switches

Often it is helpful to pseudo code your program design before you start coding. You pseudo code need not be extremely elaborate, but it helps you to think through the program structure. The following pseudo code specifies what program COBTRN3 will do.

```
Announce start of program
Open Files
Do Priming Read
Do Until End of Input File
        Move Input Fields to Output Fields
        Display Pay Values
        Write Output Record
        Read Next Input Record
End Do Until
Close Files
Announce End of Program
```

Program Listing for COBTRN3

To use input and/or output files in a COBOL program, you must declare file names and file/record descriptors. The file names are coded in the Input-Output section of the Environment Division. The file descriptors are coded in the File-Section of the Data Division.

For the Input-Output section of the Environment division, we must declare a file name and assign it to an identifier whose name corresponds to the DD name of the file in the execution JCL. For example, here's the JCL we will use to run the program. Notice there is an **EMPIFILE** DD and an **EMPOFILE** DD. Those are the input and output files respectively.

```
//USER01D JOB MSGLEVEL=(1,1),NOTIFY=&SYSUID
//*
//*  RUN A COBOL PROGRAM
//*
//STEP01  EXEC PGM=COBTRN3
//STEPLIB  DD  DSN=USER01.LOADLIB,DISP=SHR
//SYSOUT   DD  SYSOUT=*
//EMPIFILE DD DSN=USER01.EMPLOYEE.PAY,DISP=SHR
//EMPOFILE DD DSN=USER01.EMPLOYEE.PAYOUT,DISP=(OLD,KEEP,KEEP)
//SYSPRINT DD  SYSOUT=*
//SYSUDUMP DD  SYSOUT=*
//SYSOUT   DD  SYSOUT=*
```

To reference these files in our COBOL program, we will create COBOL file name variables, and then reference the DD names of the files from the JCL. Here is what we will code in the Input-Output section of the program.

```
SELECT EMPLOYEE-IN-FILE   ASSIGN TO EMPIFILE.
SELECT EMPLOYEE-OUT-FILE  ASSIGN TO EMPOFILE.
```

This above means the file we refer to in our program as EMPLOYEE-IN-FILE is the file that has DD name EMPIFILE in the JCL we use to execute the program. Similarly, the EMPLOYEE-OUT-FILE in our program refers to the file with DD name EMPOFILE in the JCL. So far, our program looks like this:

```
IDENTIFICATION DIVISION.
PROGRAM-ID. COBTRN3.

*****************************************************
*       PROGRAM USING FILE INPUT AND OUTPUT       *
*       TO REFORMAT EMPLOYEE PAY INFORMATION.     *
*****************************************************

ENVIRONMENT DIVISION.
INPUT-OUTPUT SECTION.

    FILE-CONTROL.
        SELECT EMPLOYEE-IN-FILE   ASSIGN TO EMPIFILE.
        SELECT EMPLOYEE-OUT-FILE  ASSIGN TO EMPOFILE.
```

Now we must provide file descriptors for each of the two files in the File-Section of the Data Division. Here, we will use the file name we created in the File-Control section. We will code FD (which means file descriptor), followed by the file name, and then a record structure name. Here's an example:

```
FD EMPLOYEE-IN-FILE.
01 EMPLOYEE-RECORD-IN.
    05  E-ID          PIC X(04).
    05  FILLER        PIC X(76).
```

The file descriptor above says that by default our input file will be read into structure EMPLOY-EE-RECORD-IN. The EMPLOYEE-RECORD-IN structure does not have to be detailed here if you want to use a different structure later. I've seen it done both ways, but most commonly I see a basic structure defined in the FD and a more detailed structure defined in working storage with the latter used in the actual READ statement. We will do it that way.

Before we move on, I want to suggest a couple of optional entries to use on the file descriptor. Although not required, I recommend that you include the: RECORDING MODE IS, RECORD CONTAINS X CHARS, and the DATA RECORD IS clauses. So your file descriptor for this file would be:

```
FD  EMPLOYEE-IN-FILE
    RECORDING MODE IS F
    RECORD CONTAINS 80 CHARACTERS
    DATA RECORD IS EMPLOYEE-RECORD-IN.

01 EMPLOYEE-RECORD-IN.
    05  E-ID          PIC X(04).
    05  FILLER        PIC X(76).
```

RECORDING MODE IS F – means that the format of the records is fixed. If you were using variable length records you would specify RECORDING MODE IS V. If you do not specify a recording mode, COBOL will assume mode F. This is not a problem as long as you are using fixed length records. If you are not, you'll get a run time error. In any case, it's good documentation to specify the recording mode.

RECORD CONTAINS 80 CHARACTERS – specifies that the logical record length is 80 bytes. This is helpful for the next programmer as documentation.

DATA RECORD IS EMPLOYEE-RECORD-IN – explicitly ties the file descriptor to a defined data structure. This entry can simply be a dummy structure with the correct record length. You do not have to use this structure when you actually reference the file. By default when you read or write data to a file, the designated data record structure in the FD is the one it will be read into. So if you choose to read into EMPLOYEE-RECORD-IN, you only need to code the following for the read statement:

```
READ EMPLOYEE-IN-FILE
```

However, if you want the record read into a different data structure, you must include the INTO clause in your read statement. Suppose in our program we actually want to read into a different structure named IN-EMPLOYEE-RECORD which we'll define in working storage. In that case, our READ statement will need to be:

```
READ EMPLOYEE-IN-FILE INTO IN-EMPLOYEE-RECORD
```

Ok, let's define the structure for EMPLOYEE-IN-FILE in working storage. It will be helpful to look at the actual file content. Let's say that the employee id occupies the first 4 bytes, the regular pay uses bytes 10 through 16, and the bonus pay bytes uses 19 through 24. Here is a file whose structure matches that definition. You can verify the placement of the data fields by entering COLS in the command field of the BROWSE window (this will give you a ruler of sorts).

```
BROWSE     USER01.EMPLOYEE.PAY                    Line 00000000 Col 001 080
  Command ===>                                              Scroll ===> CSR
----+----1----+----2----+----3----+----4----+----5----+----6----+----7----+----
******************************* Top of Data ***********************************
1111    8700000  670000
1122    8200000  600000
3217    6500000  550000
4175    5500000  150000
4720    8000000  250000
4836    6200000  220000
6288    7000000  200000
7459    8500000  450000
9134    7500000  250000
****************************** Bottom of Data *********************************
```

Now let's code the structure which is named IN-EMPLOYEE-RECORD. Note the use of the reserved world **FILLER** which is reused to place spacing bytes in the record.

```
01  IN-EMPLOYEE-RECORD.
    05   EMP-ID-IN      PIC X(04).
    05   FILLER         PIC X(05).
    05   REG-PAY-IN     PIC 99999V99.
    05   FILLER         PIC X(02).
    05   BON-PAY-IN     PIC 9999V99.
    05   FILLER         PIC X(54).
```

Remember that when we want to read the file into this structure we must code:

```
READ EMPLOYEE-IN-FILE INTO IN-EMPLOYEE-RECORD
```

Now let's look at defining the output file. The file descriptor is quite similar to the input file.

```
FD  EMPLOYEE-OUT-FILE
    RECORDING MODE IS F
    RECORD CONTAINS 80 CHARACTERS
    DATA RECORD IS EMPLOYEE-RECORD-OUT.

  01 EMPLOYEE-RECORD-OUT.
      05  EMP-DATA      PIC X(80).
```

Our next step is to define a detailed output record structure in working storage. Suppose we want to save some space by compressing the readable numbers into a more compact format. We can make the employee id BINARY format, and we'll use a packed decimal (COMP-3) format for our pay values. Also notice we use an implied decimal point on the money fields).

Here is our output structure:

```
01 OUT-EMPLOYEE-RECORD.
   05   EMP-ID-OUT    PIC S9(9) USAGE COMP.
   05   FILLER        PIC X(05).
   05   REG-PAY-OUT   PIC S9(6)V9(2) USAGE COMP-3.
   05   FILLER        PIC X(02).
   05   BON-PAY-OUT   PIC S9(6)V9(2) USAGE COMP-3.
   05   FILLER        PIC X(59) VALUE SPACES.
```

Finally, we need code to write a record to the output file by specifying the WRITE verb, the file name and the record structure.

```
WRITE EMPLOYEE-RECORD-OUT FROM OUT-EMPLOYEE-RECORD
```

Now let's briefly return to our pseudo code for the roadmap of the program.

> Announce start of program
> Open Files
> Do Priming Read
> Do Until End of Input file
> > Move Input Fields to Output Fields
> > Display Pay Values
> > Write Output Record
> > Read Next Input Record
> End Do Until
> Close Files
> Announce End of Program

And here is our program implementing the pseudo code elements we provided above plus some program control features.

```
IDENTIFICATION DIVISION.
PROGRAM-ID. COBTRN3.
**********************************************************
*       PROGRAM USING FILE INPUT AND OUTPUT          *
*       TO REFORMAT EMPLOYEE PAY INFORMATION.        *
**********************************************************
ENVIRONMENT DIVISION.
INPUT-OUTPUT SECTION.

    FILE-CONTROL.
        SELECT EMPLOYEE-IN-FILE   ASSIGN TO EMPIFILE.
        SELECT EMPLOYEE-OUT-FILE  ASSIGN TO EMPOFILE.

DATA DIVISION.

FILE SECTION.
FD  EMPLOYEE-IN-FILE
    RECORDING MODE IS F
    LABEL RECORDS ARE STANDARD
    RECORD CONTAINS 80 CHARACTERS
    BLOCK CONTAINS 0 RECORDS
    DATA RECORD IS EMPLOYEE-RECORD-IN.

    01 EMPLOYEE-RECORD-IN.
        05  E-ID          PIC X(04).
        05  FILLER        PIC X(76).

FD  EMPLOYEE-OUT-FILE
    RECORDING MODE IS F
    LABEL RECORDS ARE STANDARD
    RECORD CONTAINS 80 CHARACTERS
    BLOCK CONTAINS 0 RECORDS
    DATA RECORD IS EMPLOYEE-RECORD-OUT.

    01 EMPLOYEE-RECORD-OUT.
        05  EMP-DATA      PIC X(80).


WORKING-STORAGE SECTION.

    01 WS-FLAGS.
        05  SW-END-OF-FILE-SWITCH   PIC X(1) VALUE 'N'.
            88  SW-END-OF-FILE              VALUE 'Y'.
            88  SW-NOT-END-OF-FILE          VALUE 'N'.

    01 IN-EMPLOYEE-RECORD.
        05  EMP-ID-IN    PIC X(04).
```

84

```
            05  FILLER       PIC X(05).
            05  REG-PAY-IN    PIC 99999V99.
            05  FILLER       PIC X(02).
            05  BON-PAY-IN    PIC 9999V99.
            05  FILLER       PIC X(54).

        01 OUT-EMPLOYEE-RECORD.
            05  EMP-ID-OUT    PIC S9(9) USAGE COMP.
            05  FILLER       PIC X(05).
            05  REG-PAY-OUT   PIC S9(6)V9(2) USAGE COMP-3.
            05  FILLER       PIC X(02).
            05  BON-PAY-OUT   PIC S9(6)V9(2) USAGE COMP-3.
            05  FILLER       PIC X(59) VALUE SPACES.

        01 DISPLAY-EMPLOYEE-PIC.
            05  DIS-REG-PAY   PIC 99999.99.
            05  DIS-BON-PAY   PIC 9999.99.

    PROCEDURE DIVISION.

        PERFORM P100-INITIALIZATION.
        PERFORM P200-MAINLINE.
        PERFORM P300-TERMINATION.
        GOBACK.

    P100-INITIALIZATION.

        DISPLAY 'COBTRN3 - SAMPLE COBOL PROGRAM: INPUT AND OUTPUT'.
        OPEN INPUT  EMPLOYEE-IN-FILE.
        OPEN OUTPUT EMPLOYEE-OUT-FILE.
        INITIALIZE IN-EMPLOYEE-RECORD, OUT-EMPLOYEE-RECORD.

    P200-MAINLINE.

*    MAIN LOOP - READ THE INPUT FILE, LOAD THE OUTPUT
*              STRUCTURE AND WRITE THE RECORD TO OUTPUT.

        READ EMPLOYEE-IN-FILE INTO IN-EMPLOYEE-RECORD
           AT END SET SW-END-OF-FILE TO TRUE
        END-READ

        PERFORM UNTIL SW-END-OF-FILE

*        MOVE FIELDS

            MOVE EMP-ID-IN   TO EMP-ID-OUT
            MOVE REG-PAY-IN  TO REG-PAY-OUT, DIS-REG-PAY
            MOVE BON-PAY-IN  TO BON-PAY-OUT, DIS-BON-PAY

            DISPLAY ' EMP ID: '   EMP-ID-OUT
            DISPLAY ' REG PAY: '  DIS-REG-PAY
            DISPLAY ' BONUS PAY: ' DIS-BON-PAY
```

```
            WRITE EMPLOYEE-RECORD-OUT FROM OUT-EMPLOYEE-RECORD

            READ EMPLOYEE-IN-FILE INTO IN-EMPLOYEE-RECORD
               AT END SET SW-END-OF-FILE TO TRUE
            END-READ

         END-PERFORM.

      P300-TERMINATION.

         CLOSE EMPLOYEE-IN-FILE,
               EMPLOYEE-OUT-FILE.

         DISPLAY 'COBTRN3 - SUCCESSFULLY ENDED'.

   *     END OF SOURCE CODE
```

You'll see I added a couple more features such as creating display variables, and we also have a file processing control loop using a switch variable with 88 level variables. An 88 level variable specifies a particular value for another variable. In our case we have a variable named SW-END-OF-FILE-SWITCH. We've initialized it to 'N' meaning no. We declared two 88 level variables under it, namely SW-END-OF-FILE whose value is 'Y', and SW-NOT-END-OF-FILE whose value is 'N'.

```
      05  SW-END-OF-FILE-SWITCH   PIC X(1) VALUE 'N'.
          88  SW-END-OF-FILE                VALUE 'Y'.
          88  SW-NOT-END-OF-FILE            VALUE 'N'.
```

Strictly speaking you do not need both of these 88 level variables unless your code requires them. Our code only references the SW-END-OF-FILE value, so we don't need the SW-NOT-END-OF-FILE variable. We've included it here just for documentation and to aid in understanding the concept of 88 levels. Keep in mind these rules:

- You don't use a picture clause with a level 88 variable.
- An 88 level number is always declared under another level number 01-49.

There are many uses for 88 levels and they are almost always used for branching. One benefit of using 88 level variables is that it allows you to declare meaningful names for the possible values that may appear in a variable. You will likely see these a lot in legacy code. In our case, we will set the value automatically with this code:

```
      READ EMPLOYEE-IN-FILE INTO IN-EMPLOYEE-RECORD
         AT END SET SW-END-OF-FILE TO TRUE
      END-READ
```

The above means that when we encounter end of file on a read statement, we automatically set the value of our loop control variable to true which ends the loop.

```
            PERFORM UNTIL SW-END-OF-FILE
```

Now let's compile and run the program. Here is the output.

```
COBTRN3 - SAMPLE COBOL PROGRAM: INPUT AND OUTPUT
EMP ID: 000001111
REG PAY: 87000.00
BONUS PAY: 6700.00
EMP ID: 000001122
REG PAY: 82000.00
BONUS PAY: 6000.00
EMP ID: 000003217
REG PAY: 65000.00
BONUS PAY: 5500.00
EMP ID: 000004175
REG PAY: 55000.00
BONUS PAY: 1500.00
EMP ID: 000004720
REG PAY: 80000.00
BONUS PAY: 2500.00
EMP ID: 000004836
REG PAY: 62000.00
BONUS PAY: 2200.00
EMP ID: 000006288
REG PAY: 70000.00
BONUS PAY: 2000.00
EMP ID: 000007459
REG PAY: 85000.00
BONUS PAY: 4500.00
EMP ID: 000009134
REG PAY: 75000.00
BONUS PAY: 2500.00
COBTRN3 - SUCCESSFULLY ENDED
```

Now let's look at the output file. You'll notice it is not very readable because the data has been packed into compressed format.

```
BROWSE     USER01.EMPOFILE                       Line 00000000 Col 001 080
 Command ===>                                             Scroll ===> CSR
----+----1----+----2----+----3----+----4----+----5----+----6----+----7----+----8
******************************* Top of Data *********************************
...j......Á.......&..
..........e.......&..
...Þ......Í.......&..
...ø......Ø.......&..
...°......ø..........
***************************** Bottom of Data ********************************
```

You can determine the actual values by issuing the HEX command on the command line. This displays the hex value for each byte.

```
BROWSE     USER01.EMPOFILE                            Line 00000000 Col 001 080
 Command ===>                                              Scroll ===> CSR
----+----1----+----2----+----3----+----4----+----5----+----6----+----7----+----8
----+----F----+----F----+----F----+----F----+----F----+----F----+----F----+----F
----+----1----+----2----+----3----+----4----+----5----+----6----+----7----+----8
--------------------------------------------------------------------------------
****************************** Top of Data ********************************


  --------------------------------------------------------------------------
...j......Á.......&..
0009000000600000000500444444444444444444444444444444444444444444444444444444444
00C1000000500C000500C00000000000000000000000000000000000000000000000000000000000
  --------------------------------------------------------------------------
.........e.......&..
0012000000800000000500444444444444444444444444444444444444444444444444444444444
00D3000000500C000400C00000000000000000000000000000000000000000000000000000000000
  --------------------------------------------------------------------------
...Þ......Í.......&..
002A000000700000000500444444444444444444444444444444444444444444444444444444444
003E000000500C000200C00000000000000000000000000000000000000000000000000000000000
  --------------------------------------------------------------------------
```

Turn off the hex view by issuing the command HEX OFF.

I suggest you use the COBTRN3 source code as a model for Input/Output programs and customize it as you see fit. You'll most likely use this pattern quite a lot in batch programs. We'll look at file I/O for VSAM with COBOL in chapter 4. We'll look at online programs in chapter 7 when we learn COBOL with CICS.

That's it for the basics of file input/output in COBOL. In the next section we'll look at creating a report from the employee data.

Reporting

These days there are many ways to get data from a central data source, especially when much of the data is stored in relational databases. But not so many years ago most reports were generated by application programs. In fact, the last three environments I worked in still used plenty of COBOL for generating reports. So you'll almost certainly encounter reporting programs if you work in a COBOL shop.

COBOL can be used for simple or complex reports. Basically what you do is define record structures for your headers, detail lines and trailer lines (if any). Then you write an output record using the appropriate record structure. If your report tends to be more than a printed page, you'll typically use line and page counters to repeat the headers and bump up the page number at the appropriate time.

ANSI Carriage Control

At one time almost all IBM mainframe shops and customers used printers that operated with something called ANSI Carriage Control. That means you used the first byte of each report line to control advancing the paper – line feeds, skipping to the next page, etc.

Here are the values you would code in the first byte of a report line to create the desired operation for a printer that uses ANSI carriage control.

Character	Action
blank	Advance 1 line (single spacing)
1	Advance to next page (form feed)
0	Advance 2 lines (double spacing)
–	Advance 3 lines (triple spacing)
+	Do not advance any lines before printing, overstrike previous line with current line

Now having explained about ANSI carriage control, I am going to recommend that you not use it. Why? Because many printers don't use this type of carriage control anymore. This is true especially if your form of distribution is through email or by storage on a Windows or Unix server (or on a web site). Also, your users may not intend to print the report in which case the carriage control characters can be distracting as they appear on the report text as extraneous characters. This is something you will need to discover in talking with your user base.

If you decide that some sort of control is needed for establishing hearers and/or paging, you can create your own. If you want to skip a line, simply write a blank line. Or two lines or three lines as necessary. Use line and page counters to determine when headers need to be recreated. We'll show an example of that before we leave this section.

Report Program Sample

Ok, let's start on program COBTRN4 which will report data from the input file (employee id plus salary). Let's say we still want to write an output file with binary pay information like we did in COBTRN3, so when we add the report function for COBTRN4 we'll need a second output file. Let's call it REPORT-FILE-OUT and the JCL DD name will be EMPREPRT.

```
    SELECT REPORT-OUT-FILE    ASSIGN TO EMPREPRT.
```

COBTRN4 will define several structures that are "lines" to be written to the report file. Some of these are static header lines, others are detail lines where we'll fill in the employee-specific data. Here is one way of declaring these variables.

```
01 HDR-LINE-01.
    05  FILLER         PIC X(25) VALUE SPACES.
    05  FILLER         PIC X(30)
        VALUE 'EMPLOYEE ANNUAL SALARY REPORT '.
    05  FILLER         PIC X(25) VALUE SPACES.

01 HDR-LINE-02.
    05  FILLER         PIC X(25) VALUE SPACES.
    05  FILLER         PIC X(30)
        VALUE '-------------------------- '.
    05  FILLER         PIC X(25) VALUE SPACES.

01 SPC-LINE            PIC X(80) VALUE SPACES.

01 DTL-HDR01.
    05 FILLER          PIC  X(25) VALUE SPACE.
    05 FILLER          PIC  X(06) VALUE 'EMP-ID'.
    05 FILLER          PIC  X(03) VALUE SPACES.
    05 FILLER          PIC  X(08) VALUE 'REGULAR '.
    05 FILLER          PIC  X(04) VALUE SPACE.
    05 FILLER          PIC  X(08) VALUE 'BONUS    '.
    05 FILLER          PIC  X(19) VALUE SPACES.

01 DTL-HDR02.
    05 FILLER          PIC  X(25) VALUE SPACE.
    05 FILLER          PIC  X(06) VALUE '------'.
    05 FILLER          PIC  X(03) VALUE SPACES.
    05 FILLER          PIC  X(08) VALUE '--------'.
    05 FILLER          PIC  X(04) VALUE SPACE.
    05 FILLER          PIC  X(08) VALUE '--------'.
    05 FILLER          PIC  X(19) VALUE SPACES.

01 DTL-LINE.
    05 FILLER          PIC  X(27) VALUE SPACE.
    05 RPT-EMP-ID      PIC  9(04).
    05 FILLER          PIC  X(03) VALUE SPACES.
```

90

```
            05 RPT-REG-PAY    PIC  99999.99.
            05 FILLER         PIC  X(04) VALUE SPACE.
            05 RPT-BON-PAY    PIC  99999.99.
            05 FILLER         PIC  X(19) VALUE SPACES.

        01 TRLR-LINE-01.
            05  FILLER         PIC X(26) VALUE SPACES.
            05  FILLER         PIC X(30)
                VALUE ' END OF ANNUAL SALARY REPORT  '.
            05  FILLER         PIC X(24) VALUE SPACES.
```

We can use COBTRN3 as our base program, so copy it to a file member named COBTRN4. Next let's add these new report structures, and some additional logic to write out the report. Here is COBTRN4 listing including the writing of headers and trailer records. Notice that we write the headers in the Initialization routine, and the end-of-report footer in the Termination routine. Also we move the appropriate data values to variables we defined on our detail record. This is how we do reports in COBOL.

Take a good look at the code, and then we'll do one more enhancement to accommodate headers and footers.

```
        IDENTIFICATION DIVISION.
        PROGRAM-ID. COBTRN4.
        ********************************************************
        *       PROGRAM USING FILE INPUT AND OUTPUT          *
        *       TO REFORMAT EMPLOYEE PAY INFORMATION.        *
        *       ALSO PRODUCES A REPORT OF THE INFORMATION.   *
        ********************************************************
        ENVIRONMENT DIVISION.
        INPUT-OUTPUT SECTION.

           FILE-CONTROL.
              SELECT EMPLOYEE-IN-FILE   ASSIGN TO EMPIFILE.
              SELECT EMPLOYEE-OUT-FILE  ASSIGN TO EMPOFILE.
              SELECT REPORT-OUT-FILE    ASSIGN TO EMPREPRT.

        DATA DIVISION.
        FILE SECTION.
        FD  EMPLOYEE-IN-FILE
            RECORDING MODE IS F
            LABEL RECORDS ARE STANDARD
            RECORD CONTAINS 80 CHARACTERS
            BLOCK CONTAINS 0 RECORDS
            DATA RECORD IS EMPLOYEE-RECORD-IN.

        01 EMPLOYEE-RECORD-IN.
             05  E-ID         PIC X(04).
             05  FILLER       PIC X(76).
```

```
FD  EMPLOYEE-OUT-FILE
    RECORDING MODE IS F
    LABEL RECORDS ARE STANDARD
    RECORD CONTAINS 80 CHARACTERS
    BLOCK CONTAINS 0 RECORDS
    DATA RECORD IS EMPLOYEE-RECORD-OUT.

01 EMPLOYEE-RECORD-OUT.
    05  EMP-DATA     PIC X(80).

FD  REPORT-OUT-FILE
    RECORDING MODE IS F
    LABEL RECORDS ARE STANDARD
    RECORD CONTAINS 80 CHARACTERS
    BLOCK CONTAINS 0 RECORDS
    DATA RECORD IS REPORT-RECORD-OUT.

01 REPORT-RECORD-OUT.
    05  RPT-DATA     PIC X(80).

WORKING-STORAGE SECTION.

01 WS-FLAGS.
    05  SW-END-OF-FILE-SWITCH   PIC X(1) VALUE 'N'.
        88  SW-END-OF-FILE              VALUE 'Y'.
        88  SW-NOT-END-OF-FILE          VALUE 'N'.


01 IN-EMPLOYEE-RECORD.
    05  EMP-ID-IN     PIC X(04).
    05  FILLER        PIC X(05).
    05  REG-PAY-IN    PIC 99999V99.
    05  FILLER        PIC X(02).
    05  BON-PAY-IN    PIC 9999V99.
    05  FILLER        PIC X(54).

01 OUT-EMPLOYEE-RECORD.
    05  EMP-ID-OUT    PIC S9(9) USAGE COMP.
    05  FILLER        PIC X(05).
    05  REG-PAY-OUT   PIC S9(6)V9(2) USAGE COMP-3.
    05  FILLER        PIC X(02).
    05  BON-PAY-OUT   PIC S9(6)V9(2) USAGE COMP-3.
    05  FILLER        PIC X(59) VALUE SPACES.

01 DISPLAY-EMPLOYEE-PIC.
    05  DIS-REG-PAY   PIC 99999.99.
    05  DIS-BON-PAY   PIC 9999.99.

01 HDR-LINE-01.
    05  FILLER        PIC X(25) VALUE SPACES.
    05  FILLER        PIC X(30)
```

```
                   VALUE 'EMPLOYEE ANNUAL SALARY REPORT '.
           05  FILLER        PIC X(25) VALUE SPACES.

       01 HDR-LINE-02.
           05  FILLER        PIC X(25) VALUE SPACES.
           05  FILLER        PIC X(30)
               VALUE '---------------------------- '.
           05  FILLER        PIC X(25) VALUE SPACES.

       01 SPC-LINE           PIC X(80) VALUE SPACES.

       01 DTL-HDR01.
           05 FILLER         PIC  X(25) VALUE SPACE.
           05 FILLER         PIC  X(06) VALUE 'EMP-ID'.
           05 FILLER         PIC  X(03) VALUE SPACES.
           05 FILLER         PIC  X(08) VALUE 'REGULAR '.
           05 FILLER         PIC  X(04) VALUE SPACE.
           05 FILLER         PIC  X(08) VALUE 'BONUS   '.
           05 FILLER         PIC  X(19) VALUE SPACES.

       01 DTL-HDR02.
           05 FILLER         PIC  X(25) VALUE SPACE.
           05 FILLER         PIC  X(06) VALUE '------'.
           05 FILLER         PIC  X(03) VALUE SPACES.
           05 FILLER         PIC  X(08) VALUE '--------'.
           05 FILLER         PIC  X(04) VALUE SPACE.
           05 FILLER         PIC  X(08) VALUE '--------'.
           05 FILLER         PIC  X(19) VALUE SPACES.

       01 DTL-LINE.
           05 FILLER         PIC  X(27) VALUE SPACE.
           05 RPT-EMP-ID     PIC  9(04).
           05 FILLER         PIC  X(03) VALUE SPACES.
           05 RPT-REG-PAY    PIC  99999.99.
           05 FILLER         PIC  X(04) VALUE SPACE.
           05 RPT-BON-PAY    PIC  99999.99.
           05 FILLER         PIC  X(19) VALUE SPACES.

       01 TRLR-LINE-01.
           05  FILLER        PIC X(26) VALUE SPACES.
           05  FILLER        PIC X(30)
               VALUE ' END OF ANNUAL SALARY REPORT  '.
           05  FILLER        PIC X(24) VALUE SPACES.

   PROCEDURE DIVISION.

       PERFORM P100-INITIALIZATION.
       PERFORM P200-MAINLINE.
       PERFORM P300-TERMINATION.
       GOBACK.

   P100-INITIALIZATION.
```

```
        DISPLAY 'COBTRN4 - SAMPLE COBOL PROGRAM: I/O AND REPORTS '.
        OPEN INPUT  EMPLOYEE-IN-FILE,
        OPEN OUTPUT EMPLOYEE-OUT-FILE,
        OPEN OUTPUT REPORT-OUT-FILE.

        INITIALIZE IN-EMPLOYEE-RECORD, OUT-EMPLOYEE-RECORD.

        WRITE REPORT-RECORD-OUT FROM HDR-LINE-01
        WRITE REPORT-RECORD-OUT FROM HDR-LINE-02
        WRITE REPORT-RECORD-OUT FROM SPC-LINE
        WRITE REPORT-RECORD-OUT FROM DTL-HDR01
        WRITE REPORT-RECORD-OUT FROM DTL-HDR02.


    P200-MAINLINE.

*       MAIN LOOP - READ THE INPUT FILE, LOAD THE OUTPUT
*                   STRUCTURE AND WRITE THE RECORD TO OUTPUT.

        SET SW-NOT-END-OF-FILE TO TRUE.

        READ EMPLOYEE-IN-FILE INTO IN-EMPLOYEE-RECORD
           AT END SET SW-END-OF-FILE TO TRUE
        END-READ

        PERFORM UNTIL SW-END-OF-FILE

*           MOVE FIELDS

            MOVE EMP-ID-IN TO EMP-ID-OUT, RPT-EMP-ID
            MOVE REG-PAY-IN
               TO REG-PAY-OUT, RPT-REG-PAY, DIS-REG-PAY
            MOVE BON-PAY-IN
               TO BON-PAY-OUT, RPT-BON-PAY, DIS-BON-PAY

            DISPLAY ' EMP ID: '   EMP-ID-IN
                    ' REG PAY: '  DIS-REG-PAY
                    ' BONUS PAY: ' DIS-BON-PAY

            WRITE EMPLOYEE-RECORD-OUT
               FROM OUT-EMPLOYEE-RECORD

            WRITE REPORT-RECORD-OUT FROM DTL-LINE

            READ EMPLOYEE-IN-FILE
               INTO IN-EMPLOYEE-RECORD
                  AT END SET SW-END-OF-FILE TO TRUE
            END-READ

        END-PERFORM.
```

```
    P300-TERMINATION.

        WRITE REPORT-RECORD-OUT FROM SPC-LINE
        WRITE REPORT-RECORD-OUT FROM SPC-LINE
        WRITE REPORT-RECORD-OUT FROM TRLR-LINE-01

        CLOSE EMPLOYEE-IN-FILE,
              EMPLOYEE-OUT-FILE,
              REPORT-OUT-FILE.

        DISPLAY 'COBTRN4 - SUCCESSFULLY ENDED'.

*       END OF SOURCE CODE
```

When we run the program we get a report file that looks like this:

```
    EMPLOYEE ANNUAL SALARY REPORT PAGE
    ----------------------------------


    EMP-ID   REGULAR      BONUS
    ------   --------     --------
      1111   87000.00    06700.00
      1122   82000.00    06000.00
      3217   65000.00    05500.00
      4175   55000.00    01500.00
      4720   80000.00    02500.00
      4836   62000.00    02200.00
      6288   70000.00    02000.00
      7459   85000.00    04500.00
      9134   75000.00    02500.00


        END OF ANNUAL SALARY REPORT
```

We don't have much data in our example, but for a longer report you might want to use page breaks and page numbers. This can be done fairly easily with a line counter and page counter. Let's go ahead and revise our program to use the counters. If you don't need them you can just use the existing version of the program above.

Since we only have a few records for our example, let's say we want to see a maximum of 6 detail lines per page. We'll code the line and page counters, and here is the revised program listing. Notice we have broken the header writing logic into its own routine. We've added a maximum lines constant, the page counter and a line counter. When we reach maximum lines, we write a new header.

Again, you can use this as a logic model for your report programs.

```
IDENTIFICATION DIVISION.
PROGRAM-ID. COBTRN4.
*********************************************************
*       PROGRAM USING FILE INPUT AND OUTPUT         *
*       TO REFORMAT EMPLOYEE PAY INFORMATION.       *
*       ALSO PRODUCES A REPORT OF THE INFORMATION.  *
*********************************************************
ENVIRONMENT DIVISION.
INPUT-OUTPUT SECTION.

   FILE-CONTROL.
       SELECT EMPLOYEE-IN-FILE   ASSIGN TO EMPIFILE.
       SELECT EMPLOYEE-OUT-FILE  ASSIGN TO EMPOFILE.
       SELECT REPORT-OUT-FILE    ASSIGN TO EMPREPRT.

DATA DIVISION.

FILE SECTION.
FD  EMPLOYEE-IN-FILE
    RECORDING MODE IS F
    LABEL RECORDS ARE STANDARD
    RECORD CONTAINS 80 CHARACTERS
    BLOCK CONTAINS 0 RECORDS
    DATA RECORD IS EMPLOYEE-RECORD-IN.

   01 EMPLOYEE-RECORD-IN.
        05  E-ID          PIC X(04).
        05  FILLER        PIC X(76).

FD  EMPLOYEE-OUT-FILE
    RECORDING MODE IS F
    LABEL RECORDS ARE STANDARD
    RECORD CONTAINS 80 CHARACTERS
    BLOCK CONTAINS 0 RECORDS
    DATA RECORD IS EMPLOYEE-RECORD-OUT.

   01 EMPLOYEE-RECORD-OUT.
        05  EMP-DATA      PIC X(80).

FD  REPORT-OUT-FILE
    RECORDING MODE IS F
    LABEL RECORDS ARE STANDARD
    RECORD CONTAINS 80 CHARACTERS
    BLOCK CONTAINS 0 RECORDS
    DATA RECORD IS REPORT-RECORD-OUT.

   01 REPORT-RECORD-OUT.
        05  RPT-DATA      PIC X(80).

WORKING-STORAGE SECTION.

   01 WS-FLAGS.
```

```
      05  SW-END-OF-FILE-SWITCH   PIC X(1) VALUE 'N'.
          88  SW-END-OF-FILE                VALUE 'Y'.
          88  SW-NOT-END-OF-FILE            VALUE 'N'.


01 IN-EMPLOYEE-RECORD.
   05  EMP-ID-IN     PIC X(04).
   05  FILLER        PIC X(05).
   05  REG-PAY-IN    PIC 99999V99.
   05  FILLER        PIC X(02).
   05  BON-PAY-IN    PIC 9999V99.
   05  FILLER        PIC X(54).


01 OUT-EMPLOYEE-RECORD.
   05  EMP-ID-OUT    PIC S9(9) USAGE COMP.
   05  FILLER        PIC X(05).
   05  REG-PAY-OUT   PIC S9(6)V9(2) USAGE COMP-3.
   05  FILLER        PIC X(02).
   05  BON-PAY-OUT   PIC S9(6)V9(2) USAGE COMP-3.
   05  FILLER        PIC X(59) VALUE SPACES.


01 DISPLAY-EMPLOYEE-PIC.
   05  DIS-REG-PAY   PIC 99999.99.
   05  DIS-BON-PAY   PIC 9999.99.


01 HDR-LINE-01.
   05  FILLER        PIC X(25) VALUE SPACES.
   05  FILLER        PIC X(30)
       VALUE 'EMPLOYEE ANNUAL SALARY REPORT '.
   05  FILLER        PIC X(05) VALUE 'PAGE-'.
   05  R-PAGE-NO     PIC Z9.
   05  FILLER        PIC X(18) VALUE SPACES.


01 HDR-LINE-02.
   05  FILLER        PIC X(25) VALUE SPACES.
   05  FILLER        PIC X(38)
       VALUE '---------------------------------- '.
   05  FILLER        PIC X(17) VALUE SPACES.


01 SPC-LINE         PIC X(80) VALUE SPACES.


01 DTL-HDR01.
   05 FILLER         PIC  X(25) VALUE SPACE.
   05 FILLER         PIC  X(06) VALUE 'EMP-ID'.
   05 FILLER         PIC  X(03) VALUE SPACES.
   05 FILLER         PIC  X(08) VALUE 'REGULAR '.
   05 FILLER         PIC  X(04) VALUE SPACE.
   05 FILLER         PIC  X(08) VALUE 'BONUS   '.
   05 FILLER         PIC  X(19) VALUE SPACES.


01 DTL-HDR02.
   05 FILLER         PIC  X(25) VALUE SPACE.
   05 FILLER         PIC  X(06) VALUE '------'.
```

```
            05 FILLER          PIC  X(03) VALUE SPACES.
            05 FILLER          PIC  X(08) VALUE '--------'.
            05 FILLER          PIC  X(04) VALUE SPACE.
            05 FILLER          PIC  X(08) VALUE '--------'.
            05 FILLER          PIC  X(19) VALUE SPACES.

         01 DTL-LINE.
            05 FILLER          PIC  X(27) VALUE SPACE.
            05 RPT-EMP-ID      PIC  9(04).
            05 FILLER          PIC  X(03) VALUE SPACES.
            05 RPT-REG-PAY     PIC  99999.99.
            05 FILLER          PIC  X(04) VALUE SPACE.
            05 RPT-BON-PAY     PIC  99999.99.
            05 FILLER          PIC  X(19) VALUE SPACES.

         01 TRLR-LINE-01.
            05  FILLER         PIC X(26) VALUE SPACES.
            05  FILLER         PIC X(30)
                VALUE ' END OF ANNUAL SALARY REPORT  '.
            05  FILLER         PIC X(24) VALUE SPACES.

         77 C-MAX-LINES    PIC S9(9) USAGE COMP VALUE 6.

         01 ACCUMULATORS.
            05 A-PAGE-CTR      PIC S9(9) USAGE COMP VALUE 0.
            05 A-LINE-CTR      PIC S9(9) USAGE COMP VALUE 0.

     PROCEDURE DIVISION.

         PERFORM P100-INITIALIZATION.
         PERFORM P200-MAINLINE.
         PERFORM P300-TERMINATION.
         GOBACK.

     P100-INITIALIZATION.

         DISPLAY 'COBTRN4 - SAMPLE COBOL PROGRAM: I/O AND REPORTS '.
         OPEN INPUT  EMPLOYEE-IN-FILE,
         OPEN OUTPUT EMPLOYEE-OUT-FILE,
         OPEN OUTPUT REPORT-OUT-FILE.

         INITIALIZE IN-EMPLOYEE-RECORD, OUT-EMPLOYEE-RECORD.

         PERFORM P1000-WRITE-HEADERS.

     P200-MAINLINE.

     *    MAIN LOOP - READ THE INPUT FILE, LOAD THE OUTPUT
     *                STRUCTURE AND WRITE THE RECORD TO OUTPUT.

         SET SW-NOT-END-OF-FILE TO TRUE.
```

```
        READ EMPLOYEE-IN-FILE INTO IN-EMPLOYEE-RECORD
           AT END SET SW-END-OF-FILE TO TRUE
        END-READ

        PERFORM UNTIL SW-END-OF-FILE

*           MOVE FIELDS

            MOVE EMP-ID-IN TO EMP-ID-OUT, RPT-EMP-ID
            MOVE REG-PAY-IN
               TO REG-PAY-OUT, RPT-REG-PAY, DIS-REG-PAY
            MOVE BON-PAY-IN
               TO BON-PAY-OUT, RPT-BON-PAY, DIS-BON-PAY

            DISPLAY ' EMP ID: '   EMP-ID-IN
                    ' REG PAY: '  DIS-REG-PAY
                    ' BONUS PAY: '  DIS-BON-PAY

            WRITE EMPLOYEE-RECORD-OUT
               FROM OUT-EMPLOYEE-RECORD

            ADD +1 TO A-LINE-CTR
            IF A-LINE-CTR > C-MAX-LINES THEN
               PERFORM P1000-WRITE-HEADERS
            END-IF

            WRITE REPORT-RECORD-OUT FROM DTL-LINE

            READ EMPLOYEE-IN-FILE
               INTO IN-EMPLOYEE-RECORD
                  AT END SET SW-END-OF-FILE TO TRUE
            END-READ

        END-PERFORM.

    P300-TERMINATION.

        WRITE REPORT-RECORD-OUT FROM SPC-LINE
        WRITE REPORT-RECORD-OUT FROM SPC-LINE
        WRITE REPORT-RECORD-OUT FROM TRLR-LINE-01

        CLOSE EMPLOYEE-IN-FILE,
              EMPLOYEE-OUT-FILE,
              REPORT-OUT-FILE.

        DISPLAY 'COBTRN4 - SUCCESSFULLY ENDED'.

    P1000-WRITE-HEADERS.

        WRITE REPORT-RECORD-OUT FROM SPC-LINE
        ADD +1 TO A-PAGE-CTR
        MOVE A-PAGE-CTR TO R-PAGE-NO
```

```
              WRITE REPORT-RECORD-OUT FROM HDR-LINE-01
              WRITE REPORT-RECORD-OUT FROM HDR-LINE-02
              WRITE REPORT-RECORD-OUT FROM SPC-LINE
              WRITE REPORT-RECORD-OUT FROM DTL-HDR01
              WRITE REPORT-RECORD-OUT FROM DTL-HDR02.

              MOVE +0 TO A-LINE-CTR.

      *    END OF SOURCE CODE
```

When we run this revised program, here is the output:

```
COBTRN4 - SAMPLE COBOL PROGRAM: I/O AND REPORTS
 EMP ID: 1111 REG PAY: 87000.00 BONUS PAY: 6700.00
 EMP ID: 1122 REG PAY: 82000.00 BONUS PAY: 6000.00
 EMP ID: 3217 REG PAY: 65000.00 BONUS PAY: 5500.00
 EMP ID: 4175 REG PAY: 55000.00 BONUS PAY: 1500.00
 EMP ID: 4720 REG PAY: 80000.00 BONUS PAY: 2500.00
 EMP ID: 4836 REG PAY: 62000.00 BONUS PAY: 2200.00
 EMP ID: 6288 REG PAY: 70000.00 BONUS PAY: 2000.00
 EMP ID: 7459 REG PAY: 85000.00 BONUS PAY: 4500.00
 EMP ID: 9134 REG PAY: 75000.00 BONUS PAY: 2500.00
COBTRN4 - SUCCESSFULLY ENDED
```

And here is the cataloged report file:

```
EMPLOYEE ANNUAL SALARY REPORT PAGE- 1
-------------------------------------

EMP-ID    REGULAR     BONUS
------    --------    --------
  1111    87000.00    06700.00
  1122    82000.00    06000.00
  3217    65000.00    05500.00
  4175    55000.00    01500.00
  4720    80000.00    02500.00
  4836    62000.00    02200.00


EMPLOYEE ANNUAL SALARY REPORT PAGE- 2
-------------------------------------

EMP-ID    REGULAR     BONUS
------    --------    --------
  6288    70000.00    02000.00
  7459    85000.00    04500.00
  9134    75000.00    02500.00


   END OF ANNUAL SALARY REPORT
```

If you like you could include a continuation trailer record for all except the last page. This record would state something like 'REPORT CONTINUED ON NEXT PAGE'. I'm sure you'll have some other ideas of how to improve the report image and usefulness.

In concluding this section, let's remember that while COBOL may not be the IT reporting workhorse of yesteryear, it is still quite handy (especially when pulling together data from various sources). I think you'll find plenty of COBOL code that generates reports in the world of legacy IBM mainframe systems. Be prepared for it.

Calculations

Over the years, some believed COBOL to be weak for scientific and engineering applications because it was primarily developed for business applications. FORTRAN was the language of choice for science/engineering applications that needed great precision.[7] I never worked in scientific applications so I won't offer an opinion, but I can assure you that COBOL has pretty robust calculation features for most applications.

For this next program COBTRN5, we will give our employees a 10% raise based on the salary data in the employee pay file we processed earlier. Also, we'll modify the report we produced to show both the old salary and the new salary.

Calculations with GIVING

There are a few ways of performing calculations in COBOL. One basic way is to use the ADD, SUBTRACT, MULTIPLY and DIVIDE verbs with the GIVING clause. So for example we could code:

```
MULTIPLY VARIABLE1 BY 5 GIVING VARIABLE2
```

Let's look at how we could do that in our employee pay program. First we need a few arithmetic variables. We'll use packed values (COMP-3) and here they are:

```
77 REG-PAY-PKD      PIC S9(6)V9(2) USAGE COMP-3.
77 PAY-RAISE-PKD    PIC S9(6)V9(2) USAGE COMP-3.
77 NEW-PAY-PKD      PIC S9(6)V9(2) USAGE COMP-3.
77 NEW-PAY          PIC 99999.99.
```

Notice we used an implied decimal point in the packed decimal variables above, all of which are dealing with money. Now we could code:

---

7 http://science.sciencemag.org/content/303/5662/1331

```
MOVE REG-PAY-IN  TO REG-PAY-PKD
MULTIPLY REG-PAY-PKD BY 0.10 GIVING PAY-RAISE-PKD
MOVE PAY-RAISE-PKD TO PAY-RAISE
MOVE REG-PAY-PKD TO REG-PAY
ADD PAY-RAISE-PKD TO REG-PAY-PKD GIVING NEW-PAY-PKD
MOVE NEW-PAY-PKD TO NEW-PAY
DISPLAY NEW-PAY
```

This would work, and it is a very common way to code arithmetic operation. However, we have an alternative which sometimes gives us more flexibility and may be more appropriate for complex calculations. The alternative is to use the COMPUTE verb.

Calculations with COMPUTE

We could code calculations using the COMPUTE verb. This is a simple model, and here is an example:

```
COMPUTE VARIABLE2 = (VARIABLE1 * 5)
```

For our employee pay example, either method works. In our code example, we will provide a commented out version of the GIVING example, and we will actually use the COMPUTE method. That way you have a program with samples of both methods.

You can copy COBTRN4 to COBTRN5 and save the effort of recoding everything we did prior to COBTRN5. Also note that for COBTRN5 we have changed the report to include both the original and new pay.

Here is our listing.

```
IDENTIFICATION DIVISION.
PROGRAM-ID. COBTRN5.

************************************************************
*      PROGRAM USING FILE INPUT AND OUTPUT            *
*      TO REFORMAT EMPLOYEE PAY INFORMATION.          *
*      CALCULATE A 10% RAISE FOR EACH EMPLOYEE.       *
*      ALSO PRODUCES A REPORT OF THE INFORMATION.     *
************************************************************

ENVIRONMENT DIVISION.
INPUT-OUTPUT SECTION.

    FILE-CONTROL.
        SELECT EMPLOYEE-IN-FILE   ASSIGN TO EMPIFILE.
        SELECT EMPLOYEE-OUT-FILE  ASSIGN TO EMPOFILE.
        SELECT REPORT-OUT-FILE    ASSIGN TO EMPREPRT.
```

```
DATA DIVISION.

FILE SECTION.
FD  EMPLOYEE-IN-FILE
    RECORDING MODE IS F
    LABEL RECORDS ARE STANDARD
    RECORD CONTAINS 80 CHARACTERS
    BLOCK CONTAINS 0 RECORDS
    DATA RECORD IS EMPLOYEE-RECORD-IN.

   01 EMPLOYEE-RECORD-IN.
       05  E-ID          PIC X(04).
       05  FILLER        PIC X(76).

FD  EMPLOYEE-OUT-FILE
    RECORDING MODE IS F
    LABEL RECORDS ARE STANDARD
    RECORD CONTAINS 80 CHARACTERS
    BLOCK CONTAINS 0 RECORDS
    DATA RECORD IS EMPLOYEE-RECORD-OUT.

   01 EMPLOYEE-RECORD-OUT.
       05  EMP-DATA      PIC X(80).

FD  REPORT-OUT-FILE
    RECORDING MODE IS F
    LABEL RECORDS ARE STANDARD
    RECORD CONTAINS 80 CHARACTERS
    BLOCK CONTAINS 0 RECORDS
    DATA RECORD IS REPORT-RECORD-OUT.

   01 REPORT-RECORD-OUT.
       05  RPT-DATA      PIC X(80).

WORKING-STORAGE SECTION.

   01 WS-FLAGS.
       05  SW-END-OF-FILE-SWITCH   PIC X(1) VALUE 'N'.
           88  SW-END-OF-FILE              VALUE 'Y'.
           88  SW-NOT-END-OF-FILE          VALUE 'N'.

   01 IN-EMPLOYEE-RECORD.
       05  EMP-ID-IN     PIC X(04).
       05  FILLER        PIC X(05).
       05  REG-PAY-IN    PIC 99999V99.
       05  FILLER        PIC X(02).
       05  BON-PAY-IN    PIC 9999V99.
       05  FILLER        PIC X(54).

   01 OUT-EMPLOYEE-RECORD.
       05  EMP-ID-OUT    PIC S9(9) USAGE COMP.
       05  FILLER        PIC X(05).
```

```
        05  REG-PAY-OUT    PIC S9(6)V9(2) USAGE COMP-3.
        05  FILLER         PIC X(02).
        05  BON-PAY-OUT    PIC S9(6)V9(2) USAGE COMP-3.
        05  FILLER         PIC X(02).
        05  NEW-PAY-OUT    PIC S9(6)V9(2) USAGE COMP-3.
        05  FILLER         PIC X(49) VALUE SPACES.

    01 DISPLAY-EMPLOYEE-PIC.
         05  DIS-REG-PAY   PIC 99999.99.
         05  DIS-NEW-PAY   PIC 99999.99.
         05  DIS-BON-PAY   PIC 9999.99.

    01 HDR-LINE-01.
        05  FILLER         PIC X(25) VALUE SPACES.
        05  FILLER         PIC X(30)
           VALUE 'EMPLOYEE ANNUAL SALARY REPORT '.
        05  FILLER         PIC X(25) VALUE SPACES.

    01 HDR-LINE-02.
        05  FILLER         PIC X(25) VALUE SPACES.
        05  FILLER         PIC X(30)
           VALUE '----------------------------- '.
        05  FILLER         PIC X(25) VALUE SPACES.

    01 SPC-LINE           PIC X(80) VALUE SPACES.

    01 DTL-HDR01.
        05 FILLER         PIC  X(19) VALUE SPACE.
        05 FILLER         PIC  X(06) VALUE 'EMP-ID'.
        05 FILLER         PIC  X(03) VALUE SPACES.
        05 FILLER         PIC  X(08) VALUE 'REGULAR '.
        05 FILLER         PIC  X(04) VALUE SPACE.
        05 FILLER         PIC  X(08) VALUE 'BONUS   '.
        05 FILLER         PIC  X(04) VALUE SPACE.
        05 FILLER         PIC  X(08) VALUE 'NEW PAY '.
        05 FILLER         PIC  X(13) VALUE SPACES.

    01 DTL-HDR02.
        05 FILLER         PIC  X(19) VALUE SPACE.
        05 FILLER         PIC  X(06) VALUE '------'.
        05 FILLER         PIC  X(03) VALUE SPACES.
        05 FILLER         PIC  X(08) VALUE '--------'.
        05 FILLER         PIC  X(04) VALUE SPACE.
        05 FILLER         PIC  X(08) VALUE '--------'.
        05 FILLER         PIC  X(04) VALUE SPACE.
        05 FILLER         PIC  X(08) VALUE '--------'.
        05 FILLER         PIC  X(13) VALUE SPACES.

    01 DTL-LINE.
        05 FILLER         PIC  X(21) VALUE SPACE.
        05 RPT-EMP-ID     PIC  9(04).
        05 FILLER         PIC  X(03) VALUE SPACES.
```

```
        05 RPT-REG-PAY    PIC  99999.99.
        05 FILLER         PIC  X(04) VALUE SPACE.
        05 RPT-BON-PAY    PIC  99999.99.
        05 FILLER         PIC  X(04) VALUE SPACE.
        05 RPT-NEW-PAY    PIC  99999.99.
        05 FILLER         PIC  X(13) VALUE SPACES.

    01 TRLR-LINE-01.
        05  FILLER        PIC X(26) VALUE SPACES.
        05  FILLER        PIC X(30)
            VALUE ' END OF ANNUAL SALARY REPORT  '.
        05  FILLER        PIC X(24) VALUE SPACES.

    77 REG-PAY-PKD        PIC S9(6)V9(2) USAGE COMP-3.
    77 PAY-RAISE-PKD      PIC S9(6)V9(2) USAGE COMP-3.
    77 NEW-PAY-PKD        PIC S9(6)V9(2) USAGE COMP-3.
    77 NEW-PAY            PIC 99999.99.


PROCEDURE DIVISION.

    PERFORM P100-INITIALIZATION.
    PERFORM P200-MAINLINE.
    PERFORM P300-TERMINATION.
    GOBACK.


P100-INITIALIZATION.

    DISPLAY 'COBTRN5 - SAMPLE COBOL PROGRAM: CALCULATIONS '.
    OPEN INPUT  EMPLOYEE-IN-FILE,
    OPEN OUTPUT EMPLOYEE-OUT-FILE,
    OPEN OUTPUT REPORT-OUT-FILE.

    INITIALIZE IN-EMPLOYEE-RECORD, OUT-EMPLOYEE-RECORD.

    WRITE REPORT-RECORD-OUT FROM HDR-LINE-01
    WRITE REPORT-RECORD-OUT FROM HDR-LINE-02
    WRITE REPORT-RECORD-OUT FROM SPC-LINE
    WRITE REPORT-RECORD-OUT FROM DTL-HDR01
    WRITE REPORT-RECORD-OUT FROM DTL-HDR02.

P200-MAINLINE.

*   MAIN LOOP - READ THE INPUT FILE, LOAD THE OUTPUT
*               STRUCTURE AND WRITE THE RECORD TO OUTPUT.

    SET SW-NOT-END-OF-FILE TO TRUE.

    READ EMPLOYEE-IN-FILE INTO IN-EMPLOYEE-RECORD
       AT END SET SW-END-OF-FILE TO TRUE
    END-READ

    PERFORM UNTIL SW-END-OF-FILE
```

```
*          MOVE REG-PAY-IN   TO REG-PAY-PKD
*          MULTIPLY REG-PAY-PKD BY 0.10 GIVING PAY-RAISE-PKD
*          MOVE PAY-RAISE-PKD TO PAY-RAISE
*          MOVE REG-PAY-PKD TO REG-PAY
*          ADD PAY-RAISE-PKD TO REG-PAY-PKD GIVING NEW-PAY-PKD
*          MOVE NEW-PAY-PKD TO NEW-PAY
*          DISPLAY NEW-PAY

           MOVE REG-PAY-IN   TO REG-PAY-PKD
           COMPUTE NEW-PAY-PKD =
               (REG-PAY-PKD + (REG-PAY-PKD * 0.10))
           MOVE NEW-PAY-PKD TO NEW-PAY

*          MOVE FIELDS

           MOVE EMP-ID-IN    TO EMP-ID-OUT, RPT-EMP-ID
           MOVE REG-PAY-IN
           TO REG-PAY-OUT, RPT-REG-PAY, DIS-REG-PAY
           MOVE BON-PAY-IN
           TO BON-PAY-OUT, RPT-BON-PAY, DIS-BON-PAY
           MOVE NEW-PAY-PKD
           TO NEW-PAY-OUT, RPT-NEW-PAY, DIS-NEW-PAY

           DISPLAY ' EMP ID: '   EMP-ID-IN
                   ' REG PAY: '  DIS-REG-PAY
                   ' BONUS PAY: ' DIS-BON-PAY
                   ' NEW PAY  : ' DIS-NEW-PAY

           WRITE EMPLOYEE-RECORD-OUT
              FROM OUT-EMPLOYEE-RECORD

           WRITE REPORT-RECORD-OUT FROM DTL-LINE

           READ EMPLOYEE-IN-FILE
              INTO IN-EMPLOYEE-RECORD
                 AT END SET SW-END-OF-FILE TO TRUE
           END-READ

      END-PERFORM.

  P300-TERMINATION.

      WRITE REPORT-RECORD-OUT FROM SPC-LINE
      WRITE REPORT-RECORD-OUT FROM SPC-LINE
      WRITE REPORT-RECORD-OUT FROM TRLR-LINE-01

      CLOSE EMPLOYEE-IN-FILE,
            EMPLOYEE-OUT-FILE,
            REPORT-OUT-FILE.

      DISPLAY 'COBTRN5 - SUCCESSFULLY ENDED'.
```

```
    *    END OF SOURCE CODE
```

Here is the output from the program:

```
COBTRN5 - SAMPLE COBOL PROGRAM: CALCULATIONS
 EMP ID: 1111 REG PAY: 87000.00 BONUS PAY: 6700.00 NEW PAY : 95700.00
 EMP ID: 1122 REG PAY: 82000.00 BONUS PAY: 6000.00 NEW PAY : 90200.00
 EMP ID: 3217 REG PAY: 65000.00 BONUS PAY: 5500.00 NEW PAY : 71500.00
 EMP ID: 4175 REG PAY: 55000.00 BONUS PAY: 1500.00 NEW PAY : 60500.00
 EMP ID: 4720 REG PAY: 80000.00 BONUS PAY: 2500.00 NEW PAY : 88000.00
 EMP ID: 4836 REG PAY: 62000.00 BONUS PAY: 2200.00 NEW PAY : 68200.00
 EMP ID: 6288 REG PAY: 70000.00 BONUS PAY: 2000.00 NEW PAY : 77000.00
 EMP ID: 7459 REG PAY: 85000.00 BONUS PAY: 4500.00 NEW PAY : 93500.00
 EMP ID: 9134 REG PAY: 75000.00 BONUS PAY: 2500.00 NEW PAY : 82500.00
COBTRN5 - SUCCESSFULLY ENDED
```

And here is the report from the program.

```
        EMPLOYEE ANNUAL SALARY REPORT
        ----------------------------


EMP-ID    REGULAR      BONUS       NEW PAY
------    --------     --------    --------
  1111    87000.00     06700.00    95700.00
  1122    82000.00     06000.00    90200.00
  3217    65000.00     05500.00    71500.00
  4175    55000.00     01500.00    60500.00
  4720    80000.00     02500.00    88000.00
  4836    62000.00     02200.00    68200.00
  6288    70000.00     02000.00    77000.00
  7459    85000.00     04500.00    93500.00
  9134    75000.00     02500.00    82500.00

        END OF ANNUAL SALARY REPORT
```

Edits and Validation

So far the data we used for our training programs did not have any errors, and we haven't yet coded for any. Obviously in the real world you have to handle errors, both those that can cause your program to abend, as well as those that don't. The latter are sometimes worse because it can allow bad data to corrupt your client's business information.

For COBTRN6, we will start with our previous program COBTRN5 and we will change our data to include some obvious errors. Let's say we have these rules:

1. Employee id must be numeric and greater than zero

2. Regular Pay must be numeric and greater than zero
3. Bonus Pay must be numeric and greater than zero

And let's modify our pay file to create a couple of errors. In this case, the first record has a non-numeric regular pay value, and the second has a zero value for regular pay.

```
----+----1----+----2----+----3
******************************
1111    8700GGG  670000
1122    0000000  600000
3217    6500000  550000
4175    5500000  150000
4720    8000000  250000
4836    6200000  220000
6288    7000000  200000
7459    8500000  450000
9134    7500000  250000
******************************
```

We should edit these fields to ensure all values are valid. Also, once an error is encountered, we may not want to continue editing the same field, as this may possibly force an abend. For example if we check the employee id for numeric, and it is not numeric, checking the value for greater than zero will force a data exception. So we should code around these sorts of problems.

Obviously you would want to check your values for numeric first so as to avoid the data exception. But how to stop doing additional edits when an error is encountered? There are several ways to do this. A classic way is to create a label at the end of the procedure and use a GOTO statement to transfer control to the end of the procedure. That is a way that would work, but it is often discouraged as being inconsistent with structured programming.

Another way to validate selectively is to set an error flag and to include a check of the flag as part of each edit. If there are no previous errors, then perform the current edit. I prefer this method and will use it here, given that one feature of structured programming is supposed to be that you avoid the use of GOTO statements.

So you could set up an edit/validation procedure that is called from the main procedure. Your main procedure would check to see if any errors were found. Here's some pseudo code for this processing.

```
MAIN PROCEDURE
CALL ERROR CHECKING PROCEDURE
IF ERROR SWITCH IS NO THEN
   PROCESS SUCCESSFULLY EDITED RECORD
```

```
ELSE
   DO ERROR PROCESSING

ERROR CHECKING PROCEDURE
SET ERROR SWITCH TO NO

IF VAR1 IS NUMERIC THEN
   CONTINUE
ELSE
   SET ERROR SWITCH TO YES
   ASSIGN ERROR MESSAGE
END IF

IF ERROR SWITCH IS NO THEN
   IF VAR2 IS NUMERIC THEN
      CONTINUE
   ELSE
      SET ERROR SWITCH TO YES
      ASSIGN ERROR MESSAGE
   END IF
END IF

IF ERROR SWITCH IS NO THEN
   IF VAR1 IS GREATER THAN ZERO THEN
      CONTINUE
   ELSE
      SET ERROR SWITCH TO YES
      ASSIGN ERROR MESSAGE

   END IF
END IF

IF VAR1 IS GREATER THAN ZERO THEN
   IF VAR2 IS NUMERIC THEN
      CONTINUE
   ELSE
      SET ERROR SWITCH TO YES
       ASSIGN ERROR MESSAGE
   END IF
END IF
```

Of course your routine can be much more elaborate. But this is how we will implement our edits and validations in COBTRN6. First let us add a flag called SW-ERROR-SW and two 88 levels that mean yes and no:

```
01 WS-FLAGS.
    05  SW-END-OF-FILE-SWITCH    PIC X(1) VALUE 'N'.
        88  SW-END-OF-FILE                VALUE 'Y'.
        88  SW-NOT-END-OF-FILE            VALUE 'N'.

    05  SW-ERROR-SWITCH          PIC X(1) VALUE 'N'.
        88  SW-HAS-ERROR                  VALUE 'Y'.
        88  SW-NO-ERRORS                  VALUE 'N'.
```

Now let's code an error routine based on the pseudo code above. I think the only thing we're doing different than before is using the SET verb. When you set an 88 variable to TRUE you are forcing the value of the 88 level variable into its parent variable. Coding:

```
SET SW-NO-ERRORS TO TRUE
```

is equivalent to coding:

```
MOVE 'N' TO SW-ERROR-SWITCH.
```

You can do it either way. I prefer the SET method because it's a bit clearer (I think).   Ok here's our routine.

```
    P1000-EDIT-RECORD.

        SET SW-NO-ERRORS TO TRUE

        IF SW-NO-ERRORS THEN
           IF EMP-ID-IN IS NUMERIC THEN
              MOVE EMP-ID-IN TO EMP-ID-BIN
           ELSE
              SET SW-HAS-ERROR TO TRUE
              DISPLAY 'EMP ID IS NOT NUMERIC ' EMP-ID-IN
           END-IF
        END-IF

        IF SW-NO-ERRORS THEN
           IF REG-PAY-IN IS NUMERIC THEN
              MOVE REG-PAY-IN TO REG-PAY-PKD
           ELSE
              SET SW-HAS-ERROR TO TRUE
              DISPLAY 'REG PAY IS NOT NUMERIC ' REG-PAY-IN
           END-IF
        END-IF

        IF SW-NO-ERRORS THEN
           IF BON-PAY-IN IS NUMERIC THEN
              MOVE BON-PAY-IN TO BON-PAY-PKD
           ELSE
              DISPLAY 'BON PAY IS NOT NUMERIC ' BON-PAY-IN
              SET SW-HAS-ERROR TO TRUE
           END-IF
        END-IF

        IF SW-NO-ERRORS THEN
           IF EMP-ID-BIN <= 0 THEN
              SET SW-HAS-ERROR TO TRUE
              DISPLAY 'EMP ID CANNOT BE ZERO ' EMP-ID-IN
```

```
         END-IF
      END-IF

      IF SW-NO-ERRORS THEN
         IF REG-PAY-PKD <= 0 THEN
            SET SW-HAS-ERROR TO TRUE
            DISPLAY 'REG PAY CANNOT BE ZERO ' REG-PAY-IN
         END-IF
      END-IF

      IF SW-NO-ERRORS THEN
         IF BON-PAY-PKD <= 0 THEN
            SET SW-HAS-ERROR TO TRUE
            DISPLAY 'BON PAY CANNOT BE ZERO ' BON-PAY-IN
         END-IF
      END-IF.
```

Now as you can see, each error is checked only if an error has not already been found. We could compact this further by nesting our IF/THEN to three levels. For example, if our value is numeric then we can nest yet another IF to check for greater than zero.

To take the employee id field as an example:

```
IF SW-NO-ERRORS THEN
   IF EMP-ID-IN IS NUMERIC THEN
      IF EMP-ID-BIN > 0 THEN
            MOVE EMP-ID-IN TO EMP-ID-BIN
         ELSE
            SET SW-HAS-ERROR TO TRUE
            DISPLAY 'EMP ID CANNOT BE ZERO ' EMP-ID-IN
         END-IF
   ELSE
      SET SW-HAS-ERROR TO TRUE
      DISPLAY 'EMP ID IS NOT NUMERIC ' EMP-ID-IN
   END-IF
ELSE
   CONTINUE
END-IF.
```

That's somewhat more compact, and I prefer this nesting because all the edits for the employee id field are in one sentence. On the other hand, when you exceed two or three levels of nesting, the code can become difficult to read. For our COBTRN6 program we'll just nest to two levels.

Ok here is our complete program listing. Notice that in the main procedure we are checking the error switch to make sure the edit routine found no errors before mapping our data fields and writing records.

```
          IDENTIFICATION DIVISION.
          PROGRAM-ID. COBTRN6.
          *******************************************************
          *       PROGRAM USING FILE INPUT AND OUTPUT          *
          *       TO REFORMAT EMPLOYEE PAY INFORMATION.        *
          *       CALCULATE A 10% RAISE FOR EACH EMPLOYEE.     *
          *       ALSO PRODUCES A REPORT OF THE INFORMATION.   *
          *******************************************************
          ENVIRONMENT DIVISION.
          INPUT-OUTPUT SECTION.

             FILE-CONTROL.
                 SELECT EMPLOYEE-IN-FILE   ASSIGN TO EMPIFILE.
                 SELECT EMPLOYEE-OUT-FILE  ASSIGN TO EMPOFILE.
                 SELECT REPORT-OUT-FILE    ASSIGN TO EMPREPRT.

          DATA DIVISION.
          FILE SECTION.
          FD  EMPLOYEE-IN-FILE
              RECORDING MODE IS F
              LABEL RECORDS ARE STANDARD
              RECORD CONTAINS 80 CHARACTERS
              BLOCK CONTAINS 0 RECORDS
              DATA RECORD IS EMPLOYEE-RECORD-IN.

             01 EMPLOYEE-RECORD-IN.
                   05  E-ID        PIC X(04).
                   05  FILLER      PIC X(76).

          FD  EMPLOYEE-OUT-FILE
              RECORDING MODE IS F
              LABEL RECORDS ARE STANDARD
              RECORD CONTAINS 80 CHARACTERS
              BLOCK CONTAINS 0 RECORDS
              DATA RECORD IS EMPLOYEE-RECORD-OUT.

             01 EMPLOYEE-RECORD-OUT.
                   05  EMP-DATA    PIC X(80).

          FD  REPORT-OUT-FILE
              RECORDING MODE IS F
              LABEL RECORDS ARE STANDARD
              RECORD CONTAINS 80 CHARACTERS
              BLOCK CONTAINS 0 RECORDS
              DATA RECORD IS REPORT-RECORD-OUT.

             01 REPORT-RECORD-OUT.
                   05  RPT-DATA    PIC X(80).
```

```
WORKING-STORAGE SECTION.
    01 WS-FLAGS.
        05  SW-END-OF-FILE-SWITCH   PIC X(1) VALUE 'N'.
            88  SW-END-OF-FILE              VALUE 'Y'.
            88  SW-NOT-END-OF-FILE          VALUE 'N'.

        05  SW-ERROR-SWITCH          PIC X(1) VALUE 'N'.
            88  SW-HAS-ERROR                VALUE 'Y'.
            88  SW-NO-ERRORS                VALUE 'N'.

    01 IN-EMPLOYEE-RECORD.
        05  EMP-ID-IN      PIC X(04).
        05  FILLER         PIC X(05).
        05  REG-PAY-IN     PIC 99999V99.
        05  FILLER         PIC X(02).
        05  BON-PAY-IN     PIC 9999V99.
        05  FILLER         PIC X(54).

    01 OUT-EMPLOYEE-RECORD.
        05  EMP-ID-OUT     PIC S9(9) USAGE COMP.
        05  FILLER         PIC X(05).
        05  REG-PAY-OUT    PIC S9(6)V9(2) USAGE COMP-3.
        05  FILLER         PIC X(02).
        05  BON-PAY-OUT    PIC S9(6)V9(2) USAGE COMP-3.
        05  FILLER         PIC X(02).
        05  NEW-PAY-OUT    PIC S9(6)V9(2) USAGE COMP-3.
        05  FILLER         PIC X(49) VALUE SPACES.

    01 DISPLAY-EMPLOYEE-PIC.
        05  DIS-REG-PAY    PIC 99999.99.
        05  DIS-NEW-PAY    PIC 99999.99.
        05  DIS-BON-PAY    PIC 9999.99.

    01 HDR-LINE-01.
        05  FILLER         PIC X(25) VALUE SPACES.
        05  FILLER         PIC X(30)
            VALUE 'EMPLOYEE ANNUAL SALARY REPORT '.
        05  FILLER         PIC X(25) VALUE SPACES.

    01 HDR-LINE-02.
        05  FILLER         PIC X(25) VALUE SPACES.
        05  FILLER         PIC X(30)
            VALUE '--------------------------- '.
        05  FILLER         PIC X(25) VALUE SPACES.

    01 SPC-LINE          PIC X(80) VALUE SPACES.

    01 DTL-HDR01.
        05 FILLER         PIC  X(19) VALUE SPACE.
        05 FILLER         PIC  X(06) VALUE 'EMP-ID'.
        05 FILLER         PIC  X(03) VALUE SPACES.
        05 FILLER         PIC  X(08) VALUE 'REGULAR '.
```

113

```cobol
           05 FILLER          PIC  X(04) VALUE SPACE.
           05 FILLER          PIC  X(08) VALUE 'BONUS    '.
           05 FILLER          PIC  X(04) VALUE SPACE.
           05 FILLER          PIC  X(08) VALUE 'NEW PAY '.
           05 FILLER          PIC  X(13) VALUE SPACES.

       01 DTL-HDR02.
           05 FILLER          PIC  X(19) VALUE SPACE.
           05 FILLER          PIC  X(06) VALUE '------'.
           05 FILLER          PIC  X(03) VALUE SPACES.
           05 FILLER          PIC  X(08) VALUE '--------'.
           05 FILLER          PIC  X(04) VALUE SPACE.
           05 FILLER          PIC  X(08) VALUE '--------'.
           05 FILLER          PIC  X(04) VALUE SPACE.
           05 FILLER          PIC  X(08) VALUE '--------'.
           05 FILLER          PIC  X(13) VALUE SPACES.

       01 DTL-LINE.
           05 FILLER          PIC  X(21) VALUE SPACE.
           05 RPT-EMP-ID       PIC  9(04).
           05 FILLER          PIC  X(03) VALUE SPACES.
           05 RPT-REG-PAY      PIC  99999.99.
           05 FILLER          PIC  X(04) VALUE SPACE.
           05 RPT-BON-PAY      PIC  99999.99.
           05 FILLER          PIC  X(04) VALUE SPACE.
           05 RPT-NEW-PAY      PIC  99999.99.
           05 FILLER          PIC  X(13) VALUE SPACES.

       01 TRLR-LINE-01.
           05  FILLER         PIC X(26) VALUE SPACES.
           05  FILLER         PIC X(30)
               VALUE ' END OF ANNUAL SALARY REPORT  '.
           05  FILLER         PIC X(24) VALUE SPACES.

       77 EMP-ID-BIN        PIC S9(9) USAGE COMP.
       77 REG-PAY-PKD       PIC S9(6)V9(2) USAGE COMP-3.
       77 BON-PAY-PKD       PIC S9(6)V9(2) USAGE COMP-3.
       77 PAY-RAISE-PKD     PIC S9(6)V9(2) USAGE COMP-3.
       77 NEW-PAY-PKD       PIC S9(6)V9(2) USAGE COMP-3.
       77 NEW-PAY           PIC 99999.99.

   PROCEDURE DIVISION.

       PERFORM P100-INITIALIZATION.
       PERFORM P200-MAINLINE.
       PERFORM P300-TERMINATION.
       GOBACK.

   P100-INITIALIZATION.

       DISPLAY 'COBTRN6 - SAMPLE COBOL PROGRAM: CHECKING DATA '.
       OPEN INPUT  EMPLOYEE-IN-FILE,
```

```
       OPEN OUTPUT EMPLOYEE-OUT-FILE,
       OPEN OUTPUT REPORT-OUT-FILE.

       INITIALIZE IN-EMPLOYEE-RECORD, OUT-EMPLOYEE-RECORD.

       WRITE REPORT-RECORD-OUT FROM HDR-LINE-01
       WRITE REPORT-RECORD-OUT FROM HDR-LINE-02
       WRITE REPORT-RECORD-OUT FROM SPC-LINE
       WRITE REPORT-RECORD-OUT FROM DTL-HDR01
       WRITE REPORT-RECORD-OUT FROM DTL-HDR02.

   P200-MAINLINE.

*      MAIN LOOP - READ THE INPUT FILE, LOAD THE OUTPUT
*                  STRUCTURE AND WRITE THE RECORD TO OUTPUT.

       SET SW-NOT-END-OF-FILE TO TRUE.

       READ EMPLOYEE-IN-FILE INTO IN-EMPLOYEE-RECORD
          AT END SET SW-END-OF-FILE TO TRUE
       END-READ

       PERFORM UNTIL SW-END-OF-FILE

           PERFORM P1000-EDIT-RECORD

           IF SW-NO-ERRORS THEN

               COMPUTE NEW-PAY-PKD =
                   (REG-PAY-PKD + (REG-PAY-PKD * 0.10))
               MOVE NEW-PAY-PKD TO NEW-PAY
               DISPLAY 'NEW PAY ' NEW-PAY

*          MOVE FIELDS

               MOVE EMP-ID-IN   TO EMP-ID-OUT, RPT-EMP-ID
               MOVE REG-PAY-IN
               TO REG-PAY-OUT, RPT-REG-PAY, DIS-REG-PAY
               MOVE BON-PAY-IN
               TO BON-PAY-OUT, RPT-BON-PAY, DIS-BON-PAY
               MOVE NEW-PAY-PKD
               TO NEW-PAY-OUT, RPT-NEW-PAY, DIS-NEW-PAY

               DISPLAY ' EMP ID: '   EMP-ID-IN
                       ' REG PAY: '  DIS-REG-PAY
                       ' BONUS PAY: ' DIS-BON-PAY
                       ' NEW PAY  : ' DIS-NEW-PAY


               WRITE EMPLOYEE-RECORD-OUT
                  FROM OUT-EMPLOYEE-RECORD
```

```
        WRITE REPORT-RECORD-OUT FROM DTL-LINE
    ELSE
        DISPLAY '** RECORD DISCARDED **'
    END-IF

    READ EMPLOYEE-IN-FILE
        INTO IN-EMPLOYEE-RECORD
            AT END SET SW-END-OF-FILE TO TRUE
    END-READ

END-PERFORM.

P300-TERMINATION.

    WRITE REPORT-RECORD-OUT FROM SPC-LINE
    WRITE REPORT-RECORD-OUT FROM SPC-LINE
    WRITE REPORT-RECORD-OUT FROM TRLR-LINE-01

    CLOSE EMPLOYEE-IN-FILE,
          EMPLOYEE-OUT-FILE,
          REPORT-OUT-FILE.

    DISPLAY 'COBTRN6 - SUCCESSFULLY ENDED'.

P1000-EDIT-RECORD.

    SET SW-NO-ERRORS TO TRUE

    IF SW-NO-ERRORS THEN
        IF EMP-ID-IN IS NUMERIC THEN
            MOVE EMP-ID-IN TO EMP-ID-BIN
        ELSE
            SET SW-HAS-ERROR TO TRUE
            DISPLAY 'EMP ID IS NOT NUMERIC ' EMP-ID-IN
        END-IF
    END-IF

    IF SW-NO-ERRORS THEN
        IF REG-PAY-IN IS NUMERIC THEN
            MOVE REG-PAY-IN TO REG-PAY-PKD
        ELSE
            SET SW-HAS-ERROR TO TRUE
            DISPLAY 'REG PAY IS NOT NUMERIC ' REG-PAY-IN
        END-IF
    END-IF

    IF SW-NO-ERRORS THEN
        IF BON-PAY-IN IS NUMERIC THEN
            MOVE BON-PAY-IN TO BON-PAY-PKD
        ELSE
            DISPLAY 'BON PAY IS NOT NUMERIC ' BON-PAY-IN
            SET SW-HAS-ERROR TO TRUE
```

```
            END-IF
        END-IF

        IF SW-NO-ERRORS THEN
            IF EMP-ID-BIN <= 0 THEN
                SET SW-HAS-ERROR TO TRUE
                DISPLAY 'EMP ID CANNOT BE ZERO ' EMP-ID-IN
            END-IF
        END-IF

        IF SW-NO-ERRORS THEN
            IF REG-PAY-PKD <= 0 THEN
                SET SW-HAS-ERROR TO TRUE
                DISPLAY 'REG PAY CANNOT BE ZERO ' REG-PAY-IN
            END-IF
        END-IF

        IF SW-NO-ERRORS THEN
            IF BON-PAY-PKD <= 0 THEN
                SET SW-HAS-ERROR TO TRUE
                DISPLAY 'BON PAY CANNOT BE ZERO ' BON-PAY-IN
            END-IF
        END-IF.

    *    END OF SOURCE CODE
```

Here is the program output:

```
COBTRN6 - SAMPLE COBOL PROGRAM: CHECKING DATA
REG PAY IS NOT NUMERIC 8700GGG
** RECORD DISCARDED **
REG PAY CANNOT BE ZERO 0000000
** RECORD DISCARDED **
NEW PAY 71500.00
 EMP ID: 3217 REG PAY: 65000.00 BONUS PAY: 5500.00 NEW PAY  : 71500.00
NEW PAY 60500.00
 EMP ID: 4175 REG PAY: 55000.00 BONUS PAY: 1500.00 NEW PAY  : 60500.00
NEW PAY 88000.00
 EMP ID: 4720 REG PAY: 80000.00 BONUS PAY: 2500.00 NEW PAY  : 88000.00
NEW PAY 68200.00
 EMP ID: 4836 REG PAY: 62000.00 BONUS PAY: 2200.00 NEW PAY  : 68200.00
NEW PAY 77000.00
 EMP ID: 6288 REG PAY: 70000.00 BONUS PAY: 2000.00 NEW PAY  : 77000.00
NEW PAY 93500.00
EMP ID: 7459 REG PAY: 85000.00 BONUS PAY: 4500.00 NEW PAY  : 93500.00
NEW PAY 82500.00
 EMP ID: 9134 REG PAY: 75000.00 BONUS PAY: 2500.00 NEW PAY  : 82500.00
COBTRN6 - SUCCESSFULLY ENDED
```

There are many methods of doing edits and validations in a program. This sample program gives you one method. You can use it as a model unless/until you encounter something better in your shop.

Tables

There are many reasons to use internal tables in an application program. For example you may have an external file that contains valid values for a field edit. Putting this information into an internal table in the program makes it easily searchable to perform the validation.

Another example is when you want to accumulate complex statistics for raw data – a table can be perfect for this.

A third example is when you need to combine data from multiple files to create a composite file. For our training program COBTRN7 we are going to combine employee-related data elements from two files to produce a combined file that includes all the data elements. Let's look at our files in detail.

We have an employee profile file that contains employee id, last and first name, years of service and the date of the employee's last promotion.

```
----+----1----+----2----+----3----+----4----+----5----+----6----+----7----+----8
***************************** Top of Data ********************************
1111 VEREEN                 CHARLES              12 2017-01-01
1122 JENKINS                DEBORAH               5 2017-01-01
3217 JOHNSON                EDWARD                4 2017-01-01
4175 TURNBULL               FRED                  1 2016-12-01
4720 SCHULTZ                TIM                   9 2017-01-01
4836 SMITH                  SANDRA                3 2017-01-01
6288 WILLARD                JOE                   6 2016-01-01
7459 STEWART                BETTY                 7 2016-07-31
9134 FRANKLIN               BRIANNA               0 2016-10-01
***************************** Bottom of Data ********************************
```

The second file is the pay file we have been working with, and it looks like this:

```
----+----1----+----2----+----3----+----4----+----5----+----6----+----7----+----8
***************************** Top of Data ********************************
1111      8700000  670000
1122      8200000  600000
3217      6500000  550000
4175      5500000  150000
4720      8000000  250000
4836      6200000  220000
6288      7000000  200000
7459      8500000  450000
9134      7500000  250000
*****************************
```

So what we are going to do is to:

1. Declare a table in the program that includes all the fields for each employee.

2. Read the pay file and load the pay data into the table.

3. Read the employee profile file, locate the appropriate table entry by searching on employee id, and when matched load the profile data into the table.

4. Write the data in the employee table to a master file.

Our program id is now COBTRN7. Again you can copy COBTRN6 to get a baseline. We'll need file declarations for the pay file, the employee profile file, and the master file that will used for output.

```
ENVIRONMENT DIVISION.
 INPUT-OUTPUT SECTION.

    FILE-CONTROL.
        SELECT EMPPAY-IN-FILE    ASSIGN TO EMPPAYFL.
        SELECT EMPLOYEE-IN-FILE  ASSIGN TO EMPLOYIN.
        SELECT EMPLMAST-OUT-FILE ASSIGN TO EMPLMAST.

 DATA DIVISION.

 FILE SECTION.
 FD  EMPPAY-IN-FILE
     RECORDING MODE IS F
     LABEL RECORDS ARE STANDARD
     RECORD CONTAINS 80 CHARACTERS
     BLOCK CONTAINS 0 RECORDS
     DATA RECORD IS EMPPAY-RECORD-IN.

    01 EMPPAY-RECORD-IN.
         05  E-ID         PIC X(04).
         05  FILLER       PIC X(76).

 FD  EMPLOYEE-IN-FILE
     RECORDING MODE IS F
     LABEL RECORDS ARE STANDARD
     RECORD CONTAINS 80 CHARACTERS
     BLOCK CONTAINS 0 RECORDS
     DATA RECORD IS EMPLOYEE-RECORD-IN.

    01 EMPLOYEE-RECORD-IN.
         05  E-ID         PIC X(04).
         05  FILLER       PIC X(76).

 FD  EMPLMAST-OUT-FILE
     RECORDING MODE IS F
     LABEL RECORDS ARE STANDARD
```

```
        RECORD CONTAINS 85 CHARACTERS
        BLOCK CONTAINS 0 RECORDS
        DATA RECORD IS EMPLOYEE-MASTER-OUT.


    01 EMPLOYEE-MASTER-OUT.
        05  EMP-MAST-DATA PIC X(85).
```

We'll of course need record structures as well. Here are the structures for each of the three files:

```
    01 IN-EMPPAY-RECORD.
        05  EMP-ID-IN    PIC X(04).
        05  FILLER       PIC X(05).
        05  REG-PAY-IN   PIC 99999V99.
        05  FILLER       PIC X(02).
        05  BON-PAY-IN   PIC 9999V99.
        05  FILLER       PIC X(54).


    01 IN-EMPLOYEE-RECORD.
        05  EMPL-ID-IN   PIC X(04).
        05  FILLER       PIC X(01).
        05  EMPL-LNAME   PIC X(30).
        05  FILLER       PIC X(01).
        05  EMPL-FNAME   PIC X(20).
        05  FILLER       PIC X(01).
        05  EMPL-YRS-SRV PIC X(02).
        05  FILLER       PIC X(01).
        05  EMPL-PRM-DTE PIC X(10).
        05  FILLER       PIC X(10).


    01 OUT-EMPLMAST-RECORD.
        05  EMPLMAST-EMP-ID   PIC X(04).
        05  FILLER            PIC X(01) VALUE SPACES.
        05  EMPLMAST-LNAME    PIC X(30).
        05  EMPLMAST-FNAME    PIC X(20).
        05  FILLER            PIC X(01) VALUE SPACE.
        05  EMPLMAST-YRS-SRV  PIC X(02).
        05  FILLER            PIC X(01) VALUE SPACE.
        05  EMPLMAST-PRM-DTE  PIC X(10).
        05  FILLER            PIC X(01) VALUE SPACE.
        05  EMPLMAST-REG-PAY  PIC 99999V99.
        05  FILLER            PIC X(01) VALUE SPACE.
        05  EMPLMAST-BON-PAY  PIC 9999V99.
        05  FILLER            PIC X(01) VALUE SPACES.
```

Next we'll need to declare our table. For this example, we are going to assume that we know beforehand exactly how many entries we need for the table (looking at our files, we know we need 9 entries), and that is how many we will build the table with. Here's how to define it in

120

COBOL:

```
01 EMP-MASTER-TBL.
   05 EMP-DATA OCCURS 9 TIMES
      ASCENDING KEY IS EMP-ID INDEXED BY EMP-NDX.
      10 EMP-ID         PIC X(04).
      10 EMP-LAST-NAME  PIC X(30).
      10 EMP-FIRST-NAME PIC X(20).
      10 EMP-YRS-SERVICE PIC 99.
      10 EMP-PROM-DATE  PIC X(10).
      10 EMP-REG-PAY    PIC 99999V99.
      10 EMP-BON-PAY    PIC 9999V99.
```

We declared an 01 level structure, with an 05 level named EMP-DATA that "occurs" 9 times. The OCCURS clause is the key – it determines how many entries the table will have. We've also established an index for the table which is named EMP-NDX. The index makes the table searchable. And we establish that the table is indexed on EMP-ID.

To refer to an entry or element in the table, use the EMP-NDX index. You can set the index to a particular value using the SET verb. To set the index to the first entry in the table, code:

```
SET EMP-NDX TO +1
```

To bump the index to the next higher value, code:

```
SET EMP-NDX UP BY +1
```

To bump the index to the next lower value, code:

```
SET EMP-NDX DOWN BY +1
```

To search for a value in the table, use the SEARCH verb plus the AT END and WHEN clauses. For our example, we are going to read an employee profile record and use the EMPL_ID_IN value from the input file to match the EMP-ID value in the table. If we do not find the employee id in the table (which means we reach the AT END clause) we will display an error. If we do find the employee id, then we'll add the fields from the profile record to our table. Here's the logic.

```
SEARCH ALL EMP-DATA
AT END DISPLAY 'RECORD NOT FOUND'
   DISPLAY EMPL-ID-IN
WHEN EMP-ID(EMP-NDX) = EMPL-ID-IN
   MOVE EMPL-LNAME   TO EMP-LAST-NAME(EMP-NDX)
   MOVE EMPL-FNAME   TO EMP-FIRST-NAME(EMP-NDX)
   MOVE EMPL-YRS-SRV TO EMP-YRS-SERVICE(EMP-NDX)
```

```
          MOVE EMPL-PRM-DTE TO EMP-PROM-DATE(EMP-NDX)
```

Finally, when we are ready to unload the table to create the output file we'll use a loop that cycles through the table varying the EMP-NDX value from 1 to 9. The PERFORM VARYING is a very handy coding device that you'll use often.

```
     PERFORM VARYING EMP-NDX FROM +1 BY +1
        UNTIL EMP-NDX > 9

        MOVE EMP-ID (EMP-NDX)          TO EMPLMAST-EMP-ID
        MOVE EMP-LAST-NAME (EMP-NDX)   TO EMPLMAST-LNAME
        MOVE EMP-FIRST-NAME(EMP-NDX)   TO EMPLMAST-FNAME
        MOVE EMP-YRS-SERVICE(EMP-NDX)  TO EMPLMAST-YRS-SRV
        MOVE EMP-PROM-DATE(EMP-NDX)    TO EMPLMAST-PRM-DTE
        MOVE EMP-REG-PAY(EMP-NDX)      TO EMPLMAST-REG-PAY
        MOVE EMP-BON-PAY(EMP-NDX)      TO EMPLMAST-BON-PAY
        WRITE EMPLOYEE-MASTER-OUT
           FROM OUT-EMPLMAST-RECORD
     END-PERFORM.
```

So we loop through the nine entries in the table, map the values to the output variables, and then write the master record. Here is the complete program listing.

```
     IDENTIFICATION DIVISION.
     PROGRAM-ID. COBTRN7.
     ******************************************************
     *      PROGRAM USING FILE INPUT AND TABLE TO MERGE   *
     *      TWO FILES.  THE RESULTS WILL BE USED TO       *
     *      CREATE AN OUTPUT FILE AND A REPORT FILE.      *
     ******************************************************
     ENVIRONMENT DIVISION.
     INPUT-OUTPUT SECTION.

        FILE-CONTROL.
           SELECT EMPPAY-IN-FILE     ASSIGN TO EMPPAYFL.
           SELECT EMPLOYEE-IN-FILE   ASSIGN TO EMPLOYIN.
           SELECT EMPLMAST-OUT-FILE  ASSIGN TO EMPLMAST.

     DATA DIVISION.

     FILE SECTION.
     FD  EMPPAY-IN-FILE
         RECORDING MODE IS F
         LABEL RECORDS ARE STANDARD
         RECORD CONTAINS 80 CHARACTERS
         BLOCK CONTAINS 0 RECORDS
         DATA RECORD IS EMPPAY-RECORD-IN.

         01 EMPPAY-RECORD-IN.
```

```
            05  E-ID         PIC X(04).
            05  FILLER       PIC X(76).


FD  EMPLOYEE-IN-FILE
    RECORDING MODE IS F
    LABEL RECORDS ARE STANDARD
    RECORD CONTAINS 80 CHARACTERS
    BLOCK CONTAINS 0 RECORDS
    DATA RECORD IS EMPLOYEE-RECORD-IN.

    01 EMPLOYEE-RECORD-IN.
            05  E-ID         PIC X(04).
            05  FILLER       PIC X(76).


FD  EMPLMAST-OUT-FILE
    RECORDING MODE IS F
    LABEL RECORDS ARE STANDARD
    RECORD CONTAINS 85 CHARACTERS
    BLOCK CONTAINS 0 RECORDS
    DATA RECORD IS EMPLOYEE-MASTER-OUT.

    01 EMPLOYEE-MASTER-OUT.
            05  EMP-MAST-DATA PIC X(85).



WORKING-STORAGE SECTION.

    01 WS-FLAGS.
            05  SW-END-OF-FILE-SWITCH   PIC X(1) VALUE 'N'.
                88  SW-END-OF-FILE              VALUE 'Y'.
                88  SW-NOT-END-OF-FILE          VALUE 'N'.

            05  SW-ERROR-SWITCH         PIC X(1) VALUE 'N'.
                88  SW-HAS-ERROR                VALUE 'Y'.
                88  SW-NO-ERRORS                VALUE 'N'.

    01 IN-EMPPAY-RECORD.
        05  EMP-ID-IN    PIC X(04).
        05  FILLER       PIC X(05).
        05  REG-PAY-IN   PIC 99999V99.
        05  FILLER       PIC X(02).
        05  BON-PAY-IN   PIC 9999V99.
        05  FILLER       PIC X(54).

    01 IN-EMPLOYEE-RECORD.
        05  EMPL-ID-IN   PIC X(04).
        05  FILLER       PIC X(01).
        05  EMPL-LNAME   PIC X(30).
        05  FILLER       PIC X(01).
        05  EMPL-FNAME   PIC X(20).
        05  FILLER       PIC X(01).
        05  EMPL-YRS-SRV PIC X(02).
```

```cobol
           05   FILLER        PIC X(01).
           05   EMPL-PRM-DTE  PIC X(10).
           05   FILLER        PIC X(10).


       01 OUT-EMPLMAST-RECORD.
           05   EMPLMAST-EMP-ID    PIC X(04).
           05   FILLER             PIC X(01) VALUE SPACES.
           05   EMPLMAST-LNAME     PIC X(30).
           05   EMPLMAST-FNAME     PIC X(20).
           05   FILLER             PIC X(01) VALUE SPACE.
           05   EMPLMAST-YRS-SRV   PIC X(02).
           05   FILLER             PIC X(01) VALUE SPACE.
           05   EMPLMAST-PRM-DTE   PIC X(10).
           05   FILLER             PIC X(01) VALUE SPACE.
           05   EMPLMAST-REG-PAY   PIC 99999V99.
           05   FILLER             PIC X(01) VALUE SPACE.
           05   EMPLMAST-BON-PAY   PIC 9999V99.
           05   FILLER             PIC X(01) VALUE SPACES.


       01 DISPLAY-EMPLOYEE-PIC.
            05  DIS-REG-PAY   PIC 99999.99.
            05  DIS-NEW-PAY   PIC 99999.99.
            05  DIS-BON-PAY   PIC 9999.99.


       77 EMP-ID-BIN      PIC S9(9) USAGE COMP.
       77 REG-PAY-PKD     PIC S9(6)V9(2) USAGE COMP-3.
       77 BON-PAY-PKD     PIC S9(6)V9(2) USAGE COMP-3.
       77 PAY-RAISE-PKD   PIC S9(6)V9(2) USAGE COMP-3.
       77 NEW-PAY-PKD     PIC S9(6)V9(2) USAGE COMP-3.
       77 NEW-PAY         PIC 99999.99.


       01 EMP-MASTER-TBL.
          05 EMP-DATA OCCURS 9 TIMES
             ASCENDING KEY IS EMP-ID INDEXED BY EMP-NDX.
             10 EMP-ID          PIC X(04).
             10 EMP-LAST-NAME   PIC X(30).
             10 EMP-FIRST-NAME  PIC X(20).
             10 EMP-YRS-SERVICE PIC 99.
             10 EMP-PROM-DATE   PIC X(10).
             10 EMP-REG-PAY     PIC 99999V99.
             10 EMP-BON-PAY     PIC 9999V99.



   PROCEDURE DIVISION.

       PERFORM P100-INITIALIZATION.
       PERFORM P200-MAINLINE.
       PERFORM P300-TERMINATION.
       GOBACK.


   P100-INITIALIZATION.
```

124

```
         DISPLAY 'COBTRN7 - COBOL WITH TABLE HANDLING TECHNIQUES'.

         OPEN INPUT  EMPPAY-IN-FILE,
              INPUT  EMPLOYEE-IN-FILE,
              OUTPUT EMPLMAST-OUT-FILE.

         INITIALIZE IN-EMPPAY-RECORD,
                    IN-EMPLOYEE-RECORD,
                    OUT-EMPLMAST-RECORD.


    P200-MAINLINE.

         PERFORM P1000-LOAD-PAY-DATA.

         PERFORM P1100-LOAD-EMPLOYEE-DATA.

         PERFORM P1200-WRITE-EMPLOYEE-MSTR-FILE.

*        THIRD LOOP  - WRITE THE TABLE DATA TO AN OUTPUT
*                      RECORD, AND ALSO A REPORT FILE.
*

    P300-TERMINATION.

         CLOSE EMPPAY-IN-FILE,
               EMPLOYEE-IN-FILE,
               EMPLMAST-OUT-FILE.

         DISPLAY 'COBTRN7 - SUCCESSFULLY ENDED'.

*****************************************************************
*    READ THE INPUT FILE, LOAD THE PAY DATA TO THE TABLE. *
*****************************************************************

    P1000-LOAD-PAY-DATA.

         SET SW-NOT-END-OF-FILE TO TRUE.
         SET EMP-NDX TO +1

         READ EMPPAY-IN-FILE INTO IN-EMPPAY-RECORD
            AT END SET SW-END-OF-FILE TO TRUE
         END-READ

         PERFORM UNTIL SW-END-OF-FILE

            MOVE EMP-ID-IN TO EMP-ID(EMP-NDX)
            MOVE REG-PAY-IN TO EMP-REG-PAY(EMP-NDX)
            MOVE BON-PAY-IN TO EMP-BON-PAY(EMP-NDX)

            DISPLAY ' EMP ID: '   EMP-ID(EMP-NDX)
                    ' REG PAY: '  EMP-REG-PAY(EMP-NDX)
```

125

```cobol
                    ' BONUS PAY: ' EMP-BON-PAY(EMP-NDX)

        SET EMP-NDX UP BY +1

        READ EMPPAY-IN-FILE
            INTO IN-EMPPAY-RECORD
                AT END SET SW-END-OF-FILE TO TRUE
        END-READ

    END-PERFORM.

P1100-LOAD-EMPLOYEE-DATA.

    DISPLAY 'P1100-LOAD-EMPLOYEE-DATA'.

    SET SW-NOT-END-OF-FILE TO TRUE.

    READ EMPLOYEE-IN-FILE INTO IN-EMPLOYEE-RECORD
        AT END SET SW-END-OF-FILE TO TRUE
    END-READ

    PERFORM UNTIL SW-END-OF-FILE

        SEARCH ALL EMP-DATA
            AT END DISPLAY 'RECORD NOT FOUND'
                DISPLAY EMPL-ID-IN
            WHEN EMP-ID(EMP-NDX) = EMPL-ID-IN
                MOVE EMPL-LNAME    TO EMP-LAST-NAME(EMP-NDX)
                MOVE EMPL-FNAME    TO EMP-FIRST-NAME(EMP-NDX)
                MOVE EMPL-YRS-SRV  TO EMP-YRS-SERVICE(EMP-NDX)
                MOVE EMPL-PRM-DTE  TO EMP-PROM-DATE(EMP-NDX)

                DISPLAY ' LNAME     : ' EMP-LAST-NAME(EMP-NDX)
                        ' FNAME     : ' EMP-FIRST-NAME(EMP-NDX)
                        ' YRS SVC   : ' EMP-YRS-SERVICE(EMP-NDX)
                        ' PROM DTE  : ' EMP-PROM-DATE(EMP-NDX)
                        ' REG PAY   : '  EMP-REG-PAY(EMP-NDX)
                        ' BONUS PAY:' EMP-BON-PAY(EMP-NDX)

        END-SEARCH

        READ EMPLOYEE-IN-FILE
            INTO IN-EMPLOYEE-RECORD
                AT END SET SW-END-OF-FILE TO TRUE
        END-READ

    END-PERFORM.

P1200-WRITE-EMPLOYEE-MSTR-FILE.

    DISPLAY 'P1200-WRITE-EMPLOYEE-MSTR-FILE'.
```

126

```
          PERFORM VARYING EMP-NDX FROM +1 BY +1
             UNTIL EMP-NDX > 9

             MOVE EMP-ID (EMP-NDX)          TO EMPLMAST-EMP-ID
             MOVE EMP-LAST-NAME (EMP-NDX)   TO EMPLMAST-LNAME
             MOVE EMP-FIRST-NAME(EMP-NDX)   TO EMPLMAST-FNAME
             MOVE EMP-YRS-SERVICE(EMP-NDX)  TO EMPLMAST-YRS-SRV
             MOVE EMP-PROM-DATE(EMP-NDX)    TO EMPLMAST-PRM-DTE
             MOVE EMP-REG-PAY(EMP-NDX)      TO EMPLMAST-REG-PAY
             MOVE EMP-BON-PAY(EMP-NDX)      TO EMPLMAST-BON-PAY
             WRITE EMPLOYEE-MASTER-OUT
                 FROM OUT-EMPLMAST-RECORD

          END-PERFORM.


    *     END OF SOURCE CODE
```

Here's the output from our program.

```
COBTRN7 - COBOL WITH TABLE HANDLING TECHNIQUES
 EMP ID: 1111 REG PAY: 8700000 BONUS PAY: 670000
 EMP ID: 1122 REG PAY: 8200000 BONUS PAY: 600000
 EMP ID: 3217 REG PAY: 6500000 BONUS PAY: 550000
 EMP ID: 4175 REG PAY: 5500000 BONUS PAY: 150000
 EMP ID: 4720 REG PAY: 8000000 BONUS PAY: 250000
 EMP ID: 4836 REG PAY: 6200000 BONUS PAY: 220000
 EMP ID: 6288 REG PAY: 7000000 BONUS PAY: 200000
 EMP ID: 7459 REG PAY: 8500000 BONUS PAY: 450000
 EMP ID: 9134 REG PAY: 7500000 BONUS PAY: 250000
P1100-LOAD-EMPLOYEE-DATA
 LNAME    : VEREEN                    FNAME    : CHARLES         YRS S
: 8700000 BONUS PAY:670000
 LNAME    : JENKINS                   FNAME    : DEBORAH         YRS S
: 8200000 BONUS PAY:600000
 LNAME    : JOHNSON                   FNAME    : EDWARD          YRS S
: 6500000 BONUS PAY:550000
 LNAME    : TURNBULL                  FNAME    : FRED            YRS S
: 5500000 BONUS PAY:150000
 LNAME    : SCHULTZ                   FNAME    : TIM             YRS S
: 8000000 BONUS PAY:250000
 LNAME    : SMITH                     FNAME    : SANDRA          YRS S
: 6200000 BONUS PAY:220000
 LNAME    : WILLARD                   FNAME    : JOE             YRS S
: 7000000 BONUS PAY:200000
 LNAME    : STEWART                   FNAME    : BETTY           YRS S
: 8500000 BONUS PAY:450000
 LNAME    : FRANKLIN                  FNAME    : BRIANNA         YRS S
: 7500000 BONUS PAY:250000
P1200-WRITE-EMPLOYEE-MSTR-FILE
COBTRN7 - SUCCESSFULLY ENDED
```

And finally the master file looks like this:

```
----+----1----+----2----+----3----+----4----+----5----+----6----+----7----+----8-----
1111 VEREEN                  CHARLES               12 2017-01-01 8700000 670000
1122 JENKINS                 DEBORAH               05 2017-01-01 8200000 600000
3217 JOHNSON                 EDWARD                04 2017-01-01 6500000 550000
4175 TURNBULL                FRED                  01 2016-12-01 5500000 150000
4720 SCHULTZ                 TIM                   09 2017-01-01 8000000 250000
4836 SMITH                   SANDRA                03 2017-01-01 6200000 220000
6288 WILLARD                 JOE                   06 2016-01-01 7000000 200000
7459 STEWART                 BETTY                 07 2016-07-31 8500000 450000
9134 FRANKLIN                BRIANNA               00 2016-10-01 7500000 250000
```

That's it for the basics of tables. Before we leave the COBOL chapter, we should briefly visit sub programs. This is important because sub programs are handled somewhat differently in each programming language

Sub Programs

Sub programs are a way of reusing code because they allow multiple programs to perform the same logic or function by issuing a simple invocation statement. A sub program then is simply a program that can be called by another program. In COBOL a sub program is invoked with the CALL verb and the program name.

I've worked in an environment that never used sub programs, and then in another environment that nested sub programs six and seven levels deep. I hope your environment is somewhere in between. Anyway let's look at the basics of subprograms and how they work.

Subprograms can be loaded statically or dynamically. A static load means that the subprogram is linked with the calling program at compile time, hence the sub program object module is included in the final load module for the calling program. Subsequent changes to the sub program will not be reflected in the calling program until/unless the calling program is recompiled. A dynamic load means the subprogram is not statically linked with the calling program at compile time. Rather the sub program is resolved and loaded dynamically at run time. This means you get the most current version of that program from the load library. We'll show both statically and dynamically loaded sub programs with this example.

Let's create two programs: a subprogram named COBELVL and a calling program COBTRN8. The COBELVL program will accept an integer which represents the years of service for an employee. COBELVL will then return a string value representing the employee's "level" of service. The level will vary from ENTRY to ADVANCED to SENIOR depending on how many years of service. Less than one year is ENTRY, 1 year through 5 years is ADVANCED and more than 5 years is SENIOR.

We'll first need to establish a way to pass the years of service from the calling program to the sub program, and then to pass the employee level of service back to the calling program. We use the **linkage section** of the subprogram for this. Here is our declaration:

```
    LINKAGE SECTION.
    ***************************************************
    *   DECLARE THE I/O PARAMETERS FOR THE PROCEDURE
    ***************************************************

    01 LK-EMP-VARIABLES.
       10   LK-YEARS          PIC S9(9) USAGE COMP.
       10   LK-EMP-LEVEL      PIC X(10).
```

The calling program must declare a similar structure in working storage, and include that structure in the call to the subprogram. We'll see that in a few minutes when we code the calling program.

Meanwhile, the COBELVL program will use the years of service in the LK-YEARS variable. CO-BELVL will determine the correct string value for employee level, and then load the string into the LK-EMP-LEVEL variable.

So now we need to code some logic to determine the employee level of service based on years of service. We could code this using IF/THEN logic as follows:

```
    IF LK-YEARS IS LESS THAN 1 THEN
       MOVE 'ENTRY     ' TO LK-EMP-LEVEL
    ELSE
       IF LK-YEARS IS LESS THAN OR EQUAL TO 5 THEN
          MOVE 'ADVANCED  ' TO LK-EMP-LEVEL
       ELSE
          MOVE 'SENIOR    ' TO LK-EMP-LEVEL
       END-IF
    END-IF
```

This would work, but let's use a somewhat more elegant technique that implements the CASE programming construct. COBOL implements CASE using the EVALUATE verb. Here's what it looks like:

```
    EVALUATE LK-YEARS
       WHEN 0          MOVE 'ENTRY     ' TO LK-EMP-LEVEL
       WHEN 1 THRU 5   MOVE 'ADVANCED  ' TO LK-EMP-LEVEL
       WHEN OTHER      MOVE 'SENIOR    ' TO LK-EMP-LEVEL
    END-EVALUATE.
```

I like this code better. For one thing it is easier to read. It also allows for a lot more value ranges without requiring the nesting you would have to do if you used the IF/THEN. I suggest using the EVALUATE whenever you are branching on the value of a variable and there are more than two branches.

Ok, here is our program code.

```
IDENTIFICATION DIVISION.
PROGRAM-ID. COBELVL.
*******************************************************
*       THIS SUB PROGRAM WILL RETURN AN EMPLOYEE'S    *
*       JOB LEVEL (ENTRY, ADVANCED, SENIOR) BASED     *
*       ON YEARS OF SERVICE WHICH IS PASSED FROM      *
*       THE CALLING PROGRAM.                          *
*******************************************************
ENVIRONMENT DIVISION.
DATA DIVISION.
WORKING-STORAGE SECTION.

LINKAGE SECTION.
*****************************************************
*    DECLARE THE I/O PARAMETERS FOR THE PROCEDURE
*****************************************************

01 LK-EMP-VARIABLES.
    10   LK-YEARS          PIC S9(9) USAGE COMP.
    10   LK-EMP-LEVEL      PIC X(10).

PROCEDURE DIVISION USING LK-EMP-VARIABLES.

MAIN-PARA.
    DISPLAY "SAMPLE COBOL SUB-PROGRAM".

*   DETERMINE AN EMPLOYEE SERVICE LEVEL BASED ON YEARS OF SERVICE

    EVALUATE LK-YEARS
        WHEN 0          MOVE 'ENTRY     ' TO LK-EMP-LEVEL
        WHEN 1 THRU 5   MOVE 'ADVANCED  ' TO LK-EMP-LEVEL
        WHEN OTHER      MOVE 'SENIOR    ' TO LK-EMP-LEVEL
    END-EVALUATE.

    GOBACK.
```

Next we need a program to call the subprogram. Let's write COBTRN8 to read the employee master file that we created in COBTRN8. We'll retrieve the years of service from the master record, pass the years of service to the subprogram, and then display the return value. Here's our code, and notice especially how we call the subprogram COBELVL.

```cobol
 IDENTIFICATION DIVISION.
 PROGRAM-ID. COBTRN8.
**********************************************************
*      PROGRAM THAT READS A FILE AND CALLS A        *
*      SUB-PROGRAM TO CALCULATE A VALUE.            *
**********************************************************

 ENVIRONMENT DIVISION.
 INPUT-OUTPUT SECTION.

    FILE-CONTROL.
       SELECT EMPLMAST-IN-FILE   ASSIGN TO EMPMASTR.

 DATA DIVISION.

 FILE SECTION.
 FD  EMPLMAST-IN-FILE
     RECORDING MODE IS F
     LABEL RECORDS ARE STANDARD
     RECORD CONTAINS 85 CHARACTERS
     BLOCK CONTAINS 0 RECORDS
     DATA RECORD IS EMPLOYEE-RECORD-IN.

    01 EMPLMAST-RECORD-IN.
         05  E-ID          PIC X(04).
         05  FILLER        PIC X(76).

 WORKING-STORAGE SECTION.

    01 WS-FLAGS.
         05  SW-END-OF-FILE-SWITCH   PIC X(1) VALUE 'N'.
             88  SW-END-OF-FILE            VALUE 'Y'.
             88  SW-NOT-END-OF-FILE        VALUE 'N'.

    01 IN-EMPLMAST-RECORD.
         05  EMPLMAST-EMP-ID   PIC X(04).
         05  FILLER            PIC X(01) VALUE SPACES.
         05  EMPLMAST-LNAME    PIC X(30).
         05  EMPLMAST-FNAME    PIC X(20).
         05  FILLER            PIC X(01) VALUE SPACE.
         05  EMPLMAST-YRS-SRV  PIC X(02).
         05  FILLER            PIC X(01) VALUE SPACE.
         05  EMPLMAST-PRM-DTE  PIC X(10).
         05  FILLER            PIC X(01) VALUE SPACE.
         05  EMPLMAST-REG-PAY  PIC 99999V99.
         05  FILLER            PIC X(01) VALUE SPACE.
         05  EMPLMAST-BON-PAY  PIC 9999V99.
         05  FILLER            PIC X(01) VALUE SPACES.

    01 EMP-LINK-DATA.
       05 EMP-ID-BIN      PIC S9(9) USAGE COMP.
       05 EMP-LEVEL       PIC X(10).
```

```
PROCEDURE DIVISION.

    PERFORM P100-INITIALIZATION.
    PERFORM P200-MAINLINE.
    PERFORM P300-TERMINATION.
    GOBACK.

P100-INITIALIZATION.

    DISPLAY 'COBTRN8 - SAMPLE COBOL PROGRAM: SUB-PROGRAM CALL'.

    OPEN INPUT  EMPLMAST-IN-FILE.

    INITIALIZE EMP-LINK-DATA,
               IN-EMPLMAST-RECORD.

P200-MAINLINE.

*   MAIN LOOP - READ THE INPUT FILE, PASS THE YEARS OF
*              SERVICE TO SUB-PROGRAM COBELVL, AND THEN
*              DISPLAY THE RESULTS.

    READ EMPLMAST-IN-FILE INTO IN-EMPLMAST-RECORD
       AT END SET SW-END-OF-FILE TO TRUE
    END-READ

    PERFORM UNTIL SW-END-OF-FILE

       MOVE EMPLMAST-YRS-SRV TO EMP-ID-BIN
       MOVE SPACES           TO EMP-LEVEL

       CALL 'COBELVL' USING EMP-LINK-DATA

       DISPLAY ' EMP ID: '   EMPLMAST-EMP-ID
       DISPLAY ' YEARS : '   EMPLMAST-YRS-SRV
       DISPLAY ' LEVEL : '   EMP-LEVEL

       READ EMPLMAST-IN-FILE INTO IN-EMPLMAST-RECORD
          AT END SET SW-END-OF-FILE TO TRUE
       END-READ

    END-PERFORM.

P300-TERMINATION.

    CLOSE EMPLMAST-IN-FILE.
    DISPLAY 'COBTRN8 - SUCCESSFULLY ENDED'.

*   END OF SOURCE CODE
```

Now let's focus for a minute on the call to COBELVL:

```
CALL 'COBELVL' USING EMP-LINK-DATA
```

Notice a couple of things. One is that we use the EMP-LINK-DATA structure as a parameter to pass data back and forth between COBTRN8 and COBELVL. Any time we are exchanging data, it is required that the calling program pass a matching data area to the subprogram (which in turn defines that data area in its linkage section).

The other thing to notice is that we placed the sub program name in quote marks when we called it. This method of calling the program will result in a **static** link, meaning the COBELVL object module will be linked into the COBTRN8 load module when COBTRN8 is compiled and linked. So the version of COBELVL you use at compile time is what will be executed at runtime by COBTRN8 (because COBELVL is now part of the executable load module for COBTRN8).

To prove this, you can use File Manager to browse the load module. Select File Manager from your ISPF menu.

From this screen, select **Utilities.**

```
File Manager                    Primary Option Menu
Command ===>

0   Settings      Set processing options           User ID . : USER01
1   View          View data                        System ID : MATE
2   Edit          Edit data                        Appl ID . : FMN
3   Utilities     Perform utility functions        Version . : 11.1.0
4   Tapes         Tape specific functions          Terminal. : 3278
5   Disk/VSAM     Disk track and VSAM CI functions   Screen. . : 1
6   OAM           Work with OAM objects            Date. . . : 2018/02/06
7   Templates     Template and copybook utilities   Time. . . : 05:27
8   HFS           Access Hierarchical File System
9   WebSphere MQ  List, view and edit MQ data
X   Exit          Terminate File Manager
```

From this screen select LOADLIB

```
File Manager                    Utility Functions
Command ===>

0   DBCS            Set DBCS data format for print
1   Create          Create data
2   Print           Print data
3   Copy            Copy data
4   Dslist          Catalog services
5   VTOC            Work with VTOC
6   Find/Change     Search for and change data
7   AFP             Browse AFP data
8   Storage         Browse user storage
9   Printdsn        Browse File Manager print data set
10  Loadlib         Load module utility functions
11  Compare         Compare data
12  Audit trail     Print audit trail report
13  Copybook        View and Print
14  WebSphere MQ    List WebSphere MQ managers and queues
```

Now select option 1 (View).

```
File Manager                Load module utility functions
Command ===>

1   View            View load module information
2   Compare         Compare load modules
```

134

Finally, enter the library name and member name COBTRN8, **and press Enter.**

```
File Manager                   Load Module Information
Command ===>

Input:
   Data set name . . . . .  'USER01.LOADLIB'
   Member  . . . . . . . .  COBTRN8         (Blank or pattern for member list)
   Volume serial . . . . .                  (If not cataloged)


Processing Options:

   Order CSECTs by                    Output to
   1   1. Address                     1   1. Display
       2. Name                            2. Printer

   Enter "/" to select option
      YY/MM/DD date format (default: YYYY.DDD)
      Batch execution                           Advanced member selection
                                                Skip member name list
```

Notice that both COBTRN8 **and** COBELVL **are present in the** COBTRN8 **load module.**

```
 Process    Options   Help
SSSSSSSSSSSSSSSSSSSSSSSSSSSSSSSSSSSSSSSSSSSSSSSSSSSSSSSSSSSSSSSSSSSSSSSSSSSSS
 File Manager              Load Module Information        Row 00001 of 00032
Command ===>                                                     Scroll PAGE

 Load Library    USER01.LOADLIB
 Load Module     COBTRN8
      Linked on 2018.016 at 10:06:05 by PROGRAM BINDER 5695-PMB V1R1
           EPA 000000 Size 0001D80 TTR 002F10 SSI          AC 00 AM 31  RM ANY
```

Address	CSECT name	Type	Size	AMODE	RMODE	Compiler 1	Date 1	Compil ±
*		*		*	*	*	*	*
<---+->	<---+----10---+->	<-->	<---+->	<--->	<--->	<---+---->	<---+-->	<---+>
0000000	COBTRN8	SD	0000B92	MIN	ANY	COBOL V4R2	2018.016	
0000000	**COBTRN8.COBTRN8**	**LD**	**0000B92**	**MIN**	**ANY**	**COBOL V4R2**	**2018.016**	
0000B98	CEESG005	SD	0000018	MIN	ANY	PL/X V2R4	2011.074	HLASM
0000B98	CEESG005	ID	0000018	MIN	ANY	PL/X V2R4	2011.074	HLASM
0000B98	CEESG005.CEESG005	LD	0000018	MIN	ANY	PL/X V2R4	2011.074	HLASM
0000BB0	COBELVL	SD	00005B6	MIN	ANY	COBOL V4R2	2018.016	
0000BB0	**COBELVL.COBELVL**	**LD**	**00005B6**	**MIN**	**ANY**	**COBOL V4R2**	**2018.016**	
0001168	CEEBETBL	SD	0000028	MIN	ANY	HLASM V1R6	2011.077	
0001168	CEEBETBL.CEEBETBL	LD	0000028	MIN	ANY	HLASM V1R6	2011.077	

Now let's run the program and view the output.

```
COBTRN8 - SAMPLE COBOL PROGRAM: SUB-PROGRAM CALL
SAMPLE COBOL SUB-PROGRAM
```

```
EMP ID: 1111
YEARS : 12
LEVEL : SENIOR
SAMPLE COBOL SUB-PROGRAM
 EMP ID: 1122
 YEARS : 05
 LEVEL : ADVANCED
SAMPLE COBOL SUB-PROGRAM
 EMP ID: 3217
 YEARS : 04
 LEVEL : ADVANCED
SAMPLE COBOL SUB-PROGRAM
 EMP ID: 4175
 YEARS : 01
 LEVEL : ADVANCED
SAMPLE COBOL SUB-PROGRAM
 EMP ID: 4720
 YEARS : 09
LEVEL : SENIOR
SAMPLE COBOL SUB-PROGRAM
 EMP ID: 4836
 YEARS : 03
 LEVEL : ADVANCED
SAMPLE COBOL SUB-PROGRAM
 EMP ID: 6288
 YEARS : 06
 LEVEL : SENIOR
SAMPLE COBOL SUB-PROGRAM
 EMP ID: 7459
 YEARS : 07
 LEVEL : SENIOR
SAMPLE COBOL SUB-PROGRAM
 EMP ID: 9134
 YEARS : 00
 LEVEL : ENTRY
COBTRN8 - SUCCESSFULLY ENDED
```

Looks good. This version of the program, sub program and load method works.

You may know that it is also possible to call COBELVL dynamically (again, meaning the current version of the sub program is loaded at runtime). Let's look at how to do that and then we'll mention some reasons why you might want to do this.

To call a program dynamically in COBOL, instead of enclosing the sub program name in single quote marks, you assign the program name to a variable. Then you issue the call using the variable name instead of the literal program name. Specifically, let's define a 77 level character variable named PROG-NAME and assign it the value "COBELVL".

```
77 PROG-NAME  PIC X(8) VALUE 'COBELVL'.
```

Now when you issue the call, you specify the variable name instead of the actual program name, and you do not use quote marks:

```
CALL PROG-NAME USING EMP-LINK-DATA.
```

So here is our modified listing where we will call the COBELVL program dynamically.

```
        IDENTIFICATION DIVISION.
        PROGRAM-ID. COBTRN8.
       ********************************************************
       *       PROGRAM THAT READS A FILE AND CALLS A        *
       *       SUB-PROGRAM TO CALCULATE A VALUE.            *
       ********************************************************

        ENVIRONMENT DIVISION.
        INPUT-OUTPUT SECTION.

           FILE-CONTROL.
               SELECT EMPLMAST-IN-FILE   ASSIGN TO EMPMASTR.

        DATA DIVISION.

        FILE SECTION.
        FD  EMPLMAST-IN-FILE
            RECORDING MODE IS F
            LABEL RECORDS ARE STANDARD
            RECORD CONTAINS 85 CHARACTERS
            BLOCK CONTAINS 0 RECORDS
            DATA RECORD IS EMPLOYEE-RECORD-IN.

           01 EMPLMAST-RECORD-IN.
               05  E-ID          PIC X(04).
               05  FILLER        PIC X(76).

        WORKING-STORAGE SECTION.

           01 WS-FLAGS.
               05  SW-END-OF-FILE-SWITCH   PIC X(1) VALUE 'N'.
                   88  SW-END-OF-FILE              VALUE 'Y'.
                   88  SW-NOT-END-OF-FILE          VALUE 'N'.

           01 IN-EMPLMAST-RECORD.
               05   EMPLMAST-EMP-ID   PIC X(04).
               05   FILLER            PIC X(01) VALUE SPACES.
               05   EMPLMAST-LNAME    PIC X(30).
               05   EMPLMAST-FNAME    PIC X(20).
               05   FILLER            PIC X(01) VALUE SPACE.
```

```
       05  EMPLMAST-YRS-SRV  PIC X(02).
       05  FILLER            PIC X(01) VALUE SPACE.
       05  EMPLMAST-PRM-DTE  PIC X(10).
       05  FILLER            PIC X(01) VALUE SPACE.
       05  EMPLMAST-REG-PAY  PIC 99999V99.
       05  FILLER            PIC X(01) VALUE SPACE.
       05  EMPLMAST-BON-PAY  PIC 9999V99.
       05  FILLER            PIC X(01) VALUE SPACES.

   01 EMP-LINK-DATA.
       05 EMP-ID-BIN        PIC S9(9) USAGE COMP.
       05 EMP-LEVEL         PIC X(10).

   77 PROG-NAME            PIC X(8) VALUE 'COBELVL'.


PROCEDURE DIVISION.

    PERFORM P100-INITIALIZATION.
    PERFORM P200-MAINLINE.
    PERFORM P300-TERMINATION.
    GOBACK.

P100-INITIALIZATION.

    DISPLAY 'COBTRN8 - SAMPLE COBOL PROGRAM: SUB-PROGRAM CALL'.

    OPEN INPUT  EMPLMAST-IN-FILE.

    INITIALIZE EMP-LINK-DATA,
               IN-EMPLMAST-RECORD.

P200-MAINLINE.

*    MAIN LOOP - READ THE INPUT FILE, PASS THE YEARS OF
*                SERVICE TO SUB-PROGRAM COBELVL, AND THEN
*                DISPLAY THE RESULTS.

    READ EMPLMAST-IN-FILE INTO IN-EMPLMAST-RECORD
       AT END SET SW-END-OF-FILE TO TRUE
    END-READ

    PERFORM UNTIL SW-END-OF-FILE

       MOVE EMPLMAST-YRS-SRV TO EMP-ID-BIN
       MOVE SPACES           TO EMP-LEVEL

       CALL PROG-NAME USING EMP-LINK-DATA

       DISPLAY ' EMP ID: '  EMPLMAST-EMP-ID
       DISPLAY ' YEARS : '  EMPLMAST-YRS-SRV
       DISPLAY ' LEVEL : '  EMP-LEVEL
```

```
              READ EMPLMAST-IN-FILE INTO IN-EMPLMAST-RECORD
                 AT END SET SW-END-OF-FILE TO TRUE
              END-READ

           END-PERFORM.

        P300-TERMINATION.

           CLOSE EMPLMAST-IN-FILE.

           DISPLAY 'COBTRN8 - SUCCESSFULLY ENDED'.

     *     END OF SOURCE CODE
```

Now let's compile the program and look at the load module. When we view it using File Manager, you will see that the COBELVL module is no longer there. Instead it will be loaded into memory at run time.

```
 Process    Options    Help
 sssssssssssssssssssssssssssssssssssssssssssssssssssssssssssssssssssssssssssss
 File Manager              Load Module Information       Row 00001 of 00030
 Command ===>                                                    Scroll PAGE


 Load Library    USER01.LOADLIB
 Load Module     COBTRN8
        Linked on 2018.016 at 08:39:37 by PROGRAM BINDER 5695-PMB V1R1
             EPA 000000 Size 00018E0 TTR 002F02 SSI        AC 00 AM 31  RM ANY


 Address CSECT name         Type Size    AMODE RMODE Compiler 1 Date 1   Compil ±
         *                  *    *        *     *     *          *        *
 <---+-> <---+----10---+-> <--> <---+-> <---> <---> <---+----> <---+--> <---+->
 0000000 COBTRN8           SD   0000CAE MIN   ANY   COBOL V4R2 2018.016
 0000000 COBTRN8.COBTRN8   LD   0000CAE MIN   ANY   COBOL V4R2 2018.016
 0000CB0 CEESG005          SD   0000018 MIN   ANY   PL/X  V2R4 2011.074 HLASM
 0000CB0 CEESG005          ID   0000018 MIN   ANY   PL/X  V2R4 2011.074 HLASM
 0000CB0 CEESG005.CEESG005 LD   0000018 MIN   ANY   PL/X  V2R4 2011.074 HLASM
 0000CC8 CEEBETBL          SD   0000028 MIN   ANY   HLASM V1R6 2011.077
 0000CC8 CEEBETBL.CEEBETBL LD   0000028 MIN   ANY   HLASM V1R6 2011.077
 0000CF0 CEESTART          SD   00000B0 MIN   ANY   HLASM V1R6 2011.077
 0000CF0 CEESTART.CEESTART LD   00000B0 MIN   ANY   HLASM V1R6 2011.077
 0000DA0 IGZCBSO           SD   0000580 31    ANY   PL/X  V2R4 2011.074 HLASM
 0000DA0 IGZCBSO           ID   0000580 31    ANY   PL/X  V2R4 2011.074 HLASM
```

Go ahead and run it again to make sure the results are the same.

```
COBTRN8 - SAMPLE COBOL PROGRAM: SUB-PROGRAM CALL
SAMPLE COBOL SUB-PROGRAM
 EMP ID: 1111
 YEARS : 12
 LEVEL : SENIOR
SAMPLE COBOL SUB-PROGRAM
```

```
 EMP ID: 1122
 YEARS : 05
 LEVEL : ADVANCED
SAMPLE COBOL SUB-PROGRAM
 EMP ID: 3217
 YEARS : 04
 LEVEL : ADVANCED
SAMPLE COBOL SUB-PROGRAM
 EMP ID: 4175
 YEARS : 01
 LEVEL : ADVANCED
SAMPLE COBOL SUB-PROGRAM
 EMP ID: 4720
YEARS : 09
 LEVEL : SENIOR
SAMPLE COBOL SUB-PROGRAM
 EMP ID: 4836
 YEARS : 03
 LEVEL : ADVANCED
SAMPLE COBOL SUB-PROGRAM
 EMP ID: 6288
 YEARS : 06
 LEVEL : SENIOR
SAMPLE COBOL SUB-PROGRAM
 EMP ID: 7459
 YEARS : 07
 LEVEL : SENIOR
SAMPLE COBOL SUB-PROGRAM
 EMP ID: 9134
 YEARS : 00
 LEVEL : ENTRY
COBTRN8 - SUCCESSFULLY ENDED
```

The benefits of calling sub programs dynamically may be obvious. Suppose you have 10 programs that call a sub program. If you use the static approach, any change to the subprogram requires recompiling/redeploying not only the sub program, but also all 10 of the calling programs. Otherwise your changes will never be picked up.

However, if you are calling the sub program dynamically you need only recompile/redeploy the subprogram (unless the linkage structure changes). Since a dynamic call loads the current version of the sub program, you will always get the latest. [8]

---

8   There are some exceptions to this, as with IMS and CICS where programs are loaded into a different run time library on first use. In that case, you must either cycle the online environment or issue a reload command that is specific to the product and environment.

Additional Resources

We've reached the end of the COBOL chapter of this book. For additional information check out the IBM **Enterprise COBOL for z/OS Programming Guide**. Also I suggest you obtain the latest version of the IBM **Enterprise COBOL for z/OS Language Reference**. You can Google search for the latest IBM URL for these manuals and download them for free (you may need to set up a free account first).

Chapter 3 Review Questions

1. Name some elements of the COBOL IDENTIFICATION division.

2. Which clause do you use to define a table?

3. What does the INITIALIZE keyword do?

4. What is the LINKAGE SECTION used for?

5. What verb do you use to identify external files that the program will be using?

6. How do you terminate an IF/ELSE statement?

7. What is an 88 level data element used for?

8. Explain the meaning of a PIC 9v99 field.

9. If you are not certain how many entries a table should have, how would you create a variable length table?

10. In COBOL, how do you call a program statically? How about dynamically?

11. What type of picture can be used for alphanumeric data types?

12. When you open a file in I-O mode, what verb is used to update a record?

13. What are the different modes for opening a file in COBOL?

14. Explain what an EVALUATE statement is used for?

15. If you have a complex arithmetic calculation, which verb could you use to perform the calculation with a single statement?

Chapter 4 : Virtual Storage Access Method (VSAM)
Introduction
Virtual Storage Access Method (VSAM) is an IBM DASD (direct access storage device) file storage access method. It has been used for many years, including with the Multiple Virtual Storage (MVS) architecture and now in z/OS. VSAM offers four data set organizations:

- Key Sequenced Data Set (KSDS)
- Entry Sequenced Data Set (ESDS)
- Relative Record Data Set (RRDS)
- Linear Data Set (LDS)

The KSDS, RRDS and ESDS organizations all are record-based. The LDS organization uses a sequence of pages without a predefined record structure.

VSAM records are either fixed or variable length. Records are organized in fixed-size blocks called Control Intervals (CIs). The CI's are organized into larger structures called Control Areas (CAs). Control Interval sizes are measured in bytes — for example 4 kilobytes — while Control Area sizes are measured in disk tracks or cylinders. When a VSAM file is read, a complete Control Interval will be transferred to memory.

The Access Method Services utility program IDCAMS is used to define and delete VSAM data sets. In addition, you can write custom programs in COBOL, PLI and Assembler to access VSAM datasets using Data Definition (DD) statements in Job Control Language (JCL), or via dynamic allocation or in online regions such as in Customer Information Control System (CICS).

Types of VSAM Files

Key Sequence Data Set (KSDS)
This organization type is the most commonly used. Each record has one or more key fields and a record can be retrieved (or inserted) by key value. This provides random access to data. Records are of variable length. IMS uses KSDS files (as we'll see in the next chapter).

Entry Sequence Data Set (ESDS)
This organization keeps records in the order in which they were entered. Records must be accessed sequentially. This organization is used by IMS for overflow datasets.

Relative Record Data Set (RRDS)

This organization is based on record retrieval number; record 1, record 2, etc. This provides random access but the program must have a way of knowing which record number it is looking for.

Linear Data Set (LDS)

This organization is a byte-stream data set. It is rarely used by application programs.

We'll focus on KSDS files because they are the most commonly used and the most useful.

A KSDS cluster consists of following two components:

Index – the index component of the KSDS cluster is comprised of the list of key values for the records in the cluster with pointers to the corresponding records in the data component. The index component relates the key of each record to the record's relative location in the data set. When a record is added or deleted, this index is updated.

Data – the data component of the KSDS cluster contains the actual data. Each record in the data component of a KSDS cluster contains a key field with same number of characters and occurs in the same relative position in each record.

Creating VSAM Files

You create VSAM files using the IDCAMS utility. Here is the meaning of the keywords in the control statement.

NAME	The cluster name which is then extended one node for the data and index physical files. See example below.
RECSZ	The record length.
TRK	The space allocated for the file. It can be in tracks or cylinders.
FREESPACE	How much free space to leave on each control interval.
KEYS	The length and displacement of the key field.
CISZ	The Control Interval Size specified in bytes
VOLUMES	The DASD volume(s) which will physically store the data.
INDEX	The data set name of the index file.
DATA	The data set name that houses the data records.

Here is sample JCL:

```
//USER01D JOB MSGLEVEL=(1,1),NOTIFY=&SYSUID
//*
//************************************************
//* DEFINE VSAM KSDS CLUSTER
//************************************************
//JS010    EXEC PGM=IDCAMS
//SYSUDUMP DD SYSOUT=*
//SYSPRINT DD SYSOUT=*
//SYSOUT   DD SYSOUT=*
//SYSIN    DD  *
  DEFINE CLUSTER(NAME(USER01.EMPLOYEE)  -
  RECSZ(80 80)      -
  TRK(2,1)          -
  FREESPACE(5,10) -
  KEYS(4,0)         -
  CISZ(4096)        -
  VOLUMES(DEVHD1) -
  INDEXED)          -
  INDEX(NAME(USER01.EMPLOYEE.INDEX)) -
  DATA(NAME(USER01.EMPLOYEE.DATA))
/*
//SYSPRINT DD SYSOUT=*
//SYSUDUMP DD SYSOUT=*
```

This creates a catalog entry with two datasets, one for the record data and one for the index.

```
DSLIST - Data Sets Matching USER01.EMPLOYEE                 Row 1 of 11
Command ===>                                          Scroll ===> CSR

Command - Enter "/" to select action          Message       Volume
-------------------------------------------------------------------------
       USER01.EMPLOYEE                                       *VSAM*
       USER01.EMPLOYEE.DATA                                  DEVHD1
       USER01.EMPLOYEE.INDEX                                 DEVHD1
```

Loading and Unloading VSAM Files

You can add data to a VSAM KSDS in several ways:

1. Copying data from a flat file
2. Using File Manager
3. Using an application program

We'll show examples of all three. First, let us design a VSAM file. For purposes of this text book, we will be creating and maintaining a simple employee file for a fictitious company. So we'll create objects with that in mind. Here are the columns and data types for our file which we will name EMPLOYEE.

Field Name	Type
EMP_ID	Numeric 4 bytes
EMP_LAST_NAME	Character(30)
EMP_FIRST_NAME	Character(20)
EMP_SERVICE_YEARS	Numeric 2 bytes
EMP_PROMOTION_DATE	Date in format YYYY-MM-DD

Now let's say we have created a simple text file in this format. We can browse it:

```
BROWSE     USER01.EMPLOYEE.LOAD                       Line 00000000 Col 001 080
 Command ===>                                          Scroll ===> CSR
----+----1----+----2----+----3----+----4----+----5----+----6----+----7----+----8
******************************** Top of Data ********************************
3217JOHNSON                    EDWARD          042017-01-01
7459STEWART                    BETTY           072016-07-31
9134FRANKLIN                   BRIANNA         032016-10-01
4720SCHULTZ                    TIM             092017-01-01
6288WILLARD                    JOE             062016-01-01
1122JENKINS                    DEBORAH         052016-09-01
```

We can use this text file to load our VSAM file. Note however that before we load, we need to sort the records into key sequence. Otherwise IDCAMS will give us an error when we try to load. You can edit the file and on the command line issue a SORT 1 4 command (space between 1 and 4) to sort the records.

```
BROWSE     USER01.EMPLOYEE.LOAD                       Line 00000000 Col 001 080
 Command ===>                                          Scroll ===> CSR
----+----1----+----2----+----3----+----4----+----5----+----6----+----7----+----8
******************************** Top of Data ********************************
1122JENKINS                    DEBORAH         052016-09-01
3217JOHNSON                    EDWARD          042017-01-01
4720SCHULTZ                    TIM             092017-01-01
6288WILLARD                    JOE             062016-01-01
7459STEWART                    BETTY           072016-07-31
9134FRANKLIN                   BRIANNA         032016-10-01
```

Save the file to apply the changes. Now we are ready. We can use the following IDCAMS JCL to load the VSAM file. The INDATASET is our input file, and the OUTDATASET is the VSAM file. Note that we specify the VSAM file cluster name in this job, not the DATA or INDEX file names.

148

```
//USER01D JOB 'NAME',MSGLEVEL=(1,1),NOTIFY=&SYSUID
//*
//****************************************************************
//* REPRO/COPY DATA FROM PS TO VSAM KSDS
//****************************************************************
//STEP90   EXEC PGM=IDCAMS
//SYSPRINT DD SYSOUT=*
//SYSOUT   DD SYSOUT=*
//SYSUDUMP DD SYSOUT=*
//SYSIN    DD *
  REPRO -
  INDATASET (USER01.EMPLOYEE.LOAD) -
  OUTDATASET(USER01.EMPLOYEE)
/*
//*
```

Once loaded, we can view the data using the ISPF **BROWSE** function. If this doesn't work on your system, you'll need to use IBM File Manager or another tool which allows you to browse/edit VSAM files.

```
Browse           USER01.EMPLOYEE.DATA                    Top of 6
Command ===>                                             Scroll PAGE
                      Type DATA      RBA                  Format CHAR
                                          Col 1
----+----10---+----2----+----3----+----4----+----5----+----6----+----7----+----
****  Top of data  ****
1122JENKINS                DEBORAH            052016-09-01
3217JOHNSON                EDWARD             042017-01-01
4720SCHULTZ                TIM                092017-01-01
6288WILLARD                JOE                062016-01-01
7459STEWART                BETTY              072016-07-31
9134FRANKLIN               BRIANNA            032016-10-01
****  End of data  ****
```

To edit the data you will need to use a tool such as File Manager. Let's do this next.

VSAM Updates with File Manager

You can perform adds, changes and deletes to data records in File Manager. First, it will be useful if we create a file layout to assist us with viewing and updating data. Let's create a COBOL layout as follows.

```
BROWSE    USER01.COPYLIB(EMPLOYEE) - 01.00        Line 00000000 Col 001 080
  Command ===>                                            Scroll ===> CSR
********************************* Top of Data **********************************
      ************************************************************
      * COBOL DECLARATION FOR VSAM FILE EMPLOYEE                 *
      ************************************************************
      01  EMPLOYEE.
          05 EMP-ID              PIC 9(04).
```

```
                05 EMP-LAST-NAME         PIC X(30).
                05 EMP-FIRST-NAME        PIC X(20).
                05 EMP-SERVICE-YEARS     PIC 9(02).
                05 EMP-PROMOTION-DATE    PIC X(10).
                05 FILLER                PIC X(14).
```

Now let's go to File Manager. Select File Manager from your ISPF menu (it may be different on your system). Below is the main FM menu. Select the EDIT option.

```
File Manager                      Primary Option Menu
Command ===>

0    Settings       Set processing options              User ID . : USER01
1    View           View data                           System ID : MATE
2    Edit           Edit data                           Appl ID . : FMN
3    Utilities      Perform utility functions           Version . : 11.1.0
4    Tapes          Tape specific functions          Terminal. : 3278
5    Disk/VSAM      Disk track and VSAM CI functions    Screen. . : 2
6    OAM            Work with OAM objects               Date. . . : 2018/03/07
7    Templates      Template and copybook utilities     Time. . . : 02:41
8    HFS            Access Hierarchical File System
9    WebSphere MQ   List, view and edit MQ data
X    Exit           Terminate File Manager
```

Enter your file name, copybook file name, and select the processing option 1.

```
File Manager                    Edit Entry Panel
Command ===>

Input Partitioned, Sequential or VSAM Data Set, or HFS file:
   Data set/path name 'USER01.EMPLOYEE'                          +
   Member . . . . . .          (Blank or pattern for member list)
   Volume serial . .           (If not cataloged)
   Start position . .                                 +
   Record limit . . .          Record sampling
   Inplace edit . . .          (Prevent inserts and deletes)
Copybook or Template:
   Data set name  . . 'USER01.COPYLIB(EMPLOYEE)'
   Member . . . . . .          (Blank or pattern for member list)
Processing Options:
 Copybook/template   Start position type   Enter "/" to select option
 1  1. Above            1. Key               Edit template   Type (1,2,S)
    2. Previous         2. RBA               Include only selected records
    3. None             3. Record number     Binary mode, reclen 80
    4. Create dynamic   4. Formatted key     Create audit trail
```

Now you will see this screen. Notice the format is TABL which shows the data in list format. If you want to change it to show one record at a time, type over the TABL with SNGL (which means single record).

```
Edit            USER01.EMPLOYEE                          Top of 6
Command ===>                                             Scroll PAGE
    Key                 Type KSDS    RBA                 Format TABL
       EMP-ID EMP-LAST-NAME               EMP-FIRST-NAME    EMP-SERVICE-
          #2 #3                           #4                         #5
       ZD 1:4 AN 5:30                     AN 35:20            ZD 55:2
          <---> <---+----1----+----2----+----> <---+----1----+----> <->
****** **** Top of data   ****
000001   1122 JENKINS                     DEBORAH                    5
000002   3217 JOHNSON                     EDWARD                     4
000003   4720 SCHULTZ                     TIM                        9
000004   6288 WILLARD                     JOE                        6
000005   7459 STEWART                     BETTY                      7
000006   9134 FRANKLIN                    BRIANNA                    3
****** **** End of data   ****
```

Now you can edit each field on the record except the key. You cannot change the key, although you can specify a different key to bring up a different record. Let's bring up employee 6288.

```
Edit               USER01.EMPLOYEE                      Rec 1 of 6
Command ===>                                            Scroll PAGE
Key 1122                  Type KSDS     RBA 0           Format SNGL
                                               Top Line is 1    of 6
Current 01: EMPLOYEE                                    Length 80
Field              Data
EMP-ID               1122
EMP-LAST-NAME        JENKINS
EMP-FIRST-NAME       DEBORAH
EMP-SERVICE-YEARS    5
EMP-PROMOTION-DATE   2016-09-01
FILLER
***  End of record  ***
```

Now we can change this record. Let's modify the years of service by changing it to 8.

```
Edit               USER01.EMPLOYEE                      Rec 4 of 6
Command ===>                                            Scroll PAGE
Key 6288                  Type KSDS     RBA 240         Format SNGL
                                               Top Line is 1    of 6
Current 01: EMPLOYEE                                    Length 80
Field              Data
EMP-ID               6288
EMP-LAST-NAME        WILLARD
EMP-FIRST-NAME       JOE
EMP-SERVICE-YEARS    6
EMP-PROMOTION-DATE   2016-01-01
FILLER
***  End of record  ***
```

Now you can either type SAVE on the command line or simply PF3 to exit from the record. In this case, let's press PF3 to exit the Edit screen. You will be notified that the record was updated by the message on the upper right portion of the screen.

```
File Manager                   Edit Entry Panel              1 record(s) updated
Command ===>

Input Partitioned, Sequential or VSAM Data Set, or HFS file:
   Data set/path name 'USER01.EMPLOYEE'                              +
   Member . . . . . .            (Blank or pattern for member list)
   Volume serial  . .            (If not cataloged)
   Start position . .                              +
   Record limit . . .         Record sampling
   Inplace edit . . .            (Prevent inserts and deletes)
Copybook or Template:
   Data set name  . . 'USER01.COPYLIB(EMPLOYEE)'
   Member . . . . . .            (Blank or pattern for member list)
Processing Options:
 Copybook/template   Start position type   Enter "/" to select option
 1  1. Above            1. Key             Edit template    Type (1,2,S)
    2. Previous         2. RBA             Include only selected records
    3. None             3. Record number   Binary mode, reclen 80
    4. Create dynamic   4. Formatted key   Create audit trail
```

Now let's see how we can insert and delete records. Actually it is pretty simple. If you are in table mode, you just use the I(nsert) command to insert a record, or the D(elete) command to delete one. Let's add a record for employee 1111 who is Sandra Smith with 9 years of service and a promotion date of 01/01/2017. To do this, type I on the first line of detail.

```
Edit              USER01.EMPLOYEE                      Rec 1 of 6
Command ===>                                           Scroll PAGE
     Key 1122            Type KSDS      RBA 0          Format TABL
        EMP-ID EMP-LAST-NAME            EMP-FIRST-NAME  EMP-SERVICE-
           #2 #3                        #4                        #5
           ZD 1:4 AN 5:30               AN 35:20        ZD 55:2
           <---> <---+----1----+----2----+----> <---+----1----+----> <->
I00001   1122 JENKINS                   DEBORAH                   5
000002   3217 JOHNSON                   EDWARD                    4
000003   4720 SCHULTZ                   TIM                       9
000004   6288 WILLARD                   JOE                       8
000005   7459 STEWART                   BETTY                     7
000006   9134 FRANKLIN                  BRIANNA                   3
****** ****  End of data   ****
```

Now you can enter the data. You will need to scroll to the right (PF11) to add the correct
years of service and promotion date.

```
Edit              USER01.EMPLOYEE                        Rec 1 of 7
Command ===>                                             Scroll PAGE
      Key 1122                  Type KSDS     RBA 0       Format TABL
         EMP-ID EMP-LAST-NAME                  EMP-FIRST-NAME    EMP-SERVICE-
            #2 #3                               #4                        #5
            ZD 1:4 AN 5:30                      AN 35:20             ZD 55:2
            <---> <---+----1----+----2----+----> <---+----1----+----->    <->
000001    1122 JENKINS                     DEBORAH                       5
000002    1111 SMITH                       SANDRA                        0
000003    3217 JOHNSON                     EDWARD                        4
000004    4720 SCHULTZ                     TIM                           9
000005    6288 WILLARD                     JOE                           8
000006    7459 STEWART                     BETTY                         7
000007    9134 FRANKLIN                    BRIANNA                       3
****** ****   End of data   ****
```

You could also switch to SNGL mode to make it easier to enter the data on one page.

```
Edit              USER01.EMPLOYEE                        Rec 1 of 7
Command ===>                                             Scroll PAGE
Key 1111                     Type KSDS     RBA 0         Format SNGL
                                           Top Line is 1    of 6
Current 01: EMPLOYEE                                     Length 80
Field                 Data
EMP-ID                 1111
EMP-LAST-NAME         SMITH
EMP-FIRST-NAME        SANDRA
EMP-SERVICE-YEARS       9
EMP-PROMOTION-DATE    2017-01-01
FILLER
***  End of record  ***
```

Now type SAVE on the command line.

```
Edit              USER01.EMPLOYEE                        Rec 1 of 7
Command ===>      SAVE                                   Scroll PAGE
Key 1111                     Type KSDS     RBA 0         Format SNGL
                                           Top Line is 1    of 6
Current 01: EMPLOYEE                                     Length 80
Field                 Data
EMP-ID                 1111
EMP-LAST-NAME         SMITH
EMP-FIRST-NAME        SANDRA
EMP-SERVICE-YEARS       9
EMP-PROMOTION-DATE    2017-01-01
FILLER
***  End of record  ***
```

When you press Enter you can verify the record was saved.

```
Edit              USER01.EMPLOYEE                    1 record(s) updated
Command ===>                                              Scroll PAGE
Key 1111                  Type KSDS     RBA 0             Format SNGL
                                              Top Line is 1    of 6
Current 01: EMPLOYEE                                     Length 80
Field                 Data
EMP-ID                   1111
EMP-LAST-NAME         SMITH
EMP-FIRST-NAME       SANDRA
EMP-SERVICE-YEARS        9
EMP-PROMOTION-DATE   2017-01-01
FILLER
***  End of record  ***
```

Finally, to delete a record, just go to TABL mode, find the record you want to delete, and use a D action. Let's delete the record we just added.

```
Edit              USER01.EMPLOYEE                    Rec 1 of 7
Command ===>                                              Scroll PAGE
    Key 1111              Type KSDS     RBA 0             Format TABL
       EMP-ID EMP-LAST-NAME             EMP-FIRST-NAME    EMP-SERVICE-
          #2 #3                         #4                         #5
          ZD 1:4 AN 5:30                AN 35:20             ZD 55:2
          <--->  <---+----1----+----2----+---->  <---+----1----+---->  <->
D00001    1111 SMITH                    SANDRA                      9
000002    1122 JENKINS                  DEBORAH                     5
000003    3217 JOHNSON                  EDWARD                      4
000004    4720 SCHULTZ                  TIM                         9
000005    6288 WILLARD                  JOE                         8
000006    7459 STEWART                  BETTY                       7
000007    9134 FRANKLIN                 BRIANNA                     3
****** ****  End of data  ****
```

When you press Enter, the record will disappear from the list. You can either type SAVE on the command line, or simply exit the file and the delete action will be saved.

```
Edit                USER01.EMPLOYEE                       1 record(s) updated
Command ===>                                               Scroll PAGE
       Key 1122              Type KSDS      RBA 80         Format TABL
           EMP-ID EMP-LAST-NAME             EMP-FIRST-NAME    EMP-SERVICE-
               #2 #3                        #4                          #5
           ZD 1:4 AN 5:30                   AN 35:20            ZD 55:2
           <---> <---+----1----+----2----+----> <---+----1----+---->  <->
000001     1122 JENKINS                    DEBORAH                      5
000002     3217 JOHNSON                    EDWARD                       4
000003     4720 SCHULTZ                    TIM                          9
000004     6288 WILLARD                    JOE                          8
000005     7459 STEWART                    BETTY                        7
000006     9134 FRANKLIN                   BRIANNA                      3
****** ****  End of data   ****
```

Application Programming with VSAM

COBOL Program to Read Records (COBVS1)

Now it's time to use VSAM in an application program. A program to retrieve a record is not much different from reading a flat file.[9] The main difference is that with VSAM you specify the key value of the record you want to retrieve. Let's name our first program COBVS1.

In our file definition, we must reference the DD name of the VSAM cluster name. We also specify that the file is indexed, and that we will be accessing it in random mode. We specify the EMP_ID as the file key. Finally we specify a variable name that VSAM will use to return the status code from each action on the file. In VSAM we want the file status to be 00 which indicates a successful operation.

Here is our program listing. Take a few minutes to look it over. Notice that we are checking for a file status of zero which means the data operation (in this case a read) was successful.

```
        IDENTIFICATION DIVISION.
        PROGRAM-ID. COBVS1.

        ****************************************************
        *       PROGRAM TO RETRIEVE A RECORD FROM          *
        *       EMPLOYEE VSAM FILE.                        *
        ****************************************************

        ENVIRONMENT DIVISION.
        INPUT-OUTPUT SECTION.

           FILE-CONTROL.
              SELECT EMPLOYEE-VS-FILE    ASSIGN TO EMPVSFIL
              ORGANIZATION IS INDEXED
              ACCESS MODE  IS RANDOM
              RECORD KEY   IS EMP-ID
              FILE STATUS  IS EMP-FILE-STATUS.

        DATA DIVISION.

        FILE SECTION.
        FD EMPLOYEE-VS-FILE.
           01  EMPLOYEE.
               05 EMP-ID              PIC 9(04).
               05 EMP-LAST-NAME       PIC X(30).
               05 EMP-FIRST-NAME      PIC X(20).
               05 EMP-SERVICE-YEARS   PIC 9(02).
               05 EMP-PROMOTION-DATE  PIC X(10).
```

---

9 If you are not familiar with the basics of how to read a file in COBOL, you might want to check out the previous chapter. It covers basic COBOL.

```
                05 FILLER                     PIC X(14).

        WORKING-STORAGE SECTION.

            01 WS-FLAGS.
                 05  SW-END-OF-FILE-SWITCH    PIC X(1) VALUE 'N'.
                     88  SW-END-OF-FILE                 VALUE 'Y'.
                     88  SW-NOT-END-OF-FILE             VALUE 'N'.

            01  EMP-FILE-STATUS.
                     05  EMPFILE-STAT1      PIC X.
                     05  EMPFILE-STAT2      PIC X.



        PROCEDURE DIVISION.

                PERFORM P100-INITIALIZATION.
                PERFORM P200-MAINLINE.
                PERFORM P300-TERMINATION.
                GOBACK.

            P100-INITIALIZATION.

                DISPLAY 'COBVS1 - SAMPLE COBOL PROGRAM: VSAM INPUT'.
                OPEN INPUT  EMPLOYEE-VS-FILE.

                INITIALIZE EMPLOYEE.
                MOVE '3217' TO EMP-ID.

            P200-MAINLINE.

        *    READ THE INPUT FILE TO GET THE REQUESTED RECORD
        *    AND DISPLAY THE DATA VALUES

                READ EMPLOYEE-VS-FILE
                IF  EMP-FILE-STATUS = '00' THEN
        *          DISPLAY THE DATA
                   DISPLAY 'EMPLOYEE DATA IS ' EMPLOYEE
                ELSE
                   DISPLAY 'RECORD WAS NOT FOUND'.

            P300-TERMINATION.

                CLOSE EMPLOYEE-VS-FILE.

                DISPLAY 'COBVS1 - SUCCESSFULLY ENDED'.

        *    END OF SOURCE CODE
```

Compile and link (according to the procedures in your installation), and then run the program.

Here is our output:

```
SDSF OUTPUT DISPLAY USER01D  JOB05473  DSID   101 LINE 1        COLUMNS 02- 81
 COMMAND INPUT ===>                                             SCROLL ===> CSR
COBVS1 - SAMPLE COBOL PROGRAM: VSAM INPUT
EMPLOYEE DATA IS 3217JOHNSON                          EDWARD           042017-01
COBVS1 - SUCCESSFULLY ENDED
```

As you can see, we successfully retrieved the record for employee 3217. That's all there is to it. Not much different than reading a flat file, but obviously more powerful because of the random access to the data based on the record key.

COBOL Program to Add Records (COBVS2)

Now let's do a program COBVS2 to add a record. Let's add back the record for the employee we previously deleted. This is employee 1111 who is Sandra Smith with 9 years of service and a promotion date of 01/01/2017. Here's how the add program will look.

Notice we have opened the VSAM file for input and output (**I-O**). We simply load the record structure, and then do the WRITE. Also we are checking file status after opening, writing and closing the file. A list of VSAM file status codes is provided at the end of this chapter.

```
       IDENTIFICATION DIVISION.
       PROGRAM-ID. COBVS2.

      *******************************************************
      *      PROGRAM TO ADD A RECCORED TO THE            *
      *      EMPLOYEE VSAM FILE.                         *
      *******************************************************

       ENVIRONMENT DIVISION.
       INPUT-OUTPUT SECTION.

          FILE-CONTROL.
             SELECT EMPLOYEE-VS-FILE   ASSIGN TO EMPVSFIL
             ORGANIZATION IS INDEXED
             ACCESS MODE  IS RANDOM
             RECORD KEY   IS EMP-ID
             FILE STATUS  IS EMP-FILE-STATUS.

       DATA DIVISION.

       FILE SECTION.
       FD EMPLOYEE-VS-FILE.
          01  EMPLOYEE.
             05 EMP-ID              PIC 9(04).
             05 EMP-LAST-NAME       PIC X(30).
```

```cobol
          05 EMP-FIRST-NAME         PIC X(20).
          05 EMP-SERVICE-YEARS      PIC 9(02).
          05 EMP-PROMOTION-DATE     PIC X(10).
          05 FILLER                 PIC X(14).

   WORKING-STORAGE SECTION.

      01 WS-FLAGS.
          05  SW-END-OF-FILE-SWITCH   PIC X(1) VALUE 'N'.
              88  SW-END-OF-FILE                VALUE 'Y'.
              88  SW-NOT-END-OF-FILE            VALUE 'N'.

      01  EMP-FILE-STATUS.
              05  EMPFILE-STAT1     PIC X.
              05  EMPFILE-STAT2     PIC X.

   PROCEDURE DIVISION.

      PERFORM P100-INITIALIZATION.
      PERFORM P200-MAINLINE.
      PERFORM P300-TERMINATION.
      GOBACK.

   P100-INITIALIZATION.

      DISPLAY 'COBVS2 - SAMPLE COBOL PROGRAM: VSAM INSERT'.
      OPEN I-O EMPLOYEE-VS-FILE.

      IF  EMP-FILE-STATUS = '00' OR '97' THEN
         NEXT SENTENCE
      ELSE
         DISPLAY 'ERROR ON OPEN - FILE STATUS ' EMP-FILE-STATUS.

      INITIALIZE EMPLOYEE.

   P200-MAINLINE.

*    SET UP DATA ON THE RECORD STRUCTURE AND
*    THEN WRITE THE RECORD

      MOVE '1111' TO EMP-ID
      MOVE 'SMITH'      TO    EMP-LAST-NAME
      MOVE 'SANDRA'     TO    EMP-FIRST-NAME
      MOVE '09'         TO    EMP-SERVICE-YEARS
      MOVE '2017-01-01' TO    EMP-PROMOTION-DATE

      WRITE EMPLOYEE

      IF  EMP-FILE-STATUS = '00' THEN
*        DISPLAY THE DATA
         DISPLAY 'ADD SUCCESSFUL - DATA IS ' EMPLOYEE
      ELSE
```

```
          DISPLAY 'ERROR ON INSERT - FILE STATUS ' EMP-FILE-STATUS.

     P300-TERMINATION.

          CLOSE EMPLOYEE-VS-FILE.

          DISPLAY 'COBVS2 - SUCCESSFULLY ENDED'.

     *     END OF SOURCE CODE
```

Now when we compile, link and run the program we get this output.

```
COBVS2 - SAMPLE COBOL PROGRAM: VSAM INSERT
ADD SUCCESSFUL - DATA IS 1111SMITH                        SANDRA            092017-01-01
COBVS2 - SUCCESSFULLY ENDED
```

And we can verify that the record was added by checking File Manager.

```
View            USER01.EMPLOYEE                                    Top of 7
Command ===>                                                       Scroll PAGE
     Key                      Type KSDS      RBA                    Format TABL
       EMP-ID EMP-LAST-NAME                  EMP-FIRST-NAME         EMP-SERVICE-
          #2 #3                              #4                            #5
       ZD 1:4 AN 5:30                        AN 35:20               ZD 55:2
       <---> <---+----1----+----2----+----> <---+----1----+---->        <->
****** ****   Top of data   ****
000001   1111 SMITH                          SANDRA                        9
000002   1122 JENKINS                        DEBORAH                       5
000003   3217 JOHNSON                        EDWARD                        4
000004   4720 SCHULTZ                        TIM                           9
000005   6288 WILLARD                        JOE                           8
000006   7459 STEWART                        BETTY                         7
000007   9134 FRANKLIN                       BRIANNA                       3
****** ****   End of data   ****
```

COBOL Program to Update Records (COBVS3)

For COBVS3 we will update a record. To do that we must first read the record into the record structure, make modifications and then REWRITE the record. Let's say we need to change the years of service for Sandra Smith from 9 to 10. Here is a program that would do this. Note that we opened the file for I-O.

```
     IDENTIFICATION DIVISION.
     PROGRAM-ID. COBVS3.

     ****************************************************
     *       PROGRAM TO RETRIEVE AND UPDATE A RECORD    *
     *       ON THE EMPLOYEE VSAM FILE.                 *
     ****************************************************
```

161

```
ENVIRONMENT DIVISION.
INPUT-OUTPUT SECTION.

   FILE-CONTROL.
      SELECT EMPLOYEE-VS-FILE    ASSIGN TO EMPVSFIL
      ORGANIZATION IS INDEXED
      ACCESS MODE  IS RANDOM
      RECORD KEY   IS EMP-ID
      FILE STATUS  IS EMP-FILE-STATUS.

DATA DIVISION.

FILE SECTION.
FD EMPLOYEE-VS-FILE.
   01  EMPLOYEE.
       05 EMP-ID               PIC 9(04).
       05 EMP-LAST-NAME        PIC X(30).
       05 EMP-FIRST-NAME       PIC X(20).
       05 EMP-SERVICE-YEARS    PIC 9(02).
       05 EMP-PROMOTION-DATE   PIC X(10).
       05 FILLER               PIC X(14).

WORKING-STORAGE SECTION.

   01 WS-FLAGS.
       05  SW-END-OF-FILE-SWITCH   PIC X(1) VALUE 'N'.
           88  SW-END-OF-FILE              VALUE 'Y'.
           88  SW-NOT-END-OF-FILE          VALUE 'N'.

   01  EMP-FILE-STATUS.
           05  EMPFILE-STAT1     PIC X.
           05  EMPFILE-STAT2     PIC X.

PROCEDURE DIVISION.

   PERFORM P100-INITIALIZATION.
   PERFORM P200-MAINLINE.
   PERFORM P300-TERMINATION.
   GOBACK.

P100-INITIALIZATION.

   DISPLAY 'COBVS2 - SAMPLE COBOL PROGRAM: VSAM UPDATE'.
   OPEN I-O EMPLOYEE-VS-FILE.

   IF  EMP-FILE-STATUS = '00' OR '97' THEN
      NEXT SENTENCE
   ELSE
      DISPLAY 'ERROR ON OPEN - FILE STATUS ' EMP-FILE-STATUS.

   INITIALIZE EMPLOYEE.
```

```
    P200-MAINLINE.

    *     FIRST READ THE SPECIFIED RECORD.   THEN
    *     MAKE CHANGES TO THE RECORD. FINALLY
    *     REWRITE THE RECORD TO THE VSAM FILE.

        MOVE '1111' TO EMP-ID
        READ EMPLOYEE-VS-FILE

        IF  EMP-FILE-STATUS = '00' THEN
           NEXT SENTENCE
        ELSE
           DISPLAY 'ERROR ON READ - FILE STATUS ' EMP-FILE-STATUS.

        MOVE '10'          TO   EMP-SERVICE-YEARS

        REWRITE EMPLOYEE

        IF  EMP-FILE-STATUS = '00' THEN
           DISPLAY 'UPDATE SUCCESSFUL - DATA IS ' EMPLOYEE
        ELSE
           DISPLAY 'ERROR ON REWRITE - FILE STATUS ' EMP-FILE-STATUS.

    P300-TERMINATION.

        CLOSE EMPLOYEE-VS-FILE.

        DISPLAY 'COBVS3 - SUCCESSFULLY ENDED'.

    *     END OF SOURCE CODE
```

Now let's compile, link and run. Here's the output.

```
COBVS3 - SAMPLE COBOL PROGRAM: VSAM UPDATE
UPDATE SUCCESSFUL - DATA IS 1111SMITH          SANDRA              102017-01-01
COBVS3 - SUCCESSFULLY ENDED
```

And we can verify that the change took place by checking in File Manager.

```
View              USER01.EMPLOYEE                           Rec 1 of 7
 Command ===>                                               Scroll PAGE
 Key 1111                    Type KSDS     RBA 0            Format SNGL
                                           Top Line is 1    of 6
 Current 01: EMPLOYEE                                       Length 80
 Field              Data
 EMP-ID               1111
 EMP-LAST-NAME        SMITH
 EMP-FIRST-NAME       SANDRA
 EMP-SERVICE-YEARS    10
 EMP-PROMOTION-DATE   2017-01-01
 FILLER               ..............
 ***  End of record  ***
```

COBOL Program to Delete Records (COBVS4)

Now let's write program COBVS4 to delete the Sandra Smith record we just worked with. Actually it will be similar to the update program, except we don't have to first retrieve the record before deleting. And of course we will use the verb DELETE instead of REWRITE.

```
IDENTIFICATION DIVISION.
PROGRAM-ID. COBVS4.

************************************************
*      PROGRAM TO DELETE A RECORD FROM THE       *
*      EMPLOYEE VSAM FILE.                        *
************************************************

ENVIRONMENT DIVISION.
INPUT-OUTPUT SECTION.

    FILE-CONTROL.
       SELECT EMPLOYEE-VS-FILE    ASSIGN TO EMPVSFIL
       ORGANIZATION IS INDEXED
       ACCESS MODE  IS RANDOM
       RECORD KEY   IS EMP-ID
       FILE STATUS  IS EMP-FILE-STATUS.

DATA DIVISION.

FILE SECTION.
FD EMPLOYEE-VS-FILE.
    01  EMPLOYEE.
        05 EMP-ID              PIC 9(04).
        05 EMP-LAST-NAME       PIC X(30).
        05 EMP-FIRST-NAME      PIC X(20).
        05 EMP-SERVICE-YEARS   PIC 9(02).
        05 EMP-PROMOTION-DATE  PIC X(10).
        05 FILLER              PIC X(14).

WORKING-STORAGE SECTION.

    01 WS-FLAGS.
        05  SW-END-OF-FILE-SWITCH   PIC X(1) VALUE 'N'.
            88  SW-END-OF-FILE               VALUE 'Y'.
            88  SW-NOT-END-OF-FILE           VALUE 'N'.

    01  EMP-FILE-STATUS.
            05  EMPFILE-STAT1      PIC X.
            05  EMPFILE-STAT2      PIC X.
```

```
PROCEDURE DIVISION.

    PERFORM P100-INITIALIZATION.
    PERFORM P200-MAINLINE.
    PERFORM P300-TERMINATION.
    GOBACK.

P100-INITIALIZATION.

    DISPLAY 'COBVS4 - SAMPLE COBOL PROGRAM: VSAM DELETE'.
    OPEN I-O EMPLOYEE-VS-FILE.

    IF  EMP-FILE-STATUS = '00' OR '97' THEN
       NEXT SENTENCE
    ELSE
       DISPLAY 'ERROR ON OPEN - FILE STATUS ' EMP-FILE-STATUS.

    INITIALIZE EMPLOYEE.

P200-MAINLINE.

*    DELETE THE RECORD FROM THE VSAM FILE.

    MOVE '1111' TO EMP-ID
    DELETE EMPLOYEE-VS-FILE

    IF  EMP-FILE-STATUS = '00' THEN
       DISPLAY 'SUCCESSFUL DELETE OF EMPLOYEE ' EMP-ID
    ELSE
       DISPLAY 'ERROR ON DELETE - FILE STATUS ' EMP-FILE-STATUS.

P300-TERMINATION.

    CLOSE EMPLOYEE-VS-FILE.

    DISPLAY 'COBVS4 - SUCCESSFULLY ENDED'.

*    END OF SOURCE CODE
```

Here is our execution output:

```
COBVS4 - SAMPLE COBOL PROGRAM: VSAM DELETE
SUCCESS DELETE OF EMPLOYEE 1111
COBVS4 - SUCCESSFULLY ENDED
```

And we can verify that the record was deleted by checking in File Manager. As we can see, there is no longer an employee 1111.

```
View              USER01.EMPLOYEE                        Rec 1 of 6
Command ===>                                             Scroll PAGE
    Key 1122             Type KSDS      RBA 0            Format TABL
       EMP-ID EMP-LAST-NAME              EMP-FIRST-NAME   EMP-SERVICE-
          #2 #3                          #4                        #5
          ZD 1:4 AN 5:30                 AN 35:20          ZD 55:2
          <---> <---+----1----+----2----+----> <---+----1----+----->  <->
000001    1122 JENKINS                   DEBORAH                    5
000002    3217 JOHNSON                   EDWARD                     4
000003    4720 SCHULTZ                   TIM                        9
000004    6288 WILLARD                   JOE                        8
000005    7459 STEWART                   BETTY                      7
000006    9134 FRANKLIN                  BRIANNA                    3
****** **** End of data  ****
```

Let's go ahead and run the program again to check the error logic. And in fact the program does report the error.

```
COBVS4 - SAMPLE COBOL PROGRAM: VSAM DELETE
ERROR ON DELETE - FILE STATUS 23
COBVS4 - SUCCESSFULLY ENDED
```

If course, you could do more by stating that file status 23 means a requested record was not found. You could even define the various file status codes in working storage with a description (see table at the end of this chapter), and display the text as an error message.

COBOL Program to Retrieve Records Sequentially (COBVS5)

Now let's read all of the records sequentially with program COBVS5. We will need to define the file for sequential access, and we'll use a loop which will stop when we reach end of file. Note that end of file is VSAM file status code 10.

```
       IDENTIFICATION DIVISION.
       PROGRAM-ID. COBVS5.

      *********************************************************
      *      PROGRAM TO RETRIEVE RECORDS SEQUENTIALLY      *
      *      FROM THE EMPLOYEE VSAM FILE.                  *
      *********************************************************
       ENVIRONMENT DIVISION.
       INPUT-OUTPUT SECTION.

          FILE-CONTROL.
             SELECT EMPLOYEE-VS-FILE   ASSIGN TO EMPVSFIL
             ORGANIZATION IS INDEXED
             ACCESS MODE  IS SEQUENTIAL
             RECORD KEY   IS EMP-ID
             FILE STATUS  IS EMP-FILE-STATUS.
```

```
DATA DIVISION.

FILE SECTION.
FD EMPLOYEE-VS-FILE.
   01  EMPLOYEE.
       05 EMP-ID              PIC 9(04).
       05 EMP-LAST-NAME       PIC X(30).
       05 EMP-FIRST-NAME      PIC X(20).
       05 EMP-SERVICE-YEARS   PIC 9(02).
       05 EMP-PROMOTION-DATE  PIC X(10).
       05 FILLER              PIC X(14).

WORKING-STORAGE SECTION.

   01 WS-FLAGS.
       05  SW-END-OF-FILE-SWITCH   PIC X(1) VALUE 'N'.
           88  SW-END-OF-FILE              VALUE 'Y'.
           88  SW-NOT-END-OF-FILE          VALUE 'N'.

   01  EMP-FILE-STATUS.
           05  EMPFILE-STAT1      PIC X.
           05  EMPFILE-STAT2      PIC X.

PROCEDURE DIVISION.

   PERFORM P100-INITIALIZATION.
   PERFORM P200-MAINLINE.
   PERFORM P300-TERMINATION.
   GOBACK.

P100-INITIALIZATION.

   DISPLAY 'COBVS5 - SAMPLE COBOL PROGRAM: READ LOOP'.
   OPEN INPUT  EMPLOYEE-VS-FILE.

   INITIALIZE EMPLOYEE.

P200-MAINLINE.

   READ EMPLOYEE-VS-FILE
   IF  EMP-FILE-STATUS = '10' THEN
      DISPLAY 'END OF FILE ENCOUNTERED'
      SET SW-END-OF-FILE TO TRUE
   END-IF.

   IF NOT SW-END-OF-FILE THEN
      PERFORM UNTIL SW-END-OF-FILE
*        DISPLAY THE DATA VALUES
         DISPLAY 'EMP-ID               ' EMP-ID
         DISPLAY 'EMP LAST NAME        ' EMP-LAST-NAME
         DISPLAY 'EMP FIRST NAME       ' EMP-FIRST-NAME
         DISPLAY 'EMP YEARS OF SERVICE ' EMP-SERVICE-YEARS
```

167

```
            DISPLAY 'EMP PROMOTION DATE    ' EMP-PROMOTION-DATE

        READ EMPLOYEE-VS-FILE
        IF  EMP-FILE-STATUS = '10' THEN
            SET SW-END-OF-FILE TO TRUE
            DISPLAY 'END OF FILE ENCOUNTERED'
        END-IF

    END-PERFORM
ELSE
    DISPLAY 'NO RECORDS IN FILE'

END-IF.

  P300-TERMINATION.

    CLOSE EMPLOYEE-VS-FILE.

    DISPLAY 'COBVS5 - SUCCESSFULLY ENDED'.

*    END OF SOURCE CODE
```

Compile, link and run the program.

```
COBVS5 - SAMPLE COBOL PROGRAM: READ LOOP
EMP-ID                1122
EMP LAST NAME         JENKINS
EMP FIRST NAME        DEBORAH
EMP YEARS OF SERVICE 05
EMP PROMOTION DATE    2016-09-01
EMP-ID                3217
EMP LAST NAME         JOHNSON
EMP FIRST NAME        EDWARD
EMP YEARS OF SERVICE 04
EMP PROMOTION DATE    2017-01-01
EMP-ID                4720
EMP LAST NAME         SCHULTZ
EMP FIRST NAME        TIM
EMP YEARS OF SERVICE 09
EMP PROMOTION DATE    2017-01-01
EMP-ID                6288
EMP LAST NAME         WILLARD
EMP FIRST NAME        JOE
EMP YEARS OF SERVICE 08
EMP PROMOTION DATE    2016-01-01
EMP-ID                7459
EMP LAST NAME         STEWART
EMP FIRST NAME        BETTY
EMP YEARS OF SERVICE 07
EMP PROMOTION DATE    2016-07-31
```

```
EMP-ID                9134
EMP LAST NAME         FRANKLIN
EMP FIRST NAME        BRIANNA
EMP YEARS OF SERVICE 03
EMP PROMOTION DATE    2016-10-01
END OF FILE ENCOUNTERED
COBVS5 - SUCCESSFULLY ENDED
```

So this is a model you can use whenever you need to cycle through a VSAM file sequentially. It should prove useful.

Creating and Accessing Alternate Indexes

So far we've dealt with a VSAM file that has a single index which is associated with the key. Suppose however that you need another index on a file? That is, you need to randomly access your data using another field from the file? You can do this with VSAM, and you can access the data via the alternate index in application programs.

Suppose we want to add a social security number field to our EMPLOYEE file, and that we want an alternate index on it. To do this we will do the following:

1. Modify our file layout to include a social security number field named EMP-SSN.
2. Reload the EMPLOYEE VSAM file to include the social security numbers.
3. Create the alternate index which will be named EMPSSN.
4. Build and test the alternate index.

First, let's update our file layout in the EMPLOYEE copybook. Here it is with the EMP_SSN added.

```
      *******************************************************************
      * COBOL DECLARATION FOR VSAM FILE EMPLOYEE                        *
      *******************************************************************
       01  EMPLOYEE.
           05 EMP-ID              PIC 9(04).
           05 EMP-LAST-NAME       PIC X(30).
           05 EMP-FIRST-NAME      PIC X(20).
           05 EMP-SERVICE-YEARS   PIC 9(02).
           05 EMP-PROMOTION-DATE  PIC X(10).
           05 EMP-SSN             PIC X(09).
           05 FILLER              PIC X(05).
```

Then we could add the social security numbers through File Manager. Another alternative is to unload the data first into a flat file, add the social security number values to the flat file, and then scratch and recreate the VSAM file (using the revised unload file. If you want to do the unload, here is some sample JCL.

```
//USER01D JOB 'WINGATE',MSGLEVEL=(1,1),NOTIFY=&SYSUID
//*
//***********************************************************
//* UNLOAD DATA FROM VSAM KSDS TO PS DATA SET
//***********************************************************
//JS010    EXEC PGM=IDCAMS
//SYSPRINT DD SYSOUT=*
//SYSOUT   DD SYSOUT=*
//DD1      DD DSN=USER01.EMPLOYEE,DISP=SHR
//DD2      DD DSN=USER01.EMPLOYEE.UNLOAD,
//            DISP=(NEW,CATLG,DELETE),
//            SPACE=(TRK,(1,1),RLSE),
//            UNIT=SYSDA,VOL=SER=DEVHD1,
//            DCB=(DSORG=PS,RECFM=FB,LRECL=80,BLKSIZE=27920)
//SYSIN    DD  *
  REPRO -
  INFILE(DD1) -
  OUTFILE(DD2)
/*
```

I'm going to use File Manager instead. Here's the first record. I am of course adding random nine digit numbers here, not real social security numbers.

```
Edit              USER01.EMPLOYEE                    Rec 1 of 6
Command ===>                                         Scroll PAGE
Key 1122                 Type KSDS    RBA 0          Format SNGL
                                             Top Line is 1   of 7
Current 01: EMPLOYEE                                 Length 80
Field             Data
EMP-ID             1122
EMP-LAST-NAME      JENKINS
EMP-FIRST-NAME     DEBORAH
EMP-SERVICE-YEARS  5
EMP-PROMOTION-DATE 2016-09-01
EMP_SSN            034658724
FILLER
***  End of record  ***
```

Once I've finished adding SSNs, I will verify that all six records have them.

```
   Edit              USER01.EMPLOYEE                        Rec 1 of 6
   Command ===>                                             Scroll PAGE
       Key 1122                Type KSDS     RBA 0          Format TABL
         EMP-SERVICE-YEARS EMP-PROMOTION-DATE EMP_SSN     FILLER
                      #5 #6                   #7          #8
                      ZD 55:2 AN 57:10        AN 67:9     AN 76:5
                      <-> <---+---->          <---+---> <--->
   000001            5 2016-09-01             034658724
   000002            4 2017-01-01             493082938
   000003            9 2017-01-01             209482059
   000004            8 2016-01-01             030467384
   000005            7 2016-07-31             991837283
   000006            3 2016-10-01             333073948
   ****** ****   End of data   ****
```

Now, it's time to build the alternate index. First, we give it a file name and establish the other attributes. We'll give the index file name USER01.EMPLOYEE.ALX. And we will define the key as 9 bytes beginning at displacement 66. That's where the social security number is. We also indicate that it is related to the USER01.EMPLOYEE cluster.

DEFINE PATH is used to relate the alternate index to the base cluster. While defining path we specify the name of the path and the alternate index to which this path is related. This is the actual link between the VSAM cluster and the alternate index.

Finally, the BLDINDEX command is used to build the alternate index. BLDINDEX reads all the records in the VSAM indexed data set (base cluster) and extracts the data needed to build the alternate index.

```
//USER01D JOB 'WINGATE',MSGLEVEL=(1,1),NOTIFY=&SYSUID
//*
//****************************************************************
//* DEFINE ALTERNAME INDEX
//****************************************************************
//JS010    EXEC PGM=IDCAMS
//SYSPRINT DD SYSOUT=*
//SYSOUT   DD SYSOUT=*
//SYSIN    DD  *
  DEFINE AIX  -
  (NAME(USER01.EMPLOYEE.ALX) -
  RELATE(USER01.EMPLOYEE)     -
  CISZ(4096) -
  KEYS(9,66) -
  UNIQUEKEY -
  UPGRADE -
```

```
    RECORDSIZE(80,80) -
    TRK(2,1) -
    FREESPACE(10,20) -
    VOLUMES(DEVHD1)  -
    )
/*
//*
//************************************************************
//* DEFINE PATH
//************************************************************
//JS020    EXEC PGM=IDCAMS
//SYSPRINT DD SYSOUT=*
//SYSOUT   DD SYSOUT=*
//SYSIN    DD  *
  DEFINE PATH (NAME(USER01.EMPLOYEE.PATH) -
               PATHENTRY(USER01.EMPLOYEE.ALX) UPDATE
/*
//*
//************************************************************
//* BUILD INDEX
//************************************************************
//JS030    EXEC PGM=IDCAMS
//SYSPRINT DD SYSOUT=*
//SYSOUT   DD SYSOUT=*
//SYSIN    DD *
  BLDINDEX -
       INDATASET (USER01.EMPLOYEE) -
       OUTDATASET(USER01.EMPLOYEE.ALX)
/*
//*
```

COBOL Program to Read Alternate Index (COBVS6)

Now we can use this alternate index to randomly access the data using the EMP-SSN field. We'll write program COBVS6 to demonstrate this. Suppose for example we want to retrieve the record with SSN value 209482059 which is Tim Shultz. We can clone the first program COBVS1 into COBVS6. We do need to change a few things.

First our JCL must include a DD name for the PATH associated with the alternate index. When you use an alternate index, the DD name for the PATH must be the same as the DD name for the cluster except that the PATH DD name must have a 1 at the end of it. Since a DD identifier can be a maximum of 8 bytes, we must shorten the DD name of our EMPLOYEE VSAM file (in the program and JCL) to 7 bytes to so we can include a corresponding DD name for the PATH. We will shorten our cluster DD name to EMPVSFL. We can then define the PATH DD name as EMPVSFL1. Here's our JCL.

```
//USER01D JOB MSGLEVEL=(1,1),NOTIFY=&SYSUID
//*
//*  RUN A COBOL PROGRAM
//*
//STEP01  EXEC PGM=COBVS6
//STEPLIB  DD  DSN=USER01.LOADLIB,DISP=SHR
//SYSOUT   DD  SYSOUT=*
//EMPVSFL  DD DSN=USER01.EMPLOYEE,DISP=SHR
//EMPVSFL1 DD DSN=USER01.EMPLOYEE.PATH,DISP=SHR
//SYSPRINT DD  SYSOUT=*
//SYSUDUMP DD  SYSOUT=*
//SYSOUT   DD  SYSOUT=*
```

Second we need to identify the alternate key in the File Control section, and also change the **ASSIGN TO** clause to match the DD name change we made to the JCL (note that you do **not** need to do an assign statement for the PATH DD). Here is the code change. Notice the reference to the ALTERNATE KEY.

```
FILE-CONTROL.
   SELECT EMPLOYEE-VS-FILE    ASSIGN TO EMPVSFL
   ORGANIZATION     IS INDEXED
   ACCESS MODE      IS RANDOM
   RECORD KEY       IS EMP-ID
   ALTERNATE KEY    IS EMP-SSN
   FILE STATUS      IS EMP-FILE-STATUS.
```

Finally, we need to establish that the alternate key is to be used in the READ. We do this with the KEY IS clause.

```
READ EMPLOYEE-VS-FILE KEY IS EMP-SSN
```

Here is the final program listing with these features:

```
IDENTIFICATION DIVISION.
PROGRAM-ID. COBVS6.

****************************************************
*       PROGRAM TO RETRIEVE A RECORD FROM          *
*       EMPLOYEE VSAM FILE USING ALTERNATE INDEX.   *
****************************************************

ENVIRONMENT DIVISION.
INPUT-OUTPUT SECTION.

   FILE-CONTROL.
      SELECT EMPLOYEE-VS-FILE   ASSIGN TO EMPVSFL
      ORGANIZATION    IS INDEXED
```

```
            ACCESS MODE      IS RANDOM
            RECORD KEY       IS EMP-ID
            ALTERNATE KEY    IS EMP-SSN
            FILE STATUS      IS EMP-FILE-STATUS.

    DATA DIVISION.

    FILE SECTION.
    FD EMPLOYEE-VS-FILE.
        01   EMPLOYEE.
            05 EMP-ID               PIC 9(04).
            05 EMP-LAST-NAME        PIC X(30).
            05 EMP-FIRST-NAME       PIC X(20).
            05 EMP-SERVICE-YEARS    PIC 9(02).
            05 EMP-PROMOTION-DATE   PIC X(10).
            05 EMP-SSN              PIC X(09).
            05 FILLER               PIC X(05).

    WORKING-STORAGE SECTION.

        01 WS-FLAGS.
            05  SW-END-OF-FILE-SWITCH   PIC X(1) VALUE 'N'.
                88  SW-END-OF-FILE             VALUE 'Y'.
                88  SW-NOT-END-OF-FILE         VALUE 'N'.

        01  EMP-FILE-STATUS.
                05  EMPFILE-STAT1     PIC X.
                05  EMPFILE-STAT2     PIC X.

    PROCEDURE DIVISION.

        PERFORM P100-INITIALIZATION.
        PERFORM P200-MAINLINE.
        PERFORM P300-TERMINATION.
        GOBACK.

    P100-INITIALIZATION.

        DISPLAY 'COBVS6 - SAMPLE COBOL PROGRAM: VSAM ALT INDEX'.
        OPEN INPUT EMPLOYEE-VS-FILE.

        IF  EMP-FILE-STATUS = '00' OR '97' THEN
            NEXT SENTENCE
        ELSE
            DISPLAY 'ERROR ON OPEN - FILE STATUS ' EMP-FILE-STATUS.

        INITIALIZE EMPLOYEE.

    P200-MAINLINE.

*   READ THE INPUT FILE TO GET THE REQUESTED RECORD
*   AND DISPLAY THE DATA VALUES
```

174

```
        MOVE '209482059' TO EMP-SSN
        READ EMPLOYEE-VS-FILE KEY IS EMP-SSN

        IF  EMP-FILE-STATUS = '00' THEN
   *        DISPLAY THE DATA
            DISPLAY 'EMP-ID                ' EMP-ID
            DISPLAY 'EMP LAST NAME         ' EMP-LAST-NAME
            DISPLAY 'EMP FIRST NAME        ' EMP-FIRST-NAME
            DISPLAY 'EMP YEARS OF SERVICE ' EMP-SERVICE-YEARS
            DISPLAY 'EMP PROMOTION DATE    ' EMP-PROMOTION-DATE
            DISPLAY 'EMP SOCIAL SECURITY  ' EMP-SSN
        ELSE
            DISPLAY 'RECORD WAS NOT FOUND - RC = ' EMP-FILE-STATUS.

    P300-TERMINATION.

        CLOSE EMPLOYEE-VS-FILE.

        IF  EMP-FILE-STATUS = '00' THEN
            NEXT SENTENCE
        ELSE
            DISPLAY 'ERROR ON CLOSE - FILE STATUS ' EMP-FILE-STATUS.

        DISPLAY 'COBVS6 - SUCCESSFULLY ENDED'.

   *    END OF SOURCE CODE
```

Now we can compile, link and execute our program. Here is the result.

```
COBVS6 - SAMPLE COBOL PROGRAM: VSAM ALT INDEX
EMP-ID                4720
EMP LAST NAME         SCHULTZ
EMP FIRST NAME        TIM
EMP YEARS OF SERVICE 09
EMP PROMOTION DATE    2017-01-01
EMP SOCIAL SECURITY  209482059
COBVS6 - SUCCESSFULLY ENDED
```

As you can see, we successfully retrieved the record using the alternate index and specifying the KEY IS field.

Alternate keys give you tremendous flexibility when using VSAM. You can have more than one or more alternate keys on a file and you can specify more than one key in your application programs.

Other VSAM JCL

We haven't gone into much detail about the other file organizations because KSDS is the most common. However, here is some sample JCL for creating the ESDS and RRDS formats.

JCL to CREATE ESDS

```
//***********************************************************
//* DEFINE VSAM ESDS CLUSTER
//***********************************************************
//STEP30   EXEC PGM=IDCAMS
//SYSPRINT DD SYSOUT=*
//SYSOUT   DD SYSOUT=*
//SYSIN    DD  *
  DEFINE CLUSTER(NAME(USER01.TEST.ESDS.CLUSTER)  -
  RECORDSIZE(45,45)    -
  CYLINDERS(2,1)       -
  CISZ(4096)           -
  VOLUMES(DEVHD1)      -
  NONINDEXED)          -
  DATA(NAME(USER01.TEST.ESDS.DATA))
/*
//*
```

JCL to CREATE RRDS

```
//*
//***********************************************************
//* DEFINE VSAM RRDS CLUSTER
//***********************************************************
//STEP40   EXEC PGM=IDCAMS
//SYSPRINT DD SYSOUT=*
//SYSOUT   DD SYSOUT=*
//SYSIN    DD  *
  DEFINE CLUSTER(NAME(USER01.TEST.RRDS.CLUSTER)  -
  RECORDSIZE(45,45)    -
  CYLINDERS(2,1)       -
  NUMBERED)            -
  DATA(NAME(USER01.TEST.RRDS.DATA))
/*
```

JCL to LIST DATASET INFORMATION

```
//USER01L JOB 'WINGATE',MSGLEVEL=(1,1),NOTIFY=&SYSUID
//*
//*************************************************************
//* LISTCAT COMMAND
//*************************************************************
//STEP110  EXEC PGM=IDCAMS
//SYSPRINT DD SYSOUT=*
//SYSOUT   DD SYSOUT=*
//SYSIN    DD  *
     LISTCAT ENTRIES(USER01.EMPLOYEE) ALL
/*
//*


IDCAMS  SYSTEM SERVICES                                        TIME: 08:05:59
     LISTCAT ENTRIES(USER01.EMPLOYEE) ALL
CLUSTER ------- USER01.EMPLOYEE
     IN-CAT --- CATALOG.Z113.MASTER
     HISTORY
        DATASET-OWNER-----(NULL)    CREATION--------2018.064
        RELEASE----------------2    EXPIRATION------0000.000
        CA-RECLAIM---------(YES)
        EATTR------------(NULL)
        BWO STATUS--------(NULL)     BWO TIMESTAMP-----(NULL)
        BWO--------------(NULL)
      PROTECTION-PSWD-----(NULL)     RACF---------------(NO)
     ASSOCIATIONS
        DATA-----USER01.EMPLOYEE.DATA
        INDEX----USER01.EMPLOYEE.INDEX
        AIX------USER01.EMPLOYEE.ALX
   DATA ------- USER01.EMPLOYEE.DATA
     IN-CAT --- CATALOG.Z113.MASTER
     HISTORY                                                            DATA-
SET-OWNER-----(NULL)    CREATION--------2018.064
        RELEASE----------------2    EXPIRATION------0000.000
        ACCOUNT-INFO----------------------------------(NULL)
      PROTECTION-PSWD-----(NULL)     RACF---------------(NO)
     ASSOCIATIONS
        CLUSTER--USER01.EMPLOYEE
     ATTRIBUTES
        KEYLEN----------------4     AVGLRECL--------------80     BUFSPACE-------
        RKP-------------------0     MAXLRECL--------------80     EXCPEXIT-------
        SHROPTNS(1,3)  RECOVERY     UNIQUE        NOERASE     INDEXED      N
        NONSPANNED
     STATISTICS  (* - VALUE MAY BE INCORRECT)
        REC-TOTAL-------------7*    SPLITS-CI-------------0*    EXCPS----------
        REC-DELETED-----------9*    SPLITS-CA-------------0*    EXTENTS--------
        REC-INSERTED----------3*    FREESPACE-%CI----------5    SYSTEM-TIMESTAM
        REC-UPDATED----------11*    FREESPACE-%CA---------10       X'D3FD7137
        REC-RETRIEVED-------191*    FREESPC-----------45056*
     ALLOCATION
        SPACE-TYPE--------TRACK     HI-A-RBA-----------49152
        SPACE-PRI-------------1     HI-U-RBA-----------49152
         SPACE-SEC------------1
```

```
    VOLUME
        VOLSER-----------DEVHD1          PHYREC-SIZE---------4096      HI-A-RBA-------
        DEVTYPE------X'3010200F'         PHYRECS/TRK-----------12      HI-U-RBA-------
        VOLFLAG-----------PRIME          TRACKS/CA--------------1
        EXTENTS:
        LOW-CCHH-----X'00AF000E'         LOW-RBA----------------0      TRACKS---------
        HIGH-CCHH----X'00AF000E'         HIGH-RBA-----------49151
    INDEX ------ USER01.EMPLOYEE.INDEX
      IN-CAT --- CATALOG.Z113.MASTER
      HISTORY
        DATASET-OWNER-----(NULL)         CREATION--------2018.064
          RELEASE----------------2      EXPIRATION------0000.000
        PROTECTION-PSWD-----(NULL)       RACF---------------(NO)
      ASSOCIATIONS
        CLUSTER--USER01.EMPLOYEE
      ATTRIBUTES
        KEYLEN----------------4          AVGLRECL---------------0      BUFSPACE-------
        RKP-------------------0          MAXLRECL-----------4089      EXCPEXIT-------
        SHROPTNS(1,3)   RECOVERY         UNIQUE            NOERASE      NOWRITECHK
      STATISTICS  (* - VALUE MAY BE INCORRECT)
        REC-TOTAL-------------1*          SPLITS-CI--------------0*     EXCPS----------
        REC-DELETED-----------0*          SPLITS-CA--------------0*     EXTENTS--------
        REC-INSERTED----------0*          FREESPACE-%CI----------0      SYSTEM-TIMESTAM
        REC-UPDATED-----------0*          FREESPACE-%CA----------0         X'D3FD7137
        REC-RETRIEVED---------4*          FREESPC-----------45056*
      ALLOCATION
        SPACE-TYPE---------TRACK          HI-A-RBA-----------49152
        SPACE-PRI-------------1          HI-U-RBA------------4096
        SPACE-SEC-------------1
    VOLUME
        VOLSER-----------DEVHD1          PHYREC-SIZE---------4096      HI-A-RBA-------
        DEVTYPE------X'3010200F'         PHYRECS/TRK-----------12      HI-U-RBA-------
        VOLFLAG-----------PRIME          TRACKS/CA--------------1
        EXTENTS:
        LOW-CCHH-----X'00B60007'         LOW-RBA----------------0      TRACKS---------
        HIGH-CCHH----X'00B60007'         HIGH-RBA-----------49151
IDCAMS   SYSTEM SERVICES                                          TIME: 08:05:59
            THE NUMBER OF ENTRIES PROCESSED WAS:
                        AIX -------------------0
                        ALIAS -----------------0
                        CLUSTER ---------------1
                        DATA ------------------1
                        GDG -------------------0
                        INDEX -----------------1
                        NONVSAM ---------------0
                        PAGESPACE -------------0
                        PATH ------------------0
                        SPACE -----------------0
                        USERCATALOG -----------0
                        TAPELIBRARY -----------0
                        TAPEVOLUME ------------0
                        TOTAL -----------------3
            THE NUMBER OF PROTECTED ENTRIES SUPPRESSED WAS 0
IDC0001I FUNCTION COMPLETED, HIGHEST CONDITION CODE WAS 0


IDC0002I IDCAMS PROCESSING COMPLETE. MAXIMUM CONDITION CODE WAS 0
```

VSAM File Status Codes

Here is a list of the VSAM status codes you might encounter.

Code	Description
00	Operation completed successfully
02	Non-Unique Alternate Index duplicate key found
04	Invalid fixed length record
05	While performing OPEN File and file is not present
10	End of File encountered
14	Attempted to READ a relative record outside file boundary
20	Invalid Key for VSAM KSDS or RRDS
21	Sequence error while performing WRITE or changing key on REWRITE
22	Primary duplicate Key found
23	Record not found or File not found
24	Key outside boundary of file
30	Permanent I/O Error
34	Record outside file boundary
35	While performing OPEN File and file is not present
37	OPEN file with wrong mode
38	Tried to OPEN a Locked file
39	OPEN failed because of conflicting file attributes
41	Tried to OPEN a file that is already open
42	Tried to CLOSE a file that is not OPEN
43	Tried to REWRITE without READing a record first
44	Tried to REWRITE a record of a different length
46	Tried to READ beyond End-of-file
47	Tried to READ from a file that was not opened I-O or INPUT
48	Tried to WRITE to a file that was not opened I-O or OUTPUT
49	Tried to DELETE or REWRITE to a file that was not opened I-O
91	Password or authorization failed
92	Logic Error
93	Resources are not available
94	Sequential record unavailable or concurrent OPEN error
95	File Information invalid or incomplete
96	No DD statement for the file
97	OPEN successful and file integrity verified
98	File is Locked - OPEN failed
99	Record Locked - record access failed

Chapter Four Review Questions

1. What are the three types of VSAM datasets?

2. How are records stored in an ESDS (entry sequenced) dataset?

3. What VSAM feature enables you to access the records in a KSDS dataset based on a key that is different than the file's primary key?

4. What is the general purpose utility program that provides services for VSAM files?

5. Which AMS function lists information about datasets?

6. If you are mostly going to use a KSDS file for sequential access, should you define a larger or smaller control interval when creating the file?

7. What is the basic AMS command to create a VSAM file?

8. To use the REWRITE command in COBOL, the VSAM file must be opened in what mode?

9. When you define an alternate index, what is the function of the RELATE parameter?

10. When you define a path using DEFINE PATH, what does the PATHENTRY parameter do?

11. After you've defined an alternate index and path, what AMS command must you issue to actually populate the alternate index?

12. After you've created a VSAM file, if you need to add additional DASD volumes that can be used with that file, what command would you use?

13. If you want to set a VSAM file to read only status, what command would you use?

14. What are some ways you can improve the performance of a KSDS file?

15. Do primary key values in a KSDS have to be unique?

16. In the COBOL SELECT statement what organization should be specified for a KSDS file?

17. In the COBOL SELECT statement for a KSDS what are the three possibilities for ACCESS?

18. Is there a performance penalty for using an alternate index compared to using the primary key?

19. What file status code will you receive if an operation succeeded?

Chapter 5 : Information Management System (IMS) Introduction

IMS is a hierarchical database management system (DBMS) that has been around since the 1960's. Although relational DBMSs are more common now, there is still an installed base of IMS users and IBM provides robust support for it. IMS is highly tuned for transaction management and generally provides excellent performance for that environment.

This text deals with IMS-DB, the IMS database manager. IMS also has a transaction manager called IMS-DC. We will be covering IMS-DB in this volume, and IMS-DC in later volume.

There are two modes of running IMS programs. One is DLI which runs within its own address space. There is also Batch Mode Processing (BMP) which runs under the IMS online control region. The practical difference between the two concerns programs that update the database. In DLI mode, a program requires exclusive use of the database to perform updates. In BMP mode, a program does not require exclusive use of the database because it is run in the shared IMS online environment. The IMS online system "referees" the shared online environment.

Before going further I need to point out that in IMS data records are called "segments". I'll use the terms segment and record more or less interchangeably throughout the chapter. There are usually multiple segment types in an IMS database, although not always.

Designing and Creating IMS Databases

Sample System Specification

We're going to create a hierarchical database for a Human Resource system that will involve employees. In fact the database will be named EMPLOYEE and the root segment (highest level segment type) will also be named EMPLOYEE. This segment will include information such as name, years of service and last promotion date.

The EMPLOYEE segment will have a child segment that stores details about the employee's pay. The segment will be named EMPPAY and include the effective date, annual pay and bonus pay.

The EMPPAY segment will have a child segment named EMPPAYHS that includes historical details about each paycheck an employee received.

Note: there can be multiple EMPPAY segments under each EMPLOYEE segment, and there can be multiple EMPPAYHS segments under each EMPPAY segment. The following diagram depicts our EMPLOYEE database visually as a hierarchy.

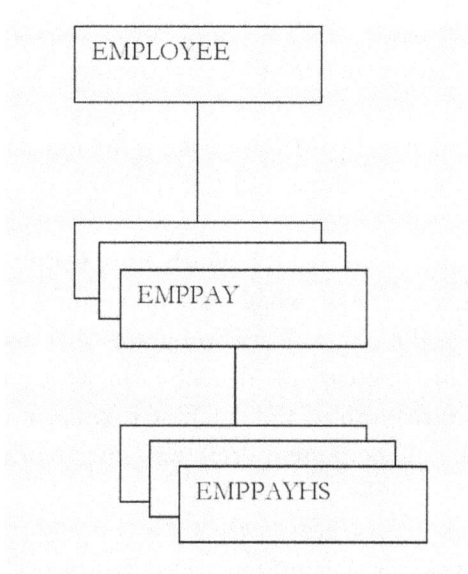

The following shows the segment structure we will be using to organize the three record types. Note that the EMP_ID key is only required on the root segment. You cannot access the child segments except through the root, so this makes sense.

EMPLOYEE Segment (key is EMP_ID).

Field Name	Type
EMP_ID	**INTEGER**
EMP_LAST_NAME	VARCHAR(30)
EMP_FIRST_NAME	VARCHAR(20)
EMP_SERVICE_YEARS	INTEGER
EMP_PROMOTION_DATE	DATE

EMPPAY segment (key is EFF_DATE which means effective date):

Field Name	Type
EFF_DATE	DATE
EMP_REGULAR_PAY	DECIMAL
EMP_BONUS_PAY	DECIMAL
EMP_SEMIMTH_PAY	DECIMAL

EMPPAYHS segment (key is PAY_DATE):

Field Name	Type
PAY_DATE	DATE
ANNUAL_PAY	DECIMAL
PAY_CHECK_AMT	DECIMAL

Having decided the content of our segment types, we can now create a record layout for each of these. We'll do this in COBOL since we will be writing our IMS programs in that language.

Here's the layout for the EMPLOYEE segment:

```
01  IO-EMPLOYEE-RECORD.
    05  EMP-ID        PIC X(04).
    05  FILLER        PIC X(01).
    05  EMPL-LNAME    PIC X(30).
    05  FILLER        PIC X(01).
    05  EMPL-FNAME    PIC X(20).
    05  FILLER        PIC X(01).
    05  EMPL-YRS-SRV  PIC X(02).
    05  FILLER        PIC X(01).
    05  EMPL-PRM-DTE  PIC X(10).
    05  FILLER        PIC X(10).
```

We've provided a bit of filler between fields, and we've left 10 bytes at the end (yes later we will be adding a field so we need some free space). Our total is 80 bytes for this segment type. The record will be keyed on **EMP-ID** which is the first four bytes of the record. We'll need this information to define the database.

Next, the EMPPAY segment layout is as follows:

```
01  IO-EMPPAY-RECORD.
    05  PAY-EFF-DATE  PIC X(8).
    05  PAY-REG-PAY   PIC S9(6)V9(2) USAGE COMP-3.
    05  PAY-BON-PAY   PIC S9(6)V9(2) USAGE COMP-3.
```

185

```
      05   SEMIMTH-PAY    PIC S9(6)V9(2) USAGE COMP-3.
      05   FILLER         PIC X(57).
```

Notice there is no `EMP-ID` field. As mentioned earlier, child segments do not need to repeat the parent segment key. The hierarchical structure of the database makes this unnecessary. The `PAY-EFF-DATE` field will be the key for the `EMPPAY` segment, and it is 8 bytes. The format will be `YYYYMMDD`.

Also notice that we padded the record with filler to total 80 bytes. We didn't have to do this. The record size is actually 23 bytes without the filler. But often it's a convenience to leave space in the IO layout for future expansion.

Finally, here is the layout for the `EMPPAYHS` segment. Out key will be `PAY-DATE` and it will be formatted as `YYYYMMDD`.

```
   01 IO-EMPPAYHS-RECORD.
      05   PAY-DATE       PIC X(8).
      05   PAY-ANN-PAY    PIC S9(6)V9(2) USAGE COMP-3.
      05   PAY-AMT        PIC S9(6)V9(2) USAGE COMP-3.
      05   FILLER         PIC X(57).
```

Now we are ready to build the data base descriptor!

Database Descriptor (DBD)

A database descriptor is required to have an IMS database. The descriptor specifies the name of the database, plus the various segment types. Typically a database administrator will create and maintain DBDs. You should still understand how to read the DBD code to understand the structure of the database.

Here's the DBD code for our EMPLOYEE database.

```
PRINT NOGEN
DBD NAME=EMPLOYEE,ACCESS=HISAM
DATASET DD1=EMPLOYEE,OVFLW=EMPLFLW
SEGM NAME=EMPLOYEE,PARENT=0,BYTES=80
FIELD NAME=(EMPID,SEQ,U),BYTES=04,START=1,TYPE=C
SEGM NAME=EMPPAY,PARENT=EMPLOYEE,BYTES=23
FIELD NAME=(EFFDATE,SEQ,U),START=1,BYTES=8,TYPE=C
SEGM  NAME=EMPPAYHS,PARENT=EMPPAY,BYTES=18
FIELD NAME=(PAYDATE,SEQ,U),START=1,BYTES=8,TYPE=C
DBDGEN
FINISH
END
```

The code above specifies the name of the database which is EMPLOYEE, as well as an access method of HISAM (Hierarchical Indexed Sequential Access Method). HISAM database records are stored in two data sets: a primary data set and an overflow data set. The primary dataset is always a VSAM KSDS and the overflow dataset is a VSAM ESDS. The ESDS dataset is used if the KSDS dataset becomes full. In that case any new records are inserted to the (overflow) ESDS dataset.

There is considerable information available about how HISAM records are stored. However, that information is frankly not very useful for application programmer duties, so we exclude it here.

Looking at the DBD code, we see that the DATASET DD1 and OVFLW keywords define the DD names of the primary cluster and the overflow dataset, respectively. We defined these values as EMPLOYEE and EMPLFLW. Later when we run batch jobs against the database, the DD name in our JCL must be EMPLOYEE for the KSDS file, and EMPLFLW for the overflow dataset.

Next, we define our segment types using the SEGM NAME= keywords. We also define the parent of each segment unless the segment is the root segment in which case we specify PARENT=0. Next you specify the length of your segment. We've defined the length as the total of the fields we mapped out earlier in the COBOL layouts.

For each segment, if you have any searchable fields (such as keys), they must be defined with the FIELD NAME= keywords. In our case, we will only specify the key fields for each segment. W specify the field name, that the records are to be ordered sequentially (SEQ), and that the field content must be unique(U). Then we specify how many bytes the field is, and it's displacement in the record. We also specify C for character data – the actual data we store can be of any type but we specify C to indicate the default type is character data.

Here's an example of defining the employee id in the DBD from above:

```
FIELD NAME=(EMPID,SEQ,U),BYTES=04,START=1,TYPE=C
```

Finally, you conclude the DBD with

```
DBDGEN
FINISH
END
```

Now you are ready to run your installation's JCL to generate a DBD. Most likely you will ask a DBA to do this. Here's the JCL I run which executes a proc named DBDGEN. It will be different for your installation.

```
//USER01D JOB MSGLEVEL=(1,1),NOTIFY=&SYSUID
//*
//PLIB     JCLLIB ORDER=SYS1.IMS.PROCLIB
//DGEN     EXEC DBDGEN,
//               MEMBER=EMPLOYEE,           <= DBD SOURCE MEMBER
//               SRCLIB=USER01.IMS.SRCLIB,  <= DBD SOURCE LIBRARY
//               DBDLIB=USER01.IMS.DBDLIB   <= DBD LIBRARY
//*
```

More information about designing, coding and generating DBDs is available on the IBM product web site.

Supporting VSAM Files

Now that we have a DBD generated, we can create the physical files for the database. Actually it can be done in either order, but you do need to know the maximum record size for all segments in order to build the IDCAMS JCL. For IMS datasets we use a VSAM key sequenced data set (KSDS).

Here is the JCL for creating our EMPLOYEE IMS database. Notice that we specify a RECORD-SIZE that is 8 bytes longer than the logical record size that we defined in the DBD. And although the key is the first logical byte of each record, we specify the key displacement at byte 6. The IMS system uses the first 5 bytes of each record, so this is required (IMS also uses the last 3 bytes, so we end up with 8 additional bytes for the record size).

Finally, note that we have a second job step to repro a dummy file to our VSAM cluster name to initialize it. This is required. Otherwise we will get an abend the first time we try to access it. Go ahead and run the JCL or ask your DBA to create the physical files.

```
//USER01D JOB MSGLEVEL=(1,1),NOTIFY=&SYSUID
//*
//*************************************************************
//* DEFINE VSAM KSDS CLUSTER FOR EMPLOYEE DATABASE
//*************************************************************
//VDEF     EXEC PGM=IDCAMS
//SYSPRINT DD SYSOUT=*
//SYSIN    DD  *
  DEFINE CLUSTER(NAME(USER01.IMS.EMPLOYEE.CLUSTER)  -
                 INDEXED                  -
                 KEYS(4,6)                -
                 RECORDSIZE(88,88)        -
                 TRACKS(2,1)              -
```

```
                CISZ(2048)                          -
                VOLUMES(DEVHD1)                      -
                )                                    -
          DATA(NAME(USER01.IMS.EMPLOYEE.DATA))
//*
//***************************************************************
//* INITIALIZE THE VSAM FILE TO PLACE EOF MARK
//***************************************************************
//VINIT     EXEC PGM=IDCAMS
//SYSPRINT DD SYSOUT=*
//INF       DD   DUMMY
//OUTF      DD   DSN=USER01.IMS.EMPLOYEE.CLUSTER,DISP=SHR
//SYSIN     DD   *
  REPRO INFILE(INF) OUTFILE(OUTF)
/*
//*
```

Next, here is the JCL for creating the overflow dataset. When your KSDS is full, new records will be placed in the overflow dataset. Again, specify RECORDSIZE 88 (not 80). You must also specify NONINDEXED to get an ESDS file.

```
//USER01D JOB MSGLEVEL=(1,1),NOTIFY=&SYSUID
//*
//***************************************************************
//* DEFINE VSAM ESDS CLUSTER FOR IMS DATA BASE
//***************************************************************
//VDEF      EXEC PGM=IDCAMS
//SYSPRINT DD SYSOUT=*
//SYSOUT   DD SYSOUT=*
//SYSIN     DD   *
  DEFINE CLUSTER(NAME(USER01.IMS.EMPLFLW.CLUSTER)   -
                NONINDEXED                          -
                RECORDSIZE(88,88)                   -
                TRACKS(2,1)                         -
                CISZ(2048)                          -
                VOLUMES(DEVHD1)                      -
                )                                    -
          DATA(NAME(USER01.IMS.EMPLFLW.DATA))
/*
```

Ok, time to move on to our next IMS entity which is a PSB.

Program Specification Block (PSB)

A Program Specification Block (PSB) is an IMS entity that specifies which segments and operations can be performed on one or more databases (using this particular PSB authority). PSBs consist of one or more Program Communication Blocks (PCB) which are logical views of a database. It is typical for each IMS application program to have a separate PSB defined for it, but this is convention, not an IMS requirement. For our programming examples we will mostly use just one PSB, but we will modify it a few times. Here is the code for the PSB that we will be using for most of the examples.

```
PRINT NOGEN
PCB    TYPE=DB,NAME=EMPLOYEE,KEYLEN=20,PROCOPT=AP
SENSEG NAME=EMPLOYEE,PARENT=0
SENSEG NAME=EMPPAY,PARENT=EMPLOYEE
SENSEG NAME=EMPPAYHS,PARENT=EMPPAY
PSBGEN LANG=COBOL,PSBNAME=EMPLOYEE
END
```

Here's the meaning of each keyword for defining the PCB.

PCB – this is where you define a pointer to your database.

TYPE - typically this is DB to indicate a database PCB which provides access to a specific database. There is also a terminal (TP) PCB that is used for teleprocessing calls in IMS DC, but we won't be doing IMS DC in this text.

NAME - identifies the database to be accessed.

PROCOPT - Processing options. This value specifies which operations can be performed such as read, update or delete. The following are the most common options:

G   Get
I   Insert
R   Replace
D   Delete
A   All Options (G, I, R, D)
L   Load Function (Initial Loading)
LS   Load Function (Loading Sequentially)
K   Key Function - Access only key of the segment.
O   Used with G option to indicate that HOLD is not allowed.
P   Path Function (Used during Path Calls)

The PROCOPT can be defined for the entire PCB or it can be more granular by applying it to specific segments. If specified at the segment level, it overrides any PROCOPT at the PCB level. In our case we have specified PROCOPT=AP for the entire PSB. That is powerful. It means All (G, I, R, D) plus authority to do path calls.

KEYLEN – specifies the length of the concatenated key. Concatenated key is the maximum length of all the segment keys added up. This needs to be calculated by adding the longest segment key in each level from top to bottom.

SENSEG means sensitive segment, which means you can access that segment via this PSB. You can specify which segments you want to access. You might not always want all segments to be accessed. In our case we do, so we define a SENSEG for each segment type.

You must then execute a PSBGEN (or ask your DBA to). Here is the JCL I use which executes a proc named PSBGEN, and note that the member name I stored the PSB source under is EMPPSB. Your JCL will be different and specific to the installation:

```
//USER01D JOB MSGLEVEL=(1,1),NOTIFY=&SYSUID
//*
//PLIB    JCLLIB ORDER=SYS1.IMS.PROCLIB
//PGEN    EXEC PSBGEN,
//            MEMBER=EMPPSB,              <= PSB SOURCE MEMBER
//            SRCLIB=USER01.IMS.SRCLIB,   <= PSB SOURCE LIBRARY
//            PSBLIB=USER01.IMS.PSBLIB    <= PSB LIBRARY
//*
```

Now you have the basic building blocks of IMS – a database descriptor (DBD), the physical VSAM files to support the database, and a PSB that provides permissions to access the data within the database via PCBs. We are ready to start programming.

IMS Application Programming Basics

The IMS Program Interface
To request IMS data services in an application program, you must call the IMS interface program for that programming language. The interface program for COBOL is CBLTDLI. This program is called with several parameters which vary depending on the operation being requested. The call also needs to tell CBLTDLI how many parameters are being passed, so we'll declare some constants in our program for that.

```
01 IMS-RET-CODES.
   05 THREE        PIC S9(9) COMP VALUE +3.
   05 FOUR         PIC S9(9) COMP VALUE +4.
   05 FIVE         PIC S9(9) COMP VALUE +5.
   05 SIX          PIC S9(9) COMP VALUE +6.
```

Let's take an example where you want to retrieve an EMPLOYEE segment from the EMPLOYEE database, and you want employee number 3217. Here is the call with appropriate parameters. We'll discuss each of these in turn.

```
CALL 'CBLTDLI' USING FOUR,
               DLI-FUNCGU,
               PCB-MASK,
               SEG-IO-AREA,
               EMP-QUALIFIED-SSA
```

The first parameter specifies the number of parameters being passed. In the case of the above call, the number would be four.

The second parameter is the call type. The following are the common IMS calls used to insert, retrieve, modify and delete data in an IMS database. There are some other calls we'll introduce later such as for checkpointing and rolling back data changes.

DLET	The Delete (DLET) call is used to remove a segment and its dependents from the database.
GN/GHN	The Get Next (GN) call is used to retrieve segments sequentially from the database. The Get Hold Next (GHN) is the hold form for a GN call.
GNP/GHNP	The Get Next in Parent (GNP) call retrieves dependents sequentially. The Get Hold Next in Parent (GHNP) call is the hold form of the GNP call.
GU/GHU	The Get Unique (GU) call is used to directly retrieve segments and to establish a starting position in the database for sequential processing. The Get Hold Unique (GHU) is the hold form for a GU call.
ISRT	The Insert (ISRT) call is used to load a database and to add one or more segments to the database. You can use ISRT to add a record to the end of a GSAM database or for an alternate PCB that is set up for IAFP processing.
REPL	The Replace (REPL) call is used to change the values of one or more fields in a segment.

It is a common practice to define a set of constants in your program that specify the value of the specific IMS calls. Here's the COBOL code to put in working storage for this purpose.

```
01 DLI-FUNCTIONS.
   05 DLI-FUNCISRT   PIC X(4)  VALUE 'ISRT'.
   05 DLI-FUNCGU     PIC X(4)  VALUE 'GU  '.
   05 DLI-FUNCGN     PIC X(4)  VALUE 'GN  '.
   05 DLI-FUNCGHU    PIC X(4)  VALUE 'GHU '.
   05 DLI-FUNCGHN    PIC X(4)  VALUE 'GHN '.
   05 DLI-FUNCGNP    PIC X(4)  VALUE 'GNP '.
   05 DLI-FUNCREPL   PIC X(4)  VALUE 'REPL'.
   05 DLI-FUNCDLET   PIC X(4)  VALUE 'DLET'.
   05 DLI-FUNCXRST   PIC X(4)  VALUE 'XRST'.
   05 DLI-FUNCCHKP   PIC X(4)  VALUE 'CHKP'.
   05 DLI-FUNCROLL   PIC X(4)  VALUE 'ROLL'.
```

As you can see from the call above and the constant definitions, we are doing a Get Unique (GU) call. The `DLI-FUNCGU` specifies it.

The next parameter is a `PCB` data area that we defined as `PCB-MASK`. This returns various information from IMS after the database call. You must define this structure in the Linkage Section of your program since it is passing data back and forth from the `CBLTDLI` interface program.

```
LINKAGE SECTION.
01 PCB-MASK.
   03 DBD-NAME        PIC X(8).
   03 SEG-LEVEL       PIC XX.
   03 STATUS-CODE     PIC XX.
   03 PROC-OPT        PIC X(4).
   03 FILLER          PIC X(4).
   03 SEG-NAME        PIC X(8).
   03 KEY-FDBK        PIC S9(5) COMP.
   03 NUM-SENSEG      PIC S9(5) COMP.
   03 KEY-FDBK-AREA.
      05 EMPLOYEE-KEY  PIC X(04).
```

One of the most important data elements in the `PCB-MASK` is the two byte status code returned by the call, the `STATUS-CODE`. A blank status code means that the call was successful. Other status codes indicate the reason why the call failed. Here is a subset of the status codes you may encounter.

IMS Status Codes

PCB Status Code	Description
AC	Hierarchic error in SSAs.
AD	Function parameter incorrect. Only applies to full-function DEQ calls.
AI	Data management OPEN error.
AJ	Incorrect parameter format in I/O area; incorrect SSA format; incorrect command used to insert a logical child segment. I/O area length in AIB is invalid; incorrect class parameter specified in Fast Path Q command code.
AK	Invalid SSA field name.
AM	Call function not compatible with processing option, segment sensitivity, transaction code, definition, or program type.
AU	SSAs too long.
DA	Segment key field or non-replaceable field has been changed.
DJ	No preceding successful GHU or GHN call or an SSA supplied at a level not retrieved.
FT	Too many SSAs on call.
GB	End of database.
GE	Segment not found.
GG	Segment contains invalid pointer.
GP	No parentage established.
II	Segment already exists.

In your application program control is passed from IMS through an entry point. Your entry point must refer to the PCBs in the order in which they have been defined in the PSB. When you code each DL/I call, you must provide the PCB you want to use for that call. Here is the entry point code at the beginning of the procedure division for this program.

```
ENTRY 'DLITCBL' USING PCB-MASK
```

The next parameter is the segment I/O area. This is where IMS returns the data segment you requested, or where you load data to be inserted/updated on an insert or replace command. For the EMPLOYEE record, we will define the I/O area in COBOL as:

```
01 IO-EMPLOYEE-RECORD.
   05  EMP-ID        PIC X(04).
   05  FILLER        PIC X(01).
   05  EMPL-LNAME    PIC X(30).
   05  FILLER        PIC X(01).
   05  EMPL-FNAME    PIC X(20).
   05  FILLER        PIC X(01).
   05  EMPL-YRS-SRV  PIC X(02).
   05  FILLER        PIC X(01).
   05  EMPL-PRM-DTE  PIC X(10).
   05  FILLER        PIC X(10).
```

The last parameter in our call is a Segment Search Argument (SSA). This is where we specify the type of segment we want and the key value. It is also possible to simply request the next record of a particular segment type without regard to key value. When we specify a key, that means we are using a "qualified" SSA. When we don't specify a key, it means we are using an unqualified SSA.

Here's the COBOL definition of the qualified and unqualified SSAs for the EMPLOYEE segment.

```
01 EMP-QUALIFIED-SSA.
   05  SEGNAME    PIC X(08) VALUE 'EMPLOYEE'.
   05  FILLER     PIC X(01) VALUE '('.
   05  FIELD      PIC X(08) VALUE 'EMPID'.
   05  OPER       PIC X(02) VALUE ' ='.
   05  EMP-ID-VAL PIC X(04) VALUE '    '.
   05  FILLER     PIC X(01) VALUE ')'.

01 EMP-UNQUALIFIED-SSA.
   05  SEGNAME    PIC X(08) VALUE 'EMPLOYEE'.
   05  FILLER     PIC X(01) VALUE ' '.
```

Both qualified and unqualified SSAs must specify the segment type or name. You specify the key for the qualified SSA in the field we've named EMP-ID-VAL. We'll show many examples of SSAs in the program examples, including the use of Boolean SSA values.

Loading an IMS Database

Ok, finally to our first program. We're going to load the IMS database with a few records from a text file (a.k.a. a flat file). Here is the data file contents:

```
----+----1----+----2----+----3----+----4----+----5----+----6----+----7----+---8
****************************** Top of Data ********************************
1111 VEREEN                   CHARLES              12 2017-01-01 937253058
1122 JENKINS                  DEBORAH              05 2017-01-01 435092366
3217 JOHNSON                  EDWARD               04 2017-01-01 397342007
4175 TURNBULL                 FRED                 01 2016-12-01 542083017
4720 SCHULTZ                  TIM                  09 2017-01-01 650450254
4836 SMITH                    SANDRA               03 2017-01-01 028374669
6288 WILLARD                  JOE                  06 2016-01-01 209883920
7459 STEWART                  BETTY                07 2016-07-31 019572830
9134 FRANKLIN                 BRIANNA              00 2016-10-01 937293598
```

As you can see, we've formatted the records exactly like we want them to be applied to the database. Of course, your input layout could be different than the IMS segment layout, but using the same layout makes it easier because you don't have to do field assignments in the program.

Now let's create a program named COBIMS1 to load the data. We'll define the input file of course. In our program, let's call the file EMPFILE. Let's assume that the DD name for the employee load file is EMPIFILE.

```
ENVIRONMENT DIVISION.
INPUT-OUTPUT SECTION.
FILE-CONTROL.
    SELECT EMPFILE ASSIGN TO EMPIFILE.

DATA DIVISION.
FILE SECTION.
FD  EMPFILE
    RECORDING MODE IS F
    RECORD CONTAINS 80 CHARACTERS.

01 INSRT-REC.
   05 SEG-IO-AREA PIC X(80).
```

We've specified a SEG-IO-AREA variable to read the input file into and to write the IMS record from. We could have used the fully detailed IO-EMPLOYEE-RECORD instead (and we will later), but I want to demonstrate the value of having your input records structured the same as the IMS segment. When you do this, it really simplifies the coding such that you can both read and write using use a one element structure like SEG-IO-AREA.

Next we'll code the working storage section with a few things including:

- An end of file switch for the loop we'll create to load the records.
- The DLI call constants.
- The Employee segment I/O structure.
- The Employee segment SSA.

```
WORKING-STORAGE SECTION.

  01 WS-FLAGS.
     05  SW-END-OF-FILE-SWITCH    PIC X(1) VALUE 'N'.
         88  SW-END-OF-FILE                VALUE 'Y'.
         88  SW-NOT-END-OF-FILE            VALUE 'N'.

  01 DLI-FUNCTIONS.
     05 DLI-FUNCISRT  PIC X(4) VALUE 'ISRT'.
     05 DLI-FUNCGU    PIC X(4) VALUE 'GU  '.
     05 DLI-FUNCGN    PIC X(4) VALUE 'GN  '.
     05 DLI-FUNCGHU   PIC X(4) VALUE 'GHU '.
     05 DLI-FUNCGNP   PIC X(4) VALUE 'GNP '.
     05 DLI-FUNCREPL  PIC X(4) VALUE 'REPL'.
     05 DLI-FUNCDLET  PIC X(4) VALUE 'DLET'.
     05 DLI-FUNCXRST  PIC X(4) VALUE 'XRST'.
     05 DLI-FUNCCKPT  PIC X(4) VALUE 'CKPT'.

  01 IO-EMPLOYEE-RECORD.
     05   EMPL-ID-IN    PIC X(04).
     05   FILLER        PIC X(01).
     05   EMPL-LNAME    PIC X(30).
     05   FILLER        PIC X(01).
     05   EMPL-FNAME    PIC X(20).
     05   FILLER        PIC X(01).
     05   EMPL-YRS-SRV  PIC X(02).
     05   FILLER        PIC X(01).
     05   EMPL-PRM-DTE  PIC X(10).
     05   FILLER        PIC X(10).


   01 EMP-UNQUALIFIED-SSA.
      05   SEGNAME     PIC X(08) VALUE 'EMPLOYEE'.
      05   FILLER      PIC X(01) VALUE ' '.

   01 EMP-QUALIFIED-SSA.
      05   SEGNAME     PIC X(08) VALUE 'EMPLOYEE'.
      05   FILLER      PIC X(01) VALUE '('.
      05   FIELD       PIC X(08) VALUE 'EMPID'.
      05   OPER        PIC X(02) VALUE ' ='.
      05   EMP-ID-VAL  PIC X(04) VALUE '    '.
      05   FILLER      PIC X(01) VALUE ')'.
```

```
01 IMS-RET-CODES.
   05 THREE            PIC S9(9) COMP VALUE +3.
   05 FOUR             PIC S9(9) COMP VALUE +4.
   05 FIVE             PIC S9(9) COMP VALUE +5.
   05 SIX              PIC S9(9) COMP VALUE +6.
```

Finally, we'll code the linkage section which includes the database PCB mask.

```
LINKAGE SECTION.
01 PCB-MASK.
   03 DBD-NAME         PIC X(8).
   03 SEG-LEVEL        PIC XX.
   03 STATUS-CODE      PIC XX.
   03 PROC-OPT         PIC X(4).
   03 FILLER           PIC X(4).
   03 SEG-NAME         PIC X(8).
   03 KEY-FDBK         PIC S9(5) COMP.
   03 NUM-SENSEG       PIC S9(5) COMP.
   03 KEY-FDBK-AREA.
      05 EMPLOYEE-KEY  PIC X(04).
```

We'll work on the procedure division next, which will complete the program. Let's talk about the actual database call. Here's what we'll use:

```
CALL 'CBLTDLI' USING FOUR,
     DLI-FUNCISRT,
     PCB-MASK,
     SEG-IO-AREA,
     EMP-UNQUALIFIED-SSA
```

This is similar to the example we gave earlier with a couple of differences. One difference of course is that we are doing an ISRT call, so we specify the constant DLI-FUNCISRT. The other difference is that we will use an unqualified SSA. On an insert operation, IMS will always establish the record key from the I/O area and therefore it does not use a qualified SSA.

To be clear, any time you are inserting a record, you will use an **unqualified** SSA at the level of the record you are inserting. So if you are inserting a root segment, you will always use an unqualified SSA. If you are inserting a child segment under a root, you will use a qualified SSA on the root segment, and then an unqualified SSA for the child segment. If this seems a bit cryptic now, it should make more sense in later examples where we use child segments and multiple SSAs.

Ok, here's our complete program code. See what you think.

```
       IDENTIFICATION DIVISION.
       PROGRAM-ID. COBIMS1.

      *******************************************************
      *   INSERT A RECORD INTO IMS EMPLOYEE DATABASE        *
      *******************************************************

       ENVIRONMENT DIVISION.
       INPUT-OUTPUT SECTION.
       FILE-CONTROL.
           SELECT EMPFILE ASSIGN TO EMPIFILE.

       DATA DIVISION.
       FILE SECTION.
       FD  EMPFILE
           RECORDING MODE IS F
           RECORD CONTAINS 80 CHARACTERS.

       01 INSRT-REC.
          05 SEG-IO-AREA PIC X(80).

      *******************************************************
      *   W O R K I N G   S T O R A G E   S E C T I O N     *
      *******************************************************

       WORKING-STORAGE SECTION.

        01 WS-FLAGS.
           05  SW-END-OF-FILE-SWITCH   PIC X(1) VALUE 'N'.
               88  SW-END-OF-FILE                VALUE 'Y'.
               88  SW-NOT-END-OF-FILE            VALUE 'N'.

       01 DLI-FUNCTIONS.
          05 DLI-FUNCISRT  PIC X(4) VALUE 'ISRT'.
          05 DLI-FUNCGU    PIC X(4) VALUE 'GU  '.
          05 DLI-FUNCGN    PIC X(4) VALUE 'GN  '.
          05 DLI-FUNCGHU   PIC X(4) VALUE 'GHU '.
          05 DLI-FUNCGNP   PIC X(4) VALUE 'GNP '.
          05 DLI-FUNCREPL  PIC X(4) VALUE 'REPL'.
          05 DLI-FUNCDLET  PIC X(4) VALUE 'DLET'.
          05 DLI-FUNCXRST  PIC X(4) VALUE 'XRST'.
          05 DLI-FUNCCKPT  PIC X(4) VALUE 'CKPT'.

       01 IN-EMPLOYEE-RECORD.
          05  EMPL-ID-IN   PIC X(04).
          05  FILLER       PIC X(01).
          05  EMPL-LNAME   PIC X(30).
          05  FILLER       PIC X(01).
          05  EMPL-FNAME   PIC X(20).
          05  FILLER       PIC X(01).
          05  EMPL-YRS-SRV PIC X(02).
          05  FILLER       PIC X(01).
```

```cobol
        05  EMPL-PRM-DTE  PIC X(10).
        05  FILLER        PIC X(10).


    01 EMP-UNQUALIFIED-SSA.
        05  SEGNAME     PIC X(08) VALUE 'EMPLOYEE'.
        05  FILLER      PIC X(01) VALUE ' '.


    01 EMP-QUALIFIED-SSA.
        05  SEGNAME     PIC X(08) VALUE 'EMPLOYEE'.
        05  FILLER      PIC X(01) VALUE '('.
        05  FIELD       PIC X(08) VALUE 'EMPID'.
        05  OPER        PIC X(02) VALUE ' ='.
        05  EMP-ID-VAL  PIC X(04) VALUE '    '.
        05  FILLER      PIC X(01) VALUE ')'.


    01 IMS-RET-CODES.
        05 THREE        PIC S9(9) COMP VALUE +3.
        05 FOUR         PIC S9(9) COMP VALUE +4.
        05 FIVE         PIC S9(9) COMP VALUE +5.
        05 SIX          PIC S9(9) COMP VALUE +6.


LINKAGE SECTION.
 01 PCB-MASK.
        03 DBD-NAME       PIC X(8).
        03 SEG-LEVEL      PIC XX.
        03 STATUS-CODE    PIC XX.
        03 PROC-OPT       PIC X(4).
        03 FILLER         PIC X(4).
        03 SEG-NAME       PIC X(8).
        03 KEY-FDBK       PIC S9(5) COMP.
        03 NUM-SENSEG     PIC S9(5) COMP.
        03 KEY-FDBK-AREA.
           05 EMPLOYEE-KEY  PIC X(04).
           05 EMPPAYHS-KEY  PIC X(08).


PROCEDURE DIVISION.

    INITIALIZE PCB-MASK
    ENTRY 'DLITCBL' USING PCB-MASK

    PERFORM P100-INITIALIZATION.
    PERFORM P200-MAINLINE.
    PERFORM P300-TERMINATION.
    GOBACK.

P100-INITIALIZATION.

    DISPLAY '** PROGRAM COBIMS1 START **'
    DISPLAY 'PROCESSING IN P100-INITIALIZATION'
    OPEN INPUT EMPFILE.
```

```
P200-MAINLINE.

    DISPLAY 'PROCESSING IN P200-MAINLINE'

    READ EMPFILE
       AT END SET SW-END-OF-FILE TO TRUE
    END-READ

    PERFORM UNTIL SW-END-OF-FILE

       CALL 'CBLTDLI' USING FOUR,
            DLI-FUNCISRT,
            PCB-MASK,
            SEG-IO-AREA,
            EMP-UNQUALIFIED-SSA

       IF STATUS-CODE = ' '
          DISPLAY 'SUCCESSFUL INSERT-REC:' SEG-IO-AREA
       ELSE
          PERFORM P400-DISPLAY-ERROR
       END-IF

       READ EMPFILE
          AT END SET SW-END-OF-FILE TO TRUE
       END-READ

    END-PERFORM.


P300-TERMINATION.

    DISPLAY 'PROCESSING IN P300-TERMINATION'

    CLOSE EMPFILE
    DISPLAY '** COBIMS1 - SUCCESSFULLY ENDED **'.

P400-DISPLAY-ERROR.

    DISPLAY 'ERROR ENCOUNTERED - DETAIL FOLLOWS'
    DISPLAY 'SEG-IO-AREA      :' SEG-IO-AREA
    DISPLAY 'DBD-NAME1:'      DBD-NAME
    DISPLAY 'SEG-LEVEL1:'     SEG-LEVEL
    DISPLAY 'STATUS-CODE:'    STATUS-CODE
    DISPLAY 'PROC-OPT1 :'     PROC-OPT
    DISPLAY 'SEG-NAME1 :'     SEG-NAME
    DISPLAY 'KEY-FDBK1 :'     KEY-FDBK
    DISPLAY 'NUM-SENSEG1:'    NUM-SENSEG
    DISPLAY 'KEY-FDBK-AREA1:' KEY-FDBK-AREA.

*    END OF SOURCE CODE
```

Now we can compile and link the program. You'll need to ask your supervisor or teammate for the compile procedure. I am using JCL as follows to execute a COBOL-IMS compile procedure:

```
//USER01D JOB MSGLEVEL=(1,1),NOTIFY=&SYSUID
//*
//* COMPILE A IMS COBOL PROGRAM
//*
//PLIB    JCLLIB ORDER=SYS1.IMS.PROCLIB
//CL      EXEC IMSCOBCL,
//              MBR=COBIMS1,                    <= COBOL PROGRAM NAME
//              SRCLIB=USER01.COBOL.SRCLIB,     <= COBOL SOURCE LIBRARY
//              COPYLIB=USER01.COPYLIB,         <= COPY BOOK LIBRARY
//              LOADLIB=USER01.IMS.LOADLIB      <= LOAD LIBRARY
```

Finally, one time only you must create and use a special PSB to load the database. The PSB can be identical to the one we already created except it must specify a PROCOPT of **LS** which means Load Sequential. Let's clone EMPPSB into member EMPPSBL:

```
        PRINT NOGEN
        PCB   TYPE=DB,NAME=EMPLOYEE,KEYLEN=20,PROCOPT=LS
        SENSEG NAME=EMPLOYEE,PARENT=0
        SENSEG NAME=EMPPAY,PARENT=EMPLOYEE
        SENSEG NAME=EMPPAYHS,PARENT=EMPPAY
        SENSEG NAME=EMPDEP,PARENT=EMPLOYEE
        PSBGEN LANG=COBOL,PSBNAME=EMPLOYEE
        END
```

Generate this PSB, and then let's execute the program. Execution JCL will look something like this (yours will be whatever you use at your installation). Note that we **MUST** include DD statements for the IMS database and its overflow dataset. Also we include the input file.

```
//USER01D JOB MSGLEVEL=(1,1),NOTIFY=&SYSUID
//*
//* TO RUN A IMS COBOL PROGRAM
//*
//PLIB    JCLLIB ORDER=SYS1.IMS.PROCLIB
//RUN     EXEC IMSCOBGO,
//              MBR=COBIMS1,                    <= COBOL PROGRAM NAME
//              LOADLIB=USER01.IMS.LOADLIB,     <= LOAD LIBRARY
//              PSB=EMPPSBL,         <= PSB NAME
//              PSBLIB=USER01.IMS.PSBLIB,       <= PSB LIBRARY
//              DBDLIB=USER01.IMS.DBDLIB        <= DBD LIBRARY
//*
//** FLAT FILES IF ANY  *********************
//GO.EMPIFILE DD DSN=USER01.EMPIFILE,DISP=SHR
//*
//** IMS DATABASES (VSAM) *******************
```

```
//GO.EMPLOYEE DD DSN=USER01.IMS.EMPLOYEE.CLUSTER,DISP=SHR
//GO.EMPLFLW DD DSN=USER01.IMS.EMPLFLW.CLUSTER,DISP=SHR
//GO.SYSPRINT DD SYSOUT=*
//GO.SYSUDUMP DD SYSOUT=*
//GO.PLIDUMP DD SYSOUT=*
```

And here are the results of the run:

```
** PROGRAM COBIMS1 START **
PROCESSING IN P100-INITIALIZATION
PROCESSING IN P200-MAINLINE
SUCCESSFUL INSERT-REC:1111 VEREEN                   CHARLES
SUCCESSFUL INSERT-REC:1122 JENKINS                  DEBORAH
SUCCESSFUL INSERT-REC:3217 JOHNSON                  EDWARD
SUCCESSFUL INSERT-REC:4175 TURNBULL                 FRED
SUCCESSFUL INSERT-REC:4720 SCHULTZ                  TIM
SUCCESSFUL INSERT-REC:4836 SMITH                    SANDRA
SUCCESSFUL INSERT-REC:6288 WILLARD                  JOE
SUCCESSFUL INSERT-REC:7459 STEWART                  BETTY
SUCCESSFUL INSERT-REC:9134 FRANKLIN                 BRIANNA
PROCESSING IN P300-TERMINATION
** COBIMS1 - SUCCESSFULLY ENDED **
```

You can browse the IMS data using whatever tool you have such as File Manager IMS. Or you can simply browse the DATA file of the VSAM data set.

```
 Browse            USER01.IMS.EMPLOYEE.DATA                      Top of 9
 Command ===>                                                    Scroll PAGE
                         Type DATA      RBA                      Format CHAR
                                            Col 1
 ----+----10---+----2----+----3----+----4----+----5----+----6----+----7----+----
 ****  Top of data   ****
 ......1111 VEREEN                   CHARLES              12 2017-01-01 93
 ......1122 JENKINS                  DEBORAH              05 2017-01-01 43
 ......3217 JOHNSON                  EDWARD               04 2017-01-01 39
 ......4175 TURNBULL                 FRED                 01 2016-12-01 54
 ......4720 SCHULTZ                  TIM                  09 2017-01-01 65
 ......4836 SMITH                    SANDRA               03 2017-01-01 02
 ......6288 WILLARD                  JOE                  06 2016-01-01 20
 ......7459 STEWART                  BETTY                07 2016-07-31 01
 ......9134 FRANKLIN                 BRIANNA              00 2016-10-01 93
 ****  End of data   ****
```

Reading a Segment (GU)

Our next program will be named COBIMS2, and the purpose is simply to retrieve a record from the EMPLOYEE database. In this case, we want the record for employee 3217.

Our basic program structure will be similar to the load program except we will need to perform a Get Unique (GU) call, and we'll use a qualified SSA. Remember our qualified SSA structure looks like this:

```
01 EMP-QUALIFIED-SSA.
    05  SEGNAME     PIC X(08) VALUE 'EMPLOYEE'.
    05  FILLER      PIC X(01) VALUE '('.
    05  FIELD       PIC X(08) VALUE 'EMPID'.
    05  OPER        PIC X(02) VALUE ' ='.
    05  EMP-ID-VAL  PIC X(04) VALUE '    '.
    05  FILLER      PIC X(01) VALUE ')'.
```

So we must load the EMP-ID-VAL variable with character value '3217'. Our call will look like this.

```
CALL 'CBLTDLI' USING FOUR,
               DLI-FUNCGU,
               PCB-MASK,
               SEG-IO-AREA,
               EMP-QUALIFIED-SSA
```

Now we can code the entire program. We don't need a loop because we are retrieving a single record. So the program is quite simple. Note that we check for a blank status code after the IMS call, and we report an error if it is not blank.

```
IDENTIFICATION DIVISION.
PROGRAM-ID. COBIMS2.

*******************************************************
*   RETRIEVE A RECORD FROM IMS EMPLOYEE DATABASE      *
*******************************************************

ENVIRONMENT DIVISION.
DATA DIVISION.

*******************************************************
*  W O R K I N G   S T O R A G E   S E C T I O N    *
*******************************************************

WORKING-STORAGE SECTION.

01 SEG-IO-AREA     PIC X(80).

01 DLI-FUNCTIONS.
    05 DLI-FUNCISRT  PIC X(4) VALUE 'ISRT'.
    05 DLI-FUNCGU    PIC X(4) VALUE 'GU  '.
    05 DLI-FUNCGN    PIC X(4) VALUE 'GN  '.
    05 DLI-FUNCGHU   PIC X(4) VALUE 'GHU '.
    05 DLI-FUNCGNP   PIC X(4) VALUE 'GNP '.
    05 DLI-FUNCREPL  PIC X(4) VALUE 'REPL'.
    05 DLI-FUNCDLET  PIC X(4) VALUE 'DLET'.
```

```
          05 DLI-FUNCXRST  PIC X(4) VALUE 'XRST'.
          05 DLI-FUNCCKPT  PIC X(4) VALUE 'CKPT'.

       01 EMP-UNQUALIFIED-SSA.
          05  SEGNAME     PIC X(08) VALUE 'EMPLOYEE'.
          05  FILLER      PIC X(01) VALUE ' '.

       01 EMP-QUALIFIED-SSA.
          05  SEGNAME     PIC X(08) VALUE 'EMPLOYEE'.
          05  FILLER      PIC X(01) VALUE '('.
          05  FIELD       PIC X(08) VALUE 'EMPID'.
          05  OPER        PIC X(02) VALUE ' ='.
          05  EMP-ID-VAL  PIC X(04) VALUE '    '.
          05  FILLER      PIC X(01) VALUE ')'.

       01 IMS-RET-CODES.
          05 THREE           PIC S9(9) COMP VALUE +3.
          05 FOUR            PIC S9(9) COMP VALUE +4.
          05 FIVE            PIC S9(9) COMP VALUE +5.
          05 SIX             PIC S9(9) COMP VALUE +6.

       LINKAGE SECTION.
       01 PCB-MASK.
          03 DBD-NAME        PIC X(8).
          03 SEG-LEVEL       PIC XX.
          03 STATUS-CODE     PIC XX.
          03 PROC-OPT        PIC X(4).
          03 FILLER          PIC X(4).
          03 SEG-NAME        PIC X(8).
          03 KEY-FDBK        PIC S9(5) COMP.
          03 NUM-SENSEG      PIC S9(5) COMP.
          03 KEY-FDBK-AREA.
             05 EMPLOYEE-ID  PIC X(04).
             05 EMPPAYHS     PIC X(08).


       PROCEDURE DIVISION.

           INITIALIZE PCB-MASK
           ENTRY 'DLITCBL' USING PCB-MASK

           PERFORM P100-INITIALIZATION.
           PERFORM P200-MAINLINE.
           PERFORM P300-TERMINATION.
           GOBACK.

       P100-INITIALIZATION.

           DISPLAY '** PROGRAM COBIMS2 START **'
           DISPLAY 'PROCESSING IN P100-INITIALIZATION'.

       P200-MAINLINE.
```

```
            DISPLAY 'PROCESSING IN P200-MAINLINE'

            MOVE '3217' TO EMP-ID-VAL

            CALL 'CBLTDLI' USING FOUR,
                            DLI-FUNCGU,
                            PCB-MASK,
                            SEG-IO-AREA,
                            EMP-QUALIFIED-SSA

        IF STATUS-CODE = '  '
            DISPLAY 'SUCCESSFUL GET CALL  '
            DISPLAY 'SEG-IO-ARE : ' SEG-IO-AREA
        ELSE
            DISPLAY 'ERROR IN FETCH :' STATUS-CODE
            PERFORM P400-DISPLAY-ERROR
        END-IF.

    P300-TERMINATION.

        DISPLAY 'PROCESSING IN P300-TERMINATION'
        DISPLAY '** COBIMS2 - SUCCESSFULLY ENDED **'.

    P400-DISPLAY-ERROR.

        DISPLAY 'ERROR ENCOUNTERED - DETAIL FOLLOWS'
        DISPLAY 'SEG-IO-AREA      :' SEG-IO-AREA
        DISPLAY 'DBD-NAME1:'       DBD-NAME
        DISPLAY 'SEG-LEVEL1:'      SEG-LEVEL
        DISPLAY 'STATUS-CODE:'     STATUS-CODE
        DISPLAY 'PROC-OPT1 :'      PROC-OPT
        DISPLAY 'SEG-NAME1 :'      SEG-NAME
        DISPLAY 'KEY-FDBK1 :'      KEY-FDBK
        DISPLAY 'NUM-SENSEG1:'     NUM-SENSEG
        DISPLAY 'KEY-FDBK-AREA1:'  KEY-FDBK-AREA.

    *    END OF SOURCE CODE
```

Now compile, link and run the program. Here is the output showing that the data was successfully retrieved.

```
** PROGRAM COBIMS2 START **
PROCESSING IN P100-INITIALIZATION
PROCESSING IN P200-MAINLINE
SUCCESSFUL GET CALL
SEG-IO-ARE : 3217 JOHNSON           EDWARD             04 2017-01-01 397342007
PROCESSING IN P300-TERMINATION
** COBIMS2 - SUCCESSFULLY ENDED **
```

Also, we need to test a case where we try to retrieve an employee number which doesn't exist. Let's modify the program to look for EMP-ID 3218 which doesn't exist. Now recompile and re-execute the program. Here's the result:

```
** PROGRAM COBIMS2 START **
PROCESSING IN P100-INITIALIZATION
PROCESSING IN P200-MAINLINE
ERROR IN FETCH :GE
ERROR ENCOUNTERED - DETAIL FOLLOWS
SEG-IO-AREA    :
DBD-NAME1:EMPLOYEE
SEG-LEVEL1:00
STATUS-CODE:GE
PROC-OPT1 :AP
SEG-NAME1 :
KEY-FDBK1 :00000
NUM-SENSEG1:00004
KEY-FDBK-AREA1:
PROCESSING IN P300-TERMINATION
** COBIMS2 - SUCCESSFULLY ENDED **
```

Excellent, we captured and reported the error. IMS returned a GE return code which means the record was not found.

You'll use GU processing anytime you have a need to access data for a particular record in the database for read-only. Here we read all the root segments. Later we will read segments lower in the database hierarchy.

Reading a Database Sequentially (GN)

Our next program COBIMS3 will read the entire database sequentially. This scenario isn't unusual (a payroll program might process the database sequentially to generate pay checks) so you'll want to have a model of how to carry it out.

Basically we are going to create a loop that will walk through the database sequentially getting each EMPLOYEE segment using Get Next (GN) calls. We'll need a switch to indicate a stopping point which will be the end of the database (IMS status code GB). We'll also use an unqualified SSA since we don't need to know the key of each record to traverse the database. Here is the code.

```
        IDENTIFICATION DIVISION.
        PROGRAM-ID. COBIMS3.

        *********************************************************
        *  WALK THROUGH THE EMPLOYEE (ROOT) SEGMENTS OF      *
```

207

```
*  THE ENTIRE EMPLOYEE IMS DATABASE.                    *
**********************************************************

ENVIRONMENT DIVISION.
INPUT-OUTPUT SECTION.
DATA DIVISION.
**********************************************************
*  W O R K I N G   S T O R A G E   S E C T I O N    *
**********************************************************

WORKING-STORAGE SECTION.
 01 WS-FLAGS.
    05  SW-END-OF-DB-SWITCH     PIC X(1) VALUE 'N'.
        88  SW-END-OF-DB                 VALUE 'Y'.
        88  SW-NOT-END-OF-DB             VALUE 'N'.


 01 DLI-FUNCTIONS.
    05 DLI-FUNCISRT  PIC X(4) VALUE 'ISRT'.
    05 DLI-FUNCGU    PIC X(4) VALUE 'GU  '.
    05 DLI-FUNCGN    PIC X(4) VALUE 'GN  '.
    05 DLI-FUNCGHU   PIC X(4) VALUE 'GHU '.
    05 DLI-FUNCGNP   PIC X(4) VALUE 'GNP '.
    05 DLI-FUNCREPL  PIC X(4) VALUE 'REPL'.
    05 DLI-FUNCDLET  PIC X(4) VALUE 'DLET'.
    05 DLI-FUNCXRST  PIC X(4) VALUE 'XRST'.
    05 DLI-FUNCCKPT  PIC X(4) VALUE 'CKPT'.

 01 IN-EMPLOYEE-RECORD.
    05  EMPL-ID-IN   PIC X(04).
    05  FILLER       PIC X(01).
    05  EMPL-LNAME   PIC X(30).
    05  FILLER       PIC X(01).
    05  EMPL-FNAME   PIC X(20).
    05  FILLER       PIC X(01).
    05  EMPL-YRS-SRV PIC X(02).
    05  FILLER       PIC X(01).
    05  EMPL-PRM-DTE PIC X(10).
    05  FILLER       PIC X(10).

 01 EMP-UNQUALIFIED-SSA.
    05  SEGNAME      PIC X(08) VALUE 'EMPLOYEE'.
    05  FILLER       PIC X(01) VALUE ' '.

 01 EMP-QUALIFIED-SSA.
    05  SEGNAME      PIC X(08) VALUE 'EMPLOYEE'.
    05  FILLER       PIC X(01) VALUE '('.
    05  FIELD        PIC X(08) VALUE 'EMPID'.
    05  OPER         PIC X(02) VALUE ' ='.
    05  EMP-ID-VAL   PIC X(04) VALUE '    '.
    05  FILLER       PIC X(01) VALUE ')'.
```

208

```cobol
01 SEG-IO-AREA      PIC X(80).

01 IMS-RET-CODES.
   05 THREE              PIC S9(9) COMP VALUE +3.
   05 FOUR               PIC S9(9) COMP VALUE +4.
   05 FIVE               PIC S9(9) COMP VALUE +5.
   05 SIX                PIC S9(9) COMP VALUE +6.

LINKAGE SECTION.
 01 PCB-MASK.
    03 DBD-NAME          PIC X(8).
    03 SEG-LEVEL         PIC XX.
    03 STATUS-CODE       PIC XX.
    03 PROC-OPT          PIC X(4).
    03 FILLER            PIC X(4).
    03 SEG-NAME          PIC X(8).
    03 KEY-FDBK          PIC S9(5) COMP.
    03 NUM-SENSEG        PIC S9(5) COMP.
    03 KEY-FDBK-AREA.
       05 EMPLOYEE-KEY  PIC X(04).
       05 EMPPAYHS-KEY  PIC X(08).

PROCEDURE DIVISION.

    INITIALIZE PCB-MASK
    ENTRY 'DLITCBL' USING PCB-MASK

    PERFORM P100-INITIALIZATION.
    PERFORM P200-MAINLINE.
    PERFORM P300-TERMINATION.
    GOBACK.

 P100-INITIALIZATION.

    DISPLAY '** PROGRAM COBIMS3 START **'
    DISPLAY 'PROCESSING IN P100-INITIALIZATION'.

*    DO INITIAL DB READ FOR FIRST EMPLOYEE RECORD

    CALL 'CBLTDLI' USING FOUR,
         DLI-FUNCGN,
         PCB-MASK,
         SEG-IO-AREA,
         EMP-UNQUALIFIED-SSA

    IF STATUS-CODE = '  ' THEN
       NEXT SENTENCE
    ELSE
       IF STATUS-CODE = 'GB' THEN
          SET SW-END-OF-DB TO TRUE
          DISPLAY 'END OF DATABASE :'
       ELSE
```

209

```
            PERFORM P400-DISPLAY-ERROR
            GOBACK
        END-IF

    END-IF.

P200-MAINLINE.

    DISPLAY 'PROCESSING IN P200-MAINLINE'

*       CHECK STATUS CODE AND FIRST RECORD

    IF SW-END-OF-DB THEN
        DISPLAY 'NO RECORDS TO PROCESS!!'
    ELSE
        PERFORM UNTIL SW-END-OF-DB
            DISPLAY 'SUCCESSFUL READ :' SEG-IO-AREA

            CALL 'CBLTDLI' USING FOUR,
                DLI-FUNCGN,
                PCB-MASK,
                SEG-IO-AREA,
                EMP-UNQUALIFIED-SSA

            IF STATUS-CODE = 'GB' THEN
                SET SW-END-OF-DB TO TRUE
                DISPLAY 'END OF DATABASE'
            ELSE
                IF STATUS-CODE NOT EQUAL SPACES THEN
                    PERFORM P400-DISPLAY-ERROR
                    GOBACK
                END-IF
            END-IF

        END-PERFORM.

    DISPLAY 'FINISHED PROCESSING IN P200-MAINLINE'.

P300-TERMINATION.

    DISPLAY 'PROCESSING IN P300-TERMINATION'
    DISPLAY '** COBIMS3 - SUCCESSFULLY ENDED **'.

P400-DISPLAY-ERROR.

    DISPLAY 'ERROR ENCOUNTERED - DETAIL FOLLOWS'
    DISPLAY 'SEG-IO-AREA      :' SEG-IO-AREA
    DISPLAY 'DBD-NAME1:'      DBD-NAME
    DISPLAY 'SEG-LEVEL1:'     SEG-LEVEL
    DISPLAY 'STATUS-CODE:'    STATUS-CODE
    DISPLAY 'PROC-OPT1 :'     PROC-OPT
    DISPLAY 'SEG-NAME1 :'     SEG-NAME
```

```
        DISPLAY 'KEY-FDBK1 :'     KEY-FDBK
        DISPLAY 'NUM-SENSEG1:'    NUM-SENSEG
        DISPLAY 'KEY-FDBK-AREA1:' KEY-FDBK-AREA.

*     END OF SOURCE CODE
```

Now let's compile and link, and then execute COBIMS3. Here's the run output.

```
** PROGRAM COBIMS3 START **
PROCESSING IN P100-INITIALIZATION
PROCESSING IN P200-MAINLINE
SUCCESSFUL READ :1111 VEREEN              CHARLES         12 201
SUCCESSFUL READ :1122 JENKINS             DEBORAH         05 201
SUCCESSFUL READ :3217 JOHNSON             EDWARD          04 201
SUCCESSFUL READ :4175 TURNBULL            FRED            01 201
SUCCESSFUL READ :4720 SCHULTZ             TIM             09 201
SUCCESSFUL READ :4836 SMITH               SANDRA          03 201
SUCCESSFUL READ :6288 WILLARD             JOE             06 201
SUCCESSFUL READ :7459 STEWART             BETTY           07 201
SUCCESSFUL READ :9134 FRANKLIN            BRIANNA         00 201
END OF DATABASE
FINISHED PROCESSING IN P200-MAINLINE
PROCESSING IN P300-TERMINATION
** COBIMS3 - SUCCESSFULLY ENDED **
```

You now have a model for any kind of sequential processing you want to do on root segments. Processing child segments is a bit more involved, but not much. We'll show an example of that later.

Updating a Segment (GHU/REPL)

In COBIMS4 we will update a record. Updating (either changing or deleting a record) always involves two steps in IMS. You must first get and lock the record you are operating on so that no other process can make updates to it. Second you issue either a REPL or DLET call.

A Get Hold Unique (GHU) call prevents any other process from making modifications to the record until you are finished with it. Similar calls are Get Hold Next (GHN) and Get Hold Next in Parent (GHNP).

For this example, let's change the promotion date on employee 9134 to Sept 1, 2016. To do that we need a GHU call with a qualified SSA that we have loaded with the employee id value of 9134.

```
        MOVE '9134' TO EMP-ID-VAL
```

Here is the GHU call, and notice we are using IO-EMPLOYEE-RECORD as our segment I/O area. This is because it has the full record layout with all the fields which makes it easy to change the

211

promotion date by field assignment.

```
CALL 'CBLTDLI' USING FOUR,
          DLI-FUNCGHU,
          PCB-MASK,
          IO-EMPLOYEE-RECORD,
          EMP-QUALIFIED-SSA
```

Once you've done the GHU call you can change the value of the promotion date.

```
MOVE '2016-09-01' TO EMPL-PRM-DTE
```

Finally you issue the REPL call. A REPL does not use any SSA since the record is already held in memory. It simply uses the segment I/O area to perform the update to the database. So you only have three parameters.

```
CALL 'CBLTDLI' USING THREE,
          DLI-FUNCREPL,
          PCB-MASK,
          IO-EMPLOYEE-RECORD
```

Here is the entire program listing for COBIMS4.

```
      IDENTIFICATION DIVISION.
      PROGRAM-ID. COBIMS4.

     ******************************************************
     *      UPDATE A RECORD FROM IMS EMPLOYEE DATABASE     *
     ******************************************************

      ENVIRONMENT DIVISION.
      DATA DIVISION.

     ******************************************************
     * W O R K I N G   S T O R A G E   S E C T I O N      *
     ******************************************************

      WORKING-STORAGE SECTION.

      01 SEG-IO-AREA      PIC X(80).

      01 IO-EMPLOYEE-RECORD.
          05  EMPL-ID-IN    PIC X(04).
          05  FILLER        PIC X(01).
          05  EMPL-LNAME    PIC X(30).
          05  FILLER        PIC X(01).
          05  EMPL-FNAME    PIC X(20).
          05  FILLER        PIC X(01).
```

```
    05  EMPL-YRS-SRV  PIC X(02).
    05  FILLER        PIC X(01).
    05  EMPL-PRM-DTE  PIC X(10).
    05  FILLER        PIC X(10).


01 DLI-FUNCTIONS.
    05 DLI-FUNCISRT  PIC X(4) VALUE 'ISRT'.
    05 DLI-FUNCGU    PIC X(4) VALUE 'GU  '.
    05 DLI-FUNCGN    PIC X(4) VALUE 'GN  '.
    05 DLI-FUNCGHU   PIC X(4) VALUE 'GHU '.
    05 DLI-FUNCGNP   PIC X(4) VALUE 'GNP '.
    05 DLI-FUNCREPL  PIC X(4) VALUE 'REPL'.
    05 DLI-FUNCDLET  PIC X(4) VALUE 'DLET'.
    05 DLI-FUNCXRST  PIC X(4) VALUE 'XRST'.
    05 DLI-FUNCCKPT  PIC X(4) VALUE 'CKPT'.


 01 EMP-UNQUALIFIED-SSA.
    05  SEGNAME     PIC X(08) VALUE 'EMPLOYEE'.
    05  FILLER      PIC X(01) VALUE ' '.


 01 EMP-QUALIFIED-SSA.
    05  SEGNAME     PIC X(08) VALUE 'EMPLOYEE'.
    05  FILLER      PIC X(01) VALUE '('.
    05  FIELD       PIC X(08) VALUE 'EMPID'.
    05  OPER        PIC X(02) VALUE ' ='.
    05  EMP-ID-VAL  PIC X(04) VALUE '    '.
    05  FILLER      PIC X(01) VALUE ')'.


 01 IMS-RET-CODES.
    05 THREE         PIC S9(9) COMP VALUE +3.
    05 FOUR          PIC S9(9) COMP VALUE +4.
    05 FIVE          PIC S9(9) COMP VALUE +5.
    05 SIX           PIC S9(9) COMP VALUE +6.

LINKAGE SECTION.
 01 PCB-MASK.
    03 DBD-NAME       PIC X(8).
    03 SEG-LEVEL      PIC XX.
    03 STATUS-CODE    PIC XX.
    03 PROC-OPT       PIC X(4).
    03 FILLER         PIC X(4).
    03 SEG-NAME       PIC X(8).
    03 KEY-FDBK       PIC S9(5) COMP.
    03 NUM-SENSEG     PIC S9(5) COMP.
    03 KEY-FDBK-AREA.
       05 EMPLOYEE-ID  PIC X(04).
       05 EMPPAYHS     PIC X(08).

PROCEDURE DIVISION.

    INITIALIZE PCB-MASK
    ENTRY 'DLITCBL' USING PCB-MASK
```

```
            PERFORM P100-INITIALIZATION.
            PERFORM P200-MAINLINE.
            PERFORM P300-TERMINATION.
            GOBACK.

        P100-INITIALIZATION.

            DISPLAY '** PROGRAM COBIMS4 START **'
            DISPLAY 'PROCESSING IN P100-INITIALIZATION'.

        P200-MAINLINE.

            DISPLAY 'PROCESSING IN P200-MAINLINE'
            MOVE '9134' TO EMP-ID-VAL

    *     AQCUIRE THE SEGMENT WITH HOLD

            CALL 'CBLTDLI' USING FOUR,
                          DLI-FUNCGHU,
                          PCB-MASK,
                          IO-EMPLOYEE-RECORD,
                          EMP-QUALIFIED-SSA

            IF STATUS-CODE = ' '
               DISPLAY 'SUCCESSFUL GET HOLD CALL  '
               DISPLAY 'IO-EMPLOYEE-RECORD : ' IO-EMPLOYEE-RECORD

    *     NOW MAKE THE CHANGE AND REPLACE THE SEGMENT

               MOVE '2016-09-01' TO EMPL-PRM-DTE

               CALL 'CBLTDLI' USING THREE,
                             DLI-FUNCREPL,
                             PCB-MASK,
                             IO-EMPLOYEE-RECORD

            IF STATUS-CODE = ' '
               DISPLAY 'SUCCESSFUL REPLACEMENT '
               DISPLAY 'IO-EMPLOYEE-RECORD : ' IO-EMPLOYEE-RECORD
            ELSE
               DISPLAY 'ERROR IN REPLACE :' STATUS-CODE
               PERFORM P400-DISPLAY-ERROR
            END-IF

            ELSE
               DISPLAY 'ERROR IN GET HOLD :' STATUS-CODE
               PERFORM P400-DISPLAY-ERROR
            END-IF.

        P300-TERMINATION.
```

```
               DISPLAY 'PROCESSING IN P300-TERMINATION'
               DISPLAY '** COBIMS4 - SUCCESSFULLY ENDED **'.

         P400-DISPLAY-ERROR.

               DISPLAY 'ERROR ENCOUNTERED - DETAIL FOLLOWS'
               DISPLAY 'SEG-IO-AREA       :' SEG-IO-AREA
               DISPLAY 'DBD-NAME1:'       DBD-NAME
               DISPLAY 'SEG-LEVEL1:'      SEG-LEVEL
               DISPLAY 'STATUS-CODE:'     STATUS-CODE
               DISPLAY 'PROC-OPT1 :'      PROC-OPT
               DISPLAY 'SEG-NAME1 :'      SEG-NAME
               DISPLAY 'KEY-FDBK1 :'      KEY-FDBK
               DISPLAY 'NUM-SENSEG1:'     NUM-SENSEG
               DISPLAY 'KEY-FDBK-AREA1:' KEY-FDBK-AREA.


    *    END OF SOURCE CODE
```

Now let's compile and link, and then run the program.

```
** PROGRAM COBIMS4 START **
PROCESSING IN P100-INITIALIZATION
PROCESSING IN P200-MAINLINE
SUCCESSFUL GET HOLD CALL
IO-EMPLOYEE-RECORD : 9134 FRANKLIN        BRIANNA        00  2016-10-01 937293598
SUCCESSFUL REPLACEMENT
IO-EMPLOYEE-RECORD : 9134 FRANKLIN        BRIANNA        00 2016-09-01 937293598
PROCESSING IN P300-TERMINATION
** COBIMS4 - SUCCESSFULLY ENDED **
```

This is the basic model for doing updates to a database segment.

Deleting a Segment (GHU/DLET)

For COBIMS5 we are going to delete a record. Basically the code is exactly the same as for COBIMS4 except we are deleting instead of updating a record. Let's delete employee 9134, the one we just updated. You can simply copy the COBIMS4 code and make modifications to turn it into a delete program.

Here's the source code.

```
     IDENTIFICATION DIVISION.
     PROGRAM-ID. COBIMS5.
     ****************************************************
     *     DELETE A RECORD FROM IMS EMPLOYEE DATABASE    *
     ****************************************************
     ENVIRONMENT DIVISION.
     DATA DIVISION.
```

215

```
**************************************************
*  W O R K I N G   S T O R A G E   S E C T I O N   *
**************************************************
WORKING-STORAGE SECTION.

01 SEG-IO-AREA      PIC X(80).
01 IO-EMPLOYEE-RECORD.
    05  EMPL-ID-IN    PIC X(04).
    05  FILLER        PIC X(01).
    05  EMPL-LNAME    PIC X(30).
    05  FILLER        PIC X(01).
    05  EMPL-FNAME    PIC X(20).
    05  FILLER        PIC X(01).
    05  EMPL-YRS-SRV  PIC X(02).
    05  FILLER        PIC X(01).
    05  EMPL-PRM-DTE  PIC X(10).
    05  FILLER        PIC X(10).


01 DLI-FUNCTIONS.
    05 DLI-FUNCISRT  PIC X(4) VALUE 'ISRT'.
    05 DLI-FUNCGU    PIC X(4) VALUE 'GU  '.
    05 DLI-FUNCGN    PIC X(4) VALUE 'GN  '.
    05 DLI-FUNCGHU   PIC X(4) VALUE 'GHU '.
    05 DLI-FUNCGNP   PIC X(4) VALUE 'GNP '.
    05 DLI-FUNCREPL  PIC X(4) VALUE 'REPL'.
    05 DLI-FUNCDLET  PIC X(4) VALUE 'DLET'.
    05 DLI-FUNCXRST  PIC X(4) VALUE 'XRST'.
    05 DLI-FUNCCKPT  PIC X(4) VALUE 'CKPT'.


 01 EMP-UNQUALIFIED-SSA.
    05  SEGNAME     PIC X(08) VALUE 'EMPLOYEE'.
    05  FILLER      PIC X(01) VALUE ' '.


 01 EMP-QUALIFIED-SSA.
    05  SEGNAME     PIC X(08) VALUE 'EMPLOYEE'.
    05  FILLER      PIC X(01) VALUE '('.
    05  FIELD       PIC X(08) VALUE 'EMPID'.
    05  OPER        PIC X(02) VALUE ' ='.
    05  EMP-ID-VAL  PIC X(04) VALUE '    '.
    05  FILLER      PIC X(01) VALUE ')'.


 01 IMS-RET-CODES.
    05 THREE         PIC S9(9) COMP VALUE +3.
    05 FOUR          PIC S9(9) COMP VALUE +4.
    05 FIVE          PIC S9(9) COMP VALUE +5.
    05 SIX           PIC S9(9) COMP VALUE +6.


LINKAGE SECTION.
 01 PCB-MASK.
    03 DBD-NAME      PIC X(8).
    03 SEG-LEVEL     PIC XX.
```

```cobol
    03 STATUS-CODE      PIC XX.
    03 PROC-OPT         PIC X(4).
    03 FILLER           PIC X(4).
    03 SEG-NAME         PIC X(8).
    03 KEY-FDBK         PIC S9(5) COMP.
    03 NUM-SENSEG       PIC S9(5) COMP.
    03 KEY-FDBK-AREA.
       05 EMPLOYEE-ID   PIC X(04).
       05 EMPPAYHS      PIC X(08).

PROCEDURE DIVISION.

    INITIALIZE PCB-MASK
    ENTRY 'DLITCBL' USING PCB-MASK

    PERFORM P100-INITIALIZATION.
    PERFORM P200-MAINLINE.
    PERFORM P300-TERMINATION.
    GOBACK.

P100-INITIALIZATION.

    DISPLAY '** PROGRAM COBIMS5 START **'
    DISPLAY 'PROCESSING IN P100-INITIALIZATION'.

P200-MAINLINE.

    DISPLAY 'PROCESSING IN P200-MAINLINE'
    MOVE '9134' TO EMP-ID-VAL

*   AQCUIRE THE SEGMENT WITH HOLD

    CALL 'CBLTDLI' USING FOUR,
                  DLI-FUNCGHU,
                  PCB-MASK,
                  IO-EMPLOYEE-RECORD,
                  EMP-QUALIFIED-SSA

    IF STATUS-CODE = '  '
       DISPLAY 'SUCCESSFUL GET HOLD CALL  '
       DISPLAY 'IO-EMPLOYEE-RECORD : ' IO-EMPLOYEE-RECORD

*   NOW DELETE THE SEGMENT

    CALL 'CBLTDLI' USING THREE,
                  DLI-FUNCDLET,
                  PCB-MASK,
                  IO-EMPLOYEE-RECORD

    IF STATUS-CODE = '  '
       DISPLAY 'SUCCESSFUL DELETION OF ' EMP-ID-VAL
    ELSE
```

217

```
                   DISPLAY 'ERROR IN DELETE :' STATUS-CODE
                   PERFORM P400-DISPLAY-ERROR
              END-IF

           ELSE
              DISPLAY 'ERROR IN GET HOLD :' STATUS-CODE
              PERFORM P400-DISPLAY-ERROR
           END-IF.

       P300-TERMINATION.

           DISPLAY 'PROCESSING IN P300-TERMINATION'
           DISPLAY '** COBIMS5 - SUCCESSFULLY ENDED **'.

       P400-DISPLAY-ERROR.

           DISPLAY 'ERROR ENCOUNTERED - DETAIL FOLLOWS'
           DISPLAY 'SEG-IO-AREA       :' SEG-IO-AREA
           DISPLAY 'DBD-NAME1:'       DBD-NAME
           DISPLAY 'SEG-LEVEL1:'      SEG-LEVEL
           DISPLAY 'STATUS-CODE:'     STATUS-CODE
           DISPLAY 'PROC-OPT1 :'      PROC-OPT
           DISPLAY 'SEG-NAME1 :'      SEG-NAME
           DISPLAY 'KEY-FDBK1 :'      KEY-FDBK
           DISPLAY 'NUM-SENSEG1:'     NUM-SENSEG
           DISPLAY 'KEY-FDBK-AREA1:' KEY-FDBK-AREA.

      *    END OF SOURCE CODE
```

Now compile, link and run:

```
** PROGRAM COBIMS5 START **
PROCESSING IN P100-INITIALIZATION
PROCESSING IN P200-MAINLINE
SUCCESSFUL GET HOLD CALL
IO-EMPLOYEE-RECORD : 9134 FRANKLIN        BRIANNA            00 2016-09-01 937293598
SUCCESSFUL DELETION OF 9134
PROCESSING IN P300-TERMINATION
** COBIMS5 - SUCCESSFULLY ENDED **
```

As you can see, the record was deleted.

Inserting Child Segments

So far we've only dealt with root segments. That was pretty straightforward. Now let's introduce child segments. In COBIMS6 we are going to create an EMPPAY segment under each EM-
PLOYEE root segment. This will be similar to how we inserted root segments except we need
to specify which root segment to insert the child segment under.

First, let's look at our input file:

```
----+----1----+----2----+----3----+----4----+----5
******************************* Top of Data ****
1111    8700000  670000  362500  20170101
1122    8200000  600000  341666  20170101
3217    6500000  550000  270833  20170101
4175    5500000  150000  229166  20170101
4720    8000000  250000  333333  20170101
4836    6200000  220000  258333  20170101
6288    7000000  200000  291666  20170101
7459    8500000  450000  354166  20170101
9134    7500000  250000  312500  20170101
```

To decrypt here a little, the file above contains employee id numbers with annual salary, annual bonus pay, twice-per-month paycheck dollar amount, and the effective date for all this information. Let's create a record structure in COBOL for this file.

```
01  IN-EMPPAY-RECORD.
    05   EMP-ID-IN      PIC X(04).
    05   FILLER         PIC X(05).
    05   REG-PAY-IN     PIC 99999V99.
    05   FILLER         PIC X(02).
    05   BON-PAY-IN     PIC 9999V99.
    05   FILLER         PIC X(02).
    05   SEMIMTH-IN     PIC 9999V99.
    05   FILLER         PIC X(02).
    05   EFF-DATE-IN    PIC X(08).
    05   FILLER         PIC X(38).
```

We'll also need an IMS I/O area for the EMPPAY segment. How about this one? We'll map data from the input record into this I/O area before we do the ISRT action. Note that we are using packed data fields for the IMS segment. This will save some space.

```
01  IO-EMPPAY-RECORD.
    05   PAY-EFF-DATE   PIC X(8).
    05   PAY-REG-PAY    PIC S9(6)V9(2) USAGE COMP-3.
    05   PAY-BON-PAY    PIC S9(6)V9(2) USAGE COMP-3.
    05   SEMIMTH-PAY    PIC S9(6)V9(2) USAGE COMP-3.
    05   FILLER         PIC X(57).
```

Finally, we need our SSA structures. We'll be using the unqualified EMPPAY SSA, but we'll go ahead and add both the qualified and unqualified SSAs to the program.

```
01  EMPPAY-UNQUALIFIED-SSA.
    05   SEGNAME    PIC X(08) VALUE 'EMPPAY  '.
    05   FILLER     PIC X(01) VALUE ' '.

01  EMPPAY-QUALIFIED-SSA.
    05   SEGNAME    PIC X(08) VALUE 'EMPPAY  '.
    05   FILLER     PIC X(01) VALUE '('.
    05   FIELD      PIC X(08) VALUE 'EFFDATE '.
```

```
        05  OPER        PIC X(02) VALUE ' ='.
        05  EFFDATE-VAL PIC X(08) VALUE '        '.
        05  FILLER      PIC X(01) VALUE ')'.
```

So given this information, our ISRT call should look like this. Notice that we use a qualified SSA for the EMPLOYEE root segment, and an unqualified SSA for the EMPPAY segment.

```
    CALL 'CBLTDLI' USING FIVE,
            DLI-FUNCISRT,
            PCB-MASK,
            IO-EMPPAY-RECORD,
            EMP-QUALIFIED-SSA
            EMPPAY-UNQUALIFIED-SSA
```

Of course we will need a loop for reading the input pay file, and we'll need code to map the input fields to the EMPPAY fields. And we must move the employee id on the input file to the EMPLOYEE qualified SSA. Finally, recall that we deleted employee 9134, but there is a record in the input file for 9134. Have we coded to handle this missing root? We'll soon see.

Here is our completed code for COBIMS6.

```
        IDENTIFICATION DIVISION.
        PROGRAM-ID. COBIMS6.

        *******************************************************
        *    INSERT EMPLOYEE PAY RECORDS INTO THE EMPLOYEE   *
        *    IMS DATABASE. ROOT KEY MUST BE SPECIFIED.       *
        *******************************************************

        ENVIRONMENT DIVISION.
        INPUT-OUTPUT SECTION.

            FILE-CONTROL.
                SELECT EMPPAY-IN-FILE   ASSIGN TO EMPPAYFL.

        DATA DIVISION.

        FILE SECTION.
        FD EMPPAY-IN-FILE
            RECORDING MODE IS F
            RECORD CONTAINS 80 CHARACTERS
            DATA RECORD IS IN-EMPPAY-RECORD.

          01 IN-EMPPAY-RECORD.
            05  EMP-ID-IN   PIC X(04).
            05  FILLER      PIC X(05).
            05  REG-PAY-IN  PIC 99999V99.
            05  FILLER      PIC X(02).
```

```
      05  BON-PAY-IN    PIC 9999V99.
      05  FILLER        PIC X(02).
      05  SEMIMTH-IN    PIC 9999V99.
      05  FILLER        PIC X(02).
      05  EFF-DATE-IN   PIC X(08).
      05  FILLER        PIC X(38).


***********************************************
*  W O R K I N G   S T O R A G E   S E C T I O N    *
***********************************************

 WORKING-STORAGE SECTION.

 01 WS-FLAGS.
    05  SW-END-OF-FILE-SWITCH   PIC X(1) VALUE 'N'.
        88  SW-END-OF-FILE                VALUE 'Y'.
        88  SW-NOT-END-OF-FILE            VALUE 'N'.


 01 IO-EMPLOYEE-RECORD.
    05  EMPL-ID-IN    PIC X(04).
    05  FILLER        PIC X(01).
    05  EMPL-LNAME    PIC X(30).
    05  FILLER        PIC X(01).
    05  EMPL-FNAME    PIC X(20).
    05  FILLER        PIC X(01).
    05  EMPL-YRS-SRV  PIC X(02).
    05  FILLER        PIC X(01).
    05  EMPL-PRM-DTE  PIC X(10).
    05  FILLER        PIC X(10).


 01 IO-EMPPAY-RECORD.
    05  PAY-EFF-DATE  PIC X(8).
    05  PAY-REG-PAY   PIC S9(6)V9(2) USAGE COMP-3.
    05  PAY-BON-PAY   PIC S9(6)V9(2) USAGE COMP-3.
    05  SEMIMTH-PAY   PIC S9(6)V9(2) USAGE COMP-3.
    05  FILLER        PIC X(57).


 01 SEG-IO-AREA    PIC X(80).

 01 DLI-FUNCTIONS.
    05 DLI-FUNCISRT PIC X(4) VALUE 'ISRT'.
    05 DLI-FUNCGU   PIC X(4) VALUE 'GU  '.
    05 DLI-FUNCGN   PIC X(4) VALUE 'GN  '.
    05 DLI-FUNCGHU  PIC X(4) VALUE 'GHU '.
    05 DLI-FUNCGNP  PIC X(4) VALUE 'GNP '.
    05 DLI-FUNCREPL PIC X(4) VALUE 'REPL'.
    05 DLI-FUNCDLET PIC X(4) VALUE 'DLET'.
    05 DLI-FUNCXRST PIC X(4) VALUE 'XRST'.
    05 DLI-FUNCCKPT PIC X(4) VALUE 'CKPT'.

  01 EMP-UNQUALIFIED-SSA.
     05  SEGNAME     PIC X(08) VALUE 'EMPLOYEE'.
```

```
        05  FILLER       PIC X(01) VALUE ' '.

    01 EMP-QUALIFIED-SSA.
        05  SEGNAME      PIC X(08) VALUE 'EMPLOYEE'.
        05  FILLER       PIC X(01) VALUE '('.
        05  FIELD        PIC X(08) VALUE 'EMPID'.
        05  OPER         PIC X(02) VALUE ' ='.
        05  EMP-ID-VAL   PIC X(04) VALUE '    '.
        05  FILLER       PIC X(01) VALUE ')'.

    01 EMPPAY-UNQUALIFIED-SSA.
        05  SEGNAME      PIC X(08) VALUE 'EMPPAY  '.
        05  FILLER       PIC X(01) VALUE ' '.

    01 EMPPAY-QUALIFIED-SSA.
        05  SEGNAME      PIC X(08) VALUE 'EMPPAY  '.
        05  FILLER       PIC X(01) VALUE '('.
        05  FIELD        PIC X(08) VALUE 'EFFDATE '.
        05  OPER         PIC X(02) VALUE ' ='.
        05  EFFDATE-VAL  PIC X(08) VALUE '        '.
        05  FILLER       PIC X(01) VALUE ')'.

    01 IMS-RET-CODES.
        05 THREE         PIC S9(9) COMP VALUE +3.
        05 FOUR          PIC S9(9) COMP VALUE +4.
        05 FIVE          PIC S9(9) COMP VALUE +5.
        05 SIX           PIC S9(9) COMP VALUE +6.


LINKAGE SECTION.
    01 PCB-MASK.
        03 DBD-NAME      PIC X(8).
        03 SEG-LEVEL     PIC XX.
        03 STATUS-CODE   PIC XX.
        03 PROC-OPT      PIC X(4).
        03 FILLER        PIC X(4).
        03 SEG-NAME      PIC X(8).
        03 KEY-FDBK      PIC S9(5) COMP.
        03 NUM-SENSEG    PIC S9(5) COMP.
        03 KEY-FDBK-AREA.
           05 EMPLOYEE-ID  PIC X(04).
           05 EMPPAYHS     PIC X(08).

PROCEDURE DIVISION.

    INITIALIZE PCB-MASK
    ENTRY 'DLITCBL' USING PCB-MASK

    PERFORM P100-INITIALIZATION.
    PERFORM P200-MAINLINE.
    PERFORM P300-TERMINATION.
    GOBACK.
```

```
P100-INITIALIZATION.

    DISPLAY '** PROGRAM COBIMS6 START **'
    DISPLAY 'PROCESSING IN P100-INITIALIZATION'
    OPEN INPUT EMPPAY-IN-FILE.

P200-MAINLINE.

    DISPLAY 'PROCESSING IN P200-MAINLINE'

    READ EMPPAY-IN-FILE
       AT END SET SW-END-OF-FILE TO TRUE
    END-READ

    PERFORM UNTIL SW-END-OF-FILE

        DISPLAY 'MAPPING FIELDS FOR EMPLOYEE ' EMP-ID-IN
        DISPLAY 'EFF-DATE-IN ' EFF-DATE-IN
        DISPLAY 'REG-PAY-IN  ' REG-PAY-IN
        DISPLAY 'BON-PAY-IN  ' BON-PAY-IN
        DISPLAY 'SEMIMTH-IN  ' SEMIMTH-IN
        MOVE EMP-ID-IN    TO EMP-ID-VAL
        MOVE EFF-DATE-IN  TO PAY-EFF-DATE
        MOVE REG-PAY-IN   TO PAY-REG-PAY
        MOVE BON-PAY-IN   TO PAY-BON-PAY
        MOVE SEMIMTH-IN   TO SEMIMTH-PAY

        CALL 'CBLTDLI' USING FIVE,
             DLI-FUNCISRT,
             PCB-MASK,
             IO-EMPPAY-RECORD,
             EMP-QUALIFIED-SSA
             EMPPAY-UNQUALIFIED-SSA

        IF STATUS-CODE = '  '
           DISPLAY 'SUCCESSFUL INSERT-REC FOR EMP: ' EMP-ID-VAL
           DISPLAY 'SUCCESSFUL INSERT-REC VALUES : '
           IO-EMPPAY-RECORD
        ELSE
           PERFORM P400-DISPLAY-ERROR
        END-IF

        READ EMPPAY-IN-FILE
           AT END SET SW-END-OF-FILE TO TRUE
        END-READ

    END-PERFORM.

P300-TERMINATION.

    DISPLAY 'PROCESSING IN P300-TERMINATION'
```

```
            CLOSE EMPPAY-IN-FILE.
            DISPLAY '** COBIMS6 - SUCCESSFULLY ENDED **'.

        P400-DISPLAY-ERROR.

            DISPLAY 'ERROR ENCOUNTERED - DETAIL FOLLOWS'
            DISPLAY 'DBD-NAME1:'      DBD-NAME
            DISPLAY 'SEG-LEVEL1:'     SEG-LEVEL
            DISPLAY 'STATUS-CODE:'    STATUS-CODE
            DISPLAY 'PROC-OPT1 :'     PROC-OPT
            DISPLAY 'SEG-NAME1 :'     SEG-NAME
            DISPLAY 'KEY-FDBK1 :'     KEY-FDBK
            DISPLAY 'NUM-SENSEG1:'    NUM-SENSEG
            DISPLAY 'KEY-FDBK-AREA1:' KEY-FDBK-AREA.

    *    END OF SOURCE CODE
```

Compile and link, and then run the program. Here is the output.

```
** PROGRAM COBIMS6 START **
PROCESSING IN P100-INITIALIZATION
PROCESSING IN P200-MAINLINE
MAPPING FIELDS FOR EMPLOYEE 1111
EFF-DATE-IN 20170101
REG-PAY-IN  8700000
BON-PAY-IN  670000
SEMIMTH-IN  362500
SUCCESSFUL INSERT-REC FOR EMP: 1111
SUCCESSFUL INSERT-REC VALUES : 20170101 g            &
MAPPING FIELDS FOR EMPLOYEE 1122
EFF-DATE-IN 20170101
REG-PAY-IN  8200000
BON-PAY-IN  600000
SEMIMTH-IN  341666
SUCCESSFUL INSERT-REC FOR EMP: 1122
SUCCESSFUL INSERT-REC VALUES : 20170101 b            %
MAPPING FIELDS FOR EMPLOYEE 3217
EFF-DATE-IN 20170101
REG-PAY-IN  6500000
BON-PAY-IN  550000
SEMIMTH-IN  270833
SUCCESSFUL INSERT-REC FOR EMP: 3217
SUCCESSFUL INSERT-REC VALUES : 20170101      &     c
MAPPING FIELDS FOR EMPLOYEE 4175
EFF-DATE-IN 20170101
REG-PAY-IN  5500000
BON-PAY-IN  150000
SEMIMTH-IN  229166
SUCCESSFUL INSERT-REC FOR EMP: 4175
SUCCESSFUL INSERT-REC VALUES : 20170101      &     %
MAPPING FIELDS FOR EMPLOYEE 4720
```

```
EFF-DATE-IN 20170101
REG-PAY-IN  8000000
BON-PAY-IN  250000
SEMIMTH-IN  333333
SUCCESSFUL INSERT-REC FOR EMP: 4720
SUCCESSFUL INSERT-REC VALUES : 20170101         &
MAPPING FIELDS FOR EMPLOYEE 4836
EFF-DATE-IN 20170101
REG-PAY-IN  6200000
BON-PAY-IN  220000
SEMIMTH-IN  258333
SUCCESSFUL INSERT-REC FOR EMP: 4836
SUCCESSFUL INSERT-REC VALUES : 20170101
MAPPING FIELDS FOR EMPLOYEE 6288
EFF-DATE-IN 20170101
REG-PAY-IN  7000000
BON-PAY-IN  200000
SEMIMTH-IN  291666
SUCCESSFUL INSERT-REC FOR EMP: 6288
SUCCESSFUL INSERT-REC VALUES : 20170101             j %
MAPPING FIELDS FOR EMPLOYEE 7459
EFF-DATE-IN 20170101
REG-PAY-IN  8500000
BON-PAY-IN  450000
SEMIMTH-IN  354166
SUCCESSFUL INSERT-REC FOR EMP: 7459
SUCCESSFUL INSERT-REC VALUES : 20170101 e     &       %
MAPPING FIELDS FOR EMPLOYEE 9134
EFF-DATE-IN 20170101
REG-PAY-IN  7500000
BON-PAY-IN  250000
SEMIMTH-IN  312500
```
ERROR ENCOUNTERED - DETAIL FOLLOWS
DBD-NAME1:EMPLOYEE
SEG-LEVEL1:00
STATUS-CODE:GE
PROC-OPT1 :AP
SEG-NAME1 :
KEY-FDBK1 :00000
NUM-SENSEG1:00004
KEY-FDBK-AREA1:   20170101
```
PROCESSING IN P300-TERMINATION
** COBIMS6 - SUCCESSFULLY ENDED **
```

The bolded text above shows that our error code caught the missing root segment and reported it. In this case we took a "soft landing" by not terminating the program. In the real world we might have forced an abend.[10] Or we might possibly have written the record to an exception report for someone to review and correct.

So now you have a model for inserting new data to child segments in an IMS database.

Reading Child Segments Sequentially (GNP)

Now let's use COBIMS7 to read back the records we just added to the database. We can traverse the database using GN for the root segments and GNP (Get Next Within Parent) calls for the children. So we'll borrow the code from COBIMS3 for walking through the root segments. And then we'll add code for retrieving GNP.

Keep in mind that we've only added a single EMPPAY child under each root segment. If there were more than one child, our code would need to allow for that. But for now, our spec will ask us to simply get a root and then get the first child under that root. Then we will display the pay information for the employee.

We already know how to traverse the root segment. So once we get a root segment, we need to take the EMP-ID returned in the IO-EMPLOYEE-RECORD and use it to set the qualified SSA for EMPLOYEE. We could use the unqualified SSA for the EMPPAY segment, but since we already know the exact key we can as easily use the qualified SSA. And we'll load the segment data into the IO-EMPPAY-RECORD I/O area. This is what our call will look like.

```
MOVE EMPL-ID-IN TO EMP-ID-VAL
MOVE '20170101' TO EFFDATE-VAL

CALL 'CBLTDLI' USING FIVE,
     DLI-FUNCGNP,
     PCB-MASK,
     IO-EMPPAY-RECORD,
     EMP-QUALIFIED-SSA,
     EMPPAY-UNQUALIFIED-SSA
```

Other than that, our program doesn't need to use any new techniques. Here is the completed program listing.

---

10 You can force an abend with a memory dump by calling LE program CEE3DMP. Details for how to do that are at the link below. We will only take soft abends in this text, so we won't abend with CEE3DMP. https://www.ibm.com/support/knowledgecenter/en/SSLT-BW_2.3.0/com.ibm.zos.v2r3.ceea100/ceea1mst78.htm

```cobol
       IDENTIFICATION DIVISION.
       PROGRAM-ID. COBIMS7.

      **********************************************************
      *   WALK THROUGH THE EMPLOYEE AND EMPPAY SEGS OF         *
      *   THE ENTIRE EMPLOYEE IMS DATABASE.                    *
      **********************************************************

       ENVIRONMENT DIVISION.
       INPUT-OUTPUT SECTION.
       DATA DIVISION.

      **********************************************************
      *   W O R K I N G   S T O R A G E   S E C T I O N    *
      **********************************************************

       WORKING-STORAGE SECTION.

        01 WS-FLAGS.
           05  SW-END-OF-DB-SWITCH    PIC X(1) VALUE 'N'.
               88  SW-END-OF-DB                 VALUE 'Y'.
               88  SW-NOT-END-OF-DB             VALUE 'N'.

        01 IO-EMPLOYEE-RECORD.
           05  EMPL-ID-IN    PIC X(04).
           05  FILLER        PIC X(01).
           05  EMPL-LNAME    PIC X(30).
           05  FILLER        PIC X(01).
           05  EMPL-FNAME    PIC X(20).
           05  FILLER        PIC X(01).
           05  EMPL-YRS-SRV  PIC X(02).
           05  FILLER        PIC X(01).
           05  EMPL-PRM-DTE  PIC X(10).
           05  FILLER        PIC X(10).

        01 IO-EMPPAY-RECORD.
           05  PAY-EFF-DATE  PIC X(8).
           05  PAY-REG-PAY   PIC S9(6)V9(2) USAGE COMP-3.
           05  PAY-BON-PAY   PIC S9(6)V9(2) USAGE COMP-3.
           05  SEMIMTH-PAY   PIC S9(6)V9(2) USAGE COMP-3.
           05  FILLER        PIC X(57).

        01 DISPLAY-EMPLOYEE-PIC.
           05  DIS-REG-PAY   PIC ZZ999.99-.
           05  DIS-BON-PAY   PIC ZZ999.99-.
           05  DIS-SMT-PAY   PIC ZZ999.99-.

        01 EMP-UNQUALIFIED-SSA.
           05  SEGNAME    PIC X(08) VALUE 'EMPLOYEE'.
           05  FILLER     PIC X(01) VALUE ' '.

        01 EMP-QUALIFIED-SSA.
```

227

```
          05   SEGNAME      PIC X(08) VALUE 'EMPLOYEE'.
          05   FILLER       PIC X(01) VALUE '('.
          05   FIELD        PIC X(08) VALUE 'EMPID'.
          05   OPER         PIC X(02) VALUE ' ='.
          05   EMP-ID-VAL   PIC X(04) VALUE '    '.
          05   FILLER       PIC X(01) VALUE ')'.

      01 EMPPAY-UNQUALIFIED-SSA.
          05   SEGNAME      PIC X(08) VALUE 'EMPPAY  '.
          05   FILLER       PIC X(01) VALUE ' '.

      01 EMPPAY-QUALIFIED-SSA.
          05   SEGNAME      PIC X(08) VALUE 'EMPPAY  '.
          05   FILLER       PIC X(01) VALUE '('.
          05   FIELD        PIC X(08) VALUE 'EFFDATE '.
          05   OPER         PIC X(02) VALUE ' ='.
          05   EFFDATE-VAL  PIC X(08) VALUE '        '.
          05   FILLER       PIC X(01) VALUE ')'.


      01 DLI-FUNCTIONS.
          05 DLI-FUNCISRT  PIC X(4) VALUE 'ISRT'.
          05 DLI-FUNCGU    PIC X(4) VALUE 'GU  '.
          05 DLI-FUNCGN    PIC X(4) VALUE 'GN  '.
          05 DLI-FUNCGHU   PIC X(4) VALUE 'GHU '.
          05 DLI-FUNCGNP   PIC X(4) VALUE 'GNP '.
          05 DLI-FUNCREPL  PIC X(4) VALUE 'REPL'.
          05 DLI-FUNCDLET  PIC X(4) VALUE 'DLET'.
          05 DLI-FUNCXRST  PIC X(4) VALUE 'XRST'.
          05 DLI-FUNCCKPT  PIC X(4) VALUE 'CKPT'.

       01 IMS-RET-CODES.
          05 THREE         PIC S9(9) COMP VALUE +3.
          05 FOUR          PIC S9(9) COMP VALUE +4.
          05 FIVE          PIC S9(9) COMP VALUE +5.
          05 SIX           PIC S9(9) COMP VALUE +6.

      LINKAGE SECTION.
       01 PCB-MASK.
          03 DBD-NAME      PIC X(8).
          03 SEG-LEVEL     PIC XX.
          03 STATUS-CODE   PIC XX.
          03 PROC-OPT      PIC X(4).
          03 FILLER        PIC X(4).
          03 SEG-NAME      PIC X(8).
          03 KEY-FDBK      PIC S9(5) COMP.
          03 NUM-SENSEG    PIC S9(5) COMP.
          03 KEY-FDBK-AREA.
             05 EMPLOYEE-KEY  PIC X(04).
             05 EMPPAYHS-KEY  PIC X(08).

      PROCEDURE DIVISION.
```

```
    INITIALIZE PCB-MASK
    ENTRY 'DLITCBL' USING PCB-MASK

    PERFORM P100-INITIALIZATION.
    PERFORM P200-MAINLINE.
    PERFORM P300-TERMINATION.
    GOBACK.

 P100-INITIALIZATION.

    DISPLAY '** PROGRAM COBIMS7 START **'
    DISPLAY 'PROCESSING IN P100-INITIALIZATION'.

*    DO INITIAL DB READ FOR FIRST EMPLOYEE RECORD

    CALL 'CBLTDLI' USING FOUR,
         DLI-FUNCGN,
         PCB-MASK,
         IO-EMPLOYEE-RECORD,
         EMP-UNQUALIFIED-SSA

    IF STATUS-CODE = '  ' THEN
       NEXT SENTENCE
    ELSE
       IF STATUS-CODE = 'GB' THEN
          SET SW-END-OF-DB TO TRUE
          DISPLAY 'END OF DATABASE :'
       ELSE
          PERFORM P400-DISPLAY-ERROR
          GOBACK
       END-IF

    END-IF.

 P200-MAINLINE.

    DISPLAY 'PROCESSING IN P200-MAINLINE'

*    CHECK STATUS CODE AND FIRST RECORD

    IF SW-END-OF-DB THEN
       DISPLAY 'NO RECORDS TO PROCESS!!'
    ELSE
       DISPLAY 'SUCCESSFUL READ :' IO-EMPLOYEE-RECORD
       PERFORM UNTIL SW-END-OF-DB
          PERFORM P500-GET-PAY-SEG

          CALL 'CBLTDLI' USING FOUR,
               DLI-FUNCGN,
               PCB-MASK,
               IO-EMPLOYEE-RECORD,
               EMP-UNQUALIFIED-SSA
```

```
            IF STATUS-CODE = 'GB' THEN
                SET SW-END-OF-DB TO TRUE
                DISPLAY 'END OF DATABASE'
            ELSE
                IF STATUS-CODE NOT EQUAL SPACES THEN
                    PERFORM P400-DISPLAY-ERROR
                ELSE
                    DISPLAY 'SUCCESSFUL READ :' IO-EMPLOYEE-RECORD
                END-IF
            END-IF

        END-PERFORM.

    DISPLAY 'FINISHED PROCESSING IN P200-MAINLINE'.

P300-TERMINATION.

    DISPLAY 'PROCESSING IN P300-TERMINATION'
    DISPLAY '** COBIMS7 - SUCCESSFULLY ENDED **'.

P400-DISPLAY-ERROR.

    DISPLAY 'PROCESSING IN P400-DISPLAY-ERROR'
    DISPLAY 'ERROR ENCOUNTERED - DETAIL FOLLOWS'
    DISPLAY 'DBD-NAME1:'       DBD-NAME
    DISPLAY 'SEG-LEVEL1:'      SEG-LEVEL
    DISPLAY 'STATUS-CODE:'     STATUS-CODE
    DISPLAY 'PROC-OPT1 :'      PROC-OPT
    DISPLAY 'SEG-NAME1 :'      SEG-NAME
    DISPLAY 'KEY-FDBK1 :'      KEY-FDBK
    DISPLAY 'NUM-SENSEG1:'     NUM-SENSEG
    DISPLAY 'KEY-FDBK-AREA1:' KEY-FDBK-AREA.

P500-GET-PAY-SEG.

    DISPLAY 'PROCESSING IN P500-GET-PAY-SEG'

    MOVE EMPL-ID-IN TO EMP-ID-VAL
    MOVE '20170101' TO EFFDATE-VAL

    CALL 'CBLTDLI' USING FIVE,
         DLI-FUNCGNP,
         PCB-MASK,
         IO-EMPPAY-RECORD,
         EMP-QUALIFIED-SSA,
         EMPPAY-QUALIFIED-SSA

    IF STATUS-CODE NOT EQUAL SPACES THEN
       PERFORM P400-DISPLAY-ERROR
    ELSE
*   MAP FIELDS
```

230

```
              MOVE PAY-REG-PAY TO DIS-REG-PAY
              MOVE PAY-BON-PAY TO DIS-BON-PAY
              MOVE SEMIMTH-PAY TO DIS-SMT-PAY
              DISPLAY 'SUCCESSFUL PAY READ :'
              DISPLAY '   EFFECTIVE DATE = ' PAY-EFF-DATE
              DISPLAY '    PAY-REG-PAY = ' DIS-REG-PAY
              DISPLAY '    PAY-BON-PAY = ' DIS-BON-PAY
              DISPLAY '    SEMIMTH-PAY = ' DIS-SMT-PAY
          END-IF.

     *    END OF SOURCE CODE
```

Once again, let's compile and link, and then run the program. Here is the output showing both root and child segments.

```
** PROGRAM COBIMS7 START **
PROCESSING IN P100-INITIALIZATION
PROCESSING IN P200-MAINLINE
SUCCESSFUL READ :1111 VEREEN              CHARLES          12 201
PROCESSING IN P500-GET-PAY-SEG
SUCCESSFUL PAY READ :
   EFFECTIVE DATE = 20170101
   PAY-REG-PAY = 87000.00
   PAY-BON-PAY =  6700.00
   SEMIMTH-PAY =  3625.00
SUCCESSFUL READ :1122 JENKINS             DEBORAH          05 201
PROCESSING IN P500-GET-PAY-SEG
SUCCESSFUL PAY READ :
   EFFECTIVE DATE = 20170101
   PAY-REG-PAY = 82000.00
   PAY-BON-PAY =  6000.00
   SEMIMTH-PAY =  3416.66
SUCCESSFUL READ :3217 JOHNSON             EDWARD           04 201
PROCESSING IN P500-GET-PAY-SEG
SUCCESSFUL PAY READ :
   EFFECTIVE DATE = 20170101
   PAY-REG-PAY = 65000.00
   PAY-BON-PAY =  5500.00
   SEMIMTH-PAY =  2708.33
SUCCESSFUL READ :4175 TURNBULL            FRED             01 201
PROCESSING IN P500-GET-PAY-SEG
SUCCESSFUL PAY READ :
   EFFECTIVE DATE = 20170101
   PAY-REG-PAY = 55000.00
   PAY-BON-PAY =  1500.00
   SEMIMTH-PAY =  2291.66
SUCCESSFUL READ :4720 SCHULTZ             TIM              09 201
PROCESSING IN P500-GET-PAY-SEG
SUCCESSFUL PAY READ :
   EFFECTIVE DATE = 20170101
   PAY-REG-PAY = 80000.00
   PAY-BON-PAY =  2500.00
   SEMIMTH-PAY =  3333.33
SUCCESSFUL READ :4836 SMITH               SANDRA           03 201
PROCESSING IN P500-GET-PAY-SEG
```

```
SUCCESSFUL PAY READ :
   EFFECTIVE DATE = 20170101
   PAY-REG-PAY = 62000.00
   PAY-BON-PAY =  2200.00
   SEMIMTH-PAY =  2583.33
SUCCESSFUL READ :6288 WILLARD                     JOE              06 201
PROCESSING IN P500-GET-PAY-SEG
SUCCESSFUL PAY READ :
   EFFECTIVE DATE = 20170101
   PAY-REG-PAY = 70000.00
   PAY-BON-PAY =  2000.00
   SEMIMTH-PAY =  2916.66
SUCCESSFUL READ :7459 STEWART                     BETTY            07 201
PROCESSING IN P500-GET-PAY-SEG
SUCCESSFUL PAY READ :
   EFFECTIVE DATE = 20170101
   PAY-REG-PAY = 85000.00
   PAY-BON-PAY =  4500.00
   SEMIMTH-PAY =  3541.66
END OF DATABASE
FINISHED PROCESSING IN P200-MAINLINE
PROCESSING IN P300-TERMINATION
** COBIMS7 - SUCCESSFULLY ENDED **
```

Inserting Child Segments Down the Hierarchy (3 levels)

Ok, I think we have a pretty good handle on the adding and retrieving of child segments. But just to be sure, let's work with the EMPPAYHS segment, adding and retrieving records. That's slightly different that what we've done already, but not much.

For COBIMS8, let's add a pay history segment EMPPAYHS for all employees using pay date January 15, 2017, and using the twice-monthly pay information from the EMPPAY segment. So we need to position ourselves at the EMPPAY child segment under each EMPLOYEE root segment, and then ISRT an EMPPAYHS segment.

I think we've covered all the techniques required to write this program. Why don't you give it a try first, and then we'll get back together and compare our code? Take a good break and then code up your version.

. . . .

Ok, I'm back with a good cup of coffee. Here's my version of the code. I added the segment I/O and SSAs for the EMPPAYHS segment. The INSERT call for the EMPPAYHS segment is as follows:

```
        CALL 'CBLTDLI' USING SIX,
              DLI-FUNCISRT,
              PCB-MASK,
              IO-EMPPAYHS-RECORD,
              EMP-QUALIFIED-SSA,
              EMPPAY-QUALIFIED-SSA,
              EMPPAYHS-UNQUALIFIED-SSA.
```

232

Should be no surprises there. Just a bit more navigation and slightly different database calls. Note that we must use qualified SSAs for EMPLOYEE and EMPPAY. Here's the full program.

```
        IDENTIFICATION DIVISION.
        PROGRAM-ID. COBIMS8.

        **********************************************************
        *    INSERT EMPLOYEE PAY HISTORY RECS INTO THE          *
        *    EMPLOYEE IMS DATABASE. THIS EXAMPLE WALKS           *
        *    THROUGH THE ROOT AND EMPPAY SEGS AND THEN           *
        *    INSERTS THE PAY HISTORY SEGMENT UNDER THE           *
        *    EMPPAY SEGMENT.                                     *
        **********************************************************

        ENVIRONMENT DIVISION.
        DATA DIVISION.

        **********************************************************
        *  W O R K I N G   S T O R A G E   S E C T I O N   *
        **********************************************************

        WORKING-STORAGE SECTION.

        01 WS-FLAGS.
            05  SW-END-OF-FILE-SWITCH   PIC X(1) VALUE 'N'.
                88  SW-END-OF-FILE               VALUE 'Y'.
                88  SW-NOT-END-OF-FILE           VALUE 'N'.
            05  SW-END-OF-DB-SWITCH     PIC X(1) VALUE 'N'.
                88  SW-END-OF-DB                 VALUE 'Y'.
                88  SW-NOT-END-OF-DB             VALUE 'N'.

        01 IO-EMPLOYEE-RECORD.
            05  EMPL-ID       PIC X(04).
            05  FILLER        PIC X(01).
            05  EMPL-LNAME    PIC X(30).
            05  FILLER        PIC X(01).
            05  EMPL-FNAME    PIC X(20).
            05  FILLER        PIC X(01).
            05  EMPL-YRS-SRV  PIC X(02).
            05  FILLER        PIC X(01).
            05  EMPL-PRM-DTE  PIC X(10).
            05  FILLER        PIC X(10).

        01 IO-EMPPAY-RECORD.
            05  PAY-EFF-DATE  PIC X(8).
            05  PAY-REG-PAY   PIC S9(6)V9(2) USAGE COMP-3.
            05  PAY-BON-PAY   PIC S9(6)V9(2) USAGE COMP-3.
            05  SEMIMTH-PAY   PIC S9(6)V9(2) USAGE COMP-3.
            05  FILLER        PIC X(57).

        01 IO-EMPPAYHS-RECORD.
```

233

```
    05  PAY-DATE      PIC X(8).
    05  PAY-ANN-PAY   PIC S9(6)V9(2) USAGE COMP-3.
    05  PAY-AMT       PIC S9(6)V9(2) USAGE COMP-3.
    05  FILLER        PIC X(62).

01 SEG-IO-AREA     PIC X(80).

01 DLI-FUNCTIONS.
    05  DLI-FUNCISRT  PIC X(4) VALUE 'ISRT'.
    05  DLI-FUNCGU    PIC X(4) VALUE 'GU  '.
    05  DLI-FUNCGN    PIC X(4) VALUE 'GN  '.
    05  DLI-FUNCGHU   PIC X(4) VALUE 'GHU '.
    05  DLI-FUNCGNP   PIC X(4) VALUE 'GNP '.
    05  DLI-FUNCREPL  PIC X(4) VALUE 'REPL'.
    05  DLI-FUNCDLET  PIC X(4) VALUE 'DLET'.
    05  DLI-FUNCXRST  PIC X(4) VALUE 'XRST'.
    05  DLI-FUNCCKPT  PIC X(4) VALUE 'CKPT'.

01 EMP-UNQUALIFIED-SSA.
    05  SEGNAME     PIC X(08) VALUE 'EMPLOYEE'.
    05  FILLER      PIC X(01) VALUE ' '.

01 EMP-QUALIFIED-SSA.
    05  SEGNAME     PIC X(08) VALUE 'EMPLOYEE'.
    05  FILLER      PIC X(01) VALUE '('.
    05  FIELD       PIC X(08) VALUE 'EMPID'.
    05  OPER        PIC X(02) VALUE ' ='.
    05  EMP-ID-VAL  PIC X(04) VALUE '    '.
    05  FILLER      PIC X(01) VALUE ')'.

01 EMPPAY-UNQUALIFIED-SSA.
    05  SEGNAME     PIC X(08) VALUE 'EMPPAY  '.
    05  FILLER      PIC X(01) VALUE ' '.

01 EMPPAY-QUALIFIED-SSA.
    05  SEGNAME     PIC X(08) VALUE 'EMPPAY  '.
    05  FILLER      PIC X(01) VALUE '('.
    05  FIELD       PIC X(08) VALUE 'EFFDATE '.
    05  OPER        PIC X(02) VALUE ' ='.
    05  EFFDATE-VAL PIC X(08) VALUE '        '.
    05  FILLER      PIC X(01) VALUE ')'.

01 EMPPAYHS-UNQUALIFIED-SSA.
    05  SEGNAME     PIC X(08) VALUE 'EMPPAYHS'.
    05  FILLER      PIC X(01) VALUE ' '.

01 IMS-RET-CODES.
    05  THREE         PIC S9(9) COMP VALUE +3.
    05  FOUR          PIC S9(9) COMP VALUE +4.
    05  FIVE          PIC S9(9) COMP VALUE +5.
    05  SIX           PIC S9(9) COMP VALUE +6.
```

```
    77 WS-PAY-DATE      PIC X(08) VALUE '20170115'.

   LINKAGE SECTION.
    01 PCB-MASK.
        03 DBD-NAME         PIC X(8).
        03 SEG-LEVEL        PIC XX.
        03 STATUS-CODE      PIC XX.
        03 PROC-OPT         PIC X(4).
        03 FILLER           PIC X(4).
        03 SEG-NAME         PIC X(8).
        03 KEY-FDBK         PIC S9(5) COMP.
        03 NUM-SENSEG       PIC S9(5) COMP.
        03 KEY-FDBK-AREA.
           05 EMPLOYEE-ID   PIC X(04).
           05 EMPPAYHS      PIC X(08).

   PROCEDURE DIVISION.

       INITIALIZE PCB-MASK
       ENTRY 'DLITCBL' USING PCB-MASK

       PERFORM P100-INITIALIZATION.
       PERFORM P200-MAINLINE.
       PERFORM P300-TERMINATION.
       GOBACK.

    P100-INITIALIZATION.

       DISPLAY '** PROGRAM COBIMS8 START **'
       DISPLAY 'PROCESSING IN P100-INITIALIZATION'.

*      DO INITIAL DB READ FOR FIRST EMPLOYEE ROOT SEGMENT

       CALL 'CBLTDLI' USING ,
            DLI-FUNCGN,
            PCB-MASK,
            IO-EMPLOYEE-RECORD,
            EMP-UNQUALIFIED-SSA

       IF STATUS-CODE = '  ' THEN
          NEXT SENTENCE
       ELSE
          IF STATUS-CODE = 'GB' THEN
             SET SW-END-OF-DB TO TRUE
             DISPLAY 'END OF DATABASE :'
          ELSE
             PERFORM P9000-DISPLAY-ERROR
             GOBACK
          END-IF

       END-IF.
```

```
P200-MAINLINE.
    DISPLAY 'PROCESSING IN P200-MAINLINE'

*    CHECK STATUS CODE AND FIRST RECORD

    IF SW-END-OF-DB THEN
        DISPLAY 'NO RECORDS TO PROCESS!!'
    ELSE
        PERFORM UNTIL SW-END-OF-DB
            DISPLAY 'SUCCESSFUL READ :' IO-EMPLOYEE-RECORD
            MOVE EMPL-ID TO EMP-ID-VAL
            PERFORM P2000-GET-EMPPAY
            IF STATUS-CODE NOT EQUAL SPACES THEN
                PERFORM P9000-DISPLAY-ERROR
                GOBACK
            ELSE
                DISPLAY 'SUCCESSFUL PAY READ :' IO-EMPPAY-RECORD
                MOVE PAY-EFF-DATE TO EFFDATE-VAL
                MOVE WS-PAY-DATE TO PAY-DATE
                MOVE PAY-REG-PAY TO PAY-ANN-PAY
                MOVE SEMIMTH-PAY TO PAY-AMT
                PERFORM P3000-INSERT-EMPPAYHS
                IF STATUS-CODE NOT EQUAL SPACES THEN
                    PERFORM P9000-DISPLAY-ERROR
                    GOBACK
                ELSE
                    DISPLAY 'SUCCESSFUL INSERT EMPPAYHS : '
                        EMP-ID-VAL
                    DISPLAY 'SUCCESSFUL INSERT VALUES   : '
                        IO-EMPPAYHS-RECORD
                END-IF

                PERFORM P1000-GET-NEXT-ROOT
                IF STATUS-CODE = 'GB' THEN
                    SET SW-END-OF-DB TO TRUE
                    DISPLAY 'END OF DATABASE'
                END-IF

            END-IF

        END-PERFORM.

    DISPLAY 'FINISHED PROCESSING IN P200-MAINLINE'.

P300-TERMINATION.

    DISPLAY 'PROCESSING IN P300-TERMINATION'
    DISPLAY '** COBIMS8 - SUCCESSFULLY ENDED **'.

P1000-GET-NEXT-ROOT.

    DISPLAY 'PROCESSING IN P1000-GET-NEXT-ROOT'.
```

```
            CALL 'CBLTDLI' USING FOUR,
                 DLI-FUNCGN,
                 PCB-MASK,
                 IO-EMPLOYEE-RECORD,
                 EMP-UNQUALIFIED-SSA.

        P2000-GET-EMPPAY.

            DISPLAY 'PROCESSING IN P2000-GET-EMPPAY'.

            CALL 'CBLTDLI' USING FIVE,
                 DLI-FUNCGNP,
                 PCB-MASK,
                 IO-EMPPAY-RECORD,
                 EMP-QUALIFIED-SSA,
                 EMPPAY-UNQUALIFIED-SSA.

        P3000-INSERT-EMPPAYHS.

            DISPLAY 'PROCESSING IN P3000-INSERT-EMPPAYHS'.

            CALL 'CBLTDLI' USING SIX,
                 DLI-FUNCISRT,
                 PCB-MASK,
                 IO-EMPPAYHS-RECORD,
                 EMP-QUALIFIED-SSA,
                 EMPPAY-QUALIFIED-SSA,
                 EMPPAYHS-UNQUALIFIED-SSA.

        P9000-DISPLAY-ERROR.

            DISPLAY 'ERROR ENCOUNTERED - DETAIL FOLLOWS'
            DISPLAY 'DBD-NAME1:'      DBD-NAME
            DISPLAY 'SEG-LEVEL1:'     SEG-LEVEL
            DISPLAY 'STATUS-CODE:'    STATUS-CODE
            DISPLAY 'PROC-OPT1 :'     PROC-OPT
            DISPLAY 'SEG-NAME1 :'     SEG-NAME
            DISPLAY 'KEY-FDBK1 :'     KEY-FDBK
            DISPLAY 'NUM-SENSEG1:'    NUM-SENSEG
            DISPLAY 'KEY-FDBK-AREA1:' KEY-FDBK-AREA.

    *     END OF SOURCE CODE
```

Now let's compile, link and run the program. Here is the output.

```
** PROGRAM COBIMS8 START **
PROCESSING IN P100-INITIALIZATION
PROCESSING IN P200-MAINLINE
SUCCESSFUL READ :1111 VEREEN                    CHARLES         12 201
PROCESSING IN P2000-GET-EMPPAY
SUCCESSFUL PAY READ :20170101 g            &
```

```
PROCESSING IN P3000-INSERT-EMPPAYHS
SUCCESSFUL INSERT EMPPAYHS : 1111
SUCCESSFUL INSERT VALUES   : 20170115 g      &
PROCESSING IN P1000-GET-NEXT-ROOT
SUCCESSFUL READ :1122 JENKINS                DEBORAH          05 201
PROCESSING IN P2000-GET-EMPPAY
SUCCESSFUL PAY READ :20170101 b          %
PROCESSING IN P3000-INSERT-EMPPAYHS
SUCCESSFUL INSERT EMPPAYHS : 1122
SUCCESSFUL INSERT VALUES   : 20170115 b        %
PROCESSING IN P1000-GET-NEXT-ROOT
SUCCESSFUL READ :3217 JOHNSON                EDWARD           04 201
PROCESSING IN P2000-GET-EMPPAY
SUCCESSFUL PAY READ :20170101       &      c
PROCESSING IN P3000-INSERT-EMPPAYHS
SUCCESSFUL INSERT EMPPAYHS : 3217
SUCCESSFUL INSERT VALUES   : 20170115      c
PROCESSING IN P1000-GET-NEXT-ROOT
SUCCESSFUL READ :4175 TURNBULL               FRED             01 201
PROCESSING IN P2000-GET-EMPPAY
SUCCESSFUL PAY READ :20170101       &      %
PROCESSING IN P3000-INSERT-EMPPAYHS
SUCCESSFUL INSERT EMPPAYHS : 4175
SUCCESSFUL INSERT VALUES   : 20170115        %
PROCESSING IN P1000-GET-NEXT-ROOT
SUCCESSFUL READ :4720 SCHULTZ                TIM              09 201
PROCESSING IN P2000-GET-EMPPAY
SUCCESSFUL PAY READ :20170101       &
PROCESSING IN P3000-INSERT-EMPPAYHS
SUCCESSFUL INSERT EMPPAYHS : 4720
SUCCESSFUL INSERT VALUES   : 20170115
PROCESSING IN P1000-GET-NEXT-ROOT
SUCCESSFUL READ :4836 SMITH                  SANDRA           03 201
PROCESSING IN P2000-GET-EMPPAY
SUCCESSFUL PAY READ :20170101
PROCESSING IN P3000-INSERT-EMPPAYHS
SUCCESSFUL INSERT EMPPAYHS : 4836
SUCCESSFUL INSERT VALUES   : 20170115
PROCESSING IN P1000-GET-NEXT-ROOT
SUCCESSFUL READ :6288 WILLARD                JOE              06 201
PROCESSING IN P2000-GET-EMPPAY
SUCCESSFUL PAY READ :20170101          j %
PROCESSING IN P3000-INSERT-EMPPAYHS
SUCCESSFUL INSERT EMPPAYHS : 6288
SUCCESSFUL INSERT VALUES   : 20170115     j %
PROCESSING IN P1000-GET-NEXT-ROOT
SUCCESSFUL READ :7459 STEWART                BETTY            07 201
PROCESSING IN P2000-GET-EMPPAY
SUCCESSFUL PAY READ :20170101 e     &      %
PROCESSING IN P3000-INSERT-EMPPAYHS
SUCCESSFUL INSERT EMPPAYHS : 7459
SUCCESSFUL INSERT VALUES   : 20170115 e      %
PROCESSING IN P1000-GET-NEXT-ROOT
END OF DATABASE
FINISHED PROCESSING IN P200-MAINLINE
PROCESSING IN P300-TERMINATION
** COBIMS8 - SUCCESSFULLY ENDED **
```

So that's how to insert a child segment under a higher level child. To make this more interesting, change the value of the pay date to January 31, 2017. Then compile and link and run again. Do this twice more using pay dates February 15, 2017 and February 28, 2017. Now we have four paychecks for each employee. We'll read all this data back in the next training program.

Read Child Segments Down the Hierarchy (3 levels)

For COBIMS9 you'll need to retrieve and display all the pay history segments for each employee. This should be fairly straightforward by now. Yes you'll need one more loop, and more navigation. But we need the practice to really drill the techniques in. Give this one a try, then take a long break and we'll compare code.

. . . .

Ok, I hope you are enjoying coding IMS in COBOL! I'll bet you got your version of the program to work without any serious problems. Let me give you my code and see what you think. Note that I have switches both for end of database and for end of EMPPAYHS segments. The latter is needed for looping through the multiple EMPPAYHS segments.

```
        IDENTIFICATION DIVISION.
        PROGRAM-ID. COBIMS9.
        ******************************************************
        *    READ AND DISPLAY EMP HISTORY RECS FROM THE      *
        *    EMPLOYEE IMS DATABASE. THIS EXAMPLE WALKS        *
        *    THROUGH THE ROOT AND EMPPAY SEGS AND THEN        *
        *    READS THE PAY HISTORY SEGMENTS UNDER THE         *
        *    EMPPAY SEGMENT.                                  *
        ******************************************************
        ENVIRONMENT DIVISION.
        DATA DIVISION.
        ******************************************************
        *  W O R K I N G    S T O R A G E    S E C T I O N   *
        ******************************************************
        WORKING-STORAGE SECTION.
        01 WS-FLAGS.
            05  SW-END-OF-FILE-SWITCH   PIC X(1) VALUE 'N'.
                88  SW-END-OF-FILE                VALUE 'Y'.
                88  SW-NOT-END-OF-FILE            VALUE 'N'.
            05  SW-END-OF-DB-SWITCH     PIC X(1) VALUE 'N'.
                88  SW-END-OF-DB                  VALUE 'Y'.
                88  SW-NOT-END-OF-DB              VALUE 'N'.
            05  SW-END-OF-EMPPAYHS-SW   PIC X(1) VALUE 'N'.
                88  SW-END-OF-EMPPAYHS            VALUE 'Y'.
                88  SW-NOT-END-OF-EMPPAYHS        VALUE 'N'.

        01 IO-EMPLOYEE-RECORD.
            05  EMPL-ID      PIC X(04).
            05  FILLER       PIC X(01).
```

```
05    EMPL-LNAME     PIC X(30).
05    FILLER         PIC X(01).
05    EMPL-FNAME     PIC X(20).
05    FILLER         PIC X(01).
05    EMPL-YRS-SRV   PIC X(02).
05    FILLER         PIC X(01).
05    EMPL-PRM-DTE   PIC X(10).
05    FILLER         PIC X(10).


01 IO-EMPPAY-RECORD.
   05    PAY-EFF-DATE   PIC X(8).
   05    PAY-REG-PAY    PIC S9(6)V9(2) USAGE COMP-3.
   05    PAY-BON-PAY    PIC S9(6)V9(2) USAGE COMP-3.
   05    SEMIMTH-PAY    PIC S9(6)V9(2) USAGE COMP-3.
   05    FILLER         PIC X(57).


01 IO-EMPPAYHS-RECORD.
   05    PAY-DATE       PIC X(8).
   05    PAY-ANN-PAY    PIC S9(6)V9(2) USAGE COMP-3.
   05    PAY-AMT        PIC S9(6)V9(2) USAGE COMP-3.
   05    FILLER         PIC X(62).


01 SEG-IO-AREA      PIC X(80).


01 DLI-FUNCTIONS.
   05 DLI-FUNCISRT PIC X(4) VALUE 'ISRT'.
   05 DLI-FUNCGU   PIC X(4) VALUE 'GU  '.
   05 DLI-FUNCGN   PIC X(4) VALUE 'GN  '.
   05 DLI-FUNCGHU  PIC X(4) VALUE 'GHU '.
   05 DLI-FUNCGNP  PIC X(4) VALUE 'GNP '.
   05 DLI-FUNCREPL PIC X(4) VALUE 'REPL'.
   05 DLI-FUNCDLET PIC X(4) VALUE 'DLET'.
   05 DLI-FUNCXRST PIC X(4) VALUE 'XRST'.
   05 DLI-FUNCCKPT PIC X(4) VALUE 'CKPT'.


01 DISPLAY-EMPPAYHS-PIC.
   05   DIS-REG-PAY   PIC ZZ999.99-.
   05   DIS-SMT-PAY   PIC ZZ999.99-.


 01 EMP-UNQUALIFIED-SSA.
    05   SEGNAME      PIC X(08) VALUE 'EMPLOYEE'.
    05   FILLER       PIC X(01) VALUE ' '.


 01 EMP-QUALIFIED-SSA.
    05   SEGNAME      PIC X(08) VALUE 'EMPLOYEE'.
    05   FILLER       PIC X(01) VALUE '('.
    05   FIELD        PIC X(08) VALUE 'EMPID'.
    05   OPER         PIC X(02) VALUE ' ='.
    05   EMP-ID-VAL   PIC X(04) VALUE '    '.
    05   FILLER       PIC X(01) VALUE ')'.


 01 EMPPAY-UNQUALIFIED-SSA.
```

```
        05   SEGNAME      PIC X(08) VALUE 'EMPPAY  '.
        05   FILLER       PIC X(01) VALUE ' '.

    01 EMPPAY-QUALIFIED-SSA.
        05   SEGNAME      PIC X(08) VALUE 'EMPPAY  '.
        05   FILLER       PIC X(01) VALUE '('.
        05   FIELD        PIC X(08) VALUE 'EFFDATE '.
        05   OPER         PIC X(02) VALUE ' ='.
        05   EFFDATE-VAL  PIC X(08) VALUE '        '.
        05   FILLER       PIC X(01) VALUE ')'.

    01 EMPPAYHS-UNQUALIFIED-SSA.
        05   SEGNAME      PIC X(08) VALUE 'EMPPAYHS'.
        05   FILLER       PIC X(01) VALUE ' '.

    01 IMS-RET-CODES.
        05 THREE          PIC S9(9) COMP VALUE +3.
        05 FOUR           PIC S9(9) COMP VALUE +4.
        05 FIVE           PIC S9(9) COMP VALUE +5.
        05 SIX            PIC S9(9) COMP VALUE +6.

    77 WS-PAY-DATE     PIC X(08) VALUE '20170228'.

LINKAGE SECTION.
 01 PCB-MASK.
        03 DBD-NAME       PIC X(8).
        03 SEG-LEVEL      PIC XX.
        03 STATUS-CODE    PIC XX.
        03 PROC-OPT       PIC X(4).
        03 FILLER         PIC X(4).
        03 SEG-NAME       PIC X(8).
        03 KEY-FDBK       PIC S9(5) COMP.
        03 NUM-SENSEG     PIC S9(5) COMP.
        03 KEY-FDBK-AREA.
           05 EMPLOYEE-ID  PIC X(04).
           05 EMPPAYHS     PIC X(08).

PROCEDURE DIVISION.

    INITIALIZE PCB-MASK
    ENTRY 'DLITCBL' USING PCB-MASK

    PERFORM P100-INITIALIZATION.
    PERFORM P200-MAINLINE.
    PERFORM P300-TERMINATION.
    GOBACK.

P100-INITIALIZATION.

    DISPLAY '** PROGRAM COBIMS9 START **'
    DISPLAY 'PROCESSING IN P100-INITIALIZATION'.
```

```
*     DO INITIAL DB READ FOR FIRST EMPLOYEE ROOT SEGMENT

      CALL 'CBLTDLI' USING FOUR,
           DLI-FUNCGN,
           PCB-MASK,
           IO-EMPLOYEE-RECORD,
           EMP-UNQUALIFIED-SSA

      IF STATUS-CODE = '  ' THEN
         DISPLAY '********************************'
      ELSE
         IF STATUS-CODE = 'GB' THEN
            SET SW-END-OF-DB TO TRUE
            DISPLAY 'END OF DATABASE :'
         ELSE
            PERFORM P9000-DISPLAY-ERROR
            GOBACK
         END-IF

      END-IF.

  P200-MAINLINE.

      DISPLAY 'PROCESSING IN P200-MAINLINE'

*     CHECK STATUS CODE AND FIRST RECORD

      IF SW-END-OF-DB THEN
         DISPLAY 'NO RECORDS TO PROCESS!!'
      ELSE
         PERFORM UNTIL SW-END-OF-DB
            DISPLAY 'SUCCESSFUL READ :' IO-EMPLOYEE-RECORD
            MOVE EMPL-ID TO EMP-ID-VAL
            PERFORM P2000-GET-EMPPAY
            IF STATUS-CODE NOT EQUAL SPACES THEN
               PERFORM P9000-DISPLAY-ERROR
               GOBACK
            ELSE
               MOVE PAY-EFF-DATE TO EFFDATE-VAL
               SET SW-NOT-END-OF-EMPPAYHS TO TRUE
               PERFORM P3000-GET-NEXT-EMPPAYHS
                  UNTIL SW-END-OF-EMPPAYHS
            END-IF

            PERFORM P1000-GET-NEXT-ROOT
            IF STATUS-CODE = 'GB' THEN
               SET SW-END-OF-DB TO TRUE
               DISPLAY 'END OF DATABASE'
            END-IF

         END-PERFORM
```

```
        END-IF.

        DISPLAY 'FINISHED PROCESSING IN P200-MAINLINE'.

P300-TERMINATION.

        DISPLAY 'PROCESSING IN P300-TERMINATION'
        DISPLAY '** COBIMS9 - SUCCESSFULLY ENDED **'.

P1000-GET-NEXT-ROOT.

        DISPLAY '*******************************'
        DISPLAY 'PROCESSING IN P1000-GET-NEXT-ROOT'.

        CALL 'CBLTDLI' USING FOUR,
             DLI-FUNCGN,
             PCB-MASK,
             IO-EMPLOYEE-RECORD,
             EMP-UNQUALIFIED-SSA.

P2000-GET-EMPPAY.

        DISPLAY 'PROCESSING IN P2000-GET-EMPPAY'.

        CALL 'CBLTDLI' USING FIVE,
             DLI-FUNCGNP,
             PCB-MASK,
             IO-EMPPAY-RECORD,
             EMP-QUALIFIED-SSA,
             EMPPAY-UNQUALIFIED-SSA.

P3000-GET-NEXT-EMPPAYHS.

        DISPLAY 'PROCESSING IN P3000-GET-NEXT-EMPPAYHS'.

        CALL 'CBLTDLI' USING SIX,
             DLI-FUNCGNP,
             PCB-MASK,
             IO-EMPPAYHS-RECORD,
             EMP-QUALIFIED-SSA,
             EMPPAY-QUALIFIED-SSA,
             EMPPAYHS-UNQUALIFIED-SSA.

             EVALUATE STATUS-CODE
                WHEN ' '
                    DISPLAY 'GOOD READ OF EMPPAYHS : '
                        EMP-ID-VAL
                    MOVE PAY-ANN-PAY TO DIS-REG-PAY
                    MOVE PAY-AMT     TO DIS-SMT-PAY
                    DISPLAY 'PAY-DATE  : ' PAY-DATE
                    DISPLAY 'PAY-ANN-PAY: ' DIS-REG-PAY
                    DISPLAY 'PAY-AMT   : ' DIS-SMT-PAY
```

```
                     WHEN 'GE'
                     WHEN 'GB'
                        SET SW-END-OF-EMPPAYHS TO TRUE
                        DISPLAY 'NO MORE PAY HISTORY SEGMENTS'
                     WHEN OTHER
                        PERFORM P9000-DISPLAY-ERROR
                        SET SW-END-OF-EMPPAYHS TO TRUE
                        GOBACK
                  END-EVALUATE.

          P9000-DISPLAY-ERROR.

              DISPLAY 'ERROR ENCOUNTERED - DETAIL FOLLOWS'
              DISPLAY 'DBD-NAME1:'      DBD-NAME
              DISPLAY 'SEG-LEVEL1:'     SEG-LEVEL
              DISPLAY 'STATUS-CODE:'    STATUS-CODE
              DISPLAY 'PROC-OPT1 :'     PROC-OPT
              DISPLAY 'SEG-NAME1 :'     SEG-NAME
              DISPLAY 'KEY-FDBK1 :'     KEY-FDBK
              DISPLAY 'NUM-SENSEG1:'    NUM-SENSEG
              DISPLAY 'KEY-FDBK-AREA1:' KEY-FDBK-AREA.


       *     END OF SOURCE CODE
```

Compile, link, run. Here is the output.

```
** PROGRAM COBIMS9 START **
PROCESSING IN P100-INITIALIZATION
*********************************
PROCESSING IN P200-MAINLINE
SUCCESSFUL READ :1111 VEREEN                    CHARLES           12 201
PROCESSING IN P2000-GET-EMPPAY
PROCESSING IN P3000-GET-NEXT-EMPPAYHS
GOOD READ OF EMPPAYHS : 1111
PAY-DATE   : 20170115
PAY-ANN-PAY: 87000.00
PAY-AMT    : 3625.00
PROCESSING IN P3000-GET-NEXT-EMPPAYHS
GOOD READ OF EMPPAYHS : 1111
PAY-DATE   : 20170130
PAY-ANN-PAY: 87000.00
PAY-AMT    : 3625.00
PROCESSING IN P3000-GET-NEXT-EMPPAYHS
GOOD READ OF EMPPAYHS : 1111
PAY-DATE   : 20170215
PAY-ANN-PAY: 87000.00
PAY-AMT    : 3625.00
PROCESSING IN P3000-GET-NEXT-EMPPAYHS
GOOD READ OF EMPPAYHS : 1111
PAY-DATE   : 20170228
PAY-ANN-PAY: 87000.00
PAY-AMT    : 3625.00
PROCESSING IN P3000-GET-NEXT-EMPPAYHS
NO MORE PAY HISTORY SEGMENTS
```

```
********************************
PROCESSING IN P1000-GET-NEXT-ROOT
SUCCESSFUL READ :1122 JENKINS                    DEBORAH          05 201
PROCESSING IN P2000-GET-EMPPAY
PROCESSING IN P3000-GET-NEXT-EMPPAYHS
GOOD READ OF EMPPAYHS : 1122
PAY-DATE   : 20170115
PAY-ANN-PAY: 82000.00
PAY-AMT    : 3416.66
PROCESSING IN P3000-GET-NEXT-EMPPAYHS
GOOD READ OF EMPPAYHS : 1122
PAY-DATE   : 20170130
PAY-ANN-PAY: 82000.00
PAY-AMT    : 3416.66
PROCESSING IN P3000-GET-NEXT-EMPPAYHS
GOOD READ OF EMPPAYHS : 1122
PAY-DATE   : 20170215
PAY-ANN-PAY: 82000.00
PAY-AMT    : 3416.66
PROCESSING IN P3000-GET-NEXT-EMPPAYHS
GOOD READ OF EMPPAYHS : 1122
PAY-DATE   : 20170228
PAY-ANN-PAY: 82000.00
PAY-AMT    : 3416.66
PROCESSING IN P3000-GET-NEXT-EMPPAYHS
NO MORE PAY HISTORY SEGMENTS
********************************
PROCESSING IN P1000-GET-NEXT-ROOT
SUCCESSFUL READ :3217 JOHNSON                     EDWARD           04 201
PROCESSING IN P2000-GET-EMPPAY
PROCESSING IN P3000-GET-NEXT-EMPPAYHS
GOOD READ OF EMPPAYHS : 3217
PAY-DATE   : 20170115
PAY-ANN-PAY: 65000.00
PAY-AMT    : 2708.33
PROCESSING IN P3000-GET-NEXT-EMPPAYHS
GOOD READ OF EMPPAYHS : 3217
PAY-DATE   : 20170130
PAY-ANN-PAY: 65000.00
PAY-AMT    : 2708.33
PROCESSING IN P3000-GET-NEXT-EMPPAYHS
GOOD READ OF EMPPAYHS : 3217
PAY-DATE   : 20170215
PAY-ANN-PAY: 65000.00
PAY-AMT    : 2708.33
PROCESSING IN P3000-GET-NEXT-EMPPAYHS
GOOD READ OF EMPPAYHS : 3217
PAY-DATE   : 20170228
PAY-ANN-PAY: 65000.00
PAY-AMT    : 2708.33
PROCESSING IN P3000-GET-NEXT-EMPPAYHS
NO MORE PAY HISTORY SEGMENTS
********************************
PROCESSING IN P1000-GET-NEXT-ROOT
SUCCESSFUL READ :4175 TURNBULL                    FRED             01 201
PROCESSING IN P2000-GET-EMPPAY
PROCESSING IN P3000-GET-NEXT-EMPPAYHS
GOOD READ OF EMPPAYHS : 4175
```

```
PAY-DATE   : 20170115
PAY-ANN-PAY: 55000.00
PAY-AMT    :  2291.66
PROCESSING IN P3000-GET-NEXT-EMPPAYHS
GOOD READ OF EMPPAYHS : 4175
PAY-DATE   : 20170130
PAY-ANN-PAY: 55000.00
PAY-AMT    :  2291.66
PROCESSING IN P3000-GET-NEXT-EMPPAYHS
GOOD READ OF EMPPAYHS : 4175
PAY-DATE   : 20170215
PAY-ANN-PAY: 55000.00
PAY-AMT    :  2291.66
PROCESSING IN P3000-GET-NEXT-EMPPAYHS
GOOD READ OF EMPPAYHS : 4175
PAY-DATE   : 20170228
PAY-ANN-PAY: 55000.00
PAY-AMT    :  2291.66
PROCESSING IN P3000-GET-NEXT-EMPPAYHS
NO MORE PAY HISTORY SEGMENTS
*********************************
PROCESSING IN P1000-GET-NEXT-ROOT
SUCCESSFUL READ :4720 SCHULTZ              TIM           09 201
PROCESSING IN P2000-GET-EMPPAY
PROCESSING IN P3000-GET-NEXT-EMPPAYHS
GOOD READ OF EMPPAYHS : 4720
PAY-DATE   : 20170115
PAY-ANN-PAY: 80000.00
PAY-AMT    :  3333.33
PROCESSING IN P3000-GET-NEXT-EMPPAYHS
GOOD READ OF EMPPAYHS : 4720
PAY-DATE   : 20170130
PAY-ANN-PAY: 80000.00
PAY-AMT    :  3333.33
PROCESSING IN P3000-GET-NEXT-EMPPAYHS
GOOD READ OF EMPPAYHS : 4720
PAY-DATE   : 20170215
PAY-ANN-PAY: 80000.00
PAY-AMT    :  3333.33
PROCESSING IN P3000-GET-NEXT-EMPPAYHS
GOOD READ OF EMPPAYHS : 4720
PAY-DATE   : 20170228
PAY-ANN-PAY: 80000.00
PAY-AMT    :  3333.33
PROCESSING IN P3000-GET-NEXT-EMPPAYHS
NO MORE PAY HISTORY SEGMENTS
*********************************
PROCESSING IN P1000-GET-NEXT-ROOT
SUCCESSFUL READ :4836 SMITH               SANDRA        03 201
PROCESSING IN P2000-GET-EMPPAY
PROCESSING IN P3000-GET-NEXT-EMPPAYHS
GOOD READ OF EMPPAYHS : 4836
PAY-DATE   : 20170115
PAY-ANN-PAY: 62000.00
PAY-AMT    :  2583.33
PROCESSING IN P3000-GET-NEXT-EMPPAYHS
GOOD READ OF EMPPAYHS : 4836
PAY-DATE   : 20170130
```

```
PAY-ANN-PAY: 62000.00
PAY-AMT    : 2583.33
PROCESSING IN P3000-GET-NEXT-EMPPAYHS
GOOD READ OF EMPPAYHS : 4836
PAY-DATE   : 20170215
PAY-ANN-PAY: 62000.00
PAY-AMT    : 2583.33
PROCESSING IN P3000-GET-NEXT-EMPPAYHS
GOOD READ OF EMPPAYHS : 4836
PAY-DATE   : 20170228
PAY-ANN-PAY: 62000.00
PAY-AMT    : 2583.33
PROCESSING IN P3000-GET-NEXT-EMPPAYHS
NO MORE PAY HISTORY SEGMENTS
********************************
PROCESSING IN P1000-GET-NEXT-ROOT
SUCCESSFUL READ :6288 WILLARD              JOE              06 201
PROCESSING IN P2000-GET-EMPPAY
PROCESSING IN P3000-GET-NEXT-EMPPAYHS
GOOD READ OF EMPPAYHS : 6288
PAY-DATE   : 20170115
PAY-ANN-PAY: 70000.00
PAY-AMT    : 2916.66
PROCESSING IN P3000-GET-NEXT-EMPPAYHS
GOOD READ OF EMPPAYHS : 6288
PAY-DATE   : 20170130
PAY-ANN-PAY: 70000.00
PAY-AMT    : 2916.66
PROCESSING IN P3000-GET-NEXT-EMPPAYHS
GOOD READ OF EMPPAYHS : 6288
PAY-DATE   : 20170215
PAY-ANN-PAY: 70000.00
PAY-AMT    : 2916.66
PROCESSING IN P3000-GET-NEXT-EMPPAYHS
GOOD READ OF EMPPAYHS : 6288
PAY-DATE   : 20170228
PAY-ANN-PAY: 70000.00
PAY-AMT    : 2916.66
PROCESSING IN P3000-GET-NEXT-EMPPAYHS
NO MORE PAY HISTORY SEGMENTS
********************************
PROCESSING IN P1000-GET-NEXT-ROOT
SUCCESSFUL READ :7459 STEWART              BETTY            07 201
PROCESSING IN P2000-GET-EMPPAY
PROCESSING IN P3000-GET-NEXT-EMPPAYHS
GOOD READ OF EMPPAYHS : 7459
PAY-DATE   : 20170115
PAY-ANN-PAY: 85000.00
PAY-AMT    : 3541.66
PROCESSING IN P3000-GET-NEXT-EMPPAYHS
GOOD READ OF EMPPAYHS : 7459
PAY-DATE   : 20170130
PAY-ANN-PAY: 85000.00
PAY-AMT    : 3541.66
PROCESSING IN P3000-GET-NEXT-EMPPAYHS
GOOD READ OF EMPPAYHS : 7459
PAY-DATE   : 20170215
PAY-ANN-PAY: 85000.00
```

```
PAY-AMT    :  3541.66
PROCESSING IN P3000-GET-NEXT-EMPPAYHS
GOOD READ OF EMPPAYHS : 7459
PAY-DATE   : 20170228
PAY-ANN-PAY: 85000.00
PAY-AMT    :  3541.66
PROCESSING IN P3000-GET-NEXT-EMPPAYHS
NO MORE PAY HISTORY SEGMENTS
********************************
PROCESSING IN P1000-GET-NEXT-ROOT
END OF DATABASE
FINISHED PROCESSING IN P200-MAINLINE
PROCESSING IN P300-TERMINATION
** COBIMS9 - SUCCESSFULLY ENDED **
```

Ok I think we've covered the root-child relationships enough. You have some models to use for most anything you'd want to do in the hierarchy. Time to move on to other topics.

Additional IMS Programming Features
Retrieve Segments Using Searchable Fields

So far all the qualified SSA retrievals we've done have been based on a segment **key**. It is also possible to retrieve IMS segments by a searchable field that is not the key. For this example with program COBIMSA we will create a new field for our EMPLOYEE record layout, and then define this field in our DBD. Then we will write a program to search based on the new EMPSSN field which is the employee social security number.

Ok, where shall we put the field? We have a 9 byte social security number field, and we have 10 bytes of filler at the end of the record. Let's use the last 9 bytes of the record. Here is our new layout.

```
01  IO-EMPLOYEE-RECORD.
    05   FILLER        PIC X(06).
    05   EMP-ID        PIC X(04).
    05   FILLER        PIC X(01).
    05   EMPL-LNAME    PIC X(30).
    05   FILLER        PIC X(01).
    05   EMPL-FNAME    PIC X(20).
    05   FILLER        PIC X(01).
    05   EMPL-YRS-SRV  PIC X(02).
    05   FILLER        PIC X(01).
    05   EMPL-PRM-DTE  PIC X(10).
    05   FILLER        PIC X(01).
    05   EMPL-SSN      PIC X(09).
```

Now let's assign EMPL-SSN values to the original flat file we used to load the database. Here it is:

```
BROWSE      USER01.EMPIFILE                      Line 00000000 Col 001 080
----+----1----+----2----+----3----+----4----+----5----+----6----+----7----+----8
 Command ===>                                          Scroll ===> CSR
****************************** Top of Data ******************************
1111 VEREEN                  CHARLES              12 2017-01-01 937253058
1122 JENKINS                 DEBORAH              05 2017-01-01 435092366
3217 JOHNSON                 EDWARD               04 2017-01-01 397342007
4175 TURNBULL                FRED                 01 2016-12-01 542083017
4720 SCHULTZ                 TIM                  09 2017-01-01 650450254
4836 SMITH                   SANDRA               03 2017-01-01 028374669
6288 WILLARD                 JOE                  06 2016-01-01 209883920
7459 STEWART                 BETTY                07 2016-07-31 019572830
9134 FRANKLIN                BRIANNA              00 2016-10-01 937293598
****************************** Bottom of Data ***************************
```

Now let's delete all existing records in the database (you can use File Manager for this as explained earlier in the chapter). Then let's run COBIMS1 to reload the database from our flat file which now includes the EMPL-SSN values. Now we can browse the database and verify that the EMPL-SSN field is populated (you will need to scroll to the right to see the EMPSSN field).

```
Browse           USER01.IMS.EMPLOYEE.CLUSTER              Top of 9
Command ===>                                              Scroll PAGE
                   Type KSDS     RBA                      Format CHAR
Key                                     Col 10
>----+----20---+----3----+----4----+----5----+----6----+----7----+----8----+---
**** Top of data ****
1 VEREEN                  CHARLES              12 2017-01-01 937253058..
2 JENKINS                 DEBORAH              05 2017-01-01 435092366..
7 JOHNSON                 EDWARD               04 2017-01-01 397342007..
5 TURNBULL                FRED                 01 2016-12-01 542083017..
0 SCHULTZ                 TIM                  09 2017-01-01 650450254..
6 SMITH                   SANDRA               03 2017-01-01 028374669..
8 WILLARD                 JOE                  06 2016-01-01 209883920..
9 STEWART                 BETTY                07 2016-07-31 019572830..
4 FRANKLIN                BRIANNA              00 2016-10-01 937293598..
```

Ok, next step. To be able to search on a field in an IMS segment, the field must be defined in the DBD. Recall our original code for the DBD is as follows:

```
PRINT NOGEN
DBD NAME=EMPLOYEE,ACCESS=HISAM
DATASET DD1=EMPLOYEE,OVFLW=EMPLFLW
SEGM NAME=EMPLOYEE,PARENT=0,BYTES=80
FIELD NAME=(EMPID,SEQ,U),BYTES=04,START=1,TYPE=C
SEGM NAME=EMPPAY,PARENT=EMPLOYEE,BYTES=23
FIELD NAME=(EFFDATE,SEQ,U),START=1,BYTES=8,TYPE=C
SEGM  NAME=EMPPAYHS,PARENT=EMPPAY,BYTES=18
FIELD NAME=(PAYDATE,SEQ,U),START=1,BYTES=8,TYPE=C
DBDGEN
FINISH
END
```

The only searchable field right now on the EMPLOYEE segment is the primary key EMPID. To make the EMPSSN field searchable we must add it to the DBD. The appropriate code is bolded below. Note that EMPSSN starts in position 72 of the record and is 9 bytes in length.

```
PRINT NOGEN
DBD NAME=EMPLOYEE,ACCESS=HISAM
DATASET DD1=EMPLOYEE,OVFLW=EMPLFLW
SEGM NAME=EMPLOYEE,PARENT=0,BYTES=80
FIELD NAME=(EMPID,SEQ,U),BYTES=04,START=1,TYPE=C
FIELD NAME=EMPSSN,START=72,BYTES=9,TYPE=C
SEGM NAME=EMPPAY,PARENT=EMPLOYEE,BYTES=23
FIELD NAME=(EFFDATE,SEQ,U),START=1,BYTES=8,TYPE=C
SEGM  NAME=EMPPAYHS,PARENT=EMPPAY,BYTES=18
FIELD NAME=(PAYDATE,SEQ,U),START=1,BYTES=8,TYPE=C
DBDGEN
FINISH
END
```

Go ahead and run the DBD gen process.

Next we can write a program to search on the EMPSSN field. We can clone the COBIMS2 program to make COBIMSA. One change we must make is to use a different qualified SSA than the one we started with. We need only change the field name in the SSA and create a value field with an appropriate specification (in this case a 9 position character field for the SSN key). Here is our new structure:

```
01 EMP-QUALIFIED-SSA-EMPSSN.
    05   SEGNAME      PIC X(08) VALUE 'EMPLOYEE'.
    05   FILLER       PIC X(01) VALUE '('.
    05   FIELD        PIC X(08) VALUE 'EMPSSN'.
    05   OPER         PIC X(02) VALUE ' ='.
    05   EMPSSN-VAL   PIC X(09) VALUE '         '.
    05   FILLER       PIC X(01) VALUE ')'.
```

Naturally you must load the EMPSSN-VAL variable with the value you are looking for. Let's use the social security number 937253058 for Charles Vereen who is employee number 1111. Here is our COBOL program source.

```
    IDENTIFICATION DIVISION.
    PROGRAM-ID. COBIMSA.

****************************************************
*    RETRIEVE A RECORD FROM IMS EMPLOYEE DATABASE  *
*    USING SEARCHABLE FIELD EMPSSN                 *
****************************************************
```

```
ENVIRONMENT DIVISION.
DATA DIVISION.

***********************************************************
*  W O R K I N G   S T O R A G E   S E C T I O N   *
***********************************************************

WORKING-STORAGE SECTION.

01 SEG-IO-AREA      PIC X(80).


01 DLI-FUNCTIONS.
    05 DLI-FUNCISRT  PIC X(4) VALUE 'ISRT'.
    05 DLI-FUNCGU    PIC X(4) VALUE 'GU  '.
    05 DLI-FUNCGN    PIC X(4) VALUE 'GN  '.
    05 DLI-FUNCGHU   PIC X(4) VALUE 'GHU '.
    05 DLI-FUNCGNP   PIC X(4) VALUE 'GNP '.
    05 DLI-FUNCREPL  PIC X(4) VALUE 'REPL'.
    05 DLI-FUNCDLET  PIC X(4) VALUE 'DLET'.
    05 DLI-FUNCXRST  PIC X(4) VALUE 'XRST'.
    05 DLI-FUNCCKPT  PIC X(4) VALUE 'CKPT'.

 01 EMP-UNQUALIFIED-SSA.
    05  SEGNAME     PIC X(08) VALUE 'EMPLOYEE'.
    05  FILLER      PIC X(01) VALUE ' '.

 01 EMP-QUALIFIED-SSA.
    05  SEGNAME     PIC X(08) VALUE 'EMPLOYEE'.
    05  FILLER      PIC X(01) VALUE '('.
    05  FIELD       PIC X(08) VALUE 'EMPID'.
    05  OPER        PIC X(02) VALUE ' ='.
    05  EMP-ID-VAL  PIC X(04) VALUE '    '.
    05  FILLER      PIC X(01) VALUE ')'.

 01 EMP-QUALIFIED-SSA-EMPSSN.
    05  SEGNAME     PIC X(08) VALUE 'EMPLOYEE'.
    05  FILLER      PIC X(01) VALUE '('.
    05  FIELD       PIC X(08) VALUE 'EMPSSN'.
    05  OPER        PIC X(02) VALUE ' ='.
    05  EMPSSN-VAL  PIC X(09) VALUE '         '.
    05  FILLER      PIC X(01) VALUE ')'.

 01 IMS-RET-CODES.
    05 THREE        PIC S9(9) COMP VALUE +3.
    05 FOUR         PIC S9(9) COMP VALUE +4.
    05 FIVE         PIC S9(9) COMP VALUE +5.
    05 SIX          PIC S9(9) COMP VALUE +6.


 LINKAGE SECTION.
 01 PCB-MASK.
```

```cobol
        03 DBD-NAME        PIC X(8).
        03 SEG-LEVEL       PIC XX.
        03 STATUS-CODE     PIC XX.
        03 PROC-OPT        PIC X(4).
        03 FILLER          PIC X(4).
        03 SEG-NAME        PIC X(8).
        03 KEY-FDBK        PIC S9(5) COMP.
        03 NUM-SENSEG      PIC S9(5) COMP.
        03 KEY-FDBK-AREA.
           05 EMPLOYEE-ID  PIC X(04).
           05 EMPPAYHS     PIC X(08).

    PROCEDURE DIVISION.

        INITIALIZE PCB-MASK
        ENTRY 'DLITCBL' USING PCB-MASK

        PERFORM P100-INITIALIZATION.
        PERFORM P200-MAINLINE.
        PERFORM P300-TERMINATION.
        GOBACK.

    P100-INITIALIZATION.

        DISPLAY '** PROGRAM COBIMSA START **'
        DISPLAY 'PROCESSING IN P100-INITIALIZATION'.

    P200-MAINLINE.

        DISPLAY 'PROCESSING IN P200-MAINLINE'

        MOVE '937253058' TO EMPSSN-VAL

        DISPLAY 'EMP-QUALIFIED-SSA-EMPSSN ' EMP-QUALIFIED-SSA-EMPSSN

        CALL 'CBLTDLI' USING FOUR,
                      DLI-FUNCGU,
                      PCB-MASK,
                      SEG-IO-AREA,
                      EMP-QUALIFIED-SSA-EMPSSN

        IF STATUS-CODE = ' '
           DISPLAY 'SUCCESSFUL GET CALL  '
           DISPLAY 'SEG-IO-ARE : ' SEG-IO-AREA
        ELSE
           DISPLAY 'ERROR IN FETCH :' STATUS-CODE
           PERFORM P400-DISPLAY-ERROR
        END-IF.

    P300-TERMINATION.

        DISPLAY 'PROCESSING IN P300-TERMINATION'
```

```
            DISPLAY '** COBIMSA - SUCCESSFULLY ENDED **'.

       P400-DISPLAY-ERROR.

            DISPLAY 'ERROR ENCOUNTERED - DETAIL FOLLOWS'
            DISPLAY 'SEG-IO-AREA     :' SEG-IO-AREA
            DISPLAY 'DBD-NAME1:'     DBD-NAME
            DISPLAY 'SEG-LEVEL1:'    SEG-LEVEL
            DISPLAY 'STATUS-CODE:'   STATUS-CODE
            DISPLAY 'PROC-OPT1 :'    PROC-OPT
            DISPLAY 'SEG-NAME1 :'    SEG-NAME
            DISPLAY 'KEY-FDBK1 :'    KEY-FDBK
            DISPLAY 'NUM-SENSEG1:'   NUM-SENSEG
            DISPLAY 'KEY-FDBK-AREA1:' KEY-FDBK-AREA.

       *    END OF SOURCE CODE
```

Again we compile, link and execute. Here's the output:

```
** PROGRAM COBIMSA START **
PROCESSING IN P100-INITIALIZATION
PROCESSING IN P200-MAINLINE
EMP-QUALIFIED-SSA-EMPSSN EMPLOYEE(EMPSSN   =937253058)
SUCCESSFUL GET CALL
SEG-IO-ARE : 1111 VEREEN            CHARLES            12 2017-01-01 937253058
PROCESSING IN P300-TERMINATION
** COBIMSA - SUCCESSFULLY ENDED **
```

As you can see, we retrieved the desired record using the EMPSSN search field. So keep in mind that you can search on fields other than the key field as long as they are defined in the DBD. If you are going to be searching on a non-indexed field often, you'll want to check with your DBA about possibly defining a secondary index.

Retrieve Segments Using Boolean SSAs

The qualified SSA retrievals we've done so far have searched using a field value that is equal to a single searchable field. It is also possible to retrieve IMS segments using other Boolean operators such as greater than or less than. Additionally, you can specify more than one operator, such as > VALUE1 and < VALUE2.

For this example with program COBIMSB we will retrieve root segments for all employees whose EMPID is greater than 3000 and less than 7000. For that we simply need to create and use a new SSA. Here it is:

```
       01 EMP-QUALIFIED-SSA-BOOL.
          05  SEGNAME    PIC X(08) VALUE 'EMPLOYEE'.
          05  FILLER     PIC X(01) VALUE '('.
```

253

```
05   FIELD       PIC X(08) VALUE 'EMPID'.
05   OPER        PIC X(02) VALUE '>='.
05   EMP-ID-VAL1 PIC X(04) VALUE '    '.
05   OPER        PIC X(01) VALUE '&'.
05   FIELD2      PIC X(08) VALUE 'EMPID'.
05   OPER2       PIC X(02) VALUE '<='.
05   EMP-ID-VAL2 PIC X(04) VALUE '    '.
05   FILLER      PIC X(01) VALUE ')'.
```

For the above we must load (or initialize) the minimum value 3000 into `EMP-ID-VAL1`, and the ceiling value 7000 into `EMP-ID-VAL2`. Then we'll call the database using the `EMP-QUALIFIED-SSA-BOOL` SSA. We'll do a loop through the database and our retrieval loop should only return those employee records that satisfy the Boolean SSA.

Note that to end our read loop, we check both for IMS status codes GB and GE. This is because the last record that satisfies the database call may not be the physical end of the database. Consequently reading beyond the end of the "result set" of your database call will result in a GE status code unless it happens to also be the end of the database. So you have to check for both GB and GE.

Here is our program source code.

```
IDENTIFICATION DIVISION.
PROGRAM-ID. COBIMSB.
*****************************************************
*   WALK THROUGH THE EMPLOYEE SEGMENTS OF THE ENTIRE *
*   EMPLOYEE IMS DATABASE USING BOOLEAN SSA.         *
*****************************************************
ENVIRONMENT DIVISION.
INPUT-OUTPUT SECTION.
DATA DIVISION.
*****************************************************
*  W O R K I N G    S T O R A G E    S E C T I O N   *
*****************************************************
WORKING-STORAGE SECTION.
  01 WS-FLAGS.
     05  SW-END-OF-DB-SWITCH     PIC X(1) VALUE 'N'.
         88  SW-END-OF-DB                  VALUE 'Y'.
         88  SW-NOT-END-OF-DB              VALUE 'N'.

  01 DLI-FUNCTIONS.
     05 DLI-FUNCISRT  PIC X(4) VALUE 'ISRT'.
     05 DLI-FUNCGU    PIC X(4) VALUE 'GU  '.
     05 DLI-FUNCGN    PIC X(4) VALUE 'GN  '.
     05 DLI-FUNCGHU   PIC X(4) VALUE 'GHU '.
     05 DLI-FUNCGNP   PIC X(4) VALUE 'GNP '.
     05 DLI-FUNCREPL  PIC X(4) VALUE 'REPL'.
```

```cobol
        05 DLI-FUNCDLET  PIC X(4) VALUE 'DLET'.
        05 DLI-FUNCXRST  PIC X(4) VALUE 'XRST'.
        05 DLI-FUNCCKPT  PIC X(4) VALUE 'CKPT'.

    01 IO-EMPLOYEE-RECORD.
        05  EMPL-ID-IN    PIC X(04).
        05  FILLER        PIC X(01).
        05  EMPL-LNAME    PIC X(30).
        05  FILLER        PIC X(01).
        05  EMPL-FNAME    PIC X(20).
        05  FILLER        PIC X(01).
        05  EMPL-YRS-SRV  PIC X(02).
        05  FILLER        PIC X(01).
        05  EMPL-PRM-DTE  PIC X(10).
        05  FILLER        PIC X(10).

     01 EMP-UNQUALIFIED-SSA.
        05  SEGNAME      PIC X(08) VALUE 'EMPLOYEE'.
        05  FILLER       PIC X(01) VALUE ' '.

     01 EMP-QUALIFIED-SSA.
        05  SEGNAME      PIC X(08) VALUE 'EMPLOYEE'.
        05  FILLER       PIC X(01) VALUE '('.
        05  FIELD        PIC X(08) VALUE 'EMPID'.
        05  OPER         PIC X(02) VALUE ' ='.
        05  EMP-ID-VAL   PIC X(04) VALUE '    '.
        05  FILLER       PIC X(01) VALUE ')'.

    01 EMP-QUALIFIED-SSA-BOOL.
        05  SEGNAME      PIC X(08) VALUE 'EMPLOYEE'.
        05  FILLER       PIC X(01) VALUE '('.
        05  FIELD        PIC X(08) VALUE 'EMPID'.
        05  OPER         PIC X(02) VALUE '>='.
        05  EMP-ID-VAL1  PIC X(04) VALUE '    '.
        05  OPER         PIC X(01) VALUE '&'.
        05  FIELD2       PIC X(08) VALUE 'EMPID'.
        05  OPER2        PIC X(02) VALUE '<='.
        05  EMP-ID-VAL2  PIC X(04) VALUE '    '.
        05  FILLER       PIC X(01) VALUE ')'.

     01 SEG-IO-AREA    PIC X(80).

     01 IMS-RET-CODES.
        05 THREE            PIC S9(9) COMP VALUE +3.
        05 FOUR             PIC S9(9) COMP VALUE +4.
        05 FIVE             PIC S9(9) COMP VALUE +5.
        05 SIX              PIC S9(9) COMP VALUE +6.

    LINKAGE SECTION.
     01 PCB-MASK.
        03 DBD-NAME         PIC X(8).
        03 SEG-LEVEL        PIC XX.
```

```
        03 STATUS-CODE      PIC XX.
        03 PROC-OPT         PIC X(4).
        03 FILLER           PIC X(4).
        03 SEG-NAME         PIC X(8).
        03 KEY-FDBK         PIC S9(5) COMP.
        03 NUM-SENSEG       PIC S9(5) COMP.
        03 KEY-FDBK-AREA.
           05 EMPLOYEE-KEY  PIC X(04).
           05 EMPPAYHS-KEY  PIC X(08).

   PROCEDURE DIVISION.

       INITIALIZE PCB-MASK
       ENTRY 'DLITCBL' USING PCB-MASK

       PERFORM P100-INITIALIZATION.
       PERFORM P200-MAINLINE.
       PERFORM P300-TERMINATION.
       GOBACK.

   P100-INITIALIZATION.

       DISPLAY '** PROGRAM COBIMSB START **'
       DISPLAY 'PROCESSING IN P100-INITIALIZATION'.
       MOVE '3000' TO EMP-ID-VAL1
       MOVE '7000' TO EMP-ID-VAL2

*      DO INITIAL DB READ FOR FIRST EMPLOYEE RECORD

       CALL 'CBLTDLI' USING FOUR,
            DLI-FUNCGN,
            PCB-MASK,
            SEG-IO-AREA,
            EMP-QUALIFIED-SSA-BOOL

       IF STATUS-CODE = '  ' THEN
          NEXT SENTENCE
       ELSE
          IF STATUS-CODE = 'GE' OR
             STATUS-CODE = 'GB' THEN
             SET SW-END-OF-DB TO TRUE
             DISPLAY 'END OF DATABASE :'
          ELSE
             PERFORM P400-DISPLAY-ERROR
             GOBACK
          END-IF

       END-IF.

   P200-MAINLINE.

       DISPLAY 'PROCESSING IN P200-MAINLINE'
```

256

```
*    CHECK STATUS CODE AND FIRST RECORD

     IF SW-END-OF-DB THEN
        DISPLAY 'NO RECORDS TO PROCESS!!'
     ELSE
        PERFORM UNTIL SW-END-OF-DB
           DISPLAY 'SUCCESSFUL READ :' SEG-IO-AREA
           CALL 'CBLTDLI' USING FOUR,
                 DLI-FUNCGN,
                 PCB-MASK,
                 SEG-IO-AREA,
                 EMP-QUALIFIED-SSA-BOOL

           IF STATUS-CODE = 'GB' OR 'GE' THEN
              SET SW-END-OF-DB TO TRUE
              DISPLAY 'END OF DATABASE'
           ELSE
              IF STATUS-CODE NOT EQUAL SPACES THEN
                 PERFORM P400-DISPLAY-ERROR
                 GOBACK
              END-IF
           END-IF

        END-PERFORM.

     DISPLAY 'FINISHED PROCESSING IN P200-MAINLINE'.

 P300-TERMINATION.

     DISPLAY 'PROCESSING IN P300-TERMINATION'
     DISPLAY '** COBIMSB - SUCCESSFULLY ENDED **'.

 P400-DISPLAY-ERROR.

     DISPLAY 'ERROR ENCOUNTERED - DETAIL FOLLOWS'
     DISPLAY 'SEG-IO-AREA      :' SEG-IO-AREA
     DISPLAY 'DBD-NAME1:'      DBD-NAME
     DISPLAY 'SEG-LEVEL1:'     SEG-LEVEL
     DISPLAY 'STATUS-CODE:'    STATUS-CODE
     DISPLAY 'PROC-OPT1 :'     PROC-OPT
     DISPLAY 'SEG-NAME1 :'     SEG-NAME
     DISPLAY 'KEY-FDBK1 :'     KEY-FDBK
     DISPLAY 'NUM-SENSEG1:'    NUM-SENSEG
     DISPLAY 'KEY-FDBK-AREA1:' KEY-FDBK-AREA.

*    END OF SOURCE CODE
```

After we compile, link and execute, here is the output. As you can see, the only employees retrieved are those whose ids fall between 3,000 and 7,000 inclusive.

```
** PROGRAM COBIMSB START **
PROCESSING IN P100-INITIALIZATION
EMP-QUALIFIED-SSA-BOOL EMPLOYEE(EMPID   >=3000&EMPID   <=7000)
PROCESSING IN P200-MAINLINE
SUCCESSFUL READ :3217 JOHNSON                    EDWARD          04 201
SUCCESSFUL READ :4175 TURNBULL                   FRED            01 201
SUCCESSFUL READ :4720 SCHULTZ                    TIM             09 201
SUCCESSFUL READ :4836 SMITH                      SANDRA          03 201
SUCCESSFUL READ :6288 WILLARD                    JOE             06 201
END OF DATABASE
FINISHED PROCESSING IN P200-MAINLINE
PROCESSING IN P300-TERMINATION
** COBIMSB - SUCCESSFULLY ENDED **
```

Extended Boolean SSAs can be very handy when you need to ready a range of values, or for any retrieval that must satisfy multiple conditions.

Command Codes

IMS command codes change and/or extend the way an IMS call works. There are about 18 command codes that serve various purposes. See the table at the end of this topic for all the command codes and what they do.

We'll do an example of the C command code. The C command code allows you to issue a qualified SSA using the concatenated key for a child segment rather than using separate SSAs for the various parent/child segments. For example suppose we want to retrieve the paycheck record of employee 3217 for pay effective January 1, 2017, and for payday February 15, 2017. The concatenated key for that is as follows:

```
32172017010120170215
```

This is the key for the root segment (3217) plus the key for the EMPPAY segment (20170101), plus the key for the EMPPAYHS segment (20170215).

To use the C command code, we must create a new SSA structure that uses both the C command code, and accommodates the concatenated key. It will look like this:

```
01 EMPPAYHS-CCODE-SSA.
    05   SEGNAME     PIC X(08) VALUE 'EMPPAYHS'.
    05   FILLER      PIC X(02) VALUE '*C'.
    05   FILLER      PIC X(01) VALUE '('.
    05   CONCATKEY   PIC X(20) VALUE SPACES.
    05   FILLER      PIC X(01) VALUE ')'.
```

Like all SSAs, our new one includes the segment name. Position 9 of the SSA will contain an asterisk (*) or blank if a command code is not being used. We put a C in position 10 to indicate we are using a concatenated key command code. We've named our concatenated key variable CONCATKEY (the name is arbitrary – you could use any name for this variable).

The CONCATKEY length is 20 bytes (4 for the employee id, and 8 each for the salary effective date and the pay date. We have initialized the concatenated key variable to the value we are looking for. You could also load it using a MOVE statement.

Ok here is the complete code for COBIMSC. It should look very familiar except for the SSA. For comparison, we will first use the regular multiple SSA method to call the 2/15 pay record. Then we will use a second call with the C command code method and a concatenated key. The results should be identical.

```
         IDENTIFICATION DIVISION.
         PROGRAM-ID. COBIMSC.

         ****************************************************
         *    READ AND DISPLAY EMP HISTORY RECORD FROM      *
         *    EMPLOYEE IMS DATABASE. THIS EXAMPLE USES A     *
         *    C COMMAND CODE TO PROVIDE THE CONCATENATED     *
         *    KEY SSA RATHER THAN A QUALIFICATION STATEMENT  *
         *    SSA (second example).                          *
         ****************************************************

         ENVIRONMENT DIVISION.
         DATA DIVISION.

         ****************************************************
         *  W O R K I N G   S T O R A G E   S E C T I O N   *
         ****************************************************

         WORKING-STORAGE SECTION.

         01 WS-FLAGS.
            05  SW-END-OF-DB-SWITCH      PIC X(1) VALUE 'N'.
                88  SW-END-OF-DB                  VALUE 'Y'.
                88  SW-NOT-END-OF-DB              VALUE 'N'.
            05  SW-END-OF-EMPPAYHS-SW    PIC X(1) VALUE 'N'.
                88  SW-END-OF-EMPPAYHS            VALUE 'Y'.
                88  SW-NOT-END-OF-EMPPAYHS        VALUE 'N'.

         01 IO-EMPLOYEE-RECORD.
            05  EMPL-ID       PIC X(04).
            05  FILLER        PIC X(01).
            05  EMPL-LNAME    PIC X(30).
            05  FILLER        PIC X(01).
            05  EMPL-FNAME    PIC X(20).
```

```cobol
       05  FILLER        PIC X(01).
       05  EMPL-YRS-SRV  PIC X(02).
       05  FILLER        PIC X(01).
       05  EMPL-PRM-DTE  PIC X(10).
       05  FILLER        PIC X(10).

   01 IO-EMPPAY-RECORD.
       05  PAY-EFF-DATE  PIC X(8).
       05  PAY-REG-PAY   PIC S9(6)V9(2) USAGE COMP-3.
       05  PAY-BON-PAY   PIC S9(6)V9(2) USAGE COMP-3.
       05  SEMIMTH-PAY   PIC S9(6)V9(2) USAGE COMP-3.
       05  FILLER        PIC X(57).

   01 IO-EMPPAYHS-RECORD.
       05  PAY-DATE      PIC X(8).
       05  PAY-ANN-PAY   PIC S9(6)V9(2) USAGE COMP-3.
       05  PAY-AMT       PIC S9(6)V9(2) USAGE COMP-3.
       05  FILLER        PIC X(62).

   01 SEG-IO-AREA     PIC X(80).

   01 IMS-RET-CODES.
      05 THREE          PIC S9(9) COMP VALUE +3.
      05 FOUR           PIC S9(9) COMP VALUE +4.
      05 FIVE           PIC S9(9) COMP VALUE +5.
      05 SIX            PIC S9(9) COMP VALUE +6.

   01 DLI-FUNCTIONS.
      05 DLI-FUNCISRT PIC X(4) VALUE 'ISRT'.
      05 DLI-FUNCGU   PIC X(4) VALUE 'GU  '.
      05 DLI-FUNCGN   PIC X(4) VALUE 'GN  '.
      05 DLI-FUNCGHU  PIC X(4) VALUE 'GHU '.
      05 DLI-FUNCGNP  PIC X(4) VALUE 'GNP '.
      05 DLI-FUNCREPL PIC X(4) VALUE 'REPL'.
      05 DLI-FUNCDLET PIC X(4) VALUE 'DLET'.
      05 DLI-FUNCXRST PIC X(4) VALUE 'XRST'.
      05 DLI-FUNCCKPT PIC X(4) VALUE 'CKPT'.

   01 DISPLAY-EMPPAYHS-PIC.
       05  DIS-REG-PAY  PIC ZZ999.99-.
       05  DIS-SMT-PAY  PIC ZZ999.99-.

    01 EMP-UNQUALIFIED-SSA.
        05  SEGNAME     PIC X(08) VALUE 'EMPLOYEE'.
        05  FILLER      PIC X(01) VALUE ' '.

    01 EMP-QUALIFIED-SSA.
        05  SEGNAME     PIC X(08) VALUE 'EMPLOYEE'.
        05  FILLER      PIC X(01) VALUE '('.
        05  FIELD       PIC X(08) VALUE 'EMPID'.
        05  OPER        PIC X(02) VALUE ' ='.
        05  EMP-ID-VAL  PIC X(04) VALUE '    '.
```

```
            05  FILLER       PIC X(01) VALUE ')'.

   01 EMPPAY-UNQUALIFIED-SSA.
       05  SEGNAME      PIC X(08) VALUE 'EMPPAY  '.
       05  FILLER       PIC X(01) VALUE ' '.

   01 EMPPAY-QUALIFIED-SSA.
       05  SEGNAME      PIC X(08) VALUE 'EMPPAY  '.
       05  FILLER       PIC X(01) VALUE '('.
       05  FIELD        PIC X(08) VALUE 'EFFDATE '.
       05  OPER         PIC X(02) VALUE ' ='.
       05  EFFDATE-VAL  PIC X(08) VALUE '         '.
       05  FILLER       PIC X(01) VALUE ')'.

   01 EMPPAYHS-UNQUALIFIED-SSA.
       05  SEGNAME      PIC X(08) VALUE 'EMPPAYHS'.
       05  FILLER       PIC X(01) VALUE ' '.

   01 EMPPAYHS-QUALIFIED-SSA.
       05  SEGNAME      PIC X(08) VALUE 'EMPPAYHS'.
       05  FILLER       PIC X(01) VALUE '('.
       05  FIELD        PIC X(08) VALUE 'PAYDATE '.
       05  OPER         PIC X(02) VALUE ' ='.
       05  PAYDATE-VAL  PIC X(08) VALUE '         '.
       05  FILLER       PIC X(01) VALUE ')'.

   01 EMPPAYHS-CCODE-SSA.
       05  SEGNAME      PIC X(08) VALUE 'EMPPAYHS'.
       05  FILLER       PIC X(02) VALUE '*C'.
       05  FILLER       PIC X(01) VALUE '('.
       05  CONCATKEY    PIC X(20) VALUE '321720170101202017 0215'.
       05  FILLER       PIC X(01) VALUE ')'.


   LINKAGE SECTION.
    01 PCB-MASK.
       03 DBD-NAME       PIC X(8).
       03 SEG-LEVEL      PIC XX.
       03 STATUS-CODE    PIC XX.
       03 PROC-OPT       PIC X(4).
       03 FILLER         PIC X(4).
       03 SEG-NAME       PIC X(8).
       03 KEY-FDBK       PIC S9(5) COMP.
       03 NUM-SENSEG     PIC S9(5) COMP.
       03 KEY-FDBK-AREA.
          05 EMPLOYEE-ID  PIC X(04).
          05 EMPPAYHS     PIC X(08).

   PROCEDURE DIVISION.

       INITIALIZE PCB-MASK
       ENTRY 'DLITCBL' USING PCB-MASK
```

```
        PERFORM P100-INITIALIZATION.
        PERFORM P200-MAINLINE.
        PERFORM P300-TERMINATION.
        GOBACK.

P100-INITIALIZATION.

        DISPLAY '** PROGRAM COBIMSC START **'
        DISPLAY 'PROCESSING IN P100-INITIALIZATION'.


P200-MAINLINE.

        DISPLAY 'PROCESSING IN P200-MAINLINE'

        MOVE '3217'      TO EMP-ID-VAL
        MOVE '20170101' TO EFFDATE-VAL
        MOVE '201700215' TO PAYDATE-VAL

        CALL 'CBLTDLI' USING SIX,
            DLI-FUNCGU,
            PCB-MASK,
            IO-EMPPAYHS-RECORD,
            EMP-QUALIFIED-SSA,
            EMPPAY-QUALIFIED-SSA,
            EMPPAYHS-QUALIFIED-SSA.

        EVALUATE STATUS-CODE
           WHEN ' '
              DISPLAY 'GOOD READ OF EMPPAYHS : '
                 EMP-ID-VAL
              MOVE PAY-ANN-PAY TO DIS-REG-PAY
              MOVE PAY-AMT     TO DIS-SMT-PAY
              DISPLAY 'PAY-DATE   : ' PAY-DATE
              DISPLAY 'PAY-ANN-PAY: ' DIS-REG-PAY
              DISPLAY 'PAY-AMT    : ' DIS-SMT-PAY
           WHEN 'GE'
           WHEN 'GB'
              DISPLAY 'PAY HISTORY SEGMENT NOT FOUND'
           WHEN OTHER
              PERFORM P9000-DISPLAY-ERROR
              GOBACK
         END-EVALUATE.

        DISPLAY 'NOW CALLING THE 2/15/2017 REC USING C COMMAND CODE'

        CALL 'CBLTDLI' USING FOUR,
            DLI-FUNCGU,
            PCB-MASK,
            IO-EMPPAYHS-RECORD,
            EMPPAYHS-CCODE-SSA.
```

```
      EVALUATE STATUS-CODE
         WHEN ' '
            DISPLAY 'GOOD READ OF EMPPAYHS : '
               EMP-ID-VAL
            MOVE PAY-ANN-PAY TO DIS-REG-PAY
            MOVE PAY-AMT     TO DIS-SMT-PAY
            DISPLAY 'PAY-DATE   : ' PAY-DATE
            DISPLAY 'PAY-ANN-PAY: ' DIS-REG-PAY
            DISPLAY 'PAY-AMT    : ' DIS-SMT-PAY
         WHEN 'GE'
         WHEN 'GB'
            DISPLAY 'PAY HISTORY SEGMENT NOT FOUND'
         WHEN OTHER
            PERFORM P9000-DISPLAY-ERROR
            GOBACK
       END-EVALUATE.

       DISPLAY 'FINISHED PROCESSING IN P200-MAINLINE'.

   P300-TERMINATION.

       DISPLAY 'PROCESSING IN P300-TERMINATION'
       DISPLAY '** COBIMSC - SUCCESSFULLY ENDED **'.

   P9000-DISPLAY-ERROR.

       DISPLAY 'ERROR ENCOUNTERED - DETAIL FOLLOWS'
       DISPLAY 'DBD-NAME1:'     DBD-NAME
       DISPLAY 'SEG-LEVEL1:'    SEG-LEVEL
       DISPLAY 'STATUS-CODE:'   STATUS-CODE
       DISPLAY 'PROC-OPT1 :'    PROC-OPT
       DISPLAY 'SEG-NAME1 :'    SEG-NAME
       DISPLAY 'KEY-FDBK1 :'    KEY-FDBK
       DISPLAY 'NUM-SENSEG1:'   NUM-SENSEG
       DISPLAY 'KEY-FDBK-AREA1:' KEY-FDBK-AREA.

   *    END OF SOURCE CODE
```

Ok, once again we compile, link and execute. Here is our output.

```
** PROGRAM COBIMSC START **
PROCESSING IN P100-INITIALIZATION
PROCESSING IN P200-MAINLINE
FIRST CALL THE 2/15/2017 PAY REC WITH 3 SSA METHOD
GOOD READ OF EMPPAYHS : 3217
PAY-DATE   : 20170215
PAY-ANN-PAY: 65000.00
PAY-AMT    : 2708.33
NOW CALLING THE 2/15/2017 REC USING C COMMAND CODE
GOOD READ OF EMPPAYHS : 3217
```

```
PAY-DATE    : 20170215
PAY-ANN-PAY: 65000.00
PAY-AMT     :  2708.33
FINISHED PROCESSING IN P200-MAINLINE
PROCESSING IN P300-TERMINATION
** COBIMSC - SUCCESSFULLY ENDED **
```

Command codes can be very useful when you need the features they offer. Check out the following table of the command codes and how they are used. This information is from the IBM product web site. [11] You'll find more detail about each command code there as well.

Summary of Command Codes

Command Code	Description
A	Clear positioning and start the call at the beginning of the database.
C	Use the concatenated key of a segment to identify the segment.
D	Retrieve or insert a sequence of segments in a hierarchic path using only one call, instead of using a separate (path) call for each segment.
F	Back up to the first occurrence of a segment under its parent when searching for a particular segment occurrence. Disregarded for a root segment.
G	Prevent randomization or the calling of the HALDB Partition Selection exit routine and search the database sequentially.
L	Retrieve the last occurrence of a segment under its parent.
M	Move a subset pointer to the next segment occurrence after your current position. (Used with DEDBs only.)
N	Designate segments that you do not want replaced when replacing segments after a Get Hold call. Typically used when replacing a path of segments.
O	Either field names or both segment position and lengths can be contained in the SSA qualification for combine field position.
P	Set parentage at a higher level than what it usually is (the lowest-level SSA of the call).
Q	Reserve a segment so that other programs cannot update it until you have finished processing and updating it.
R	Retrieve the first segment occurrence in a subset. (Used with DEDBs only.)
S	Unconditionally set a subset pointer to the current position. (Used with DEDBs only.)

11 https://www.ibm.com/support/knowledgecenter/en/SSEPH2_13.1.0/com.ibm.ims13.doc.apr/ims_cmdcodref.htm

Command Code	Description
U	Limit the search for a segment to the dependents of the segment occurrence on which position is established.
V	Use the hierarchic level at the current position and higher as qualification for the segment.
W	Set a subset pointer to your current position, if the subset pointer is not already set. (Used with DEDBs only.)
Z	Set a subset pointer to 0, so it can be reused. (Used with DEDBs only.)
-	NULL. Use an SSA in command code format without specifying the command code. Can be replaced during execution with the command codes that you want.

Committing and Rolling Back Changes

Let's look at how we commit updated data to the database. This is not difficult to do using checkpoint calls. Using checkpoint **restart** is somewhat more involved, especially for running in DLI mode where you must use a log file. We'll provide examples of both checkpointing and checkpoint restarting. It will be better if we take it in two chunks with two programs, so that's what we'll do.

For COBIMSD our objective is to delete all the records in the database. We use the same walk-through-the-database code we used in COBIMS3 except we will use GHN to do the walking, and we will add a DLET call after each GHN to delete the root segment. Note: all child segments are automatically deleted when a root segment is deleted. In fact the principle is even broader - all children under a parent segment are deleted if the parent segment is deleted.

We will also set up checkpointing to show its usage. We will need to do four things before checkpointing can work.

1. Change the PSB to include an IO-PCB

2. Add an XRST call before any data related IMS calls are done

3. Add CHKP calls at specified intervals

4. Add code to reset database position after a checkpoint

Modifying the PSB to Add An IO-PCB

We have to back up a bit to make a fundamental change to our PSB. In order to issue IMS service commands like CHKP (as opposed to database retrieval or update commands) you must use a special PCB called the IO-PCB. Programs that run in BMP mode are always defined to use an IO-PCB, but those that run in DLI mode by default do not have to use an IO-PCB (unless they are doing IMS service calls).

Since we have only been running in DLI mode and not issuing IMS service calls, we didn't define our PSB to include an IOPCB. Since we must now use an IO-PCB to use CHKP calls, let's modify our PSB accordingly. The change is very simple and involves adding a **CMPAT=Y** clause after the PSBNAME= clause. Let's create a separate PSB named EMPPSBZ. It will be a clone of the EMPPSB except for the CMPAT=Y. Here is the code:

```
PRINT NOGEN
PCB    TYPE=DB,NAME=EMPLOYEE,KEYLEN=20,PROCOPT=AP
SENSEG NAME=EMPLOYEE,PARENT=0
SENSEG NAME=EMPPAY,PARENT=EMPLOYEE
SENSEG NAME=EMPPAYHS,PARENT=EMPPAY
SENSEG NAME=EMPDEP,PARENT=EMPLOYEE
PSBGEN LANG=COBOL,PSBNAME=EMPLOYEE,CMPAT=YES
END
```

Let's save this as member EMPPSBZ in our library and run the PSBGEN process.

So what practical effect does this have if we use the EMPPSBZ PSB to run a program? Basically this PSB **implicitly** includes an IO-PCB, meaning you don't see an IO-PCB defined in the PSB, but it must be the first PCB pointer in the linkage between your program and IMS. Since we defined the PSB this way, you **must** handle the IO-PCB in your program by:

- Including a structure for the IO-PCB.
- Including the IO-PCB structure name in the ENTRY statement in the procedure division.

Here is our new IO-PCB structure:

```
01 IO-PCB.
   05 FILLER           PICTURE X(10).
   05 IO-STATUS-CODE   PICTURE XX.
   05 FILLER           PICTURE X(20).
```

And here is the change to the ENTRY coded in the procedure division. Notice it now includes both the IO-PCB and the PCB-MASK structures.

```
ENTRY 'DLITCBL' USING IO-PCB, PCB-MASK
```

You MUST put the IO-PCB first in the parameter list before any database PCBs. The database PCBs that follow should be in the same order that they are defined in the PSB. Now we can move on to doing the restart call.

Adding an XRST Call to Initialization Routine

Now we need to include an XRST (Extended Restart Facility) call to check for restart. Don't worry that we won't actually be restarting with this program yet (the reason is because we aren't logging our changes yet – be patient, we'll get there in the next program). The XRST call is part of the procedure that we need to do symbolic checkpoints and eventually perform IMS restarts, so we include it here. [12]

First, add these structures and variables to your working storage section.

```
01 XRST-IOAREA.
   05 XRST-ID      PIC X(08) VALUE SPACES.
   05 FILLER       PIC X(04) VALUE SPACES.

77 IO-AREALEN      PIC S9(9) USAGE IS BINARY VALUE 12.

77 CHKP-ID         PIC X(08) VALUE 'IMSD    '.

77 CHKP-NBR        PIC 999   VALUE ZERO.
77 CHKP-COUNT      PIC S9(9) USAGE IS BINARY VALUE ZERO.

01 CHKP-MESSAGE.
   05 FILLER               PIC X(24) VALUE
      'COBIMSD  CHECK POINT NO:'.
   05 CHKP-MESS-NBR        PIC 999     VALUE ZERO.
   05 FILLER               PIC X(15)   VALUE ',AT INPUT REC#:'.
   05 CHKP-MESS-REC        PIC ZZZZZ9  VALUE SPACES.
   05 FILLER               PIC X(10)   VALUE ',AT EMP#:'.
   05 CHKP-MESS-EMP        PIC X(08)   VALUE SPACES.

01 IMS-CHKP-AREA-LTH.
   05 LEN                 PIC S9(9) USAGE IS BINARY VALUE +7.
```

---

12 In this text we will only deal with symbolic checkpoints. IMS also offers basic checkpoints, but these do not work with the extended restart facility (the XRST call and automated repositions, etc), so with basic checkpoints your program must do 100% of the code to perform a restart. Consequently basic checkpoints are of limited value and I don't deal with them in this text.

267

```
01 IMS-CHKP-AREA.
   05 CHKP-EMP-ID      PIC X(04) VALUE SPACES.
   05 CHKP-NBR-LAST    PIC 999   VALUE 0.
```

Second, add this code at the beginning of your Initialization paragraph.

```
* CHECK FOR RESTART

    CALL 'CBLTDLI' USING SIX,
         DLI-FUNCXRST,
         PCB-MASK,
         IO-AREALEN,
         XRST-IOAREA,
         IMS-CHKP-AREA-LTH,
         IMS-CHKP-AREA

    IF STATUS-CODE NOT EQUAL SPACES THEN
       PERFORM P9000-DISPLAY-ERROR
       GOBACK
    END-IF

    IF XRST-ID NOT EQUAL SPACES THEN
       MOVE CHKP-NBR-LAST TO CHKP-NBR
       DISPLAY '*** COBIMSD IMS RESTART ***'
       DISPLAY '*  LAST CHECK POINT :' XRST-ID
       DISPLAY '*  EMPLOYEE NUMBER  :' CHKP-EMP-ID
    ELSE
       DISPLAY '****** COBIMSD IMS NORMAL START ***'
       PERFORM P8000-TAKE-CHECKPOINT
    END-IF.
```

This code checks to see if our execution is being run as a restart. If it is, then we announce that it is a restart. If it is not, we announce a normal start. That's all we need to do with XRST right now. Later we will add code to perform the various restart actions, and we'll explain the parameters at that time.

Adding the CHKP Call

Now let's add code for taking a checkpoint. We'll code a separate procedure for this. The required parameters for the call are the CHKP function, the IO-PCB structure, the length of an IO area that contains the checkpoint id, the IO area that contains the checkpoint id, the length of the checkpoint area, and the checkpoint area structure. The latter is where you save anything you want to save for restart, such as the last processed EMP-ID, record counters and anything else you want to save for a restart. Here is the code for doing the checkpoint call.

```
P8000-TAKE-CHECKPOINT.
    DISPLAY 'PROCESSING IN P8000-TAKE-CHECKPOINT'
```

```
         ADD +1               TO CHKP-NBR
         MOVE CHKP-NBR         TO CHKP-NBR-LAST
         MOVE CHKP-NBR-LAST TO CHKP-ID(6:3)
         MOVE EMP-ID           TO CHKP-EMP-ID

         CALL 'CBLTDLI' USING SIX,
              DLI-FUNCCHKP,
              IO-PCB,
              IO-AREALEN,
              CHKP-ID,
              IMS-CHKP-AREA-LTH,
              IMS-CHKP-AREA

         IF IO-STATUS-CODE NOT EQUAL SPACES THEN
            DISPLAY 'TOOK AN ERROR DOING THE CHECKPOINT'
            DISPLAY 'IO-STATUS-CODE ' IO-STATUS-CODE
            PERFORM P9000-DISPLAY-ERROR
            PERFORM P9000-DISPLAY-ERROR
            GOBACK
         ELSE
            MOVE 0 TO CHKP-COUNT
            MOVE CHKP-NBR         TO CHKP-MESS-NBR
            MOVE CHKP-EMP-ID      TO CHKP-MESS-EMP
            DISPLAY CHKP-MESSAGE
         END-IF.
```

One final note: the third parameter in the CHKP call (the IO area length) is not actually used by IMS, but it must still be included for backward compatibility. You need only define a variable for it in the program.

Adding Code to Reposition in the Database After Checkpoint

Finally, we must create code to reposition the database after taking a checkpoint. The reason is that the checkpoint call causes the database position to be lost. If you continue GHN calls at this point without reestablishing your database position, you'll get an error.

So what we'll do is to ensure we have the next record to process and we'll include that in the checkpoint IO area that we are going to save. So our code will:

> DLET a record
> Read the next record and capture the employee id
> If it is time to take a checkpoint then
>> Take a check point using the captured employee id that was just read
>> Reposition in the database using the captured employee id

The reposition code is as follows. Notice it is using a qualified SSA to get the exact record that is needed to reposition. Of course we must use a qualified SSA, and the EMP-ID that was retrieved in the GHN call before we took the checkpoint.

```
P1000-RESET-POSITION.

    DISPLAY 'PROCESSING IN P1000-RESET-POSITION'

    CALL 'CBLTDLI' USING FOUR,
        DLI-FUNCGHU,
        PCB-MASK,
        IO-EMPLOYEE-RECORD,
        EMP-QUALIFIED-SSA

    IF STATUS-CODE NOT EQUAL SPACES THEN
        PERFORM P9000-DISPLAY-ERROR
        GOBACK
    ELSE
        DISPLAY 'SUCCESSFUL REPOSITION AT EMP ID ' EMP-ID.
```

Ok, now we've performed all four items that will enable us to commit data updates by taking checkpoints at some interval. Let's make our record interval 5. So we have eight records in the database, and we'll take a checkpoints as follows:

- At the beginning of the program.

- After each 5 records have been processed.

- At the end of the program.

Here is our complete program code for COBIMSD. As mentioned earlier, we haven't completed the code yet for a restart. But we now have the functionality to commit our data changes with the checkpoint call.

```
IDENTIFICATION DIVISION.
PROGRAM-ID. COBIMSD.

******************************************************
*   WALK THROUGH THE EMPLOYEE (ROOT) SEGMENTS OF     *
*   THE ENTIRE EMPLOYEE DATABASE. DELETE ALL RECORDS.*
******************************************************

ENVIRONMENT DIVISION.
INPUT-OUTPUT SECTION.
DATA DIVISION.

******************************************************
*   W O R K I N G   S T O R A G E   S E C T I O N    *
```

270

```
 ********************************************************

 WORKING-STORAGE SECTION.

  01 WS-FLAGS.
     05  SW-END-OF-DB-SWITCH     PIC X(1) VALUE 'N'.
         88  SW-END-OF-DB                  VALUE 'Y'.
         88  SW-NOT-END-OF-DB              VALUE 'N'.


 01 DLI-FUNCTIONS.
     05 DLI-FUNCISRT  PIC X(4) VALUE 'ISRT'.
     05 DLI-FUNCGU    PIC X(4) VALUE 'GU  '.
     05 DLI-FUNCGN    PIC X(4) VALUE 'GN  '.
     05 DLI-FUNCGHU   PIC X(4) VALUE 'GHU '.
     05 DLI-FUNCGHN   PIC X(4) VALUE 'GHN '.
     05 DLI-FUNCGNP   PIC X(4) VALUE 'GNP '.
     05 DLI-FUNCREPL  PIC X(4) VALUE 'REPL'.
     05 DLI-FUNCDLET  PIC X(4) VALUE 'DLET'.
     05 DLI-FUNCXRST  PIC X(4) VALUE 'XRST'.
     05 DLI-FUNCCHKP  PIC X(4) VALUE 'CHKP'.


 01 IO-EMPLOYEE-RECORD.
     05  EMP-ID        PIC X(04).
     05  FILLER        PIC X(01).
     05  EMPL-LNAME    PIC X(30).
     05  FILLER        PIC X(01).
     05  EMPL-FNAME    PIC X(20).
     05  FILLER        PIC X(01).
     05  EMPL-YRS-SRV  PIC X(02).
     05  FILLER        PIC X(01).
     05  EMPL-PRM-DTE  PIC X(10).
     05  FILLER        PIC X(10).


  01 EMP-UNQUALIFIED-SSA.
     05  SEGNAME     PIC X(08) VALUE 'EMPLOYEE'.
     05  FILLER      PIC X(01) VALUE ' '.


  01 EMP-QUALIFIED-SSA.
     05  SEGNAME     PIC X(08) VALUE 'EMPLOYEE'.
     05  FILLER      PIC X(01) VALUE '('.
     05  FIELD       PIC X(08) VALUE 'EMPID'.
     05  OPER        PIC X(02) VALUE ' ='.
     05  EMP-ID-VAL  PIC X(04) VALUE '    '.
     05  FILLER      PIC X(01) VALUE ')'.


  01 SEG-IO-AREA    PIC X(80).


  01 IMS-RET-CODES.
     05 ONE            PIC S9(9) COMP VALUE +1.
     05 TWO            PIC S9(9) COMP VALUE +2.
     05 THREE          PIC S9(9) COMP VALUE +3.
     05 FOUR           PIC S9(9) COMP VALUE +4.
```

```
       05 FIVE              PIC S9(9) COMP VALUE +5.
       05 SIX               PIC S9(9) COMP VALUE +6.


 01 XRST-IOAREA.
    05 XRST-ID      PIC X(08) VALUE SPACES.
    05 FILLER       PIC X(04) VALUE SPACES.

 77 IO-AREALEN      PIC S9(9) USAGE IS BINARY VALUE 12.

 77 CHKP-ID         PIC X(08) VALUE 'IMSD    '.

 77 CHKP-NBR        PIC 999   VALUE ZERO.
 77 CHKP-COUNT      PIC S9(9) USAGE IS BINARY VALUE ZERO.

 01 CHKP-MESSAGE.
    05 FILLER               PIC X(24) VALUE
       'COBIMSD  CHECK POINT NO:'.
    05 CHKP-MESS-NBR        PIC 999     VALUE ZERO.
    05 FILLER               PIC X(15)   VALUE ',AT INPUT REC#:'.
    05 CHKP-MESS-REC        PIC ZZZZZ9  VALUE SPACES.
    05 FILLER               PIC X(10)   VALUE ',AT EMP#:'.
    05 CHKP-MESS-EMP        PIC X(08)   VALUE SPACES.

 01 IMS-CHKP-AREA-LTH.
    05 LEN                  PIC S9(9) USAGE IS BINARY VALUE +7.

 01 IMS-CHKP-AREA.
    05 CHKP-EMP-ID    PIC X(04) VALUE SPACES.
    05 CHKP-NBR-LAST  PIC 999   VALUE 0.

LINKAGE SECTION.

 01 IO-PCB.
    05 FILLER           PICTURE X(10).
    05 IO-STATUS-CODE   PICTURE XX.
    05 FILLER           PICTURE X(20).

 01 PCB-MASK.
    03 DBD-NAME      PIC X(8).
    03 SEG-LEVEL     PIC XX.
    03 STATUS-CODE   PIC XX.
    03 PROC-OPT      PIC X(4).
    03 FILLER        PIC X(4).
    03 SEG-NAME      PIC X(8).
    03 KEY-FDBK      PIC S9(5) COMP.
    03 NUM-SENSEG    PIC S9(5) COMP.
    03 KEY-FDBK-AREA.
       05 EMPLOYEE-KEY PIC X(04).
       05 EMPPAYHS-KEY PIC X(08).
```

```
PROCEDURE DIVISION.

    INITIALIZE IO-PCB PCB-MASK
    ENTRY 'DLITCBL' USING IO-PCB, PCB-MASK

    PERFORM P100-INITIALIZATION.
    PERFORM P200-MAINLINE.
    PERFORM P300-TERMINATION.
    GOBACK.

 P100-INITIALIZATION.

    DISPLAY '** PROGRAM COBIMSD START **'
    DISPLAY 'PROCESSING IN P100-INITIALIZATION'.

* CHECK FOR RESTART

    CALL 'CBLTDLI' USING SIX,
         DLI-FUNCXRST,
         PCB-MASK,
         IO-AREALEN,
         XRST-IOAREA,
         IMS-CHKP-AREA-LTH,
         IMS-CHKP-AREA

    IF STATUS-CODE NOT EQUAL SPACES THEN
       PERFORM P9000-DISPLAY-ERROR
       GOBACK
    END-IF

    IF XRST-ID NOT EQUAL SPACES THEN
       MOVE CHKP-NBR-LAST TO CHKP-NBR
       DISPLAY '*** COBIMSD IMS RESTART ***'
       DISPLAY '*  LAST CHECK POINT :' XRST-ID
       DISPLAY '*  EMPLOYEE NUMBER  :' CHKP-EMP-ID
    ELSE
       DISPLAY '****** COBIMSD IMS NORMAL START ***'
       PERFORM P8000-TAKE-CHECKPOINT
    END-IF.

*    DO INITIAL DB READ FOR FIRST EMPLOYEE RECORD

    CALL 'CBLTDLI' USING FOUR,
         DLI-FUNCGHN,
         PCB-MASK,
         IO-EMPLOYEE-RECORD,
         EMP-UNQUALIFIED-SSA

    IF STATUS-CODE = '  ' THEN
       NEXT SENTENCE
    ELSE
       IF STATUS-CODE = 'GB' THEN
```

```cobol
            SET SW-END-OF-DB TO TRUE
            DISPLAY 'END OF DATABASE :'
        ELSE
            PERFORM P9000-DISPLAY-ERROR
            GOBACK
        END-IF

    END-IF.

 P200-MAINLINE.

    DISPLAY 'PROCESSING IN P200-MAINLINE'

*   CHECK STATUS CODE AND FIRST RECORD

    IF SW-END-OF-DB THEN
        DISPLAY 'NO RECORDS TO PROCESS!!'
    ELSE

        PERFORM UNTIL SW-END-OF-DB

            CALL 'CBLTDLI' USING THREE,
                DLI-FUNCDLET,
                PCB-MASK,
                IO-EMPLOYEE-RECORD

            IF STATUS-CODE NOT EQUAL SPACES THEN
                PERFORM P9000-DISPLAY-ERROR
                GOBACK
            ELSE
                DISPLAY 'SUCCESSFUL DELETE OF EMPLOYEE ' EMP-ID
            END-IF

*   GET THE NEXT RECORD

            CALL 'CBLTDLI' USING FOUR,
                DLI-FUNCGHN,
                PCB-MASK,
                IO-EMPLOYEE-RECORD,
                EMP-UNQUALIFIED-SSA

            IF STATUS-CODE = 'GB' THEN
                SET SW-END-OF-DB TO TRUE
                DISPLAY 'END OF DATABASE'
            ELSE
                IF STATUS-CODE NOT EQUAL SPACES THEN
                    PERFORM P9000-DISPLAY-ERROR
                    SET SW-END-OF-DB TO TRUE
                    GOBACK
                ELSE
                    DISPLAY 'SUCCESSFUL GET HOLD :'
                        IO-EMPLOYEE-RECORD
```

```
                    MOVE EMP-ID TO EMP-ID-VAL
                    ADD +1 TO CHKP-COUNT
                    IF CHKP-COUNT GREATER THAN OR EQUAL TO 5
                        PERFORM P8000-TAKE-CHECKPOINT
                        PERFORM P1000-RESET-POSITION
                    END-IF
                END-IF
            END-IF

        END-PERFORM.
    DISPLAY 'FINISHED PROCESSING IN P200-MAINLINE'.

P300-TERMINATION.

    DISPLAY 'PROCESSING IN P300-TERMINATION'
    ADD +1 TO CHKP-COUNT
    PERFORM P8000-TAKE-CHECKPOINT
    DISPLAY '** COBIMSD - SUCCESSFULLY ENDED **'.

P1000-RESET-POSITION.

    DISPLAY 'PROCESSING IN P1000-RESET-POSITION'

    CALL 'CBLTDLI' USING FOUR,
            DLI-FUNCGHU,
            PCB-MASK,
            IO-EMPLOYEE-RECORD,
            EMP-QUALIFIED-SSA

    IF STATUS-CODE NOT EQUAL SPACES THEN
        PERFORM P9000-DISPLAY-ERROR
        GOBACK
    ELSE
        DISPLAY 'SUCCESSFUL REPOSITION AT EMP ID ' EMP-ID.

P8000-TAKE-CHECKPOINT.

    DISPLAY 'PROCESSING IN P8000-TAKE-CHECKPOINT'

    ADD +1               TO CHKP-NBR
    MOVE CHKP-NBR        TO CHKP-NBR-LAST
    MOVE CHKP-NBR-LAST TO CHKP-ID(6:3)
    MOVE EMP-ID          TO CHKP-EMP-ID

    CALL 'CBLTDLI' USING SIX,
            DLI-FUNCCHKP,
            IO-PCB,
            IO-AREALEN,
            CHKP-ID,
            IMS-CHKP-AREA-LTH,
            IMS-CHKP-AREA
```

```
        IF IO-STATUS-CODE NOT EQUAL SPACES THEN
            DISPLAY 'TOOK AN ERROR DOING THE CHECKPOINT'
            DISPLAY 'IO-STATUS-CODE ' IO-STATUS-CODE
            PERFORM P9000-DISPLAY-ERROR
            GOBACK
        ELSE
            MOVE 0 TO CHKP-COUNT
            MOVE CHKP-NBR        TO CHKP-MESS-NBR
            MOVE CHKP-EMP-ID     TO CHKP-MESS-EMP
            DISPLAY CHKP-MESSAGE
        END-IF.

    P9000-DISPLAY-ERROR.

        DISPLAY 'ERROR ENCOUNTERED - DETAIL FOLLOWS'
        DISPLAY 'SEG-IO-AREA      :' SEG-IO-AREA
        DISPLAY 'DBD-NAME1:'       DBD-NAME
        DISPLAY 'SEG-LEVEL1:'      SEG-LEVEL
        DISPLAY 'STATUS-CODE:'     STATUS-CODE
        DISPLAY 'PROC-OPT1 :'      PROC-OPT
        DISPLAY 'SEG-NAME1 :'      SEG-NAME
        DISPLAY 'KEY-FDBK1 :'      KEY-FDBK
        DISPLAY 'NUM-SENSEG1:'     NUM-SENSEG
        DISPLAY 'KEY-FDBK-AREA1:'  KEY-FDBK-AREA.

*       END OF SOURCE CODE
```

At this point, we can compile and link, and then run the program. Make sure your JCL specifies the EMPPSBZ PSB or you'll get an error.

```
** PROGRAM COBIMSD START **
PROCESSING IN P100-INITIALIZATION
****** COBIMSD IMS NORMAL START ***
PROCESSING IN P8000-TAKE-CHECKPOINT
COBIMSD  CHECK POINT NO:001,AT INPUT REC#:        ,AT EMP#:
PROCESSING IN P200-MAINLINE
SUCCESSFUL DELETE OF EMPLOYEE 1111
SUCCESSFUL GET HOLD :1122 JENKINS                 DEBORAH           05
SUCCESSFUL DELETE OF EMPLOYEE 1122
SUCCESSFUL GET HOLD :3217 JOHNSON                 EDWARD            04
SUCCESSFUL DELETE OF EMPLOYEE 3217
SUCCESSFUL GET HOLD :4175 TURNBULL                FRED              01
SUCCESSFUL DELETE OF EMPLOYEE 4175
SUCCESSFUL GET HOLD :4720 SCHULTZ                 TIM               09
SUCCESSFUL DELETE OF EMPLOYEE 4720
SUCCESSFUL GET HOLD :4836 SMITH                   SANDRA            03
PROCESSING IN P8000-TAKE-CHECKPOINT
COBIMSD  CHECK POINT NO:002,AT INPUT REC#:        ,AT EMP#: 4836
PROCESSING IN P1000-RESET-POSITION
SUCCESSFUL REPOSITION AT EMP ID 4836
SUCCESSFUL DELETE OF EMPLOYEE 4836
SUCCESSFUL GET HOLD :6288 WILLARD                 JOE               06
```

276

```
SUCCESSFUL DELETE OF EMPLOYEE 6288
SUCCESSFUL GET HOLD :7459 STEWART                        BETTY              07
SUCCESSFUL DELETE OF EMPLOYEE 7459
END OF DATABASE
FINISHED PROCESSING IN P200-MAINLINE
PROCESSING IN P300-TERMINATION
PROCESSING IN P8000-TAKE-CHECKPOINT
COBIMSD   CHECK POINT NO:003,AT INPUT REC#:      ,AT EMP#: 7459
** COBIMSD - SUCCESSFULLY ENDED **
```

We now have an empty database. You can verify this by looking in your File Manager IMS if you have it, or you can try browsing the DATA file of the KSDS. Since it is empty, you'll get an error.

```
VSAM POINT RC X"08", Error Code X"20"
VSAM GET RC X"08", Error Code X"58"
Function terminated

***
```

We have shown we can commit updates to the database at some interval. In a real production environment we would not checkpoint every 5 records. More likely we would checkpoint at 500 records or 1,000 records or 2,000 records. You don't want to lock your data for too long, so find a record interval that commits at about once a minute, or whatever your DBA recommends.

Performing Checkpoint Restart

At this point, we've successfully committed data using checkpoints. However, we have not yet demonstrated how to perform a restart using the extended restart facility (XRST). To do that, we need to introduce IMS logging.

Using the IMS Log

To allow for IMS restartability, you must log all the transactions and checkpoints you take. When you stop the program (or when IMS stops it for an abend), your data modifications (ISRT, REPL, DLET) are automatically backed out to the last checkpoint. So typically, you will want to fix whatever the problem was, and then restart your program from the last checkpoint.

In your execution JCL for running IMS programs, there should be two DD statements that are probably dummied out.[13] The IEFRDER DD should definitely be there, and the IMSLOGR may be there (it is only referenced on restart so it might not be).

---

13 This discussion pertains to running a program in DLI mode. If you are running a program in BMP mode, you don't need these DDs because the program runs in the IMS online space which has its own transaction log.

```
//IMSLOGR   DD  DUMMY
//IEFRDER   DD  DUMMY
```

Here's what these are used for when they are not dummied out (when actual file names are specified):

- IMSLOGR – the previous (existing) generation of IMS log file created for your DLI execution.

- IEFRDER – the new generation of the IMS log file created for your DLI execution to log any updates to the database performed by your program.

You'll want to create a generation data group for your IMS log file, and then define these DDs to use the 0 and +1 generation of this data set. I created USER01.IMSLOG with 5 generations, and I created an empty first generation. Next, I have un-dummied the IMSLOGR and IEFRDER DD's by coding the new log file as follows:

```
//IMSLOGR   DD  DSN=USER01.IMSLOG(+0),
//          DISP=SHR
//IEFRDER   DD  DSN=USER01.IMSLOG(+1),
//          DISP=(NEW,CATLG,CATLG),
//          UNIT=SYSDA,
//          SPACE=(TRK,(1,1),RLSE),
//          DCB=(RECFM=VB,BLKSIZE=4096,
//          LRECL=4092,BUFNO=2)
```

Now if you specify a checkpoint value when you restart your program, IMS will scan the 0 generation of the IMS log to pick up the information from the last checkpoint. In our case, this information includes the employee id that we read before issuing the last checkpoint. You can then use that employee id key to reposition in the database.

Specifying a Checkpoint ID on Restart

You can specify the checkpoint id in the PARM value of the execute statement for your program. This is a positional parameter, so it must be placed correctly in the PARM sequence. Here is the JCL and I'm putting a sample checkpoint id at the right place in the PARM.

```
//GO      EXEC PGM=DFSRRC00,REGION=4M,
//       PARM=(DLI,&MBR,&PSB,7,0000,,0,'CHKP0003',N,0,0,,,N,N,,N,)
```

Restart Example

We need to reload the database now before we can do a restart example (remember we deleted all the records in the database earlier). You can run your COBIMS1 to do this. Although the database is empty, it is not brand new. So you can use `PSB EMPPSB` instead of `EMPPSBL`. In fact you'll get an error (AI status code) if you use the `EMPPSBL`, so make sure you use `EMPPSB`. When finished, verify that we have nine records in the database.

```
Browse              USER01.IMS.EMPLOYEE.DATA                    Top of 9
Command ===>                                            Scroll PAGE
                        Type DATA       RBA             Format CHAR
                                         Col 1
----+----10---+----2----+----3----+----4----+----5----+----6----+----7----+----
****  Top of data  ****
......1111 VEREEN               CHARLES            12 2017-01-01 93
......1122 JENKINS              DEBORAH            05 2017-01-01 43
......3217 JOHNSON              EDWARD             04 2017-01-01 39
......4175 TURNBULL             FRED               01 2016-12-01 54
......4720 SCHULTZ              TIM                09 2017-01-01 65
......4836 SMITH                SANDRA             03 2017-01-01 02
......6288 WILLARD              JOE                06 2016-01-01 20
......7459 STEWART              BETTY              07 2016-07-31 01
......9134 FRANKLIN             BRIANNA            00 2016-10-01 93
****  End of data  ****
```

For our example, we will create a new program `COBIMSE` and it will delete all the records in the database as we did with `COBIMSD`. We will checkpoint at 5 record intervals. You can start by copying `COBIMSD` to create `COBIMSE`. There will be two differences between `COBIMSD` and `COBIMSE`. One is that `COBIMSE` will intentionally cause a rollback when we encounter employee 7459 (this is just to simulate an abend type error). The rollback will back out all changes made since the last checkpoint.

The other difference is that we will code restart logic in `COBIMSE` to reposition to the appropriate employee id in the data to continue processing on a restart. In between run 1 and run 2 of `COBIMSE`, the only change we will make to the program is to not do the rollback when it gets to employee id 7459. We're simulating a "problem" to cause the rollback, then we solve the cause of the rollback and restart the program.

If you copy `COBIMSD` to create `COBIMSE`, you only need to make a few changes. First, let's create some new procedures. One procedure will get the first root in the database. We've been doing that in `P100-INITIALIZATION`, but now on a restart we need to call the reset position procedure instead. Separating these functions into separate procedures makes the code easier to read. Let's do this:

```
    P1000-GET-FIRST-ROOT.

        CALL 'CBLTDLI' USING ,
            DLI-FUNCGHN,
```

```
                PCB-MASK,
                IO-EMPLOYEE-RECORD,
                EMP-UNQUALIFIED-SSA

       IF STATUS-CODE = '  ' THEN
          NEXT SENTENCE
       ELSE
          IF STATUS-CODE = 'GB' THEN
             SET SW-END-OF-DB TO TRUE
             DISPLAY 'END OF DATABASE :'
          ELSE
             PERFORM P9000-DISPLAY-ERROR
             GOBACK
          END-IF.
```

Next let's rename `P1000-RESET-POSITION` to `P2000-RESET-POSITION`. That will keep the code more orderly.

Finally, let's add the procedure to perform the rollback.

```
       P3000-ROLLBACK.

          DISPLAY 'PROCESSING IN P3000-ROLLBACK'.

          CALL 'CBLTDLI' USING ONE,
                DLI-FUNCROLL.
```

Now let's modify the initialization logic to handle either a normal start or a restart. On a normal start we'll get the first root in the database. On a restart we'll reposition at the `EMP-ID` saved in the checkpoint that we are using to do the restart.

```
          CALL 'CBLTDLI' USING SIX,
                DLI-FUNCXRST,
                PCB-MASK,
                IO-AREALEN,
                XRST-IOAREA,
                IMS-CHKP-AREA-LTH,
                IMS-CHKP-AREA

       IF STATUS-CODE NOT EQUAL SPACES THEN
          PERFORM P9000-DISPLAY-ERROR
          GOBACK
       END-IF

       IF XRST-ID NOT EQUAL SPACES THEN
          SET SW-IMS-RESTART TO TRUE
          MOVE CHKP-NBR-LAST TO CHKP-NBR
          DISPLAY '*** COBIMSE IMS RESTART ***'
          DISPLAY '*  LAST CHECK POINT :' XRST-ID
```

280

```
          DISPLAY '*  EMPLOYEE NUMBER  :' CHKP-EMP-ID
      ELSE
          DISPLAY '****** COBIMSE IMS NORMAL START ***'
          PERFORM P8000-TAKE-CHECKPOINT
      END-IF.

 *    DO INITIAL DB READ FOR FIRST EMPLOYEE RECORD
 *    OR REPOSITION IF AN IMS RESTART.

      IF SW-IMS-RESTART THEN
          MOVE CHKP-EMP-ID TO EMP-ID-VAL
          PERFORM P2000-RESET-POSITION
      ELSE
          PERFORM P1000-GET-FIRST-ROOT

      END-IF.
```

The value of XRST-ID will be non-blank if we are doing a restart. In that case we will turn on the SW-IMS-RESTART switch. Otherwise we will branch to take the initial checkpoint. Now if the SW-IMS-RESTART is true, it means this is a restart so we load the employee id from the checkpoint area into the qualified EMPLOYEE qualified SSA value, and then we call the procedure to reset database position to where it was at that checkpoint.

If the value of XRST-ID is blank, then we are **not** doing a restart. In this case, we call the procedure to get the first root.

Finally, let's add a temporary statement to the execution loop. After a successful GHN, check to see if we have employee id 7459, and if so call the rollback procedure. We will only do this on the first run of the program so as to force a rollback.

```
      DISPLAY 'SUCCESSFUL GET HOLD :'
          IO-EMPLOYEE-RECORD
      MOVE EMP-ID TO EMP-ID-VAL
      ADD +1 TO CHKP-COUNT
      IF CHKP-COUNT GREATER THAN OR EQUAL TO 5
          PERFORM P8000-TAKE-CHECKPOINT
          PERFORM P2000-RESET-POSITION
      END-IF
      IF EMP-ID = '7459'
          PERFORM P3000-ROLLBACK
          GOBACK
      END-IF
```

So here is our complete code listing. Review it carefully to be sure you understand what is happening.

```
IDENTIFICATION DIVISION.
PROGRAM-ID. COBIMSE.

*********************************************************
*  WALK THROUGH THE EMPLOYEE (ROOT) SEGMENTS OF      *
*  THE ENTIRE EMPLOYEE IMS DATABASE, AND ROLL BACK   *
*  CHNGES WHEN A PARTICULAR CONDITION IS ENCOUNTERED *
*********************************************************
ENVIRONMENT DIVISION.
INPUT-OUTPUT SECTION.
DATA DIVISION.
*********************************************************
*  W O R K I N G   S T O R A G E   S E C T I O N    *
*********************************************************
WORKING-STORAGE SECTION.
 01 WS-FLAGS.
     05  SW-END-OF-DB-SWITCH      PIC X(1) VALUE 'N'.
         88  SW-END-OF-DB                  VALUE 'Y'.
         88  SW-NOT-END-OF-DB              VALUE 'N'.
     05  SW-IMS-RESTART-SW        PIC X(1) VALUE 'N'.
         88  SW-IMS-RESTART                VALUE 'Y'.
         88  SW-NOT-IMS-RESTART            VALUE 'N'.

 01 DLI-FUNCTIONS.
     05 DLI-FUNCISRT  PIC X(4) VALUE 'ISRT'.
     05 DLI-FUNCGU    PIC X(4) VALUE 'GU  '.
     05 DLI-FUNCGN    PIC X(4) VALUE 'GN  '.
     05 DLI-FUNCGHU   PIC X(4) VALUE 'GHU '.
     05 DLI-FUNCGHN   PIC X(4) VALUE 'GHN '.
     05 DLI-FUNCGNP   PIC X(4) VALUE 'GNP '.
     05 DLI-FUNCREPL  PIC X(4) VALUE 'REPL'.
     05 DLI-FUNCDLET  PIC X(4) VALUE 'DLET'.
     05 DLI-FUNCXRST  PIC X(4) VALUE 'XRST'.
     05 DLI-FUNCCHKP  PIC X(4) VALUE 'CHKP'.
     05 DLI-FUNCROLL  PIC X(4) VALUE 'ROLL'.

 01 IO-EMPLOYEE-RECORD.
     05  EMP-ID        PIC X(04).
     05  FILLER        PIC X(01).
     05  EMPL-LNAME    PIC X(30).
     05  FILLER        PIC X(01).
     05  EMPL-FNAME    PIC X(20).
     05  FILLER        PIC X(01).
     05  EMPL-YRS-SRV  PIC X(02).
     05  FILLER        PIC X(01).
     05  EMPL-PRM-DTE  PIC X(10).
     05  FILLER        PIC X(10).

 01 EMP-UNQUALIFIED-SSA.
     05  SEGNAME    PIC X(08) VALUE 'EMPLOYEE'.
     05  FILLER     PIC X(01) VALUE ' '.
```

```
01  EMP-QUALIFIED-SSA.
    05  SEGNAME     PIC X(08) VALUE 'EMPLOYEE'.
    05  FILLER      PIC X(01) VALUE '('.
    05  FIELD       PIC X(08) VALUE 'EMPID'.
    05  OPER        PIC X(02) VALUE ' ='.
    05  EMP-ID-VAL  PIC X(04) VALUE '    '.
    05  FILLER      PIC X(01) VALUE ')'.

01  SEG-IO-AREA    PIC X(80).

01  IMS-RET-CODES.
    05 ONE          PIC S9(9) COMP VALUE +1.
    05 TWO          PIC S9(9) COMP VALUE +2.
    05 THREE        PIC S9(9) COMP VALUE +3.
    05 FOUR         PIC S9(9) COMP VALUE +4.
    05 FIVE         PIC S9(9) COMP VALUE +5.
    05 SIX          PIC S9(9) COMP VALUE +6.

01  XRST-IOAREA.
    05  XRST-ID    PIC X(08) VALUE SPACES.
    05  FILLER     PIC X(04) VALUE SPACES.

77  IO-AREALEN     PIC S9(9) USAGE IS BINARY VALUE 12.

77  CHKP-ID        PIC X(08) VALUE 'IMSE-   '.

77  CHKP-NBR       PIC 999    VALUE ZERO.
77  CHKP-COUNT     PIC S9(9) USAGE IS BINARY VALUE ZERO.

01  CHKP-MESSAGE.
    05 FILLER              PIC X(24) VALUE
       'COBIMSE  CHECK POINT NO:'.
    05 CHKP-MESS-NBR       PIC 999      VALUE ZERO.
    05 FILLER              PIC X(15)  VALUE '    ,AT REC#:'.
    05 FILLER              PIC X(10)  VALUE ' ,AT EMP#:'.
    05 CHKP-MESS-EMP       PIC X(04)  VALUE SPACES.

01 IMS-CHKP-AREA-LTH.
   05 LEN              PIC S9(9) USAGE IS BINARY VALUE +7.

01 IMS-CHKP-AREA.
   05 CHKP-EMP-ID      PIC X(04) VALUE SPACES.
   05 CHKP-NBR-LAST    PIC 999   VALUE 0.

LINKAGE SECTION.

01 IO-PCB.
   05 FILLER           PICTURE X(10).
   05 IO-STATUS-CODE   PICTURE XX.
   05 FILLER           PICTURE X(20).

01 PCB-MASK.
```

```cobol
        03 DBD-NAME         PIC X(8).
        03 SEG-LEVEL        PIC XX.
        03 STATUS-CODE      PIC XX.
        03 PROC-OPT         PIC X(4).
        03 FILLER           PIC X(4).
        03 SEG-NAME         PIC X(8).
        03 KEY-FDBK         PIC S9(5) COMP.
        03 NUM-SENSEG       PIC S9(5) COMP.
        03 KEY-FDBK-AREA.
           05 EMPLOYEE-KEY  PIC X(04).
           05 EMPPAYHS-KEY  PIC X(08).

    PROCEDURE DIVISION.

        INITIALIZE IO-PCB PCB-MASK
        ENTRY 'DLITCBL' USING IO-PCB, PCB-MASK

        PERFORM P100-INITIALIZATION.
        PERFORM P200-MAINLINE.
        PERFORM P300-TERMINATION.
        GOBACK.

    P100-INITIALIZATION.

        DISPLAY '** PROGRAM COBIMSE START **'
        DISPLAY 'PROCESSING IN P100-INITIALIZATION'.

   * CHECK FOR RESTART

        CALL 'CBLTDLI' USING SIX,
             DLI-FUNCXRST,
             PCB-MASK,
             IO-AREALEN,
             XRST-IOAREA,
             IMS-CHKP-AREA-LTH,
             IMS-CHKP-AREA

        IF STATUS-CODE NOT EQUAL SPACES THEN
           PERFORM P9000-DISPLAY-ERROR
           GOBACK
        END-IF

        IF XRST-ID NOT EQUAL SPACES THEN
           SET SW-IMS-RESTART TO TRUE
           MOVE CHKP-NBR-LAST TO CHKP-NBR
           DISPLAY '*** COBIMSE IMS RESTART ***'
           DISPLAY '*   LAST CHECK POINT :' XRST-ID
           DISPLAY '*   EMPLOYEE NUMBER  :' CHKP-EMP-ID
        ELSE
           DISPLAY '****** COBIMSE IMS NORMAL START ***'
           PERFORM P8000-TAKE-CHECKPOINT
        END-IF.
```

284

```
*     DO INITIAL DB READ FOR FIRST EMPLOYEE RECORD
*     OR REPOSITION IF AN IMS RESTART.

      IF SW-IMS-RESTART THEN
         MOVE CHKP-EMP-ID TO EMP-ID-VAL
         PERFORM P2000-RESET-POSITION
      ELSE
         PERFORM P1000-GET-FIRST-ROOT

      END-IF.

 P200-MAINLINE.

      DISPLAY 'PROCESSING IN P200-MAINLINE'

*     CHECK STATUS CODE AND FIRST RECORD

      IF SW-END-OF-DB THEN
         DISPLAY 'NO RECORDS TO PROCESS!!'
      ELSE

         PERFORM UNTIL SW-END-OF-DB

            CALL 'CBLTDLI' USING THREE,
                DLI-FUNCDLET,
                PCB-MASK,
                IO-EMPLOYEE-RECORD

            IF STATUS-CODE NOT EQUAL SPACES THEN
               PERFORM P9000-DISPLAY-ERROR
               GOBACK
            ELSE
               DISPLAY 'SUCCESSFUL DELETE OF EMPLOYEE ' EMP-ID
               MOVE EMP-ID TO CHKP-EMP-ID
            END-IF

*     GET THE NEXT RECORD

            CALL 'CBLTDLI' USING FOUR,
                DLI-FUNCGHN,
                PCB-MASK,
                IO-EMPLOYEE-RECORD,
                EMP-UNQUALIFIED-SSA

            IF STATUS-CODE = 'GB' THEN
               SET SW-END-OF-DB TO TRUE
               DISPLAY 'END OF DATABASE'
            ELSE
               IF STATUS-CODE NOT EQUAL SPACES THEN
                  PERFORM P9000-DISPLAY-ERROR
                  SET SW-END-OF-DB TO TRUE
```

```
                   GOBACK
              ELSE
                 DISPLAY 'SUCCESSFUL GET HOLD :'
                     IO-EMPLOYEE-RECORD
                 MOVE EMP-ID TO EMP-ID-VAL
                 ADD +1 TO CHKP-COUNT
                 IF CHKP-COUNT GREATER THAN OR EQUAL TO 5
                    PERFORM P8000-TAKE-CHECKPOINT
                    PERFORM P2000-RESET-POSITION
                 END-IF
                 IF EMP-ID = '7459'
                    PERFORM P3000-ROLLBACK
                 END-IF
              END-IF
          END-IF

       END-PERFORM.

    DISPLAY 'FINISHED PROCESSING IN P200-MAINLINE'.

P300-TERMINATION.

    DISPLAY 'PROCESSING IN P300-TERMINATION'
    ADD +1 TO CHKP-COUNT
    PERFORM P8000-TAKE-CHECKPOINT
    DISPLAY '** COBIMSE - SUCCESSFULLY ENDED **'.

P1000-GET-FIRST-ROOT.

    CALL 'CBLTDLI' USING FOUR,
         DLI-FUNCGHN,
         PCB-MASK,
         IO-EMPLOYEE-RECORD,
         EMP-UNQUALIFIED-SSA

    IF STATUS-CODE = '   ' THEN
       NEXT SENTENCE
    ELSE
       IF STATUS-CODE = 'GB' THEN
          SET SW-END-OF-DB TO TRUE
          DISPLAY 'END OF DATABASE :'
       ELSE
          PERFORM P9000-DISPLAY-ERROR
          GOBACK
       END-IF.

P2000-RESET-POSITION.

    DISPLAY 'PROCESSING IN P2000-RESET-POSITION'

    CALL 'CBLTDLI' USING ,
         DLI-FUNCGHU,
```

```
                PCB-MASK,
                IO-EMPLOYEE-RECORD,
                EMP-QUALIFIED-SSA

          IF STATUS-CODE NOT EQUAL SPACES THEN
             PERFORM P9000-DISPLAY-ERROR
             GOBACK
          ELSE
             DISPLAY 'SUCCESSFUL REPOSITION AT EMP ID ' EMP-ID.


   P3000-ROLLBACK.

          DISPLAY 'PROCESSING IN P3000-ROLLBACK'.

          CALL 'CBLTDLI' USING ONE,
                DLI-FUNCROLL.


   P8000-TAKE-CHECKPOINT.

          DISPLAY 'PROCESSING IN P8000-TAKE-CHECKPOINT'
          ADD +1           TO CHKP-NBR
          MOVE CHKP-NBR      TO CHKP-NBR-LAST
          MOVE CHKP-NBR-LAST TO CHKP-ID(6:3)
          DISPLAY 'CHECKPOINT ID IS ' CHKP-ID
          MOVE EMP-ID        TO CHKP-EMP-ID

          CALL 'CBLTDLI' USING SIX,
                DLI-FUNCCHKP,
                IO-PCB,
                IO-AREALEN,
                CHKP-ID,
                IMS-CHKP-AREA-LTH,
                IMS-CHKP-AREA

          IF IO-STATUS-CODE NOT EQUAL SPACES THEN
             DISPLAY 'TOOK AN ERROR DOING THE CHECKPOINT'
             DISPLAY 'IO-STATUS-CODE ' IO-STATUS-CODE
             PERFORM P9000-DISPLAY-ERROR
             GOBACK
          ELSE
             MOVE 0 TO CHKP-COUNT
             MOVE CHKP-NBR         TO CHKP-MESS-NBR
             MOVE CHKP-EMP-ID      TO CHKP-MESS-EMP
             DISPLAY CHKP-MESSAGE
          END-IF.


   P9000-DISPLAY-ERROR.

          DISPLAY 'ERROR ENCOUNTERED - DETAIL FOLLOWS'
          DISPLAY 'SEG-IO-AREA     :' SEG-IO-AREA
          DISPLAY 'DBD-NAME1:'      DBD-NAME
          DISPLAY 'SEG-LEVEL1:'     SEG-LEVEL
```

```
          DISPLAY 'STATUS-CODE:'    STATUS-CODE
          DISPLAY 'PROC-OPT1 :'     PROC-OPT
          DISPLAY 'SEG-NAME1 :'     SEG-NAME
          DISPLAY 'KEY-FDBK1 :'     KEY-FDBK
          DISPLAY 'NUM-SENSEG1:'    NUM-SENSEG
          DISPLAY 'KEY-FDBK-AREA1:' KEY-FDBK-AREA.
     *    END OF SOURCE CODE
```

Compile and link, then run the program. The program will abend with IMS user code U0778 because of the ROLL call.[14] Here is the output:

```
** PROGRAM COBIMSE START **
PROCESSING IN P100-INITIALIZATION
****** COBIMSE IMS NORMAL START ***
PROCESSING IN P8000-TAKE-CHECKPOINT
CHECKPOINT ID IS IMSE-001
COBIMSE   CHECK POINT NO:001     ,AT REC#:  ,AT EMP#:
PROCESSING IN P200-MAINLINE
SUCCESSFUL DELETE OF EMPLOYEE 1111
SUCCESSFUL GET HOLD :1122 JENKINS            DEBORAH           05
SUCCESSFUL DELETE OF EMPLOYEE 1122
SUCCESSFUL GET HOLD :3217 JOHNSON            EDWARD            04
SUCCESSFUL DELETE OF EMPLOYEE 3217
SUCCESSFUL GET HOLD :4175 TURNBULL           FRED              01
SUCCESSFUL DELETE OF EMPLOYEE 4175
SUCCESSFUL GET HOLD :4720 SCHULTZ            TIM               09
SUCCESSFUL DELETE OF EMPLOYEE 4720
SUCCESSFUL GET HOLD :4836 SMITH              SANDRA            03
PROCESSING IN P8000-TAKE-CHECKPOINT
CHECKPOINT ID IS IMSE-002
COBIMSE   CHECK POINT NO:002     ,AT REC#:   ,AT EMP#:4836
PROCESSING IN P2000-RESET-POSITION
SUCCESSFUL REPOSITION AT EMP ID 4836
SUCCESSFUL DELETE OF EMPLOYEE 4836
SUCCESSFUL GET HOLD :6288 WILLARD            JOE               06
SUCCESSFUL DELETE OF EMPLOYEE 6288
SUCCESSFUL GET HOLD :7459 STEWART            BETTY             07
PROCESSING IN P3000-ROLLBACK
```

At this point, we can verify that the first 5 records got deleted, and we can also verify that after the last checkpoint, all deleted records were backed (meaning they are still on the database).

```
   Browse          USER01.IMS.EMPLOYEE.DATA              Top of 4
   Command ===>                                          Scroll PAGE
                          Type DATA    RBA               Format CHAR
                                        Col 1
   ----+----10---+----2----+----3----+----4----+----5----+----6----+----7----+----
   ****  Top of data   ****
   ......4836 SMITH                   SANDRA        03 2017-01-01 02
   ......6288 WILLARD                 JOE           06 2016-01-01 20
   ......7459 STEWART                 BETTY         07 2016-07-31 01
   ......9134 FRANKLIN                BRIANNA       00 2016-10-01 93
```

14 If you prefer not to take a hard abend, instead of issuing the ROLL IMS call you can issue ROLB. ROLB backs out changes the same as ROLL, but ROLB returns control to the application program instead of abending.

The next step is to remove the code in COBIMSE that forced the rollback, then restart the program. Go ahead and remove or comment out the code, recompile and then we'll set up our restart JCL.

The PARM should look like this. Note that the IMSE-002 is the last successful checkpoint in the prior run. You can verify this by looking at the output from the previous run. Here is our restart parm override:

```
//GO       EXEC PGM=DFSRRC00,REGION=4M,
//      PARM=(DLI,&MBR,&PSB,7,0000,,0,'IMSE-002',N,0,0,,,N,N,,N,)
```

Now run the program, and here is the output.

```
** PROGRAM COBIMSE START **
PROCESSING IN P100-INITIALIZATION
*** COBIMSE IMS RESTART ***
*   LAST CHECK POINT :IMSE-002
*   EMPLOYEE NUMBER   :4836
PROCESSING IN P2000-RESET-POSITION
SUCCESSFUL REPOSITION AT EMP ID 4836
PROCESSING IN P200-MAINLINE
SUCCESSFUL DELETE OF EMPLOYEE 4836
SUCCESSFUL GET HOLD :6288 WILLARD              JOE            06
SUCCESSFUL DELETE OF EMPLOYEE 6288
SUCCESSFUL GET HOLD :7459 STEWART              BETTY          07
SUCCESSFUL DELETE OF EMPLOYEE 7459
SUCCESSFUL GET HOLD :9134 FRANKLIN             BRIANNA        00
SUCCESSFUL DELETE OF EMPLOYEE 9134
END OF DATABASE
FINISHED PROCESSING IN P200-MAINLINE
PROCESSING IN P300-TERMINATION
PROCESSING IN P8000-TAKE-CHECKPOINT
CHECKPOINT ID IS IMSE-003
COBIMSE  CHECK POINT NO:003     ,AT REC#:  ,AT EMP#:9134
** COBIMSE - SUCCESSFULLY ENDED **
```

We correctly restarted at employee id 4836, and then processed in GHN mode from there on. This is what should have happened. Now the database is empty, which we can confirm by trying to browse it.

```
VSAM POINT RC X"08", Error Code X"20"
VSAM GET RC X"08", Error Code X"58"
Function terminated
***
```

You now have a basic model for doing checkpoint restart. Frankly, checkpoint restart is done somewhat differently in each of the major environments I've worked in. Typically larger companies use third party products (such as BMC tools) to keep track of checkpoints and facilitate recovery. You may need to learn a bit more to use the third party products. The examples I've provided, although plain vanilla, work fine without any third party products.

That pretty well wraps up basic IMS programming. There are plenty of other features you can use, but that will depend on your work environment. Every shop and application is different. Good luck with it, and enjoy!

IMS Programming Guidelines

Consider the COBOL code examples in this text to be my own guidelines for coding IMS programs. Different shops use different standards and there is no "true" standard.

Chapter 5 Review Questions

1. What is the name of the interface program you call from a COBOL program to perform IMS operations?

2. Here are some IMS return codes. Explain briefly what each of them means: blank, GE, GB, II.

3. What is an SSA?

4. Briefly explain these entities: DBD, PSB, PCB?

5. What is the use of CMPAT parameter in PSB ?

6. In IMS, what is the difference between a key field and a search field?

7. What does PROCOPT mean in a PCB?

8. The different PROCOPTs and their meaning are:

9. What are the four basic parameters of a DLI retrieval call?

10. What are Qualified SSA and Unqualified SSA?

11. Which PSB parameter in a PSBGEN specifies the language in which the application program is written?

12. What does SENSEG stand for and how is it used in a PCB?

13. What storage mechanism/format is used for IMS index databases?

14. What are the DL/I commands to add, change and remove a segment?

15. What return code will you receive from IMS if the DL/I call was successful?

16. If you want to retrieve the last occurrence of a child segment under its parent, what command code could you use?

17. When would you use a GU call?

18. When would you use a GHU call?

19. What is the difference between running an IMS program as DLI and BMP ?

20. When would you use a GNP call?

21. Which IMS call is used to restart an abended program?

22. How do you establish parentage on a segment occurrence?

23. What is a checkpoint?

24. How do you update the primary key of an IMS segment?

25. Do you need to use a qualified SSA with REPL/DLET calls?

Chapter 6 : Database 2 (DB2)

Database 2 (DB2) is IBM's flagship relational database management system (DBMS). It was introduced for IBM mainframe computers in 1983. If you are working on a mainframe it is most likely you will be using DB2 for data management. Consequently we devote almost half of this text book to DB2 application development.

Basic z/OS Tools for DB2

Before we get into DB2 development activities, I want to introduce you to the environment we'll be working in. If you've used DB2 on the mainframe, you're almost certainly familiar with these tools and this will be a quick review. But if you have little or no exposure to DB2 on z/OS, we need to make sure you are familiar with how to access it and use the basic tools available.

DB2 Interactive

You'll do much of your DB2 work in DB2 Interactive which is now typically is titled **DB2 Primary Option Menu.** Regardless of which shop you work in, there should be a menu option on the ISPF main menu to get to DB2. It may be called DB2 or some other name with DB2 in it. On my system, the option is called DB2 and the description is DB2 Primary Menu. Select whichever option is on your main menu for DB2.

```
   Menu  Utilities  Compilers  Options  Status  Help
  ─────────────────────────────────────────────────────────────────────
                           ISPF Primary Option Menu
   Option ===>

   0   Settings      Terminal and user parameters        User ID . : HRUSER
   1   View          Display source data or listings     Time. . . : 21:19
   2   Edit          Create or change source data        Terminal. : 3278
   3   Utilities     Perform utility functions           Screen. . : 1
   4   Foreground    Interactive language processing     Language. : ENGLISH
   5   Batch         Submit job for language processing  Appl ID . : ISR
   6   Command       Enter TSO or Workstation commands   TSO logon : MATPROC
   7   Dialog Test   Perform dialog testing              TSO prefix: HRUSER
   10  SCLM          SW Configuration Library Manager    System ID : MATE
   11  Workplace     ISPF Object/Action Workplace        MVS acct. : MT529
   12  DITTO         DITTO/ESA for MVS                   Release . : ISPF 6.0
   13  FMN           File Manager
   15  DB2           DB2 Primary Menu
   17  QMF           DB2 Query Management Facility
   S   SDSF          Spool Search and Display Facility

        Enter X to Terminate using log/list defaults
```

This is the DB2 main menu. Select the first option, which is SPUFI. SPUFI is an acronym for **SQL Processing Using File Input**.

Select option 1.

```
                       DB2I PRIMARY OPTION MENU          SSID: DB2X
COMMAND ===>

Select one of the following DB2 functions and press ENTER.

   1   SPUFI                (Process SQL statements)
   2   DCLGEN               (Generate SQL and source language declarations)
   3   PROGRAM PREPARATION  (Prepare a DB2 application program to run)
   4   PRECOMPILE           (Invoke DB2 precompiler)
   5   BIND/REBIND/FREE     (BIND, REBIND, or FREE plans or packages)
   6   RUN                  (RUN an SQL program)
   7   DB2 COMMANDS         (Issue DB2 commands)
   8   UTILITIES            (Invoke DB2 utilities)
   D   DB2I DEFAULTS        (Set global parameters)
   Q   QMF                  (Query Management Facility
   X   EXIT                 (Leave DB2I)



PRESS:                      END to exit     HELP for more information
```

You'll see the following screen.

```
                        SPUFI                          SSID: DB2X
===>

Enter the input data set name:        (Can be sequential or partitioned)
  1   DATA SET NAME ... ===> 'HRUSER.SPUFI.CNTL(EXECSQL)'
  2   VOLUME SERIAL ... ===>          (Enter if not cataloged)
  3   DATA SET PASSWORD ===>          (Enter if password protected)

Enter the output data set name:       (Must be a sequential data set)
  4   DATA SET NAME ... ===> 'HRUSER.SPUFI.OUT'

Specify processing options:
  5   CHANGE DEFAULTS   ===> NO       (Y/N - Display SPUFI defaults panel?)
  6   EDIT INPUT ...... ===> YES      (Y/N - Enter SQL statements?)
  7   EXECUTE ......... ===> YES      (Y/N - Execute SQL statements?)
  8   AUTOCOMMIT ...... ===> YES      (Y/N - Commit after successful run?)
  9   BROWSE OUTPUT ... ===> YES      (Y/N - Browse output data set?)

For remote SQL processing:
 10   CONNECT LOCATION  ===>


PRESS:  ENTER to process    END to exit         HELP for more information
```

This is the place you can specify an input file which will contain the SQL, DDL, DML or DCL statements you wish to execute. We'll explain DDL, DML and DCL shortly. For now just think of SPUFI as the place you can run SQL.

I also recommend that you specify these processing options: `NO` for `CHANGE DEFAULTS`, and `YES` for `EDIT INPUT`, `EXECUTE`, `AUTOMCOMMIT` and `BROWSE OUTPUT`. You must also specify an output file to capture the results from your statements. If these files do not already exist you must allocate them.

When you press `ENTER` your input dataset will open and you can type the statements that you want to execute. In the example below, I've coded a `SELECT` statement to retrieve all records from a sample table named `EMPLOYEE`.

```
File  Edit  Edit_Settings  Menu  Utilities  Compilers  Test  Help


EDIT       HRUSER.SPUFI.CNTL(EXECSQL) - 01.00         Columns 00001 00072
Command ===>                                          Scroll ===> PAGE
****** *************************** Top of Data ***************************
000001 SELECT * FROM EMPLOYEE;
****** ************************** Bottom of Data **************************
```

When you press PF3, and then press ENTER again, the output dataset is shown with the results from the query.

```
  Menu  Utilities  Compilers  Help
ssssssssssssssssssssssssssssssssssssssssssssssssssssssssssssssssssssssssss
BROWSE     HRUSER.SPUFI.OUT                       Line 00000000 Col 001 080
Command ===>                                          Scroll ===> PAGE
****************************** Top of Data ********************************
---------+---------+---------+---------+---------+---------+---------+---------+
SELECT * FROM EMPLOYEE;
---------+---------+---------+---------+---------+---------+---------+---------+
     EMPNO   NAME
---------+---------+---------+---------+---------+---------+---------+---------+
      100   SMITH
      200   JONES
DSNE610I NUMBER OF ROWS DISPLAYED IS 2
DSNE616I STATEMENT EXECUTION WAS SUCCESSFUL, SQLCODE IS 100
---------+---------+---------+---------+---------+---------+---------+---------+
---------+---------+---------+---------+---------+---------+---------+---------+
DSNE617I COMMIT PERFORMED, SQLCODE IS 0
DSNE616I STATEMENT EXECUTION WAS SUCCESSFUL, SQLCODE IS 0
---------+---------+---------+---------+---------+---------+---------+---------+
DSNE601I SQL STATEMENTS ASSUMED TO BE BETWEEN COLUMNS 1 AND 72
DSNE620I NUMBER OF SQL STATEMENTS PROCESSED IS 1
DSNE621I NUMBER OF INPUT RECORDS READ IS 2
DSNE622I NUMBER OF OUTPUT RECORDS WRITTEN IS 18
****************************** Bottom of Data *****************************
```

You will use SPUFI often unless your shop has adopted another tool such as **IBM Data Studio.**

DCLGEN

DCLGEN is an IBM utility that generates SQL data structures (table definition and host variables) for a table or view. DCLGEN stores the structure in a PDS member and then the PDS member can be included in a PLI or COBOL program by issuing an EXEC SQL INCLUDE statement. Put another way, DCLGEN generates table declarations (hence the name DCLGEN). Don't worry if this doesn't make sense yet. We'll generate and use these DCLGEN structures when we start writing programs.

Here's an example of running a DCLGEN for a table. From the DB2 Primary Option menu, select option 2 for DCLGEN.

```
                          DB2I PRIMARY OPTION MENU          SSID: DB2X
COMMAND ===>
Select one of the following DB2 functions and press ENTER.

   1   SPUFI                 (Process SQL statements)
   2   DCLGEN                (Generate SQL and source language declarations)
   3   PROGRAM PREPARATION   (Prepare a DB2 application program to run)
   4   PRECOMPILE            (Invoke DB2 precompiler)
   5   BIND/REBIND/FREE      (BIND, REBIND, or FREE plans or packages)
   6   RUN                   (RUN an SQL program)
   7   DB2 COMMANDS          (Issue DB2 commands)
   8   UTILITIES             (Invoke DB2 utilities)
   D   DB2I DEFAULTS         (Set global parameters)
   Q   QMF                   (Query Management Facility
   X   EXIT                  (Leave DB2I)




PRESS:                       END to exit     HELP for more information
```

Now enter the DB2 table name, owner and the partitioned data set and member name to place the DCLGEN output into. In the example below we have a table named EMP_PAY_CHECK owned by HRSCHEMA. We want the output of the DCLGEN to be placed in member EMPPAYCK of partitioned dataset HRUSER.DCLGEN.COBOL.

Once you've entered the required information, press ENTER.

```
                           DCLGEN                           SSID: DB2X
===>

Enter table name for which declarations are required:
 1   SOURCE TABLE NAME ===> EMP_PAY_CHECK


 2   TABLE OWNER ..... ===> HRSCHEMA


 3   AT LOCATION ..... ===>                              (Optional)
Enter destination data set:        (Can be sequential or partitioned)
 4   DATA SET NAME ... ===> 'HRUSER.DCLGEN.COBOL(EMPPAYCK)'
 5   DATA SET PASSWORD ===>         (If password protected)
Enter options as desired:
 6   ACTION .......... ===> ADD     (ADD new or REPLACE old declaration)
 7   COLUMN LABEL .... ===> NO      (Enter YES for column label)
 8   STRUCTURE NAME .. ===>                             (Optional)
 9   FIELD NAME PREFIX ===>                             (Optional)
10   DELIMIT DBCS .... ===> YES     (Enter YES to delimit DBCS identifiers)
11   COLUMN SUFFIX ... ===> NO      (Enter YES to append column name)
12   INDICATOR VARS .. ===> NO      (Enter YES for indicator variables)
13   ADDITIONAL OPTIONS===> NO      (Enter YES to change additional options

PRESS: ENTER to process    END to exit      HELP for more information
```

Next, you will receive a message indicating the DCLGEN has succeeded.

DSNE905I EXECUTION COMPLETE, MEMBER EMPPAYCK ADDED
***

Now you can browse the PDS member EMPPAYCK to see the resulting structures. The first structure will declare the DB2 table definition, and the other structure will declare COBOL host variables that correspond to the table definition.

```
     ******************************************************************
     * DCLGEN TABLE(HRSCHEMA.EMP_PAY_CHECK)                          *
     *        LIBRARY(HRSCHEMA.DCLGEN.COBOL(EMPPAYCK))               *
     *        LANGUAGE(COBOL)                                      *
     *        QUOTE                                               *
     * ... IS THE DCLGEN COMMAND THAT MADE THE FOLLOWING STATEMENTS  *
     ******************************************************************
          EXEC SQL DECLARE HRSCHEMA.EMP_PAY_CHECK TABLE
          ( EMP_ID                      INTEGER NOT NULL,
            EMP_REGULAR_PAY             DECIMAL(8, 2) NOT NULL,
            EMP_SEMIMTH_PAY             DECIMAL(8, 2) NOT NULL
          ) END-EXEC.
     ******************************************************************
```

```
* COBOL DECLARATION FOR TABLE HRSCHEMA.EMP_PAY_CHECK          *
*********************************************************************
  01  DCLEMP-PAY-CHECK.
      10 EMP-ID              PIC S9(9) USAGE COMP.
      10 EMP-REGULAR-PAY     PIC S9(6)V9(2) USAGE COMP-3.
      10 EMP-SEMIMTH-PAY     PIC S9(6)V9(2) USAGE COMP-3.
*********************************************************************
* THE NUMBER OF COLUMNS DESCRIBED BY THIS DECLARATION IS 3     *
*********************************************************************
```

Again, the above structure can be included in your application program by simply issuing:

```
EXEC SQL
    INCLUDE EMPPAYCK
END-EXEC
```

Once you've included the DCLGEN structure in your application program, you have host variables declared for every column in the table, so you don't need to code these variables yourself.

DB2I Defaults

Your shop should have standard settings for DB2I defaults. If you are studying on your own, I recommend setting some defaults. First select option D from the DB2 Primary Option menu:

```
                        DB2I PRIMARY OPTION MENU          SSID: DB2X
COMMAND ===>

Select one of the following DB2 functions and press ENTER.

  1   SPUFI                (Process SQL statements)
  2   DCLGEN               (Generate SQL and source language declarations)
  3   PROGRAM PREPARATION  (Prepare a DB2 application program to run)
  4   PRECOMPILE           (Invoke DB2 precompiler)
  5   BIND/REBIND/FREE     (BIND, REBIND, or FREE plans or packages)
  6   RUN                  (RUN an SQL program)
  7   DB2 COMMANDS         (Issue DB2 commands)
  8   UTILITIES            (Invoke DB2 utilities)
  D   DB2I DEFAULTS        (Set global parameters)
  Q   QMF                  (Query Management Facility)
  X   EXIT                 (Leave DB2I)




PRESS:                    END to exit     HELP for more information
```

Select the following options, specifying your correct DB2 subsystem identifier as the DB2 Name (ask your system admin if you are not sure – I have used DB2X as mine and that represents the subsystem identifier on my system).

If you are developing in COBOL, specify IBMCOB as the application language. Otherwise specify PLI or whichever language you are using. Our programming examples will be in COBOL.

```
                         DB2I DEFAULTS PANEL 1
 COMMAND ===>

 Change defaults as desired:

  1  DB2 NAME ............. ===> DB2X       (Subsystem identifier)
  2  DB2 CONNECTION RETRIES ===> 0          (How many retries for DB2 connection)
  3  APPLICATION LANGUAGE   ===> IBMCOB     (ASM, C, CPP, IBMCOB, FORTRAN, PLI)
  4  LINES/PAGE OF LISTING  ===> 60         (A number from 5 to 999)
  5  MESSAGE LEVEL ........ ===> I          (Information, Warning, Error, Severe)
  6  SQL STRING DELIMITER   ===> DEFAULT    (DEFAULT, ' or ")
  7  DECIMAL POINT ........ ===> .          (. or ,)
  8  STOP IF RETURN CODE >= ===> 8          (Lowest terminating return code)
  9  NUMBER OF ROWS ....... ===> 20         (For ISPF Tables)
 10  AS USER                ===>            (Userid to associate with the trusted
                                             connection)




 PRESS:  ENTER to process    END to cancel           HELP for more information
```

BATCH UTILITIES

DB2 provides some batch utility programs you can use. The two most useful of these are DSNTIAUL and DSNSTEP2 (and DSNSTEP4 if you have large numbers of rows to unload because it uses multi-row fetch). You can use these instead of SPUFI to operate on data (or issue other statements) in a batch mode.

DSNTIAUL

DSNTIAUL is a utility most often used to unload a DB2 table into a flat file. Here is a sample JCL to run it, and of course you must adjust file names to your own environment. In the SYSIN file you specify the name of a table to unload. In this example we will unload the EMP_PAY table (HRSCHEMA is a schema qualifier – we will talk about that later).

301

```
//HRSCHEMAD JOB MSGLEVEL=(1,1),NOTIFY=&SYSUID
//*
//*  COBOL DB2 RUN JCL
//*
//STEP01   EXEC PGM=IKJEFT01,
//            DYNAMNBR=20,REGION=4096K
//STEPLIB  DD  DISP=SHR,DSN=DSNTST.DBAX.SDSNEXIT   ß YOUR LIBRARIES
//         DD  DISP=SHR,DSN=DSNTST.SDSNLOAD        ß YOUR LIBRARIES
//SYSPRINT DD  SYSOUT=*
//SYSTSPRT DD  SYSOUT=*
//SYSPUNCH DD  SYSOUT=*
//SYSUDUMP DD  SYSOUT=*
//SYSTSIN  DD  *
DSN SYSTEM (DB2X   )
  RUN PROGRAM   (DSNTIAUL) -
      PLAN      (DSNTIAUL) -
      LIBRARY   ('DSNTST.DBAX.RUNLIB.LOAD')          ß YOUR SYS LOAD LIBRARY
END
/*
//SYSIN    DD *
 HRSCHEMA.EMP_PAY
//SYSOUT   DD  SYSOUT=*
```

The above unloads the entire EMP_PAY table. Our second example allows you to use SQL to select the data you want to unload. Notice that it includes **PARMS('SQL')** as a RUN parameter. The SQL is coded in the SYSIN file:

```
//HRSCHEMAD JOB MSGLEVEL=(1,1),NOTIFY=&SYSUID
//*
//STEP01   EXEC PGM=IKJEFT01,
//            DYNAMNBR=20,REGION=4096K
//STEPLIB  DD  DISP=SHR,DSN=DSNTST.DB2X.SDSNEXIT
//         DD  DISP=SHR,DSN=DSNTST.SDSNLOAD
//SYSPRINT DD  SYSOUT=*
//SYSTSPRT DD  SYSOUT=*
//SYSPUNCH DD  SYSOUT=*
//SYSUDUMP DD  SYSOUT=*
//SYSTSIN  DD  *
DSN SYSTEM (DB2X   )
  RUN PROGRAM   (DSNTIAUL) -
      PLAN      (DSNTIAUL) -
    PARMS('SQL') -
      LIBRARY   ('DSNTST.DB2X.RUNLIB.LOAD')
END
/*
//SYSIN    DD *
 SELECT
 EMP_ID, EMP_LAST_NAME, EMP_FIRST_NAME
 FROM HRSCHEMA.EMPLOYEE
 WHERE EMP_SERVICE_YEARS > 1
//SYSOUT   DD  SYSOUT=*
```

The above example shows that you can specify which columns to include in your output, as well as selecting which rows to unload.

DSNTEP2

The DSNTEP2 sample program allows you to execute dynamic SQL to select or update data. The JCL looks like this, and the example is reading data from the EMP_PAY_CHECK table:

```
//HRSCHEMA2 JOB MSGLEVEL=(1,1),NOTIFY=&SYSUID
//*
//STEP01   EXEC PGM=IKJEFT01,
//             DYNAMNBR=20,REGION=4096K
//STEPLIB  DD  DISP=SHR,DSN=DSNTST.DB2X.SDSNEXIT
//         DD  DISP=SHR,DSN=DSNTST.SDSNLOAD
//         DD  DISP=SHR,DSN=DSNTST.DB2X.RUNLIB.LOAD
//SYSPRINT DD  SYSOUT=*
//SYSTSPRT DD  SYSOUT=*
//SYSPUNCH DD  SYSOUT=*
//SYSUDUMP DD  SYSOUT=*
//SYSTSIN  DD  *
DSN SYSTEM (DB2X   )
  RUN PROGRAM (DSNTEP2) PLAN (DSNTEP2)
END
/*
//SYSIN      DD *
 SELECT * FROM HRSCHEMA.EMP_PAY_CHECK
 ORDER BY EMP_ID
//SYSOUT   DD  SYSOUT=*
```

You could also run an update or delete action with DSNTEP2. Just change the SYSIN control file content. Here's an example:

```
        UPDATE HRSCHEMA.EMPLOYEE
        SET EMP_SERVICE_YEARS
           = EMP_SERVICE_YEARS + 1
```

If you are not already using these sample programs, I suggest you learn more about them and use them. They are great utilities and will save you some custom coding effort!

Data Definition Language

Before rushing into programming it is a good idea to understand how to create and maintain the basic DB2 objects. We'll look at the properties of the various objects (tables, indexes, views). We'll look over the various data types that DB2 provides as well. Finally we'll look at the DB2 catalog and the information it provides.

A DB2 database is a collection of objects including tablespaces, tables, indexes, views, triggers, stored procedures and sequences. Generally a database is concerned with a single domain such as marketing, accounting, shipping and receiving, etc. Within the database there are objects such as tables, indexes, views, etc.

In many shops a DBA creates and maintains the database objects. I don't expect you to know every nuance of each database object, but you need to know the basic Data Definition Language. So let's look at how to create and maintain the basic database objects.

For purposes of this text book, we will be creating and maintaining a simple human relations system for a fictitious company. So we'll create objects with that in mind.

DATABASES

CREATE

You can create a database with the CREATE DATABASE statement and you can assign options such as bufferpool, index bufferpool, storage group and CCSID. The required syntax to create the database is:

```
CREATE DATABASE <database name>
```

However you would normally specify names for a storage group, bufferpool and index bufferpool. Here's an example:

```
CREATE DATABASE DB1
STOGROUP DS1
BUFFERPOOL BP1
INDEXBP BP2;
```

You may not have security access to create a DB2 database, even a test database. This depends on your shop and whether or not it allows application developers to create database objects. If not, then you can ask a DBA to create the objects for you.

For our purposes, let's assume that you do have security and we'll go ahead and create our HR database. Let's name the database DBHR and we'll specify a bufferpool named BPHR, an index bufferpool named IBPHR, and a storage group called SGHR. Finally we need to choose a CCSID (conceptually this is similar to a codepage) from ASCII, EBCDIC or UNICODE. We'll talk more about CCSIDs in the subsection on tables. For now, let's go ahead and specify UNICODE as the CCSID for our database.

The DDL to create our HR database is as follows:

```
CREATE DATABASE DBHR
STOGROUP SGHR
BUFFERPOOL BPHR
INDEXBP IBPHR
CCSID UNICODE;
```

ALTER

If you need to change anything about the database in the future, you can use the ALTER command. Suppose for example we want to change the default bufferpool. You could issue this DDL to change the default bufferpool in the DBHR database to BPHR2.

```
ALTER DATABASE DBHR
BUFFERPOOL BPHR2;
```

DROP

Most database objects can be removed/deleted by issuing the DROP command. The syntax to delete a database is very simple:

```
DROP DATABASE <databasename>
```

You could DROP the DBHR database by simply issuing this command:

```
DROP DATABASE DBHR;
```

We won't drop the DBHR database now because we are going to add some more objects to it in this chapter.

TABLESPACES

A tablespace is a layer between the physical containers that hold data and the logical database. In essence, a tablespace defines storage areas into which DB2 objects may be placed and maintained.

DB2 11 supports the following types of tables spaces:

- Universal
- Segmented
- Partitioned
- EA enabled
- Large Object (LOB)

- XML
- Simple (cannot be created in DB2 11 but still supported if already exists)

Tablespace Type	Description
Universal table spaces	A universal table space is a combination of partitioned and segmented table space schemes.
EA-enabled table spaces/index spaces	Table spaces and index spaces that are enabled for extended addressability is EA-enabled.
Large object table spaces	LOB table spaces (also known as auxiliary table spaces) hold large object data, such as graphics, video, or large text strings. If your data does not fit entirely within a data page, you can define one or more columns as LOB columns.
XML table spaces	An XML table space stores an XML table.
Partitioned (non-universal) table spaces (deprecated)	A table space that is partitioned stores a single table. DB2 divides the table space into partitions.
Simple table spaces (deprecated)	A simple table space is neither partitioned nor segmented. The creation of new simple table spaces is not supported in DB2 11. However, DB2 can still use existing simple table spaces.
Segmented (non-universal) table spaces (deprecated)	A table space that is segmented is useful for storing more than one table, especially relatively small tables. The pages hold segments, and each segment holds records from only one table.

Let us create a universal tablespace for our HR database, and we'll name it TSHR (tablespace HR). We'll specify storage group SGHR, provide the primary and secondary space allocations as 50 and 20, do locking at the page level and use bufferpool BPHR2. Don't worry if these values do not make perfect sense. DBAs typically take care of tablespsaces.

Meanwhile, we'll create and use our new tablespace TSHR throughout this text book as a container for tables and other objects. Here's the DDL:

```
CREATE TABLESPACE TSHR
    IN DBHR
    USING STOGROUP SGHR
      PRIQTY 50
      SECQTY 20
    LOCKSIZE PAGE
    BUFFERPOOL BPHR2;
```

One other operation we will perform before moving on to tables is to create a schema. A schema is a qualifier used for logically grouping and owning objects. In our case we will create a schema named HRSCHEMA and then use that to group our tables, indexes, views, etc. The schema must have an owner and let's assume we can have it owned by system authorized id DBA001 (you can substitute your own logon id).

```
CREATE SCHEMA HRSCHEMA
AUTHORIZATION DBA001;   ← this should be your id
```

Now you can create database objects such as tables, view, indexes and sequences and specify schema HRSCHEMA as the qualifier. For example you could create an EMPLOYEE table as HRSCHEMA.EMPLOYEE. If you do not specify a schema, DB2 assumes a default or current schema which is often your logon-id. Obviously if you are working as part of a group, a common schema name is a better alternative.

TABLES

As I'm sure you are aware, a table is the basic structure and container for DB2 data. Let's summarize the different types of tables in DB2 11, and then we'll generate sample tables for discussion.

Table Types

Type	Description
Archive	An archive table is a table that stores data that was deleted from another table. The other table is called an archive-enabled table.
Auxiliary	An auxiliary table is used to store Large Object (LOB) data that is linked to another base table.
Base	A table structure which physically persists records.
Clone	A table that is structurally identical to a base table.
History	A table that is used to store historical versions of rows from the associated system-period temporal table.
Materialized Query	A materialized query table basically stores the result set of a query. It is typically used to store aggregate results from one or more other tables.
Result	A non-persistent table that contains a set of rows that DB2 selects or generates, directly or indirectly, from one or more base tables or views in response to an SQL statement.
Temporal	A temporal table is one that keeps track of "versions" of data over time and allows you to query data according to the time frame.
Temporary	A table that is created and exists only for the duration of a session.
XML	A special table that holds only XML data.

DDL for Tables

Now let's look at the basic DDL that is used to manipulate tables. As with other DB2 objects, we use the CREATE, ALTER and DROP statements to create, change and delete tables respectively.

308

CREATE

The basic syntax to create a DB2 table specifies the table name, column specifications and the tablespace into which the table is to be created.

```
CREATE TABLE <tablename>
(field specifications)
IN <tablespace>
```

For an example, let's create the first table for our HR application. Here are the columns and data types for our table which we will name EMPLOYEE.

Field Name	Type	Attributes
EMP_ID	INTEGER	NOT NULL, PRIMARY KEY
EMP_LAST_NAME	VARCHAR(30)	NOT NULL
EMP_FIRST_NAME	VARCHAR(20)	NOT NULL
EMP_SERVICE_YEARS	INTEGER	NOT NULL, DEFAULT IS ZERO
EMP_PROMOTION_DATE	DATE	

The table can be created with the following DDL:

```
CREATE TABLE HRSCHEMA.EMPLOYEE(
    EMP_ID INT NOT NULL,
    EMP_LAST_NAME VARCHAR(30) NOT NULL,
    EMP_FIRST_NAME VARCHAR(20) NOT NULL,
    EMP_SERVICE_YEARS INT NOT NULL WITH DEFAULT 0,
    EMP_PROMOTION_DATE DATE,
    PRIMARY KEY(EMP_ID)) IN TSHR;
```

While we haven't talked about indexes yet, we will need to create a unique index to support the primary key. Otherwise when we try to access the table, our SQL will fail. Let's create an index now and then we'll talk more about indexes in the next sub-section.

```
CREATE UNIQUE INDEX NDX_EMPLOYEE
ON EMPLOYEE (EMP_ID);
```

Before we move on, let's create a couple more tables that we will use later. Let's say we need an EMP_PAY table to store the employee's annual pay amount, and an EMP_PAY_CHECK table that will be used to cut pay checks on the first and fifteen of the month. Here's the DDL for these:

```
CREATE TABLE HRSCHEMA.EMP_PAY(
EMP_ID INT NOT NULL,
EMP_REGULAR_PAY DECIMAL (8,2) NOT NULL,
```

```
EMP_BONUS_PAY DECIMAL    (8,2))
IN TSHR;

CREATE TABLE HRSCHEMA.EMP_PAY_CHECK(
EMP_ID INT NOT NULL,
EMP_REGULAR_PAY  DECIMAL (8,2) NOT NULL,
EMP_SEMIMTH_PAY DECIMAL (8,2) NOT NULL)
IN TSHR;
```

ALTER

You can change various aspects of a table using the ALTER command. ALTER is often used to add an index or additional columns. Here is an example:

```
ALTER TABLE HRSCHEMA.EMPLOYEE
ADD COLUMN EMP_PROFILE XML;
```

At this point we won't run this DDL because we want to do these operations later after some explanation. For now, just be aware that the way to change a DB2 table is to use the ALTER statement.

DROP

You can remove a table by issuing the DROP command.

```
DROP TABLE <table name>
```

Note: You cannot drop a table for which a trigger is still defined. You must first drop the trigger.

Base Tables

The most common type of table in DB2 is a base table. This is your typical table created with the CREATE TABLE statement. A sample is our original DDL used for the EMPLOYEE table:

```
CREATE TABLE HRSCHEMA.EMPLOYEE(
EMP_ID INT NOT NULL,
EMP_LAST_NAME VARCHAR(30) NOT NULL,
EMP_FIRST_NAME VARCHAR(20) NOT NULL,
EMP_SERVICE_YEARS INT NOT NULL WITH DEFAULT 0,
EMP_PROMOTION_DATE DATE,
PRIMARY KEY(EMP_ID))
IN TSHR;
```

Result Tables

A result table is called that because it is the result set of a query. It is not persistent. To show an example let's first add a record to the EMPLOYEE table. We haven't reviewed DML yet, but let's go ahead and do an INSERT with the following:

```
INSERT INTO HRSCHEMA.EMPLOYEE
(EMP_ID,
 EMP_LAST_NAME,
 EMP_FIRST_NAME,
 EMP_SERVICE_YEARS,
 EMP_PROMOTION_DATE)
VALUES (3217,
'JOHNSON',
'EDWARD',
4,
'01/01/2017');
```

Now run the following query:

```
SELECT EMP_ID, EMP_LAST_NAME,
EMP_FIRST_NAME
FROM HRSCHEMA.EMPLOYEE
WHERE EMP_ID = 3217;
---------+---------+---------+---------+---------+---
    EMP_ID   EMP_LAST_NAME        EMP_FIRST_NAME
---------+---------+---------+---------+---------+---
    3217   JOHNSON              EDWARD
DSNE610I NUMBER OF ROWS DISPLAYED IS 1
```

The displayed data is a result table. After the query is run and the data is displayed, you cannot reference or change the result table, i.e., it is not persistent.

Clone Tables

In some situations you may need to work with a copy of a base table, and then at some point switch the copy for the original. A clone table is useful for this purpose. A clone table is structurally identical to the base table. It is created by ALTERING the original. To take an example, let's clone the EMPLOYEE table:

```
ALTER TABLE HRSCHEMA.EMPLOYEE
ADD CLONE HRSCHEMA.EMPLOYEE_CLONE;
```

Now you can load the clone table with data and manipulate it in whatever fashion you need to. Later you can switch the tables using the EXCHANGE command:

```
EXCHANGE DATA BETWEEN TABLE HRSCHEMA.EMPLOYEE
AND HRSCHEMA.EMPLOYEE_CLONE;
```

In actuality, no data is moved. Instead, the names on the physical tables are just switched behind the scenes. In the example, the original EMPLOYEE base table has now become EM-PLOYEE_CLONE and the EMPLOYEE_CLONE table has become EMPLOYEE. To switch them back, issue the EXCHANGE command again.

```
EXCHANGE DATA BETWEEN TABLE HRSCHEMA.EMPLOYEE
AND HRSCHEMA.EMPLOYEE_CLONE;
```

Note: You cannot clone a table unless that table exists in a universal table space (UTS).

Archive Tables

An archive table is a table that stores data that was deleted from another table called an archive-enabled table. When a row is deleted from the archive-enabled table, DB2 automatically adds the row to the archive table. When you query the archive-enabled table, you can specify whether or not to include archived records or not. We'll look at these features in an example.

Assume we want to delete some records from our EMPLOYEE table and we want to automatically archive the deleted records to a new table named EMPLOYEE_ARCHIVE. One way to set up and define the archive table with exactly the same column definitions as EMPLOYEE is to use the LIKE clause with the CREATE statement:

```
CREATE TABLE HRSCHEMA.EMPLOYEE_ARCHIVE
LIKE HRSCHEMA.EMPLOYEE
IN TSHR;
```

To enable archiving of deleted records from table EMPLOYEE to the new table EMPLOYEE_AR-CHIVE, you would execute the following:

```
ALTER TABLE EMPLOYEE ENABLE ARCHIVE USE EMPLOYEE_ARCHIVE;
```

To automatically archive records, set the global variable SYSIBMADM.MOVE_TO_ARCHIVE to Y or E. This value indicates whether deleting a record from an archive-enabled table should store a copy of the deleted record in the archive table. The values are:

- Y - store a copy of the deleted record, and also make any attempted insert/update operation against the archive table an error.

- E - store a copy of the deleted record.

- N- do not store a copy of the deleted record.

In the future when you query the EMPLOYEE table you can choose to include or exclude the archived records in a given session. To do this, your package must first be bound with the ARCHIVESENSITIVE(YES) bind option. Then the package/program should set the GET_AR-CHIVE global variable to Y (the default is N). At this point, any query against the archive-enabled table during this session will automatically include data from the corresponding archive table. This is very handy for when you want to do historical research.

In our EMPLOYEE example, suppose we have a package EMP001 that is bound with ARCHIVESENSITIVE(YES). Suppose further that the program issues this SQL:

```
SET SYSIBMADM.GET_ARCHIVE = 'Y'
```

Now any query we issue during this session against the EMPLOYEE table will automatically return any qualifying rows from both the EMPLOYEE and EMPLOYEE_ARCHIVE tables. For example:

```
SELECT EMP_ID, EMP_LAST_NAME, EMPL_FIRST_NAME
FROM EMPLOYEE
ORDER BY EMP_ID
```

If the package needs to revert to only picking up data from the EMPLOYEE table, it can simply issue the SQL:

```
SET SYSIBMADM.GET_ARCHIVE = 'N'
```

Some design advantages of an archive table are:

1. Your historical data is managed automatically. You don't need to manually or programmatically move older data to a separate table.

2. The scope of your query is controlled using a global variable. Consequently you can modify your query results to include or exclude the archive table data and you don't have to change the SQL statement (only the global variable value).

3. Older rows that are less often retrieved can be stored in a separate archive table which could potentially be located on a cheaper device.

Auxiliary Tables

An auxiliary table is used to store Large Object (LOB) data that is linked to another base table. The best way to understand this is by example. Suppose that we decide to add an employee photo column to the EMPLOYEE table.

Now let's add an employee photo column as a 5 megabyte BLOB. The DDL we would use is as follows:

```
ALTER TABLE EMPLOYEE
ADD EMP_PHOTO BLOB(5M);
```

Now we must create the auxiliary table to store the LOB data. It must be created in the new LOB tablespace, and our DDL must specify that it will store column EMP_PHOTO from table EMPLOYEE. We also need a unique index on the auxiliary table.

If we do not already have an LOB table space, we can create one as follows:

```
CREATE LOB TABLESPACE EMP_PHOTO_TS
  IN DBHR LOG NO;
```

While not mandatory, it is good practice to avoid logging the LOB data, as this can slow performance considerably for large amounts of LOB data. Of course if the data is mission critical you may need to log it for recoverability purposes if no other recovery method exists.

Now we are ready to create the auxiliary table. The syntax for this type of table is:

```
CREATE AUX TABLE <table name>
IN <LOB table space>
STORES (the base table name>
COLUMN <column in the base table>;
```

Here's the DDL to create our new auxiliary table to support the EMP_PHOTO column:

```
CREATE AUX TABLE EMP_PHOTOS_TAB
      IN EMP_PHOTO_TS
      STORES EMPLOYEE
      COLUMN (EMP_PHOTO);

CREATE UNIQUE INDEX XEMP_PHOTO
  ON EMP_PHOTOS_TAB;
```

Now we would need to rerun the DCLGEN on the EMPLOYEE table. You'll notice that the DCL-GEN now specifies the EMP_PHOTO as a BLOB type.

```
SQL TYPE IS BLOB (5M) EMP_PHOTO;
```

Our update program can now load the photo data by defining a host variable into which you load the binary photo data, and then you use that host variable in the SQL as follows (assume we have declared PHOTO-DATA as our host variable):

```
UPDATE EMPLOYEE SET EMP_PHOTO =:PHOTO-DATA
WHERE EMP_ID = :EMP_ID;
```

Materialized Query Tables

A materialized query table basically stores the results of an SQL query. It holds the aggregate results from querying one or more other tables or views. MQTs are often used to improve performance for certain aggregation queries by providing pre-computed results. Consequently, MQTs are often used in analytic or data warehousing environments.

MQTs are either system-maintained or user maintained. For a system maintained table, the data must be updated using the REFRESH TABLE statement. A user-maintained MQT can be updated using the LOAD utility, and also the UPDATE, INSERT, and DELETE SQL statements.

Let's do an example of an MQT that summarizes monthly payroll. Assume we have a source table named EMP_PAY_HIST which will be a history of each employee's salary for each paycheck. The table requirements are summarized as follows:

Column Name	Definition
EMP_ID	Numeric
EMP_PAY_DATE	Date
EMP_PAY_AMT	Decimal(8,2)

The DDL for the table is as follows:

```
CREATE TABLE EMP_PAY_HIST(
EMP_ID              INT NOT NULL,
EMP_PAY_DATE        DATE NOT NULL,
EMP_PAY_AMT         DECIMAL (8,2) NOT NULL)
IN TSHR;
```

Before we can use the table for an MQT we need to add some records to it. Let's run the following DDL to do that:

```
INSERT INTO HRSCHEMA.EMP_PAY_HIST
VALUES (3217,'01/15/2017',2291.66);

INSERT INTO HRSCHEMA.EMP_PAY_HIST
VALUES (3217,'01/31/2017',2291.66);

INSERT INTO HRSCHEMA.EMP_PAY_HIST
VALUES (3217,'02/15/2017',2291.66);

INSERT INTO HRSCHEMA.EMP_PAY_HIST
VALUES (3217,'02/28/2017',2291.66);

INSERT INTO HRSCHEMA.EMP_PAY_HIST
VALUES (7459,'01/15/2017',3333.33);

INSERT INTO HRSCHEMA.EMP_PAY_HIST
VALUES (7459,'01/31/2017',3333.33);

INSERT INTO HRSCHEMA.EMP_PAY_HIST
VALUES (7459,'02/15/2017',3333.33);

INSERT INTO HRSCHEMA.EMP_PAY_HIST
VALUES (7459,'02/28/2017',3333.33);
```

Now let's select the data in the EMP_PAY_HIST table is as follows:

```
     SELECT * FROM HRSCHEMA.EMP_PAY_HIST;
---------+---------+---------+---------+
    EMP_ID  EMP_PAY_DATE  EMP_PAY_AMT
---------+---------+---------+---------+
      3217  2017-01-15       2291.66
      3217  2017-01-31       2291.66
      3217  2017-02-15       2291.66
      3217  2017-02-28       2291.66
      7459  2017-01-15       3333.33
      7459  2017-01-31       3333.33
      7459  2017-02-15       3333.33
      7459  2017-02-28       3333.33
DSNE610I NUMBER OF ROWS DISPLAYED IS 8
```

Finally, let's assume we regularly need an aggregated total of each employee's year to date pay. We could do this with a materialized query table. Let's build the query that will summarize the employee pay from the beginning of the year to current date:

```
   SELECT EMP_ID, SUM(EMP_PAY_AMT) AS EMP_PAY_YTD
   FROM HRSCHEMA.EMP_PAY_HIST
   GROUP BY EMP_ID
```

```
   ORDER BY EMP_ID;
---------+---------+---------+---------+---------
    EMP_ID              EMP_PAY_YTD
---------+---------+---------+---------+---------
     3217                9166.64
     7459               13333.32
DSNE610I NUMBER OF ROWS DISPLAYED IS 2
```

Using the model query above, let's create the MQT using the following DDL, and we'll make it a system managed table:

```
CREATE TABLE EMP_PAY_TOT (EMP_ID, EMP_PAY_YTD) AS
(SELECT EMP_ID, SUM(EMP_PAY_AMT) AS EMP_PAY_YTD
FROM HRSCHEMA.EMP_PAY_HIST
GROUP BY EMP_ID)
DATA INITIALLY DEFERRED
REFRESH DEFERRED
MAINTAINED BY SYSTEM
ENABLE QUERY OPTIMIZATION;
```

We can now populate the table by issuing the REFRESH TABLE statement as follows:

```
REFRESH TABLE HRSCHEMA.EMP_PAY_TOT
```

Finally we can query data from the MQT as follows:

```
SELECT * FROM HRSCHEMA.EMP_PAY_TOT;

---------+---------+---------+---------+---------+--------
    EMP_ID              EMP_PAY_YTD
---------+---------+---------+---------+---------+--------
     3217                9166.64
     7459               13333.32
DSNE610I NUMBER OF ROWS DISPLAYED IS 2
```

Again, the benefit of an MQT is that you can generate the query results once, and then the results can be queried many times by many users without the overhead of regenerating the results each time. Also, to update the table you need only issue the REFRESH command. These benefits may seem trivial for a small query, but it can save a great deal of time and CPU cycles for complex queries involving multiple tables and large amounts of aggregated data.

Temporal Tables

Temporal tables were introduced to DB2 z/OS in version 10. Briefly, a temporal table is one that keeps track of "versions" of data over time and allows you to query data according to the time frame. We won't spend too much time here going into the details because it's more an intermediate topic. We cover temporal tables in the companion book DB2 11 for z/OS: Intermediate Training for Application Developers. For now we'll just put forth the basic concepts.

Business Time

Business time concerns data that is valid in a business sense for some period of time. Let's go back to our employee application. An employee's pay typically changes over time. Besides wanting to know current salary, there are scenarios under which an HR department or supervisor might need to know what pay rate was in effect for an employee at some time in the past. We might also need to allow for cases where the employee terminated for some period of time and then returned. Or maybe they took a non-paid leave of absence.

This is the concept of business time and it can be fairly complex depending on the business rules required by the application. It basically means a period of time during which the data is accurate from a business standpoint or according to a business rule. You could think of the event or condition as having an effective date and discontinue date.

A table can only have one business time period. When a BUSINESS_TIME period is defined for a table, DB2 generates a check constraint in which the end column value must be greater than the begin column value.

System Time

System time simply means the time during which a piece of data is in the database, i.e., when the data was added, changed or deleted. Sometimes it is important to know this information. For example a user might enter an employee's salary change on a certain date but the effective date of the salary change might be earlier or later than the date it was entered into the system. The system time simply records when the data was actually entered into the system, changed in the system, or deleted from the system. An audit trail application or table often has a time-stamp that can be considered system time.

Bitemporal Support

In some cases you may need to support both business and system time in the same table. DB2 supports this and it is called bi-temporal support.

History Table

A table that is used to store historical versions of rows from the associated system-period temporal table.

Temporary Table

Sometimes you may need the use of a table for the duration of a session but no longer than that. For example you may have a programming situation where it is convenient to load a temporary table for these operations:

- To join the data in the temporary table with another table
- To store intermediate results to be queried later in the program
- To load data from a flat file into a relational format

In all cases, it is assumed that you only need a temporary table for the duration of a session (or iteration of a program) since temporary tables are dropped as soon as the session ends.

Temporary tables are created using either the CREATE statement or the DECLARE statement. We'll enumerate their differences below.

```
CREATE GLOBAL TEMPORARY TABLE
EMP_INFO(
EMP_ID    INT,
EMP_LNAME  VARCHAR(30),
EMP_FNAME  VARCHAR(30));


DECLARE GLOBAL TEMPORARY TABLE
EMP_INFO(
EMP_ID    INT,
EMP_LNAME  VARCHAR(30),
EMP_FNAME  VARCHAR(30));
```

CREATED Temporary Tables

Created temporary tables:

- Have an entry in the system catalog (SYSIBM.SYSTABLES)
- Cannot have indexes
- Their columns cannot use default values (except NULL)
- Cannot have constraints
- Cannot be used with DB2 utilities
- Cannot be used with the UPDATE statement
- If DELETE is used at all, it will delete all rows from the table
- Do not provide for locking or logging

DECLARED Temporary Tables

A declared temporary table offers some advantages over created temporary tables.

- Can have indexes and check constraints

- Can use the UPDATE statement

- Can do positioned deletes

So declared temporary tables offer more flexibility than created temporary tables. However, when a session ends, DB2 will automatically delete the table definition. So if you want a table definition that persists in the DB2 catalog for future use, you would need to use a created temporary table.

Things to remember about temporary tables:

- Use temporary tables when you need the data only for the duration of the session.

- Created temporary tables can provide excellent performance because they do not use locking or logging.

- Declared temporary tables can also be very efficient because you can choose not to log, and they only allow limited locking.

- The schema for a temporary table is always SESSION.

- If you create a temporary table and you wish to replace any existing temporary table that has the same name, use the WITH REPLACE clause.

- If you create a temporary table from another table using the LIKE clause, the temporary table will NOT have any unique constraints, foreign key constraints, triggers, indexes, table partitioning keys, or distribution keys from the original table.

XML Table

An XML table is a special table that holds only XML data. When you create a table with an XML column, DB2 implicitly creates a new XML table space with a new DB2 table to store the XML data. Here's an example of adding an employee profile XML column to the employee table we created earlier:

```
ALTER TABLE HRSCHEMA.EMPLOYEE
ADD COLUMN EMP_PROFILE XML;
```

The preceding DDL automatically creates a table space and table for the XML column. DB2 creates the XML table space and table in the same database as the table that defined the XML column (the base table). The XML table space is in the Unicode UTF-8 encoding scheme.

INDEXES

Indexes are structures that provide a means of quickly locating a record in a table. One of the main reasons for having indexes is that they improve performance when accessing data randomly. There are other reasons as well which we'll explore now.

Benefits of Indexes

Indexes are beneficial in three ways:

1. Indexes improve performance in that it is typically faster to use the index row locator to navigate to a specific row than to do a table scan (except in cases of very small tables).

2. Unique indexes ensure uniqueness of record keys (either primary or secondary).

3. Clustered indexes enable data to be organized in the base table to optimize sequential access and processing.

Types of Indexes

The types of indexes available in DB2 11 are listed below and detailed on the DB2 product documentation web site: [15]

Type	Description
Unique index	Ensures that each row contains a unique value in the column upon which the index is built.
Primary index	Indexes the primary key column of the table. When you create a table and assign a primary key, you must create a unique index on the column before you can use the table.
Secondary index	Any index that is not a primary index.
Clustering index	An index that forces a logical grouping by clustering the data based on a defined sequence. A table can only have one clustering index.
Expression-based index	An index that is defined based on a general expression.
Partitioned index	An index that is itself physically partitioned.
Partitioning index (PI)	An index that corresponds to the columns that partition the table. These columns are called the partitioning key and are specified in the PARTITION BY clause of the CREATE TABLE statement.
Secondary index	Depending on the context, a secondary index can mean one of the following two things: • An index that is not a partitioning index. • An index that is not a primary index.
Data partitioned secondary index (DPSI)	A partitioned index that is not a partitioning index. These indexes are also called partitioned secondary indexes (PSIs).
Nonpartitioned secondary index (NPSI)	An index that is not partitioned or partitioning. These indexes are also called nonpartitioned indexes (NPIs).
Multi-piece index	A nonpartitioned index that has multiple data sets. The data sets do not correspond to data partitions. Use a multi-piece index to spread a large index across multiple data sets and thus reduce the physical I/O contention on the index.
XML index	An index that uses a particular XML pattern expression to index paths and values in XML documents that are stored in a single XML column.

---

15   https://www.ibm.com/support/knowledgecenter/en/SSEPEK_11.0.0/intro/src/tpc/db2z_typesofindexes.html

Examples of Indexes
Unique primary

Recall that we defined the EMPLOYEE table with a primary key. We also created a unique primary index for the EMPLOYEE table right after we created it. Before you can use a table with a primary key, you must create a unique index on the column or columns being used for the primary key.

This is the table:

```
CREATE TABLE HRSCHEMA.EMPLOYEE(
    EMP_ID INT NOT NULL,
    EMP_LAST_NAME VARCHAR(30) NOT NULL,
    EMP_FIRST_NAME VARCHAR(20) NOT NULL,
    EMP_SERVICE_YEARS INT NOT NULL WITH DEFAULT 0,
    EMP_PROMOTION_DATE DATE,
    PRIMARY KEY(EMP_ID));
```

And this is the unique index we created to support the primary key.

```
CREATE UNIQUE INDEX HRSCHEMA.NDX_EMPLOYEE
    ON HRSCHEMA.EMPLOYEE (EMP_ID);
```

Unique

While a table can only have one primary key, it can have more than one unique index. Suppose that we added a social security column to the employee table. For data security reasons (and simply to reduce the possibility of keying errors), we might want to create a unique index on the social security number column.

The DDL to add the social security number column to the table and create a unique index on it is as follows:

```
ALTER TABLE HRSCHEMA.EMPLOYEE
ADD COLUMN EMP_SSN CHAR(09);

CREATE UNIQUE INDEX HRSCHEMA.NDX_SSN
ON HRSCHEMA.EMPLOYEE (EMP_SSN);
```

Now it will be impossible to add a second record with the same EMP_SSN value into the table. It will also be efficient to search by EMP_SSN because it is indexed.

Clustering

A clustering index forces data to be grouped and ordered according to index sequence. One case in which a clustering index is helpful is when sequential processing is called for. Suppose we create an electronic funds transfer system with a table called EMP_PAY_EFT. We will use the EMP_PAY_CHECK records to generate table entries for the EMP_PAY_EFT table on pay day.

Given the above, we might like to physically locate the data according to pay date so the records for a pay cycle will be close together (otherwise our pay check program may have to jump all over the table to get the needed records for the pay cycle).

Here's the DDL we used to create the EMP_PAY_EFT table.

```
CREATE TABLE HRSCHEMA.EMP_PAY_EFT
(EMP_ID INT NOT NULL,
EMP_PAY_DATE    DATE NOT NULL,
EMP_REGULAR_PAY  DECIMAL (8,2) NOT NULL,
EMP_SEMIMTH_PAY DECIMAL (8,2) NOT NULL)
IN TSHR;
```

We can ensure that the data is clustered as we want it by creating this index:

```
CREATE INDEX NDX_EMP_EFT
ON EMP_PAY_EFT (EMP_PAY_DATE, EMP_ID)
CLUSTER;
```

Now DB2 will attempt to order all records in the EMP_PAY_EFT table by EMP_PAY_DATE and EMP_ID. This should minimize the physical I/O required to obtain the pay data into memory.

Note: A new feature in DB2 11 allows you to prevent the entry of NULL values in an index by including the clause EXCLUDE NULL KEYS when you create the index. If you do not specify that you want to exclude null keys, then DB2 will allow null values in the index (assuming the indexed field is not defined with NOT NULL).

Data Types

DB2 supports the following data types. You should be familiar with all of these so that you can choose the best data type for your purpose. The information below plus comprehensive information about data types is available on the DB2 product documentation web site. [16]

---

16 http://www.ibm.com/support/knowledgecenter/en/SSEPEK_11.0.0/intro/src/tpc/db2z_datatypes.html

String data types

DB2 supports several types of string data: character strings, graphic strings, and binary strings.

CHARAC-TER(n)	Fixed-length character strings with a length of n bytes. n must be greater than 0 and not greater than 255. The default length is 1.
VARCHAR(n)	Varying-length character strings with a maximum length of n bytes. n must be greater than 0 and less than a number that depends on the page size of the table space. The maximum length is 32704.
GRAPHIC(n)	Fixed-length graphic strings that contain n double-byte characters. n must be greater than 0 and less than 128. The default length is 1.
VARGRAPH-IC(n)	Varying-length graphic strings. The maximum length, n, must be greater than 0 and less than a number that depends on the page size of the table space. The maximum length is 16352.
BINARY(n)	Fixed-length or varying-length binary strings with a length of n bytes. n must be greater than 0 and not greater than 255. The default length is 1.
VARBINA-RY(n)	Varying-length binary strings with a length of n bytes. The length of n must be greater than 0 and less than a number that depends on the page size of the table space. The maximum length is 32704.

Numeric Data Types

DB2 supports several types of numeric data types, each of which has its own characteristics.

SMALLINT	A small integer is binary integer in the range of -32768 to +32767.
INTEGER or INT	Large integers. A large integer is binary integer with a precision of 31 bits. The range is -2147483648 to +2147483647.
BIGINT	Big integers. A big integer is a binary integer with a precision of 63 bits. The range of big integers is -9223372036854775808 to +9223372036854775807.
DECIMAL or NUMERIC	A decimal number is a packed decimal number with an implicit decimal point. The position of the decimal point is determined by the precision and the scale of the number. The scale, which is the number of digits in the fractional part of the number, cannot be negative or greater than the precision. The maximum precision is 31 digits.

DECFLOAT	A decimal floating-point value is an IEEE 754r number with a decimal point. The position of the decimal point is stored in each decimal floating-point value. The maximum precision is 34 digits.
REAL	A single-precision floating-point number is a short floating-point number of 32 bits. The range of single-precision floating-point numbers is approximately -7.2E+75 to 7.2E+75. In this range, the largest negative value is about -5.4E-79, and the smallest positive value is about 5.4E-079.
DOUBLE	A double-precision floating-point number is a long floating-point number of 64-bits. The range of double-precision floating-point numbers is approximately -7.2E+75 to 7.2E+75. In this range, the largest negative value is about -5.4E-79, and the smallest positive value is about 5.4E-79.

Date, time, and timestamp data types

DATE	A date is a three-part value representing a year, month, and day.
TIME	A time is a three-part value representing a time of day in hours, minutes, and seconds.
TIMESTAMP	A timestamp is a seven-part value representing a date and time by year, month, day, hour, minute, second, and microsecond.

XML data type

The XML data type is used to define columns of a table that store XML values. This pureXML data type provides the ability to store well-formed XML documents in a database.

Large object data types

You can use large object data types to store audio, video, images, and other files that are larger than 32 KB.

Character large objects (CLOBs)	Use the CLOB data type to store SBCS or mixed data, such as documents that contain single character set. Use this data type if your data is larger (or might grow larger) than the VARCHAR data type permits.
Double-byte character large objects (DBCLOBs)	Use the DBCLOB data type to store large amounts of DBCS data, such as documents that use a DBCS character set.

Binary large objects (BLOBs)	Use the BLOB data type to store large amounts of non-character data, such as pictures, voice, and mixed media.

ROWID data type

You use the ROWID data type to uniquely and permanently identify rows in a DB2 subsystem.

Distinct types

A *distinct type* is a user-defined data type that is based on existing built-in DB2 data types. Here are a couple of examples:

```
CREATE DISTINCT TYPE US_DOLLAR AS DECIMAL (9,2);
CREATE DISTINCT TYPE CANADIAN_DOLLAR AS DECIMAL (9,2);
```

Default values for data types

For columns of...	Data types	Default
Numbers	SMALLINT, INTEGER, BIGINT, DECIMAL, NUMERIC, REAL, DOUBLE, DECFLOAT, or FLOAT	0
Fixed-length strings	CHAR or GRAPHIC	Blanks
	BINARY	Hexadecimal zeros
Varying-length strings	VARCHAR, CLOB, VARGRAPHIC, DBCLOB, VARBINARY, or BLOB	Empty string
Dates	DATE	CURRENT DATE
Times	TIME	CURRENT TIME
Timestamps	TIMESTAMP	CURRENT TIMESTAMP
ROWIDs	ROWID	DB2-generated

CONSTRAINTS

Types of constraints

There are basically four types of constraints, as follows:

1. Unique

2. Referential

3. Check

4. NULL

A UNIQUE constraint requires that the value in a particular field in a table be unique for each record.

A REFERENTIAL constraint enforces relationships between tables. For example you can define a referential constraint between an EMPLOYEE table and a DEPARTMENT table, such that a DEPT field in the EMPLOYEE table can only contain a value that matches a key value in the DEPARTMENT table.

A CHECK constraint establishes some condition on a field, such as the stored value must be >= 10.

A NOT NULL constraint on a column establishes that each record must have a non-null value for this column.

Unique Constraints

A unique constraint is a rule that the values of a key are valid only if they are unique in a table. In the case of the EMPLOYEE table we have been working with, you can create a unique index as follows:

```
CREATE UNIQUE INDEX NDX_EMPLOYEE
     ON EMPLOYEE (EMP_ID);
```

Trying to add a duplicate key value results in an error.

Check Constraints

A check constraint is a rule that specifies the values that are allowed in one or more columns of every row of a base table. It establishes some condition on a column, such as the stored value must be >= 10. The real worth of check constraints is that some business rules such as edits and validations can be stored in and performed by the database manager instead of by application programs.

When you try to INSERT or UPDATE a table record, the column values are evaluated against the check constraint rules (if any) for that table. If the data follows the rules, then the action is permitted; otherwise the action will fail with a constraint violation.

Let's define some check constraints on the EMP_DATA table such that the employee number must be between zero and 9999. Also the employee's age must be between 18 and 99. We'll create the table and then alter it to add the two constraints:

```
CREATE TABLE HRSCHEMA.EMP_DATA
(EMP_ID    INT,
EMP_LNAME   VARCHAR(30),
EMP_FNAME   VARCHAR(20),
EMP_AGE     INT)
IN TSHR;
---------+---------+---------+---------+---------+--------
DSNE616I STATEMENT EXECUTION WAS SUCCESSFUL, SQLCODE IS 0
```

Now let's add the constraint on employee id.

```
ALTER TABLE HRSCHEMA.EMP_DATA
ADD CONSTRAINT  X_EMPID
CHECK (EMP_ID BETWEEN 0 AND 9999);
---------+---------+---------+---------+---------+--------
DSNE616I STATEMENT EXECUTION WAS SUCCESSFUL, SQLCODE IS 0
```

Finally, we'll add the constraint on age.

```
ALTER TABLE HRSCHEMA.EMP_DATA
ADD CONSTRAINT  X_AGE
CHECK (EMP_AGE >= 18);
---------+---------+---------+---------+---------+--------
DSNE616I STATEMENT EXECUTION WAS SUCCESSFUL, SQLCODE IS 0
```

Now let's try this insert:

```
INSERT INTO HRSCHEMA.EMP_DATA
VALUES
(17888,
 'BROWN',
 'WILLIAM',
 17);
------+---------+---------+---------+---------+---------+---------+-
DSNT408I SQLCODE = -545, ERROR:  THE REQUESTED OPERATION IS NOT ALLOWED
BECAUSE A ROW DOES NOT SATISFY THE CHECK CONSTRAINT X_EMPID
```

It turns out that we mis-keyed the employee id. It should be 1788 instead of 17888. The value 17888 exceeds 9999 which is the maximum limit for employee id according to the check constraint. So DB2 prevented the insert action.

Let's fix the employee id and try again:

```
INSERT INTO HRSCHEMA.EMP_DATA
VALUES
(1788,
 'BROWN',
 'WILLIAM',
 17);
---------+---------+---------+---------+---------+---------+--------
DSNT408I SQLCODE = -545, ERROR:  THE REQUESTED OPERATION IS NOT ALLOWED
BECAUSE A ROW DOES NOT SATISFY THE CHECK CONSTRAINT X_AGE
```

Now we have violated the age constraint. Obviously 17 is less than 18 which is the defined lower limit on the employee age. So let's fix the age and we can see the row is accepted now.

```
INSERT INTO HRSCHEMA.EMP_DATA
VALUES
(1788,
 'BROWN',
 'WILLIAM',
 18);
---------+---------+---------+---------+----
DSNE615I NUMBER OF ROWS AFFECTED IS 1
```

Check constraints can be a very powerful way of building and centralizing business logic into the database itself. No application programming is required to implement a constraint (other than trapping and handling errors generated by the DBMS). There is great consistency in using check constraints because you can't forget to code them in a program (or code them inconsistently across multiple programs). They are applied by the DBMS regardless of which program or ad hoc process attempts the data modification.

Referential Constraints

A referential constraint is the rule that the non-null values of a foreign key are valid only if they also appear as a key in a parent table. The table that contains the parent key is called the parent table of the referential constraint, and the table that contains the foreign key is a dependent of that table. Referential integrity ensures data integrity by using primary and foreign key relationships between tables.

In DB2 you define a referential constraint by specifying in the child table a column which references a column in a parent table. For example, in a company you could have a DEPARTMENT table with column DEPT_CODE, and the EMP_DATA table that includes a column DEPT that represents the department code an employee is assigned to. The rule would be that you cannot have a value in the DEPT column of the EMP_DATA table that does not have a corresponding

DEPT_CODE value in the DEPARTMENT table. You can think of this as a parent and child relationship between the DEPARTMENT table and the EMP_DATA table. DEPARTMENT is the parent table and EMP_DATA is the child table.

Let's create the DEPARTMENT table and also add the DEPT column to the EMP_DATA table. We'll also add a row to the DEPARTMENT table and update the corresponding row(s) in the EMP_DATA table to match

```
CREATE TABLE HRSCHEMA.DEPARTMENT
(DEPT_CODE   CHAR(04) NOT NULL,
 DEPT_NAME   VARCHAR (20) NOT NULL,
 PRIMARY KEY(DEPT_CODE)) IN TSHR;

CREATE UNIQUE INDEX HRSCHEMA.NDX_DEPT_CODE
      ON HRSCHEMA.DEPARTMENT (DEPT_CODE);

INSERT INTO HRSCHEMA.DEPARTMENT
VALUES ('DPTA','DEPARTMENT A');


ALTER TABLE HRSCHEMA.EMP_DATA
ADD DEPT CHAR(04);

UPDATE HRSCHEMA.EMP_DATA
SET DEPT_CODE = 'DPTA';
```

Adding a Foreign Key Relationship

Now let's add the foreign key relationship by performing an ALTER action on the child table. Here is the DDL for our example:

```
ALTER TABLE HRSCHEMA.EMP_DATA
   FOREIGN KEY FK_DEPT_EMP (DEPT)
      REFERENCES HRSCHEMA.DEPARTMENT(DEPT_CODE) ;

---------+---------+---------+---------+---------+---------+---------+-
DSNT404I SQLCODE = 162, WARNING:  TABLE SPACE DBHR.TSHR HAS BEEN PLACED
         IN CHECK PENDING
```

The constraint was successfully built, but before you can continue you must deal with the CHECK PENDING status on your tablespace. The CHECK PENDING status means the table is possibly in an inconsistent state because of the new constraint, and it must be checked. To clear the CHECK PENDING status you must issue a CHECK DATA command. In our case, the command will be:

```
CHECK DATA TABLESPACE DBHR.TSHR;
```

Once the CHECK DATA finishes, your tablespace is taken out of check pending and you can continue, provided there were no errors.

Now if you try to update an EMP_DATA record with a DEPT value that does not have a DEPT_CODE with the same value as the DEPT value you are using, you'll get an SQL error -530 which means a violation of a foreign key.

```
UPDATE HRSCHEMA.EMP_DATA
SET DEPT = 'DPTB'

---------+---------+---------+---------+---------+---------+--------
DSNT408I SQLCODE = -530, ERROR:  THE INSERT OR UPDATE VALUE OF FOREIGN KEY
FK_DEPT_EMP IS INVALID
```

And here is the full explanation of the error.

```
-530
THE INSERT OR UPDATE VALUE OF FOREIGN KEY constraint-name IS INVALID
```

Explanation

```
An insert or update operation attempted to place a value in a foreign
key of the object table; however, this value was not equal to some value
of the parent key of the parent table.

When a row is inserted into a dependent table, the insert value of a
foreign key must be equal to the value of the parent key of some row of
the parent table in the associated relationship.

When the value of the foreign key is updated, the update value of a for-
eign key must be equal to the value of the parent key of some row of the
parent table of the associated relationship.
```

We know now that the parent table DEPARTMENT does not have DEPT_CODE DPTB in it, and it must be added before the EMP_DATA record can be updated.

Deleting a Record from the Parent Table

Now let's talk about what happens if you want to delete a record from the parent table. Assuming no EMP_DATA records are linked to the DEPARTMENT record, deleting that record may be fine. But what if you are trying to delete a DEPARTMENT record whose DEPT_CODE is referenced by one or more records in the EMP_DATA table?

Let's look at a record in the table:

```
SELECT EMP_ID, DEPT
FROM HRSCHEMA.EMP_DATA
WHERE EMP_ID = 1788;
```

```
--------+---------+------
    EMP_ID  DEPT
--------+---------+------
    1788  DPTA
```

Ok, we know that the DEPT_CODE in use is DPTA. Now let's try to delete DPTA from the DE-PARTMENT table.

```
DELETE FROM HRSCHEMA.DEPARTMENT
WHERE DEPT_CODE = 'DPTA';
---------+---------+---------+---------+---------+---------+---------+-
DSNT408I SQLCODE = -532, ERROR:  THE RELATIONSHIP FK_DEPT_EMP RESTRICTS THE DELE-
TION OF ROW WITH RID X'0000002201'
```

As you can see, when we try to remove the DEPT_CODE from the DEPARTMENT table, we will get a -532 SQLCODE telling us our SQL is in violation of the referential constraint. That's probably what we want, i.e., to have an error flagged. But there are some other options for handling the situation.

You can specify the action that will take place upon deleting a parent record by including an ON DELETE clause in the foreign key definition. If no ON DELETE clause is present, or if the ON DELETE RESTRICT clause is used, then the parent record cannot be deleted unless all child records referencing that parent record are first deleted (the child reference to that parent record can also be changed to some other existing parent record). RESTRICT is most commonly used with ON DELETE (or just omitting the ON DELETE clause which has the same effect). This is the case above.

Here are the two other options:

If **ON DELETE CASCADE** is specified, then any rows in the child table that correspond to the deleted parent record will also be deleted. Wow, that is probably not what we want! But there may be cases where this function is useful. Possibly if a certain product is discontinued then you might want to delete all pending SHIPPING table entries for it. I can't think of many other needs for this, but be aware that this option is available.

If **ON DELETE SET NULL** is specified, then the foreign key field will be set to NULL for corresponding child rows that reference the parent record that is being deleted.

Let's redefine our constraint to use this last option:

```
ALTER TABLE HRSCHEMA.EMP_DATA DROP CONSTRAINT FK_DEPT_EMP;
```

333

```
COMMIT WORK;

ALTER TABLE HRSCHEMA.EMP_DATA
    FOREIGN KEY FK_DEPT_EMP (DEPT)
        REFERENCES HRSCHEMA.DEPARTMENT (DEPT_CODE)
            ON DELETE SET NULL;

---------+---------+---------+---------+---------+---------+--------
DSNT404I SQLCODE = 162, WARNING:  TABLE SPACE DBHR.TSHR HAS BEEN PLACED IN CHECK PENDING
```

Go ahead and run the CHECK DATA to clear the CHECK PENDING condition.
Now try deleting the DPTA record from the DEPARTMENT table:

```
        DELETE FROM DEPARTMENT
        WHERE DEPT_CODE = 'DPTA';
        ---------+---------+---------+--------
        DSNE615I NUMBER OF ROWS AFFECTED IS 1
```

We see the delete is successful. So now let's check and see if the DEPT value for the child
record has been set to NULL.

```
        SELECT EMP_ID, DEPT
        FROM HRSCHEMA.EMP_DATA
        WHERE EMP_ID = 1788;

        ---------+---------+---------+--------
            EMP_ID  DEPT
        ---------+---------+---------+--------
             1788  ----
        DSNE610I NUMBER OF ROWS DISPLAYED IS 1
```

And in fact the DEPT column has been set to NULL.

Referential constraints are a very powerful and necessary function to ensure data integrity in a
database. Be sure to keep this in mind when you design your systems.

Not Null Constraints

A NOT NULL constraint on a column requires that when you add or update a record, you must
specify a non null value for that column. If you do not specify a non null value, you will get
an error.

Let's take an example of trying to add an EMPLOYEE record without specifying a value for one
of the columns. In this case, let's leave off the EMP_FIRST_NAME value which is defined in
the table as NOT NULL.

```
INSERT INTO HRSCHEMA.EMPLOYEE
(EMP_ID,
 EMP_LAST_NAME,
 EMP_PROMOTION_DATE)

VALUES (7420,
 'JACKSHIRT',
 '09/01/2016')

---------+---------+---------+---------+---------+---------+---------+---
DSNT408I SQLCODE = -407, ERROR:  AN UPDATE, INSERT, OR SET VALUE IS NULL, BUT
THE OBJECT COLUMN EMP_FIRST_NAME CANNOT CONTAIN NULL VALUES
DSNT418I SQLSTATE   = 23502 SQLSTATE RETURN CODE
DSNT415I SQLERRP    = DSNXODM SQL PROCEDURE DETECTING ERROR
DSNT416I SQLERRD    = 12 0  0  -1  0  0 SQL DIAGNOSTIC INFORMATION
DSNT416I SQLERRD    = X'0000000C'  X'00000000'  X'00000000'  X'FFFFFFFF'
         X'00000000'  X'00000000' SQL DIAGNOSTIC INFORMATION
```

Defining a column with the NOT NULL attribute ensures that no record can be added to the table with a NULL value in this column. Defining this requirement in the table enforces the requirement no matter which application or user is processing the data. Consequently, the rule is centralized and does not depend on program logic enforcing it.

Encoding Schemes

A code page is a table or map that denotes the character set used for encoding a particular set of characters. An IBM term related to code page is CCSID (**C**oded **C**haracter **S**et **Id**entifier) which is a number that identifies an implementation of a code page. Every CCSID belongs to an encoding scheme such as ASCII, EBCDIC or UNICODE.

CCSID values are used in DB2 for z/OS. When you need to specify a certain encoding of data, you provide the CCSID. You can specify a CCSID for an object when you create it. You can also set a CCSID for a program using the ENCODING bind option.

DB2 for z/OS uses three default CCSIDs, one for each major encoding scheme (EBCDIC, ASCII and UNICODE).

To determine the subsystem encoding scheme, use this query:

```
SELECT  GETVARIABLE('SYSIBM.ENCODING_SCHEME')
  FROM SYSIBM.SYSDUMMY1;
---------+---------+---------+---------+--------

---------+---------+---------+---------+--------
UNICODE
DSNE610I NUMBER OF ROWS DISPLAYED IS 1
```

You can determine the application encoding scheme in your environment by issuing this query:

```
SELECT GETVARIABLE('SYSIBM.APPLICATION_ENCODING_SCHEME')
FROM SYSIBM.SYSDUMMY1;
---------+---------+---------+---------+---------+---------+-

---------+---------+---------+---------+---------+---------+-
UNICODE
```

To determine the CCSID for ASCII, use this query.

```
SELECT  GETVARIABLE('SYSIBM.SYSTEM_ASCII_CCSID')
   FROM SYSIBM.SYSDUMMY1;
---------+---------+---------+---------+---------+-

---------+---------+---------+---------+---------+-
819,65534,65534
DSNE610I NUMBER OF ROWS DISPLAYED IS 1
```

You can determine the CCSID for EBCDIC with this query:

```
SELECT  GETVARIABLE('SYSIBM.SYSTEM_EBCDIC_CCSID')
   FROM SYSIBM.SYSDUMMY1;
---------+---------+---------+---------+---------+-

---------+---------+---------+---------+---------+-
1047,65534,65534
```

A table is typically created with the same CCSID as the tablespace into which it is placed. If a CCSID is not specified when a tablespace is created, the DB2 subsystem CCSID (specified when the instance is set up) is used.

Sequencies And Identities

Some scenarios require an auto-generated sequence that can be used to uniquely identify a record. DB2 provides two methods of doing this: sequences and identity columns. Although both provide auto-generated numbers, they work differently. The main difference between an identity field and a sequence is that the former is a column contained in a specific table – it cannot be shared across objects. A sequence is itself a separate database object and can be used to generate numbers for multiple objects. Make sure you are clear on this difference!

We'll look at examples of both a sequence and an identity column on a table.

Sequences

A sequence generates a sequential set of numbers. You define a sequence with a starting value and you increment it with the NEXTVAL function. You can also obtain the most recent previous value with the PREVVAL function.

The basic syntax for creating a sequence is:

```
CREATE SEQUENCE <name of sequence>
START WITH <start value>
INCREMENT BY <increment value>
NO CYCLE   <reuse old values?>
```

Now let's create a sequence for our employee table. Suppose we want to use this sequence to generate new employee numbers and that we want to start with employee number 1001. The following DDL would accomplish this:

```
CREATE SEQUENCE HRSCHEMA.EMPSEQ
START WITH 1001
INCREMENT BY 1
NO CYCLE ;
```

The NO CYCLE means we will not reuse numbers when the maximum for the data type is reached. You could explicitly create the sequence as SMALLINT, INTEGER, BIGINT, or DECIMAL with a scale of zero. If you don't specify a data type, the default is INTEGER.

Now we could add a record to the EMPLOYEE table without specifying the employee number. Instead we would use the sequence number with the NEXTVAL option. Here's how we do it:

```
INSERT INTO HRSCHEMA.EMPLOYEE
VALUES (NEXT VALUE FOR HRSCHEMA.EMPSEQ,
'HENDERSON',
'JOHN',
1,
'12/01/2016');
---------+---------+---------+---------+---------+---------+--
DSNE615I NUMBER OF ROWS AFFECTED IS 1
DSNE616I STATEMENT EXECUTION WAS SUCCESSFUL, SQLCODE IS 0
---------+---------+---------+---------+---------+---------+--
```

Now let's select the record with key 1001.

```
SELECT EMP_ID, EMP_LAST_NAME
FROM HRSCHEMA.EMPLOYEE
WHERE EMP_ID = 1001;
```

```
---------+---------+---------+---------+---------+---------+----
     EMP_ID  EMP_LAST_NAME
---------+---------+---------+---------+---------+---------+----
      1001  HENDERSON
DSNE610I NUMBER OF ROWS DISPLAYED IS 1
DSNE616I STATEMENT EXECUTION WAS SUCCESSFUL, SQLCODE IS 100
---------+---------+---------+---------+---------+---------+----
```

You can also change a sequence after it is created by using the ALTER statement. For example you could change the increment of the EMPSEQ sequence from 1 to 2 as follows:

```
ALTER SEQUENCE HRSCHEMA.EMPSEQ
INCREMENT BY 2;
```

You can delete a sequence by issuing the DROP command.

```
DROP SEQUENCE HRSCHEMA.EMPSEQ
```

Identity Columns

Now let's perform a similar setup with another table and this time we will use an identity column on the table. The identity column is so named because it allows you to uniquely identify a record. It provides a unique key. For this example we'll create a different table call EMPLOYE2.

```
CREATE TABLE HRSCHEMA.EMPLOYE2(
EMP_ID SMALLINT GENERATED ALWAYS AS IDENTITY
   (START WITH 1001,
    INCREMENT BY 1,
    NOCYCLE),
EMP_LAST_NAME VARCHAR(30) NOT NULL,
EMP_FIRST_NAME VARCHAR(20) NOT NULL,
EMP_SERVICE_YEARS INT
NOT NULL WITH DEFAULT 0,
EMP_PROMOTION_DATE DATE)
IN TSHR;
---------+---------+---------+---------+---------+---------+--
DSNE616I STATEMENT EXECUTION WAS SUCCESSFUL, SQLCODE IS 0
---------+---------+---------+---------+---------+---------+--
```

Now let's insert a row into the EMPLOYE2 table:

```
INSERT INTO HRSCHEMA.EMPLOYE2
  (EMP_LAST_NAME,
   EMP_FIRST_NAME,
   EMP_SERVICE_YEARS,
   EMP_PROMOTION_DATE)
VALUES
  ('JOHNSON',
```

338

```
        'BILL',
         1,
        '12/01/2016');
---------+---------+---------+---------+---------+--------
DSNE615I NUMBER OF ROWS AFFECTED IS 1
DSNE616I STATEMENT EXECUTION WAS SUCCESSFUL, SQLCODE IS 0
---------+---------+---------+---------+---------+--------
```

Notice that we did not specify any value for the EMP_ID. That is because it's an identity field and DB2 will generate the value. Now we can query the table and see the contents:

```
    SELECT EMP_ID, EMP_LAST_NAME
    FROM HRSCHEMA.EMPLOYE2;
---------+---------+---------+---------+-
EMP_ID   EMP_LAST_NAME
---------+---------+---------+---------+-
   1001  JOHNSON
DSNE610I NUMBER OF ROWS DISPLAYED IS 1
```

As you can see, we get the same results with the identity field that we got earlier using a sequence. The main difference is that an identity column takes care of generating the value without any prompting. Whereas with a sequence you must specify the sequence name and request the next value.

The other difference between sequences and identity columns is that sequences are a separate object in the database, i.e., they can be used for generating numbers independently of any table. But an identify column is always tied to a single table.

Views

A view is a virtual table that is based on a SELECT query against a base table or another view. Views can include more than one table (or other view), which means they can include the results of a join.

DDL for Views

CREATE

The basic syntax to create a view is as follows:

```
CREATE VIEW <name of view>
AS
SELECT <columns>
FROM <table>
WHERE <condition>
```

You can use a view not only to select data, but also to insert new records into a base table. When you add or update records using a view, you have a choice of enforcing the view definition (meaning you can only those insert records that match the view definition), or you allow inserting of records that don't match the view definition. Using the WITH CHECK OPTION when creating the view ensures that a record inserted via a view is consistent with the view definition.

For example, let's go back to our EMPLOYE2 table. Let's say we want a view named EMP_SENIOR that shows us data from EMPLOYE2 only for senior employees, meaning employees with at least 5 years of service. We will also allow records to be inserted to the EMPLOYE2 table via the view. Here is the view definition:

```
CREATE VIEW HRSCHEMA.EMP_SENIOR AS
SELECT
EMP_ID,
EMP_LAST_NAME,
EMP_FIRST_NAME,
EMP_SERVICE_YEARS,
EMP_PROMOTION_DATE
FROM HRSCHEMA.EMPLOYE2
WHERE EMP_SERVICE_YEARS >= 5;
---------+---------+---------+---------+---------+---------
DSNE616I STATEMENT EXECUTION WAS SUCCESSFUL, SQLCODE IS 0
```

Now let's insert a couple of records using this view.

```
INSERT INTO HRSCHEMA.EMP_SENIOR
  (EMP_LAST_NAME,
   EMP_FIRST_NAME,
   EMP_SERVICE_YEARS,
   EMP_PROMOTION_DATE)
VALUES
  ('FORD',
   'JAMES',
   7,
   '10/01/2015');
---------+---------+---------+---------+---------+---------+---
DSNE615I NUMBER OF ROWS AFFECTED IS 1
DSNE616I STATEMENT EXECUTION WAS SUCCESSFUL, SQLCODE IS 0
```

This record is for an employee with 5 or more years of service. We got a successful insert which we can reconfirm by querying the table using the same view. Good.

```
SELECT EMP_ID,
EMP_LAST_NAME,
EMP_FIRST_NAME
FROM HRSCHEMA.EMP_SENIOR;
---------+---------+---------+---------+---------+---------+
EMP_ID  EMP_LAST_NAME        EMP_FIRST_NAME
---------+---------+---------+---------+---------+---------+
  1002  FORD                 JAMES
DSNE610I NUMBER OF ROWS DISPLAYED IS 1
```

Now let's insert a record that does not fit the view definition. In this case, let's add a record for which the employee has only 2 years service:

```
INSERT INTO HRSCHEMA.EMP_SENIOR
  (EMP_LAST_NAME,
   EMP_FIRST_NAME,
   EMP_SERVICE_YEARS,
   EMP_PROMOTION_DATE)
VALUES
  ('BUFORD',
   'HOLLAND',
   2,
   '07/31/2016');
---------+---------+---------+---------+---------+-----
DSNE615I NUMBER OF ROWS AFFECTED IS 1
```

Interestingly, DB2 allowed the record to be added to the table via the view even though the data did not match the view definition. We cannot know that fact by querying the view because querying the view only shows us data which conforms to the view definition. Notice that only the record for the person with 5 or more years' service is returned by the view.

```
SELECT EMP_ID,
EMP_LAST_NAME,
EMP_FIRST_NAME
 FROM HRSCHEMA.EMP_SENIOR;
---------+---------+---------+---------+------
EMP_ID  EMP_LAST_NAME        EMP_FIRST_NAME
---------+---------+---------+---------+------
  1002  FORD                 JAMES
DSNE610I NUMBER OF ROWS DISPLAYED IS 1
```

But when we query the base table, we see the new record was in fact added:

```
   EMP_ID   EMP_LAST_NAME        EMP_FIRST_NAME        EMP_SERVICE_YEARS
---------+---------+---------+---------+---------+---------+---------+-
   1001   JOHNSON              BILL                                  1
   1002   FORD                 JAMES                                 7
   1003   BUFORD               HOLLAND                               2
DSNE610I NUMBER OF ROWS DISPLAYED IS 3
```

Maybe this is ok if it fits your overall business rules. However, if you want to prevent records that do not conform to the view definition from being inserted into the table using that view, you must define the view using the WITH CHECK OPTION clause. Let's drop the view and recreate it that way.

```
DROP VIEW HRSCHEMA.EMP_SENIOR;
COMMIT WORK;

CREATE VIEW HRSCHEMA.EMP_SENIOR
AS
SELECT
EMP_ID,
EMP_LAST_NAME,
EMP_FIRST_NAME,
EMP_SERVICE_YEARS,
EMP_PROMOTION_DATE
FROM HRSCHEMA.EMPLOYE2
WHERE EMP_SERVICE_YEARS >= 5
WITH CHECK OPTION;
---------+---------+---------+---------+---------+---------
DSNE616I STATEMENT EXECUTION WAS SUCCESSFUL, SQLCODE IS 0
```

Now let's try to insert two more records, first an employee with more 5 or more years of service.

```
INSERT INTO HRSCHEMA.EMP_SENIOR
  (EMP_LAST_NAME,
   EMP_FIRST_NAME,
   EMP_SERVICE_YEARS,
   EMP_PROMOTION_DATE)

  VALUES
  ('JACKSON',
   'MARLO',
   8,
   '06/30/2015');
---------+---------+---------+---------
DSNE615I NUMBER OF ROWS AFFECTED IS 1
```

342

This still works which is fine. Now let's try one with less than 5 years of service. In this case the insert fails as we can see:

```
INSERT INTO HRSCHEMA.EMP_SENIOR
  (EMP_LAST_NAME,
   EMP_FIRST_NAME,
   EMP_SERVICE_YEARS,
   EMP_PROMOTION_DATE)
VALUES
  ('TARKENTON',
  'QUINCY',
   3,
  '09/30/2015');

---------+---------+---------+---------+---------+---------+---------+-----
DSNT408I SQLCODE = -161, ERROR:  THE INSERT OR UPDATE IS NOT ALLOWED BECAUSE A RE-
SULTING ROW DOES NOT SATISFY THE VIEW DEFINITION
DSNT418I SQLSTATE   = 44000 SQLSTATE RETURN CODE
DSNT415I SQLERRP    = DSNXRSVW SQL PROCEDURE DETECTING ERROR
DSNT416I SQLERRD    = -160 0  0  -1  0  0 SQL DIAGNOSTIC INFORMATION
DSNT416I SQLERRD    = X'FFFFFF60'  X'00000000'  X'00000000'  X'FFFFFFFF'
          X'00000000'  X'00000000' SQL DIAGNOSTIC INFORMATION
---------+---------+---------+---------+---------+---------+---------+-----
```

So this is how to create a view that will disallow any insert or update that does not conform to the view definition.

ALTER

In general views cannot be changed. They must be dropped and recreated. One exception is that you can regenerate a view from the existing definition. The basic syntax to do this is:

```
ALTER VIEW <name of view> REGENERATE
```

Here's an example of regenerating the EMPLOYEE_SENIOR view.

```
ALTER VIEW HRSCHEMA.EMP_SENIOR REGENERATE;
```

DROP

Finally, you can delete a view by issuing the DROP command.

```
DROP VIEW <view name>
```

Read-Only Views

The following are some conditions under which a view is automatically read-only, meaning you cannot insert, update or delete any rows using a view that includes any of these conditions.

- The first SELECT clause includes the keyword DISTINCT.
- The first SELECT clause contains an aggregate function.
- The first FROM clause identifies multiple tables or views.
- The outer fullselect includes a GROUP BY clause.

Views for Security

A view is a classic way to restrict a subset of data columns to a specific set of users who are allowed to see or manipulate those columns. If you are going to use views for security you must make sure that users do not have direct access to the base tables, i.e. that they only access the data via view(s). Otherwise they could potentially circumvent access by view.

Let's look back at our employee table. Suppose we add a column for the employee's Social Security number. That is obviously a very private piece of information that not everyone should see. Our business rule will be that users HRUSER01, HRUSER02 and HRUSER99 are the only ones who should be able to view Social Security numbers. All other users and/or groups are not allowed to see the content of this column, but they can access all the other columns.

We can begin implementing this by adding the new column to the base table:

```
ALTER TABLE HRSCHEMA.EMPLOYEE
ADD COLUMN EMP_SSN CHAR(09);
```

Now let's update this column for employee 3217:

```
UPDATE HRSCHEMA.EMPLOYEE
SET EMP_SSN = '238297536'
WHERE EMP_ID = 3217;
```

Now let's create two views, one of which includes the EMP_SSN column, and the other of which does not:

```
CREATE VIEW HRSCHEMA.EMPLOYEE_ALL
AS SELECT
EMP_ID,
EMP_LAST_NAME,
EMP_FIRST_NAME,
EMP_SERVICE_YEARS,
EMP_PROMOTION_DATE,
```

```
EMP_PROFILE
FROM HRSCHEMA.EMPLOYEE;

CREATE VIEW HRSCHEMA.EMPLOYEE_HR
AS SELECT
EMP_ID,
EMP_LAST_NAME,
EMP_FIRST_NAME,
EMP_SERVICE_YEARS,
EMP_PROMOTION_DATE,
EMP_PROFILE,
EMP_SSN
FROM HRSCHEMA.EMPLOYEE;
```

Finally, issue the appropriate grants.

```
GRANT SELECT ON HRSCHEMA.EMPLOYEE_ALL TO PUBLIC;

GRANT SELECT on HRSCHEMA.EMPLOYEE_HR
TO HRUSER01, HRUSER02, HRUSER99;
```

At this point, assuming we are only accessing data through views, the three HR users are the only users able to access the EMP_SSN column. Other users cannot access the EMP_SSN column because it is not included in the EMPLOYEE_ALL view that they have access to. To prove this, let's run some queries:

```
SELECT EMP_ID, EMP_SSN
FROM HRSCHEMA.EMPLOYEE_ALL
WHERE EMP_ID = 3217;
---------+---------+---------+---------+---------+---------+---------+-----
DSNT408I SQLCODE = -206, ERROR:  EMP_SSN IS NOT VALID IN THE CONTEXT WHERE IT IS US
ED
```

If you are using the EMPLOYEE_ALL view, you do not have access to the SSN because it is not a column in that view. However, if you are one of the HR users, you will be able to access the EMP_SSN column using the other view, EMPLOYEE_HR:

```
SELECT EMP_ID, EMP_SSN
FROM HRSCHEMA.EMPLOYEE_HR
WHERE EMP_ID = 3217;
---------+---------+------
    EMP_ID  EMP_SSN
---------+---------+------
      3217  238297536
```

Notes to remember:

- A view can reference one or more other views

- When a view is deleted, any dependent view is marked inoperative. Referencing an inoperative view results in an error.

- A view will be automatically regenerated if the a field size is increased in the base table upon which the view is based.

As we conclude the DDL section of this study guide, I strongly recommend that you **remove all the objects we have created** thus far. This will enable us to begin with a clean slate for the next section on Data Manipulation Language. The easiest way to remove all the objects is to simply drop the database and then recreate it.

```
DROP DATABASE DBHR;
```

Then recreate the database, tablespace and schema that we will use for the HR application. The following is the DDL to do that, and of course it must be customized to your physical environment.

```
CREATE DATABASE DBHR
STOGROUP SGHR
BUFFERPOOL BPHR
INDEXBP IBPHR
CCSID UNICODE;

CREATE TABLESPACE TSHR
    IN DBHR
    USING STOGROUP SGHR
      PRIQTY 50
      SECQTY 20
    LOCKSIZE PAGE
    BUFFERPOOL BPHR2;


CREATE SCHEMA HRSCHEMA
AUTHORIZATION DBA001;    ←   This should be your DB2 id, whatever it is.
```

The DB2 Catalog

The DB2 system catalog is a wealth of information for research and problem solving. The catalog consists of a set of tables containing information about the various objects in DB2 (tablespaces, tables, indexes, views, packages, etc). The schema for the system catalog is SYS-

IBM. A complete list of the catalog tables is provided here with general information about what each table contains.

List of Catalog Tables

The catalog tables are listed here and in the IBM product documentation: [17]

Table	Description
SYSIBM.SYSCONTEXT	Contains one row for each trusted context.
SYSIBM.SYSCONTEXTAUTHIDS	Contains the authorization ID under which a trusted context can be used.
SYSIBM.IPLIST	Contains a row for each IP address that corresponds to a remote DRDA server
SYSIBM.IPNAMES	Contains a row for each remote DRDA server that DB2 can access using TCP/IP
SYSIBM.LOCATIONS	Contains a row for every accessible remote server
SYSIBM.LULIST	Contains a row for each real LU name that is associated with the dummy LU name for a data sharing group
SYSIBM.LUNAMES	Contains a row for each remote SNA client or server
SYSIBM.USERNAMES	Contains a row for each authorization ID to be translated or checked as it is sent to or from this DB2
SYSIBM.SYSDUMMY1	Contains one row. This table is used by SQL statements that do not use any tables but must specify a table name.
SYSIBM.SYSRESAUTH	Records the privileges that are held by users over buffer pools, storage groups, table spaces, and collections
SYSIBM.SYSROUTINES_SRC	Contains one or more rows for the source code for each generated routine
SYSIBM.SYSJARDATA	Contains the contents of the JAR file for each Java stored procedure
SYSIBM.SYSJARCLASS_SOURCE	Contains the source code for a Java stored procedure
SYSIBM.SYSROLES	Contains one row for each role
SYSIBM.SYSSEQUENCES	Contains one row for each identity column
SYSIBM.SYSSEQUENCEAUTH	Records the privileges that users hold on sequences
SYSIBM.SYSINDEXSTATS	Contains one row for each partition of a partitioning index or data-partitioned secondary index (DPSI)
SYSIBM.SYSAUDITPOLICIES	Each row represents an audit policy.

---

17 https://www.ibm.com/support/knowledgecenter/SSEPEK_11.0.0/cattab/src/tpc/db2z_catalogtablesintro.html

Table	Description
SYSIBM.SYSAUXRELS	Contains one row for each auxiliary table for a LOB column
SYSIBM.SYSCHECKS	Contains one row for each table check constraint
SYSIBM.SYSCOLUMNS	Contains one row for every column of each table and view
SYSIBM.SYSCONSTDEP	Records dependencies on check constraints or user-defined defaults for a column
SYSIBM.SYSCONTROLS	Contains one row for each row permission and column mask.
SYSIBM.SYSDATATYPES	Contains one row for each distinct data type
SYSIBM.SYSDATABASE	Contains one row for each database, except for database DSNDB01
SYSIBM.SYSDBAUTH	Records the privileges held by users over databases
SYSIBM.SYSCOLAUTH	Records the UPDATE privileges that are held by users on individual columns of a table or view
SYSIBM.SYSFOREIGNKEYS	Contains one row for every column of every foreign key
SYSIBM.SYSINDEXES	Contains one row for every index
SYSIBM.SYSKEYS	Contains one row for each column of an index key
SYSIBM.SYSKEYCOLUSE table	Contains a row for every column in a unique constraint (primary key or unique key)
SYSIBM.SYSPACKAUTH table	Contains the privileges that are held by users over packages
SYSIBM.SYSPACKDEP table	Records the dependencies of packages on local tables, views, synonyms, table spaces, indexes, aliases, and triggers
SYSIBM.SYSPACKAGE table	Contains one row for each package
SYSIBM.SYSPACKLIST table	Contains one row for every package list entry for a plan
SYSIBM.SYSPACKSTMT table	Contains one row for every SQL statement that belongs to a package
SYSIBM.SYSPKSYSTEM table	Contain one row of system information for each package
SYSIBM.SYSPLANAUTH table	Records the privileges that are held by users over application plans
SYSIBM.SYSPLANDEP table	Records the dependencies of plans on tables, views, aliases, synonyms, table spaces, and indexes
SYSIBM.SYSPLAN table	Contains one row for each application plan
SYSIBM.SYSPLSYSTEM table	Contains one row of system information for each plan
SYSIBM.SYSPARMS table	Contains a row for each parameter of a routine or a row for each column of a table that is passed as a parameter to a routine
SYSIBM.SYSQUERYPLAN table	Contains the plan hint information for the queries in the SYSIBM.SYSQUERY table

Table	Description
SYSIBM.SYSROUTINEAUTH	Records the privileges that users hold on routines
SYSIBM.SYSRELS table	Contains one row for every relationship
SYSIBM.SYSROUTINES table	Contains a row for every routine
SYSIBM.SYSSCHEMAAUTH table	Contains one or more rows for each grantee of a privilege on a schema
SYSIBM.SYSSTOGROUP table	Contains one row for each storage group
SYSIBM.SYSSTMT table	Contains one or more rows for each SQL statement of each DBRM
SYSIBM.SYSSYNONYMS table	Contains one row for each synonym of a table or view
SYSIBM.SYSTABLES table	Contains one row for each table, view, and alias
SYSIBM.SYSTABAUTH table	Records the privileges that are held by users on tables, views, and triggers
SYSIBM.SYSTRIGGERS table	Contains one row for each trigger
SYSIBM.SYSTABLESPACE table	Contains one row for each table space
SYSIBM.SYSVARIABLES table	Contains one row for each global variable that has been created.
SYSIBM.SYSVIEWS table	Contains one or more rows for each view
SYSIBM.SYSVOLUMES table	Contains one row for each volume of each storage group
SYSIBM.SYSVIEWDEP table	Records the dependencies of views on tables and other views
SYSIBM.SYSUSERAUTH table	Records the system privileges that are held by users
SYSIBM.SYSXMLRELS table	Contains one row for each XML table that is created for an XML column

Table and View Properties Stored in the System Catalog

I think it is good to mention here that you can discover various properties of tables, indexes and views by querying the system catalog. For example, if you want to know what type of tables you have associated with owner HRSCHEMA, you could query them this way:

```
SELECT NAME, TYPE, DBNAME, TSNAME
FROM SYSIBM.SYSTABLES
WHERE CREATOR = 'HRSCHEMA'
ORDER BY TYPE, NAME
---------+---------+---------+---------+---------+---------+------
NAME              TYPE  DBNAME                TSNAME
---------+---------+---------+---------+---------+---------+------
EMP_INFO          G     DSNDB06               SYSTSTAB
XEMPLOYEE         P     DBHR                  XEMP0000
DEPARTMENT        T     DBHR                  HRTS
DEPTMENT          T     DBHR                  HRTS
EMPLOYEE          T     DBHR                  HRTS
```

349

```
EMPLOYEE_NEW          T    DBHR           HRTS
EMPRECOG              T    DBHR           HRTS
EMP_DATA              T    DBHR           HRTS
EMP_DATA_X            T    DBHR           HRTS
EMP_PAY               T    DBHR           HRTS
EMP_PAY_CHECK         T    DBHR           HRTS
EMP_PAY_HIST          T    DBHR           HRTS
EMP_PAY_HST           T    DBHR           HRTS
EMP_PAY_X             T    DBHR           HRTS
PLAN_TABLE            T    DBHR           HRTS
EMPLOYEE_ALL          V    DBHR           HRTS
EMPLOYEE_HR           V    DBHR           HRTS
EMP_PROFILE_PAY       V    DBHR           HRTS
```

The meaning of the TYPE column is as follows:

A	Alias
C	Clone table
D	Accelerator-only table
G	Created global temporary table (note: declared global temporary is not in the catalog)
H	History table
M	Materialized query table
P	Table that was implicitly created for XML columns
R	Archive table
T	Table
V	View
X	Auxiliary table

You can also get information about views from the system catalog. For example, if you want to know what type of views you have, you could query the information this way:

```
SELECT NAME, TYPE
FROM SYSIBM.SYSVIEWS
WHERE CREATOR = 'HRSCHEMA'
ORDER BY TYPE, NAME
---------+---------+--------
NAME                 TYPE
---------+---------+--------
EMPLOYEE_ALL          V
EMPLOYEE_HR           V
EMP_PROFILE_PAY       V
```

Again, view types are as follows:

F	SQL function
M	Materialized query table
V	View

Data Manipulation Language

Overview

In this section we will explore DML (data Manipulation Language) which includes both SQL and several related topics.

Data Manipulation Language (DML) is used to add, change and delete data in a DB2 table. DML is one of the most basic and essential skills you must have as a DB2 professional. In this section we'll look at the five major DML statements: INSERT, UPDATE, DELETE, MERGE and SELECT.

XML data access and processing is another skill that you need to be familiar with. DB2 includes an XML data type and various functions for accessing and processing XML data. I'll assume you have a basic understanding of XML, but we'll do a quick review anyway. Then we'll look at some examples of creating an XML column, populating it, modifying it and manipulating it using XML functions such as XQuery.

Special registers allow you to access detailed information about the DB2 instance settings as well as certain session information. CURRENT DATE is an example of a special register. You can access special registers in SPUFI or in an application program and then use the information as needed.

Built-in functions can be used in SQL statements to return a result based on an argument. Think of these as productivity tools in that they can be used to replace custom coded functionality in an application program and thereby simplify development and maintenance. Whether your role is application developer, DBA or business services professional, the DB2 built-in functions can save you time if you know what they are and how to use them.

Database, Tablespace and Schema Conventions

Throughout this book we will be using a database called DBHR which is a database for a fictitious human relations department in a company. The main tablespace we will us is HRTS. Finally, our default schema will be HRSCHEMA. In some cases we will explicitly specify the schema in our DDL or SQL. If we don't explicitly specify a schema, it means we have defined the HRSCHEMA schema as the CURRENT SCHEMA so we don't need to specify it.

If you are following along and creating examples on your own system, you may of course use whatever database and schema is available to you on your system. If you want the basic DDL to create the objects named above, it is as follows:

```
CREATE DATABASE DBHR
STOGROUP SGHR
BUFFERPOOL BPHR
INDEXBP IBPHR
CCSID UNICODE;


CREATE TABLESPACE TSHR
      IN DBHR
      USING STOGROUP SGHR
        PRIQTY 50
        SECQTY 20
      LOCKSIZE PAGE
      BUFFERPOOL BPHR2;


CREATE SCHEMA HRSCHEMA
AUTHORIZATION DBA001;   ←    This should be your DB2 id, whatever it is.
```

DML SQL Statements

Data Manipulation Language (DML) is at the core of working with relational databases. You need to be very comfortable with DML statements: INSERT, UPDATE, DELETE, MERGE and SELECT. We'll cover the syntax and use of each of these. For purposes of this section, let's plan and create a very simple table. Here are the columns and data types for our table which we will name EMPLOYEE.

Field Name	Type	Attributes
EMP_ID	INTEGER	NOT NULL, PRIMARY KEY
EMP_LAST_NAME	VARCHAR(30)	NOT NULL
EMP_FIRST_NAME	VARCHAR(20)	NOT NULL
EMP_SERVICE_YEARS	INTEGER	NOT NULL, DEFAULT IS ZERO
EMP_PROMOTION_DATE	DATE	

The table can be created with the following DDL:

```
CREATE TABLE HRSCHEMA.EMPLOYEE(
EMP_ID INT NOT NULL,
EMP_LAST_NAME VARCHAR(30) NOT NULL,
EMP_FIRST_NAME VARCHAR(20) NOT NULL,
EMP_SERVICE_YEARS INT NOT NULL WITH DEFAULT 0,
EMP_PROMOTION_DATE DATE,
PRIMARY KEY(EMP_ID)) ;
```

We also need to create a unique index to support the primary key:

```
CREATE UNIQUE INDEX NDX_EMPLOYEE
      ON EMPLOYEE (EMP_ID);
```

INSERT Statement

The INSERT statement adds one or more rows to a table. There are three forms of the IN-SERT statement and you need to know the syntax of each of these.

1. Insert via Values
2. Insert via Select
3. Insert via FOR N ROWS

Insert Via Values

There are actually two sub-forms of the insert by values. One form explicitly names the target columns and the other does not. Generally when inserting a record you explicitly specify the target fields, followed by a VALUES clause that includes the actual values to apply to the new record. Let's use our EMPLOYEE table for this example:

```
INSERT INTO EMPLOYEE
(EMP_ID,
 EMP_LAST_NAME,
 EMP_FIRST_NAME,
 EMP_SERVICE_YEARS,
 EMP_PROMOTION_DATE)

VALUES (3217,
 'JOHNSON',
 'EDWARD',
 4,
 '01/01/2017')
```

Note that the values must be ordered in the same sequence that the columns are named in the INSERT query.

A second sub-form of the INSERT statement via values is to omit the target fields and simply provide the VALUES clause. You can do this only if your values clause includes values for ALL the columns in the correct positional order as defined in the table.

Here's an example of this second sub-form of insert via values:

```
INSERT INTO EMPLOYEE
VALUES (7459,
 'STEWART',
```

```
'BETTY',
7,
'07/31/2016')
```

Some additional rules to remember for the INSERT statement:

- You can define a column as having a default value. If a column is defined with a DEFAULT value, and you want to assign that default value to the column, then you can simply specify the DEFAULT keyword for that column in the values clause. DB2 will automatically populate the field with its default value.

- Each column that is defined as NOT NULL must have an entry in the values clause. If the NOT NULL column has a DEFAULT value attribute, you can specify DEFAULT in the values clause. If you do not specify DEFAULT, then you must assign a specific value for this column in the values clause or an error will result.

- If a column is not defined as NOT NULL (or if it is explicitly defined as NULL meaning NULL values are allowed), and you don't want to assign a value to that column, the you must specify NULL for the column in the values clause.

Note that EMP_ID is defined as a primary key on the table. If you try inserting a row for which the primary key already exists, you will receive a -803 error SQL code (more on this later when we discuss table objects in detail).

Here's an example of specifying the DEFAULT value for the EMP_SERVICE_YEARS column, and the NULL value for the EMP_PROMOTION_DATE.

```
INSERT INTO EMPLOYEE
(EMP_ID,
EMP_LAST_NAME,
EMP_FIRST_NAME,
EMP_SERVICE_YEARS,
EMP_PROMOTION_DATE)

VALUES (9134,
'FRANKLIN',
'ROSEMARY',
DEFAULT,
NULL);
```

When you define a column using WITH DEFAULT, you do not necessarily have to specify the actual default value in your DDL. DB2 provides implicit default values for most data types and if you just specify WITH DEFAULT and no specific value, the implicit default value will be used.

In the EMPLOYEE table we specified WITH DEFAULT 0 for the employee's service years. However, the implicit default value here is also zero because the column is defined as INTEGER. So we could have simply specified WITH DEFAULT and it would have the same result as specifying WITH DEFAULT 0.

The following table denotes the default values for the various data types.

Default Values for DB2 Data Types

This table appeared earlier in the book, but I include it again here because we are talking specifically about using defaults.

For columns of	Type	Default
Numbers	SMALLINT, INTEGER, BIGINT, DECIMAL, NUMERIC, REAL, DOUBLE, DECFLOAT, or FLOAT	0
Fixed-length strings	CHAR or GRAPHIC	Blanks
	BINARY	Hexadecimal zeros
Varying-length strings	VARCHAR, CLOB, VARGRAPHIC, DBCLOB, VARBINARY, or BLOB	Empty string
Dates	DATE	CURRENT DATE
Times	TIME	CURRENT TIME
Timestamps	TIMESTAMP	CURRENT TIMESTAMP
ROWIDs	ROWID	DB2-generated

Before moving on to the Insert via Select option, let's take a look at the data we have in the table so far.

```
SELECT
EMP_ID,
EMP_LAST_NAME,
EMP_FIRST_NAME,
EMP_SERVICE_YEARS,
EMP_PROMOTION_DATE
FROM EMPLOYEE
ORDER BY EMP_ID;
```

```
-------+---------+---------+---------+---------+---------+---------+--------------
  EMP_ID  EMP_LAST_NAME   EMP_FIRST_NAME   EMP_SERVICE_YEARS  EMP_PROMOTION_DATE
-------+---------+---------+---------+---------+---------+---------+--------------
    3217  JOHNSON         EDWARD                          4  2017-01-01
    7459  STEWART         BETTY                           7  2016-01-01
    9134  FRANKLIN        ROSEMARY                        0  ----------

DSNE610I NUMBER OF ROWS DISPLAYED IS 3
```

Insert via Select

You can use a SELECT query to extract data from one table and load it to another. You can even include literals or built in functions in the SELECT query in lieu of column names (if you need them). Let's do an example.

Suppose you work in HR and you have an employee recognition request table named EMPRE-COG. This table is used to generate/store recognition requests for employees who have been promoted during a certain time frame. Once the request is fulfilled, the date completed will be populated by HR in a separate process. The table specification is as follows:

Field Name	Type	Attributes
EMP_ID	INTEGER	NOT NULL
EMP_PROMOTION_DATE	DATE	NOT NULL
EMP_RECOG_RQST_DATE	DATE	NOT NULL WITH DEFAULT
EMP_RECOG_COMP_DATE	DATE	

The DDL to create the table is as follows:

```
CREATE TABLE EMPRECOG(
EMP_ID INT NOT NULL,
EMP_PROMOTION_DATE DATE NOT NULL,
EMP_RECOG_RQST_DATE DATE
NOT NULL WITH DEFAULT,
EMP_RECOG_COMP_DATE DATE)
IN TSHR;
```

Your objective is to load this table with data from the EMPLOYEE table for any employee whose promotion date occurs during the current month. The selection criteria could be expressed as:

```
SELECT
EMP_ID,
EMP_PROMOTION_DATE
FROM EMPLOYEE
WHERE MONTH(EMP_PROMOTION_DATE)
 = MONTH(CURRENT DATE);
```

To use this SQL in an `INSERT` statement on the `EMPRECOG` table, you would need to add another column for the request date (`EMP_RECOG_RQST_DATE`). Let's use the `CURRENT DATE` function to insert today's date. Now our select statement looks like this:

```
SELECT
EMP_ID,
EMP_PROMOTION_DATE,
CURRENT DATE AS RQST_DATE
FROM EMPLOYEE
WHERE MONTH(EMP_PROMOTION_DATE)
    = MONTH(CURRENT DATE)
```

Assuming we are running the SQL on January 10, 2017 we should get the following results:

```
---------+---------+---------+---------+---------+
    EMP_ID  EMP_PROMOTION_DATE  RQST_DATE
---------+---------+---------+---------+---------+
    3217  2017-01-01            2017-01-10
DSNE610I NUMBER OF ROWS DISPLAYED IS 1
```

Finally, let's create the `INSERT` statement for the `EMPRECOG` table. Since our query does not include the `EMP_RQST_COMP_DATE` (assume that the **request complete** column will be populated by another HR process when the request is complete), we must specify the target column names we are populating. Otherwise we will get a mismatch between the number of columns we are loading and the number in the table.

Professional Note: In circumstances where you have values for all the table's columns, you don't have to include the column names. You could just use the `INSERT INTO` and `SELECT` statement. But it is handy to include the target column names, even when you don't have to. It makes the DML more self-documenting and helpful for the next developer. This is a good habit to develop – thinking of the next person that will maintain your code.

Here is our SQL:

```
INSERT INTO EMPRECOG
(EMP_ID,
 EMP_PROMOTION_DATE,
 EMP_RECOG_RQST_DATE)
 SELECT
 EMP_ID,
 EMP_PROMOTION_DATE,
 CURRENT DATE AS RQST_DATE
 FROM EMPLOYEE
 WHERE MONTH(EMP_PROMOTION_DATE)
  = MONTH(CURRENT DATE)
```

If you are following along and running the examples, you may notice it doesn't work if the real date is not a January 2017 date. You can make this one work by specifying the comparison date as 1/1/2017. So your query would be:

```
INSERT INTO EMPRECOG
(EMP_ID,
 EMP_PROMOTION_DATE,
 EMP_RECOG_RQST_DATE)
 SELECT
 EMP_ID,
 EMP_PROMOTION_DATE,
 CURRENT DATE AS RQST_DATE
 FROM EMPLOYEE
 WHERE MONTH(EMP_PROMOTION_DATE)
  = MONTH('01/01/2017')
```

After you run the SQL, query the EMPRECOG table, and you can see the result:

```
SELECT * FROM EMPRECOG;
---------+---------+---------+---------+---------+---------+---------+---
    EMP_ID  EMP_PROMOTION_DATE  EMP_RECOG_RQST_DATE  EMP_RECOG_COMP_DATE
---------+---------+---------+---------+---------+---------+---------+---
      3217  2017-01-01          2017-01-10           ------------------
DSNE610I NUMBER OF ROWS DISPLAYED IS 1
```

The above is what we expect. Only one of the employees has a promotion date in January, 2017. This employee has been added to the EMPRECOG table with request date of January 10 and a NULL recognition completed date.

Insert via FOR N ROWS

The third form of the INSERT statement is used to insert multiple rows with a single statement. You can do this with an internal program table and host variables. We haven't talked yet about embedded SQL but we'll do a sample program now in COBOL, and I'll assume you know the COBOL language (if not, review the COBOL chapter).

We'll use our EMPLOYEE table and insert two new rows using the INSERT via FOR N ROWS. Note that we define our host variables with OCCURS 2 TIMES to create arrays, and then we load the arrays with data before we do the INSERT statement. Also notice the **FOR 2 ROWS** clause at the end of the SQL statement. You could also have an array with more than two rows. And the number of rows you insert using FOR X ROWS can be less than the actual array size.

```
       IDENTIFICATION DIVISION.
       PROGRAM-ID. COBEMP1.

      *********************************************************
      *      PROGRAM USING DB2 INSERT FOR MULTIPLE ROWS    *
      *********************************************************

       ENVIRONMENT DIVISION.
       DATA DIVISION.
       WORKING-STORAGE SECTION.


           EXEC SQL
             INCLUDE SQLCA
           END-EXEC.

           EXEC SQL
             INCLUDE EMPLOYEE
           END-EXEC.

           01 HV-EMP-VARIABLES.
           10  HV-ID             PIC S9(9) USAGE COMP OCCURS 2 TIMES.
           10  HV-LAST-NAME      PIC X(30) OCCURS 2 TIMES.
           10  HV-FIRST-NAME     PIC X(20) OCCURS 2 TIMES.
           10  HV-SERVICE-YEARS  PIC S9(9) USAGE COMP OCCURS 2 TIMES.
           10  HV-PROMOTION-DATE PIC X(10) OCCURS 2 TIMES.

       PROCEDURE DIVISION.

       MAIN-PARA.
           DISPLAY "SAMPLE COBOL PROGRAM: MULTIPLE ROW INSERT".

      *   LOAD THE EMPLOYEE ARRAY

           MOVE +4720            TO HV-ID (1).
           MOVE 'SCHULTZ'        TO HV-LAST-NAME(1).
           MOVE 'TIM'            TO HV-FIRST-NAME(1).
           MOVE +9              TO HV-SERVICE-YEARS(1).
           MOVE '01/01/2017'     TO HV-PROMOTION-DATE(1).

           MOVE +6288            TO HV-ID (2).
           MOVE 'WILLARD'        TO HV-LAST-NAME(2).
           MOVE 'JOE'            TO HV-FIRST-NAME(2).
           MOVE +6              TO HV-SERVICE-YEARS(2).
           MOVE '01/01/2016'     TO HV-PROMOTION-DATE(2).

      *   LOAD THE EMPLOYEE TABLE

           EXEC SQL
               INSERT INTO HRSCHEMA.EMPLOYEE
```

```
            (EMP_ID,
             EMP_LAST_NAME,
             EMP_FIRST_NAME,
             EMP_SERVICE_YEARS,
             EMP_PROMOTION_DATE)

            VALUES
            (:HV-ID,
             :HV-LAST-NAME,
             :HV-FIRST-NAME,
             :HV-SERVICE-YEARS,
             :HV-PROMOTION-DATE)

            FOR 2 ROWS

        END-EXEC.

        STOP RUN.
```

An additional option for the multiple row INSERT is to specify ATOMIC or NOT ATOMIC. Specifying ATOMIC means that if any of the row operations fails, any successful row operations are rolled back. It's all or nothing. This may be what you want, but that will depend on your program design and how you plan to handle any failed rows.

```
        EXEC SQL
            INSERT INTO HRSCHEMA.EMPLOYEE
            (EMP_ID,
             EMP_LAST_NAME,
             EMP_FIRST_NAME,
             EMP_SERVICE_YEARS,
             EMP_PROMOTION_DATE)

            VALUES
            (:HV-ID,
             :HV-LAST-NAME,
             :HV-FIRST-NAME,
             :HV-SERVICE-YEARS,
             :HV-PROMOTION-DATE)

            FOR 2 ROWS
            ATOMIC

        END-EXEC.

        STOP RUN.
```

Before you can run this program, it must be pre-compiled, compiled, link-edited and bound. How you do this depends on the shop. Typically you will either submit a JCL from your own library, or you will use a set of online panels to run the steps automatically. Check with a fellow programmer or system admin in your environment for the details of how to do this.

If you are renting a mainframe id with Mathru Technologies, they will provide you with the JCL to compile, bind and run your program.

On the `ATOMIC` option, you can specify `NOT ATOMIC CONTINUE ON SQLEXCEPTION`. In this case any successful row operations are still applied to the table, and any unsuccessful ones are not. The unsuccessfully inserted rows are discarded. The key point here is that `NOT ATOMIC` means the unsuccessful inserts do not cause the entire query to fail. Make sure to remember this point!

Note: You can also `INSERT` to an underlying table via a view. The syntax is exactly the same as for inserting to a table. This topic will be considered in a later section.

UPDATE Statement

The `UPDATE` statement is pretty straightforward. It changes one or more records based on specified conditions. There are two forms of the `UPDATE` statement:

1. The Searched Update
2. The Positioned Update

Searched Update

The searched update is performed on records that meet a certain search criteria using a `WHERE` clause. The basic form and syntax you need to know for the searched update is:

```
UPDATE <TABLENAME>
SET FIELDNAME = <VALUE>
WHERE <CONDITION>
```

For example, recall that we left the promotion date for employee 9134 with a `NULL` value. Now let's say we want to update the promotion date to October 1, 2016. We could use this SQL to do that:

```
UPDATE EMPLOYEE
SET EMP_PROMOTION_DATE = '10/01/2016'
WHERE EMP_ID = 9134;
```

If you have more than one column to update, you must use a comma to separate the column names. For example, let's update both the promotion date and the first name of the employee.

We'll make the first name Brianna and the promotion date 10/1/2016.

```
UPDATE EMPLOYEE
SET EMP_PROMOTION_DATE = '10/01/2016',
    EMP_FIRST_NAME = 'BRIANNA'
WHERE EMP_ID = 9134;
```

Another sub-form of the UPDATE statement to be aware of is UPDATE without a WHERE clause. For example, to set the **EMP_RECOG_COMP_DATE** field to January 31, 2017 for every row in the EMPRECOG table, you could use this statement:

```
UPDATE EMPRECOG
SET EMP_RECOG_COMP_DATE = '01/31/2017';
```

Obviously you should be very careful using this form of UPDATE, as it will set the column value(s) you specify for every row in the table. This is normally not what you want, but it could be useful in cases where you need to initialize one or more fields for all rows of a relatively small table.

Positioned Update

The positioned update is an update based on a cursor in an application program. Let's continue with our EMPLOYEE table examples by creating an update DB2 program that will generate a result set based on a cursor and then update a set of records.

We need to specially set up test data for our example, so if you are following along, execute the following query:

```
UPDATE EMPLOYEE
SET EMP_LAST_NAME = LOWER(EMP_LAST_NAME)
WHERE
EMP_LAST_NAME IN ('JOHNSON', 'STEWART', 'FRANKLIN');
```

Now here is the current content of our EMPLOYEE table:

```
SELECT EMP_ID, EMP_LAST_NAME, EMP_FIRST_NAME
FROM EMPLOYEE
ORDER BY EMP_ID;


---------+---------+---------+---------+---------+---------+----
    EMP_ID  EMP_LAST_NAME                       EMP_FIRST_NAME
---------+---------+---------+---------+---------+---------+----
    3217  johnson                             EDWARD
    4720  SCHULTZ                             TIM
```

```
6288   WILLARD                    JOE
7459   stewart                    BETTY
9134   franklin                   BRIANNA
```

As you can see we have some last names that are in lower case. Further, assume that we have decided we want to store all names in upper case. So we have to correct the lowercase data. We want to check all records in the EMPLOYEE table and if the last name is in lower case, we want to change it to upper case. We also want to report the name (both before and after correction) of the corrected records.

To accomplish our objective we'll define and open a cursor on the EMPLOYEE table. We can specify a WHERE clause that limits the result set to only those records where the EMP_LAST_NAME contains lower case characters. After we find them, we will change the case and replace the records.

To code a solution, first we need to identify the rows that include lower case letters in EMP_LAST_NAME. We can do this using the DB2 UPPER function. We'll compare the current contents of EMP_LAST_NAME to the value of UPPER(EMP_LAST_NAME) and if the results are not identical, we know that the row in question has lower case characters and needs to be changed. Our result set should include all rows where these two values are not identical. So our SQL would be:

```
SELECT EMP_ID, EMP_LAST_NAME
FROM HRSCHEMA.EMPLOYEE
WHERE EMP_LAST_NAME <> UPPER(EMP_LAST_NAME);
```

Once our FETCH statement has loaded the last name value into the host variable EMP-LAST-NAME, we can use the COBOL UPPER-CASE function to convert it from lower to uppercase.

```
MOVE FUNCTION UPPER-CASE (EMP-LAST-NAME) TO EMP-LAST-NAME
```

With this approach in mind, we are now ready to write the complete COBOL program. We will define and open the cursor, cycle through the result set using FETCH, modify the data and then do the UPDATE action specifying the current record of the cursor. That is what is meant by a positioned update – the cursor is positioned on the record to be changed, hence you do not need to specify a more elaborate WHERE clause in the UPDATE. Only the **WHERE CURRENT OF <cursor name>** clause need be specified. Also we will include the FOR UPDATE clause in our cursor definition to tell DB2 that our intent is to update the data we retrieve.

The program code follows:

```
IDENTIFICATION DIVISION.
PROGRAM-ID. COBEMP2.

**********************************************************
*        PROGRAM USING DB2 CURSOR HANDLING              *
**********************************************************

ENVIRONMENT DIVISION.
DATA DIVISION.
WORKING-STORAGE SECTION.

    EXEC SQL
      INCLUDE SQLCA
    END-EXEC.

    EXEC SQL
      INCLUDE EMPLOYEE
    END-EXEC.

    EXEC SQL
        DECLARE EMP-CURSOR CURSOR FOR
        SELECT EMP_ID, EMP_LAST_NAME
        FROM EMPLOYEE
        WHERE EMP_LAST_NAME <> UPPER(EMP_LAST_NAME)
        FOR UPDATE OF EMP_LAST_NAME
    END-EXEC.


PROCEDURE DIVISION.

MAIN-PARA.
    DISPLAY "SAMPLE COBOL PROGRAM: UPDATE USING CURSOR".

    EXEC SQL
        OPEN EMP-CURSOR
    END-EXEC.

    DISPLAY 'OPEN CURSOR SQLCODE: ' SQLCODE.

    PERFORM FETCH-CURSOR
      UNTIL SQLCODE NOT EQUAL 0.

    EXEC SQL
        CLOSE EMP-CURSOR
    END-EXEC.

    DISPLAY 'CLOSE CURSOR SQLCODE: ' SQLCODE.

    STOP RUN.
```

```
FETCH-CURSOR.

    EXEC SQL
        FETCH EMP-CURSOR INTO :EMP-ID, :EMP-LAST-NAME
    END-EXEC.

    IF SQLCODE = 0
       DISPLAY 'BEFORE CHANGE   ', EMP-LAST-NAME
       MOVE FUNCTION UPPER-CASE (EMP-LAST-NAME)
          TO EMP-LAST-NAME
       EXEC SQL
          UPDATE EMPLOYEE
          SET EMP_LAST_NAME = :EMP-LAST-NAME
          WHERE CURRENT OF EMP-CURSOR
       END-EXEC
    END-IF.

    IF SQLCODE = 0
       DISPLAY 'AFTER CHANGE    ', EMP-LAST-NAME
    END-IF.
```

To avoid redundancy, from this point I will assume that you will pre-compile, compile, link-edit and bind your programs using whatever procedures are used in your shop. I won't mention those steps again. I will just assume that you perform them before you run the program.

Here is the output from running our COBOL program:

```
SAMPLE COBOL PROGRAM: UPDATE USING CURSOR
OPEN CURSOR SQLCODE: 0000000000
BEFORE CHANGE    johnson
AFTER CHANGE     JOHNSON
BEFORE CHANGE    stewart
AFTER CHANGE     STEWART
BEFORE CHANGE    franklin
AFTER CHANGE     FRANKLIN
CLOSE CURSOR SQLCODE: 0000000000
```

This method of using a positioned cursor update is something you will use often, particularly when you do not know your result set beforehand, and anytime you need to examine the content of the record before you perform the update.

DELETE Statement

The DELETE statement is also pretty straightforward. It removes one or more records from the table based on specified conditions. As with the UPDATE statement, there are two forms of the DELETE statement:

1. The Searched Delete
2. The Positioned Delete

Searched DELETE

The searched delete is performed on records that meet a certain criteria, i.e., based on a WHERE clause. The basic form and syntax you need to remember for the searched DELETE is:

```
DELETE FROM <TABLENAME>
WHERE <CONDITION>
```

For example, we might want to remove the record for the employee with id 9134. We could use this SQL to do that:

```
DELETE FROM EMPLOYEE WHERE EMP_ID = 9134;
```

Another sub-form of the DELETE statement to be aware of is the DELETE without a WHERE clause. For example, to remove all records from the EMPRECOG table, use this statement:

```
DELETE FROM EMPRECOG;
```

Be very careful using this form of DELETE, as it will remove every record from the target table. This is normally not what you want, but it could be useful in cases where you need to initialize a relatively small table to empty.

Positioned Delete

The positioned DELETE is similar to the positioned UPDATE. It is a DELETE based on a cursor position in an application program. Let's create a DB2 program that will delete records based on a cursor. We'll have it delete any record where the employee has not received a promotion – don't feel bad for them, remember we're just using the example to illustrate a coding point!

Before we can proceed, we need to add a record to the EMPLOYEE table because currently we have no records that lack a promotion date. So we will add one.

```
INSERT INTO EMPLOYEE
VALUES (1122, 'JENKINS', 'DEBORAH', 5, NULL);
```

At this time, we have a single record in the table for which the promotion data is NULL, which is employee 1122, Deborah Jenkins:

```
SELECT
EMP_ID,
```

```
EMP_LAST_NAME,
EMP_FIRST_NAME,
EMP_PROMOTION_DATE
FROM EMPLOYEE
ORDER BY EMP_ID;
---------+---------+---------+---------+---------+---------+---------+---
     EMP_ID  EMP_LAST_NAME           EMP_FIRST_NAME          EMP_PROMOTION_DATE
---------+---------+---------+---------+---------+---------+---------+---
       1122  JENKINS                 DEBORAH                 ----------
       3217  JOHNSON                 EDWARD                  2017-01-01
       4720  SCHULTZ                 TIM                     2017-01-01
       6288  WILLARD                 JOE                     2016-01-01
       7459  STEWART                 BETTY                   2016-07-31
DSNE610I NUMBER OF ROWS DISPLAYED IS 5
```

The SQL for our cursor should look like this:

```
SELECT
EMP_ID,
FROM EMPLOYEE
WHERE EMP_PROMOTION_DATE IS NULL
FOR UPDATE
```

We'll include the FOR UPDATE clause with our cursor to ensure DB2 knows our intention is to use the cursor to delete the records we retrieve. In case you are wondering, there is no FOR DELETE clause. The FOR UPDATE clause covers both updates and deletes.

Our program code will look like this:

```
IDENTIFICATION DIVISION.
PROGRAM-ID. COBEMP3.

*****************************************************
*       PROGRAM USING DB2 CURSOR HANDLING AND DELETE *
*****************************************************

ENVIRONMENT DIVISION.
DATA DIVISION.
WORKING-STORAGE SECTION.

    EXEC SQL
      INCLUDE SQLCA
    END-EXEC.

    EXEC SQL
      INCLUDE EMPLOYEE
    END-EXEC.

    EXEC SQL
```

```
            DECLARE EMP-CURSOR CURSOR FOR
            SELECT EMP_ID
            FROM HRSCHEMA.EMPLOYEE
            WHERE EMP_PROMOTION_DATE IS NULL
            FOR UPDATE
       END-EXEC.

 PROCEDURE DIVISION.
 MAIN-PARA.
       DISPLAY "SAMPLE COBOL PROGRAM: UPDATE USING CURSOR".

       EXEC SQL
            OPEN EMP-CURSOR
       END-EXEC.

       DISPLAY 'OPEN CURSOR SQLCODE: ' SQLCODE.

       PERFORM FETCH-CURSOR
         UNTIL SQLCODE NOT EQUAL 0.

       EXEC SQL
            CLOSE EMP-CURSOR
       END-EXEC.

       DISPLAY 'CLOSE CURSOR SQLCODE: ' SQLCODE.
       STOP RUN.

 FETCH-CURSOR.

       EXEC SQL
            FETCH EMP-CURSOR INTO :EMP-ID, :EMP-LAST-NAME
       END-EXEC.

       IF SQLCODE = 0
          EXEC SQL
             DELETE HRSCHEMA.EMPLOYEE
             WHERE CURRENT OF EMP-CURSOR
          END-EXEC

       END-IF.
       IF SQLCODE = 0
          DISPLAY 'DELETED EMPLOYEE ', EMP-ID
       END-IF.
```

The output from the program looks like this:

```
SAMPLE COBOL PROGRAM: DELETE USING CURSOR
OPEN CURSOR SQLCODE: 0000000000
DELETED EMPLOYEE 000001122
CLOSE CURSOR SQLCODE: 0000000000
```

As with the positioned update statement, the positioned delete is something you will use when you do not know your result set beforehand, or when you have to first examine the content of the record and then decide whether or not to delete it.

MERGE Statement

The `MERGE` statement updates a target table or view using specified input data. Rows that already exist in the target table are updated as specified by the input source, and rows that do not exist in the target are inserted using data from that same input source.

So what problem does the merge solve? It adds/updates records for a table from a data source when you don't know whether the row already exists in the table or not. An example could be if you are updating data in your table based on a flat file you receive from another system, department or even another company. Assuming the other system does not send you an action code (add, change or delete), you won't know whether to use the `INSERT` or `UPDATE` statement.

One way of handling this situation is to first try doing an `INSERT` and if you get a -803 SQL error code, then you know the record already exists. In that case you would then need to do an `UPDATE` instead. Or you could first try doing an `UPDATE` and then if you received an `SQLCODE` +100, you would know the record does not exist and you would need to do an `INSERT`. This solution works, but it inevitably wastes some DB2 calls and could potentially slow down performance.

A more elegant solution is the `MERGE` statement. We'll look at an example of this below. You'll notice the example is a pretty long SQL statement, but don't be put off by that. The SQL is only slightly longer than the combined `INSERT` and `UPDATE` statements you would have needed to use otherwise.

Single Row Merge Using Values

Let's go back to our `EMPLOYEE` table for this example. Let's say we have employee information for Deborah Jenkins whom we previously deleted, and now we want to apply her information back to the table. This information is being fed to us from another system which also supplied an `EMP_ID`, but we don't know whether that `EMP_ID` already exists in our `EMPLOYEE` table or not. So let's use the `MERGE` statement:

```
MERGE INTO EMPLOYEE AS T
USING
(VALUES (1122,
'JENKINS',
'DEBORAH',
```

```
           5,
           NULL))
           AS S
           (EMP_ID,
            EMP_LAST_NAME,
            EMP_FIRST_NAME,
            EMP_SERVICE_YEARS,
            EMP_PROMOTION_DATE)
           ON S.EMP_ID = T.EMP_ID

      WHEN MATCHED
         THEN UPDATE
            SET T.EMP_LAST_NAME       = S.EMP_LAST_NAME,
                T.EMP_FIRST_NAME      = S.EMP_FIRST_NAME,
                T.EMP_SERVICE_YEARS   = S.EMP_SERVICE_YEARS,
                T.EMP_PROMOTION_DATE  = S.EMP_PROMOTION_DATE

      WHEN NOT MATCHED
         THEN INSERT
            VALUES (S.EMP_ID,
             S.EMP_LAST_NAME,
             S.EMP_FIRST_NAME,
             S.EMP_SERVICE_YEARS,
             S.EMP_PROMOTION_DATE);
```

Note that the existing EMPLOYEE table is given with a T qualifier and the new information is given with S as the qualifier (these qualifiers are arbitrary – you can use anything you want). We are matching the new information to the table based on employee id. When the specified employee id is matched to an employee id on the table, an update is performed using the S values, i.e., the new information. If it is not matched to an existing record, then an insert is performed – again based on the S values.

To see that our MERGE action was successful, let's take another look at our EMPLOYEE table.

```
SELECT
EMP_ID,
EMP_LAST_NAME,
EMP_FIRST_NAME,
EMP_PROMOTION_DATE
FROM EMPLOYEE
ORDER BY EMP_ID;
---------+---------+---------+---------+---------+---------+---------+---------+
    EMP_ID  EMP_LAST_NAME       EMP_FIRST_NAME       EMP_PROMOTION_DATE
---------+---------+---------+---------+---------+---------+---------+---------+
      1122  JENKINS             DEBORAH              ----------
      3217  JOHNSON             EDWARD               2017-01-01
      4720  SCHULTZ             TIM                  2017-01-01
      6288  WILLARD             JOE                  2016-01-01
      7459  STEWART             BETTY                2016-07-31
DSNE610I NUMBER OF ROWS DISPLAYED IS 5
```

371

Merge Using HOST Variables

You can also do a merge in an application program using host variables. For this example, let's create a new table and a new program. The table will be EMP_PAY and it will include the base and bonus pay for each employee identified by employee id. Here are the columns we need to define.

Field Name	Type	Attributes
EMP_ID	INTEGER	NOT NULL
EMP_REGULAR_PAY	DECIMAL	NOT NULL
EMP_BONUS	DECIMAL	

The DDL would look like this:

```
CREATE TABLE EMP_PAY(
EMP_ID INT NOT NULL,
EMP_REGULAR_PAY DECIMAL (8,2) NOT NULL,
EMP_BONUS_PAY DECIMAL   (8,2));
```

Next, let's add a few records:

```
INSERT INTO HRSCHEMA.EMP_PAY
VALUES (3217, 80000.00, 4000);

INSERT INTO HRSCHEMA.EMP_PAY
VALUES (7459, 80000.00, 4000);

INSERT INTO HRSCHEMA.EMP_PAY
VALUES (9134, 70000.00, NULL);
```

Now the current data in the table is as follows:

```
SELECT * FROM EMP_PAY;
---------+---------+---------+---------+----
    EMP_ID  EMP_REGULAR_PAY  EMP_BONUS_PAY
---------+---------+---------+---------+----
      3217         80000.00          4000.00
      7459         80000.00          4000.00
      9134         70000.00     -------------
```

Let's create an update file for the employees where some of the data is for brand new employees and some is for updating existing employees. We'll have the program read the file and use the input data with a MERGE statement to update the table. Here's the content of the file with the three fields, EMP_ID, EMP_REGULAR_PAY and EMP_BONUS_PAY:

```
----+----1----+----2----+----3---
3217      65000.00  5500.00
7459      85000.00  4500.00
9134      75000.00  2500.00
4720      80000.00  2500.00
6288      70000.00  2000.00
```

Looking at these records we know we will need to update three records that are already on the table, and we need to add two that don't currently exist on the table.

Here is sample code for a MERGE program that is based on reading the above input file and applying the data to the EMP_PAY table. It differs from the single row insert example only in that we are using host variables for the update data rather than using hard coded values. The power of the MERGE statement should be getting clearer to you now.

```
      IDENTIFICATION DIVISION.
      PROGRAM-ID. COBEMP4.
     ******************************************************
     *       PROGRAM USING DB2 MERGE WITH HOST VARIABLES  *
     ******************************************************
      ENVIRONMENT DIVISION.
      INPUT-OUTPUT SECTION.

         FILE-CONTROL.
            SELECT EMPLOYEE-FILE   ASSIGN TO EMPFILE.

      DATA DIVISION.
      FILE SECTION.
      FD  EMPLOYEE-FILE
          RECORDING MODE IS F
          LABEL RECORDS ARE STANDARD
          RECORD CONTAINS 80 CHARACTERS
          BLOCK CONTAINS 0 RECORDS.

         01 EMPLOYEE-RECORD.
             05  E-ID          PIC X(04).
             05  FILLER        PIC X(76).

      WORKING-STORAGE SECTION.

         EXEC SQL
           INCLUDE SQLCA
         END-EXEC.

         EXEC SQL
           INCLUDE EMPPAY
         END-EXEC.
```

```
       01 WS-FLAGS.
           05   SW-END-OF-FILE-SWITCH    PIC X(1) VALUE 'N'.
                88   SW-END-OF-FILE               VALUE 'Y'.
                88   SW-NOT-END-OF-FILE           VALUE 'N'.

       01 IN-EMPLOYEE-RECORD.
           05   EMPLOYEE-ID   PIC X(04).
           05   FILLER        PIC X(05).
           05   REGULAR-PAY   PIC 99999.99.
           05   FILLER        PIC X(02).
           05   BONUS-PAY     PIC 9999.99.
           05   FILLER        PIC X(54).

   PROCEDURE DIVISION.

   MAIN-PARA.
       DISPLAY "SAMPLE COBOL PROGRAM: UPDATE USING MERGE".

       OPEN INPUT EMPLOYEE-FILE.

*   MAIN LOOP - READ THE INPUT FILE, LOAD HOST VARIABLES
*               AND CALL THE MERGE ROUTINE.

       PERFORM UNTIL SW-END-OF-FILE

           READ EMPLOYEE-FILE INTO IN-EMPLOYEE-RECORD
               AT END SET SW-END-OF-FILE TO TRUE
           END-READ

           IF SW-END-OF-FILE
               CLOSE EMPLOYEE-FILE
           ELSE
               MOVE EMPLOYEE-ID TO  EMP-ID
               MOVE REGULAR-PAY TO  EMP-REGULAR-PAY
               MOVE BONUS-PAY   TO  EMP-BONUS-PAY
               PERFORM A1000-MERGE-RECORD
           END-IF

       END-PERFORM.

       STOP RUN.

   A1000-MERGE-RECORD.

       EXEC SQL

           MERGE INTO EMP_PAY AS TARGET
           USING (VALUES(:EMP-ID,
           :EMP-REGULAR-PAY,
           :EMP-BONUS-PAY))
           AS SOURCE(EMP_ID,
           EMP_REGULAR_PAY,
```

```
     EMP_BONUS_PAY)
     ON TARGET.EMP_ID = SOURCE.EMP_ID

     WHEN MATCHED THEN UPDATE
        SET TARGET.EMP_REGULAR_PAY
              = SOURCE.EMP_REGULAR_PAY,
           TARGET.EMP_BONUS_PAY
              = SOURCE.EMP_BONUS_PAY

     WHEN NOT MATCHED THEN INSERT
        (EMP_ID,
         EMP_REGULAR_PAY,
         EMP_BONUS_PAY)
         VALUES
        (SOURCE.EMP_ID,
         SOURCE.EMP_REGULAR_PAY,
         SOURCE.EMP_BONUS_PAY)

   END-EXEC.

   IF SQLCODE = 0
      DISPLAY 'RECORD MERGED SUCCESSFULLY', EMP-ID
   ELSE
      DISPLAY 'ERROR - SQLCODE = ', SQLCODE, EMP-ID
   END-IF.
```

Here are the results from the program.

```
SAMPLE COBOL PROGRAM: UPDATE USING MERGE
RECORD MERGED SUCCESSFULLY  000003217
RECORD MERGED SUCCESSFULLY  000007459
RECORD MERGED SUCCESSFULLY  000009134
RECORD MERGED SUCCESSFULLY  000004720
RECORD MERGED SUCCESSFULLY  000006288
```

Again the power of the MERGE statement is that you do not need to know whether a record already exists when you apply the data to the table. The program logic is simplified – there is no trial and error to determine whether or not the record exists.

SELECT Statement

SELECT is the main statement you will use to retrieve data from a table or view. The basic syntax for the select statement is:

```
SELECT            <column names>
FROM              <table or view name>
WHERE             <condition>
ORDER BY          <column name or number to sort by>
```

375

Let's return to our EMPLOYEE table for an example:

```
SELECT EMP_ID, EMP_LAST_NAME, EMP_FIRST_NAME
FROM HRSCHEMA.EMPLOYEE
WHERE EMP_ID = 3217;

---------+---------+---------+---------+---------+-----
    EMP_ID  EMP_LAST_NAME           EMP_FIRST_NAME
---------+---------+---------+---------+---------+-----
      3217  JOHNSON                 EDWARD
DSNE610I NUMBER OF ROWS DISPLAYED IS 1
```

You can also change the column heading on the result set by specifying <column name> AS <literal>. For example:

```
SELECT EMP_ID AS "EMPLOYEE NUMBER",
EMP_LAST_NAME AS "EMPLOYEE LAST NAME",
EMP_FIRST_NAME AS "EMPLOYEE FIRST NAME"
FROM HRSCHEMA.EMPLOYEE
WHERE EMP_ID = 3217 ;
---------+---------+---------+---------+---------+---------+---
EMPLOYEE NUMBER  EMPLOYEE LAST NAME   EMPLOYEE FIRST NAME
---------+---------+---------+---------+---------+---------+---
        3217  JOHNSON                 EDWARD
DSNE610I NUMBER OF ROWS DISPLAYED IS 1
```

Now let's look at some clauses that will further qualify the rows that are returned.

WHERE CONDITION

There are quite a lot of options for the WHERE condition. In fact, you can use multiple where conditions by specifying AND and OR clauses. Be aware of the equality operators which are:

=	Equal to
<>	Not equal to
>	Greater than
>=	Greater than or equal to
<	Less than
<=	Less than or equal to

Let's look at some various examples of WHERE conditions.

OR

```
SELECT EMP_ID, EMP_LAST_NAME, EMP_FIRST_NAME
FROM EMPLOYEE
WHERE EMP_ID = 3217 OR EMP_ID = 9134;
---------+---------+---------+---------+---------+---------
    EMP_ID  EMP_LAST_NAME         EMP_FIRST_NAME
---------+---------+---------+---------+---------+---------
      3217  JOHNSON               EDWARD
      9134  FRANKLIN              BRIANNA
DSNE610I NUMBER OF ROWS DISPLAYED IS 2
```

AND

```
SELECT EMP_ID,
EMP_LAST_NAME,
EMP_FIRST_NAME,
EMP_PROMOTION_DATE
FROM HRSCHEMA.EMPLOYEE
WHERE (EMP_SERVICE_YEARS > 1)
   AND (EMP_PROMOTION_DATE > '12/31/2016')


---------+---------+---------+---------+---------+---------+---------
    EMP_ID  EMP_LAST_NAME       EMP_FIRST_NAME        EMP_PROMOTION_DATE
---------+---------+---------+---------+---------+---------+---------
      3217  JOHNSON             EDWARD                2017-01-01
      4720  SCHULTZ             TIM                   2017-01-01
DSNE610I NUMBER OF ROWS DISPLAYED IS 2
```

IN

You can specify that the column value must be present in a specified collection of values, either those you code in the SQL explicitly or a collection that is a result of a query. Let's look at an example of specifying specific EMP_IDs.

```
SELECT EMP_ID,
EMP_LAST_NAME,
EMP_FIRST_NAME
FROM HRSCHEMA.EMPLOYEE
WHERE EMP_ID IN (3217, 9134);
---------+---------+---------+---------+---------+
    EMP_ID  EMP_LAST_NAME         EMP_FIRST_NAME
---------+---------+---------+---------+---------+
      3217  JOHNSON               EDWARD
      9134  FRANKLIN              BRIANNA
DSNE610I NUMBER OF ROWS DISPLAYED IS 2
```

377

Now let's provide a listing of employees who are in the EMPLOYEE table but are NOT in the EMP_PAY table yet. This example shows us two new techniques, use of the NOT keyword and use of a sub-select to create a collection result set. First, let's add a couple of records to the EMPLOYEE table:

```
INSERT INTO EMPLOYEE
(EMP_ID,
EMP_LAST_NAME,
EMP_FIRST_NAME,
EMP_SERVICE_YEARS,
EMP_PROMOTION_DATE)

VALUES (3333,
'FORD',
'JAMES',
7,
'10/01/2015');

INSERT INTO EMPLOYEE
(EMP_ID,
EMP_LAST_NAME,
EMP_FIRST_NAME,
EMP_SERVICE_YEARS,
EMP_PROMOTION_DATE)

VALUES (7777,
'HARRIS',
'ELISA',
2,
NULL);
```

Now let's run our mismatch query:

```
SELECT EMP_ID,
EMP_LAST_NAME,
EMP_FIRST_NAME
FROM EMPLOYEE
WHERE EMP_ID
NOT IN (SELECT EMP_ID FROM EMP_PAY);
---------+---------+---------+---------+---------+---------+---------+-
    EMP_ID  EMP_LAST_NAME       EMP_FIRST_NAME
---------+---------+---------+---------+---------+---------+---------+-
      3333  FORD                JAMES
      7777  HARRIS              ELISA
DSNE610I NUMBER OF ROWS DISPLAYED IS 2
```

By the way you can also use the **EXCEPT** clause to identify rows in one table that have no counterpart in the other. For example, suppose we want the employee ids of any employee who

378

has not received a paycheck. You could quickly identify them with this SQL:

```
SELECT EMP_ID
FROM EMPLOYEE
EXCEPT (SELECT EMP_ID FROM EMP_PAY);
---------+---------+---------+---------+---------+---
     EMP_ID
---------+---------+---------+---------+---------+---
        3333
        7777
DSNE610I NUMBER OF ROWS DISPLAYED IS 2
```

One limitation of the EXCEPT clause is that the two queries have to match exactly, so you could not bring back a column from EMPLOYEE that does not also exist in the EMP_PAY table. Still the EXCEPT is useful in some cases, especially where you need to identify discrepancies between tables using a single column.

BETWEEN

The BETWEEN clause allows you to specify a range of values inclusive of the start and end value you provide. Here's an example where we want to retrieve the employee id and pay rate for all employees whose pay rate is between 60,000 and 85,000 annually.

```
SELECT EMP_ID,
EMP_REGULAR_PAY
FROM EMP_PAY
WHERE EMP_REGULAR_PAY
BETWEEN 60000 AND 85000;
---------+---------+---------+---------+----
     EMP_ID  EMP_REGULAR_PAY
---------+---------+---------+---------+----
        3217          65000.00
        7459          85000.00
        9134          75000.00
        4720          80000.00
        6288          70000.00
DSNE610I NUMBER OF ROWS DISPLAYED IS 5
```

LIKE

You can use the LIKE predicate to select values that match a pattern. For example, let's choose all rows for which the last name begins with the letter B. The % character is used as a wild card for any string value or character. So in this case we are retrieving every record for which the EMP_FIRST_NAME starts with the letter B.

```
SELECT EMP_ID,
EMP_LAST_NAME,
```

```
EMP_FIRST_NAME
FROM HRSCHEMA.EMPLOYEE
WHERE EMP_FIRST_NAME LIKE 'B%'
---------+---------+---------+---------+---------+---------
    EMP_ID  EMP_LAST_NAME        EMP_FIRST_NAME
---------+---------+---------+---------+---------+---------
      7459  STEWART              BETTY
      9134  FRANKLIN             BRIANNA
DSNE610I NUMBER OF ROWS DISPLAYED IS 2
```

DISTINCT

Use the DISTINCT operator when you want to eliminate duplicate values. To illustrate this, let's create a couple of new tables. The first is called EMP_PAY_CHECK and we will use to store a calculated bi-monthly pay amount for each employee based on their annual salary. The DDL to create EMP_PAY_CHECK is a s follows:

```
CREATE TABLE EMP_PAY_CHECK(
EMP_ID INT NOT NULL,
EMP_REGULAR_PAY  DECIMAL (8,2) NOT NULL,
EMP_SEMIMTH_PAY DECIMAL (8,2) NOT NULL)
IN HRTS;
```

Now let's insert some data into the EMP_PAY_CHECK table by calculating a twice monthly pay check:

```
INSERT INTO EMP_PAY_CHECK
(SELECT EMP_ID,
EMP_REGULAR_PAY,
EMP_REGULAR_PAY / 24 FROM EMP_PAY);
```

Let's look at the results:

```
SELECT *
FROM HRSCHEMA.EMP_PAY_CHECK;
---------+---------+---------+---------+---------+--
    EMP_ID  EMP_REGULAR_PAY  EMP_SEMIMTH_PAY
---------+---------+---------+---------+---------+--
      3217        65000.00          2708.33
      7459        85000.00          3541.66
      9134        75000.00          3125.00
      4720        80000.00          3333.33
      6288        70000.00          2916.66
DSNE610I NUMBER OF ROWS DISPLAYED IS 5
```

We now know how much each employee should make in their pay check. The next step is to create a history table of each pay check the employee receives. First we'll create the table and then we'll load it with data.

```
CREATE TABLE EMP_PAY_HIST(
EMP_ID INT NOT NULL,
EMP_PAY_DATE  DATE NOT NULL,
EMP_PAY_AMT   DECIMAL (8,2) NOT NULL)
IN HRTS;
```

We can load the history table by creating pay checks for the first four pay periods of the year like this:

```
INSERT INTO EMP_PAY_HIST
SELECT EMP_ID,
 '01/15/2017',
 EMP_SEMIMTH_PAY
 FROM EMP_PAY_CHECK;

INSERT INTO EMP_PAY_HIST
SELECT EMP_ID,
 '01/31/2017',
 EMP_SEMIMTH_PAY
 FROM EMP_PAY_CHECK;

INSERT INTO EMP_PAY_HIST
SELECT EMP_ID,
 '02/15/2017',
 EMP_SEMIMTH_PAY
 FROM EMP_PAY_CHECK;

INSERT INTO EMP_PAY_HIST
SELECT EMP_ID,
 '02/28/2017',
 EMP_SEMIMTH_PAY
 FROM EMP_PAY_CHECK;
```

Now we can look at the history table content which is as follows:

```
    SELECT * from HRSCHEMA.EMP_PAY_HIST;
---------+---------+---------+---------+------
    EMP_ID  EMP_PAY_DATE  EMP_PAY_AMT
---------+---------+---------+---------+------
      3217  2017-01-15       2708.33
      7459  2017-01-15       3541.66
      9134  2017-01-15       3125.00
      4720  2017-01-15       3333.33
      6288  2017-01-15       2916.66
      3217  2017-01-31       2708.33
      7459  2017-01-31       3541.66
      9134  2017-01-31       3125.00
```

```
4720   2017-01-31          3333.33
6288   2017-01-31          2916.66
3217   2017-02-15          2708.33
7459   2017-02-15          3541.66
9134   2017-02-15          3125.00
4720   2017-02-15          3333.33
6288   2017-02-15          2916.66
3217   2017-02-28          2708.33
7459   2017-02-28          3541.66
9134   2017-02-28          3125.00
4720   2017-02-28          3333.33
6288   2017-02-28          2916.66
DSNE610I NUMBER OF ROWS DISPLAYED IS 20
```

If you want a list of all employees who got a paycheck during the month of February, you would need to eliminate the duplicate entries because there are two for each employee. You could accomplish that with this SQL:

```
SELECT DISTINCT EMP_ID
FROM HRSCHEMA.EMP_PAY_HIST
WHERE MONTH(EMP_PAY_DATE) = '02'
---------+---------+---------+---------+-----
     EMP_ID
---------+---------+---------+---------+-----
      3217
      4720
      6288
      7459
      9134
DSNE610I NUMBER OF ROWS DISPLAYED IS 5
```

The DISTINCT operator ensures that only unique records are selected based on the columns you are returning. This is important because if you included additional columns in the results, any value that makes the record unique will also make it **NOT** a duplicate.

Let's add the payment date to our query and see the results:

```
SELECT DISTINCT EMP_ID, EMP_PAY_DATE
FROM HRSCHEMA.EMP_PAY_HIST
WHERE MONTH(EMP_PAY_DATE) = '02'

---------+---------+---------+---------+----
     EMP_ID  EMP_PAY_DATE
---------+---------+---------+---------+----
      3217   2017-02-15
      3217   2017-02-28
      4720   2017-02-15
```

```
               4720   2017-02-28
               6288   2017-02-15
               6288   2017-02-28
               7459   2017-02-15
               7459   2017-02-28
               9134   2017-02-15
               9134   2017-02-28
DSNE610I NUMBER OF ROWS DISPLAYED IS 10
```

Since the combination of the employee id and payment date makes each record unique, you'll get multiple rows for each employee. So you must be careful in using DISTINCT to ensure that the structure of your query is really what you want.

FETCH FIRST X ROWS ONLY

You can limit your result set by using the FETCH FIRST X ROWS ONLY clause. For example, suppose you just want the employee id and names of the first four records from the employee table. You can code it as follows:

```
SELECT EMP_ID,
EMP_LAST_NAME,
EMP_FIRST_NAME
FROM HRSCHEMA.EMPLOYEE
FETCH FIRST 4 ROWS ONLY
---------+---------+---------+---------+---------+-
    EMP_ID  EMP_LAST_NAME        EMP_FIRST_NAME
---------+---------+---------+---------+---------+-
      3217  JOHNSON              EDWARD
      7459  STEWART              BETTY
      9134  FRANKLIN             BRIANNA
      4720  SCHULTZ              TIM
DSNE610I NUMBER OF ROWS DISPLAYED IS 4
```

Keep in mind that when you order the results you may get different records. For example if you order by last name, you would get this result:

```
SELECT EMP_ID,
EMP_LAST_NAME,
EMP_FIRST_NAME
FROM HRSCHEMA.EMPLOYEE
ORDER BY EMP_LAST_NAME
FETCH FIRST 4 ROWS ONLY


---------+---------+---------+---------+---------+-
    EMP_ID  EMP_LAST_NAME        EMP_FIRST_NAME
---------+---------+---------+---------+---------+-
      3333  FORD                 JAMES
      9134  FRANKLIN             BRIANNA
      7777  HARRIS               ELISA
      3217  JOHNSON              EDWARD
DSNE610I NUMBER OF ROWS DISPLAYED IS 4
```

SUBQUERY

A subquery is essentially a query within a query. Suppose for example we want to list the employee or employees who make the largest salary in the company. You can use a subquery to determine the maximum salary, and then use that value in the WHERE clause.

```
SELECT EMP_ID, EMP_REGULAR_PAY
FROM EMP_PAY
WHERE EMP_REGULAR_PAY
   = (SELECT MAX(EMP_REGULAR_PAY)
       FROM EMP_PAY);
---------+---------+---------+---------+----
   EMP_ID  EMP_REGULAR_PAY
---------+---------+---------+---------+----
    7459        85000.00
DSNE610I NUMBER OF ROWS DISPLAYED IS 1
```

What if there is more than one employee who makes the highest salary? Let's bump two people up to 85000 (and 4500 bonus) and see.

```
UPDATE EMP_PAY
SET EMP_REGULAR_PAY = 85000.00,
    EMP_BONUS_PAY = 4500
WHERE EMP_ID IN (4720,9134);
```

Here are the results:

```
SELECT * FROM EMP_PAY;
---------+---------+---------+---------+----
   EMP_ID  EMP_REGULAR_PAY  EMP_BONUS_PAY
---------+---------+---------+---------+----
    3217        65000.00         5500.00
    7459        85000.00         4500.00
    9134        85000.00         4500.00
    4720        85000.00         4500.00
    6288        70000.00         2000.00
DSNE610I NUMBER OF ROWS DISPLAYED IS 5
```

Now let's see if our subquery still works:

```
SELECT EMP_ID, EMP_REGULAR_PAY
FROM EMP_PAY
WHERE EMP_REGULAR_PAY
   = (SELECT MAX(EMP_REGULAR_PAY)
       FROM EMP_PAY);
---------+---------+---------+---------+----
   EMP_ID  EMP_REGULAR_PAY
---------+---------+---------+---------+----
    7459        85000.00
    9134        85000.00
```

```
      4720         85000.00
DSNE610I NUMBER OF ROWS DISPLAYED IS 3
```

The query pulls all three of the highest paid employees. Subqueries are very powerful in that any value you can produce via a subquery can be substituted into a main query as selection or exclusion criteria.

GROUP BY

You can summarize data using the GROUP BY clause. For example, let's determine how many distinct employee salary rates there are and how many employees are paid those amounts.

```
SELECT EMP_REGULAR_PAY,
  COUNT(*) AS "HOW MANY"
  FROM EMP_PAY
  GROUP BY EMP_REGULAR_PAY


---------+---------+---------+---------+-
EMP_REGULAR_PAY     HOW MANY
---------+---------+---------+---------+-
     65000.00            1
     70000.00            1
     85000.00            3
DSNE610I NUMBER OF ROWS DISPLAYED IS 3
```

ORDER BY

You can sort the display into ascending or descending sequence using the ORDER BY clause. To take the query we were just using for the group-by, let's present the data in descending sequence:

```
SELECT EMP_REGULAR_PAY,
  COUNT(*) AS "HOW MANY"
  FROM EMP_PAY
  GROUP BY EMP_REGULAR_PAY
  ORDER BY EMP_REGULAR_PAY DESC


---------+---------+---------+---------+-----
EMP_REGULAR_PAY     HOW MANY
---------+---------+---------+---------+-----
     85000.00            3
     70000.00            1
     65000.00            1
DSNE610I NUMBER OF ROWS DISPLAYED IS 3
```

HAVING

You could also use the GROUP BY with a HAVING clause that limits the results to only those groups that meet another condition. Let's specify that the group must have more than one employee in it to be included in the results.

```
SELECT EMP_REGULAR_PAY,
  COUNT(*) AS "HOW MANY"
  FROM EMP_PAY
  GROUP BY EMP_REGULAR_PAY
  HAVING COUNT(*) > 1
  ORDER BY EMP_REGULAR_PAY DESC

---------+---------+---------+---------+-
EMP_REGULAR_PAY     HOW MANY
---------+---------+---------+---------+-
       85000.00              3
DSNE610I NUMBER OF ROWS DISPLAYED IS 1
```

Or if you want pay rates that have only one employee you could specify the count 1.

```
SELECT EMP_REGULAR_PAY,
  COUNT(*) AS "HOW MANY"
  FROM EMP_PAY
  GROUP BY EMP_REGULAR_PAY
  HAVING COUNT(*) = 1
  ORDER BY EMP_REGULAR_PAY DESC

---------+---------+---------+---------+------
EMP_REGULAR_PAY     HOW MANY
---------+---------+---------+---------+------
       70000.00              1
       65000.00              1
DSNE610I NUMBER OF ROWS DISPLAYED IS 2
```

Before we move on, let's reset our two employees to whom we gave a temporary raise. Otherwise our EMP_PAY and EMP_PAY_CHECK tables will not be in sync.

```
UPDATE EMP_PAY
SET EMP_REGULAR_PAY = 80000.00,
    EMP_BONUS_PAY = 2500
WHERE EMP_ID = 4720;

UPDATE EMP_PAY
SET EMP_REGULAR_PAY = 75000.00,
    EMP_BONUS_PAY = 2500
WHERE EMP_ID = 9134;
```

Now our `EMP_PAY` table is restored:

```
   SELECT * FROM EMP_PAY;
---------+---------+---------+---------+-------
    EMP_ID  EMP_REGULAR_PAY  EMP_BONUS_PAY
---------+---------+---------+---------+-------
      3217          65000.00          5500.00
      7459          85000.00          4500.00
      9134          75000.00          2500.00
      4720          80000.00          2500.00
      6288          70000.00          2000.00
DSNE610I NUMBER OF ROWS DISPLAYED IS 5
```

CASE Expressions

In some situations you may need to code more complex conditional logic into your queries. Assume we have a requirement to report all employees according to seniority. We've invented the classifications ENTRY, ADVANCED and SENIOR. We want to report those who have less than a year service as ENTRY, employees who have a year or more service but less than 5 years as ADVANCED, and all employees with 5 years or more service as SENIOR. Here is a sample query that performs this using a `CASE` expression:

```
SELECT EMP_ID,
EMP_LAST_NAME,
EMP_FIRST_NAME,
CASE
   WHEN EMP_SERVICE_YEARS  < 1 THEN 'ENTRY'
   WHEN EMP_SERVICE_YEARS  < 5 THEN 'ADVANCED'
   ELSE 'SENIOR'
END CASE
FROM HRSCHEMA.EMPLOYEE;


---------+---------+---------+---------+---------+---------+------
    EMP_ID  EMP_LAST_NAME       EMP_FIRST_NAME      CASE
---------+---------+---------+---------+---------+---------+------
      3217  JOHNSON             EDWARD              SENIOR
      7459  STEWART             BETTY               SENIOR
      9134  FRANKLIN            BRIANNA             ENTRY
      4720  SCHULTZ             TIM                 SENIOR
      6288  WILLARD             JOE                 SENIOR
      3333  FORD                JAMEs               SENIOR
      7777  HARRIS              ELISA               ADVANCED
DSNE610I NUMBER OF ROWS DISPLAYED IS 7
```

You'll notice that the column heading for the case result is `CASE`. If you want to use a more meaningful column heading, then instead of closing the `CASE` statement with `END CASE`, close it with `END AS <some literal>`. So if we want to call the result of the `CASE` expression an employee's "LEVEL", code it this way:

```
SELECT EMP_ID,
EMP_LAST_NAME,
EMP_FIRST_NAME,
CASE
   WHEN EMP_SERVICE_YEARS  < 1 THEN 'ENTRY'
   WHEN EMP_SERVICE_YEARS  < 5 THEN 'ADVANCED'
   ELSE 'SENIOR'
END AS LEVEL
FROM HRSCHEMA.EMPLOYEE ;
```

```
---------+---------+---------+---------+---------+---------+-------
     EMP_ID  EMP_LAST_NAME      EMP_FIRST_NAME      LEVEL
---------+---------+---------+---------+---------+---------+-------
     3217  JOHNSON             EDWARD              SENIOR
     7459  STEWART             BETTY               SENIOR
     9134  FRANKLIN            BRIANNA             ENTRY
     4720  SCHULTZ             TIM                 SENIOR
     6288  WILLARD             JOE                 SENIOR
     3333  FORD                JAMEs               SENIOR
     7777  HARRIS              ELISA               ADVANCED
DSNE610I NUMBER OF ROWS DISPLAYED IS 7
```

JOINS

Now let's look at some cases where we need to pull data from more than one table. To do this we can use a join. Before we start running queries I want to add one row to the EMP_PAY_CHECK table. This is needed to make some of the joins work later, so bear with me.

```
INSERT INTO EMP_PAY_CHECK
VALUES
(7033,
77000.00,
77000 / 24);
```

Now our EMP_PAY_CHECK has these rows.

```
   SELECT * FROM EMP_PAY_CHECK;
---------+---------+---------+---------+------
   EMP_ID  EMP_REGULAR_PAY  EMP_SEMIMTH_PAY
---------+---------+---------+---------+------
     3217        65000.00          2708.33
     7459        85000.00          3541.66
     9134        75000.00          3125.00
     4720        80000.00          3333.33
     6288        70000.00          2916.66
     7033        77000.00          3208.00
DSNE610I NUMBER OF ROWS DISPLAYED IS 6
```

388

Inner joins

An inner join combines each row of one table with matching rows of the other table, keeping only the rows in which the join condition is true. You can join more than two tables but keep in mind that the more tables you join, the more record I/O is required and this could be a performance consideration. When I say a "performance consideration" I do not mean it is necessarily a problem. I mean it is one factor of many to keep in mind when designing an application process.

Let's do an example of a join. Assume we want a report that includes employee id, first and last names and pay rate for each employee. To accomplish this we need data from both the EMPLOYEE and the EMP_PAY tables. We can match the tables on EMP_ID which is the column they have in common.

We can perform our join either implicitly or with the JOIN verb (explicitly). In the first example will do the join implicitly by specifying we want to include rows for which the EMP_ID in the EMPLOYEE table matches the EMP_ID in the EMP_PAY table. The join is specified by the equality in the WHERE condition: WHERE A.EMP_ID = B.EMP_ID.

```
SELECT A.EMP_ID,
A.EMP_LAST_NAME,
A.EMP_FIRST_NAME,
B.EMP_REGULAR_PAY
FROM HRSCHEMA.EMPLOYEE A, HRSCHEMA.EMP_PAY B
WHERE A.EMP_ID = B.EMP_ID
ORDER BY EMP_ID
```

```
---------+---------+---------+---------+---------+---------+--------+---
    EMP_ID  EMP_LAST_NAME       EMP_FIRST_NAME          EMP_REGULAR_PAY
---------+---------+---------+---------+---------+---------+--------+---
      3217  JOHNSON             EDWARD                         65000.00
      4720  SCHULTZ             TIM                            80000.00
      6288  WILLARD             JOE                            70000.00
      7459  STEWART             BETTY                          85000.00
      9134  FRANKLIN            BRIANNA                        75000.00
DSNE610I NUMBER OF ROWS DISPLAYED IS 5
```

Notice that in the SQL the column names are prefixed with a tag that is associated with the table being referenced. This is needed in all cases where the column being referenced exists in both tables (using the same column name). In this case, if you do not specify the qualifying tag, you will get an error that your column name reference is ambiguous, i.e., DB2 does not know which column from which table you are referencing.

Moving on, you can use an explicit join by specifying the JOIN or INNER JOIN verbs. This is actually a best practice because it helps keep the query clearer for those developers who follow you, especially as your queries get more complex.

```
SELECT A.EMP_ID,
A.EMP_LAST_NAME,
A.EMP_FIRST_NAME,
B.EMP_REGULAR_PAY
FROM HRSCHEMA.EMPLOYEE A
INNER JOIN
HRSCHEMA.EMP_PAY B
ON A.EMP_ID = B.EMP_ID
ORDER BY EMP_ID
---------+---------+---------+---------+---------+---------+---------+---
    EMP_ID  EMP_LAST_NAME    EMP_FIRST_NAME      EMP_REGULAR_PAY
---------+---------+---------+---------+---------+---------+---------+---
      3217  JOHNSON          EDWARD                     65000.00
      4720  SCHULTZ          TIM                        80000.00
      6288  WILLARD          JOE                        70000.00
      7459  STEWART          BETTY                      85000.00
      9134  FRANKLIN         BRIANNA                    75000.00
DSNE610I NUMBER OF ROWS DISPLAYED IS 5
```

Finally let's do a join with three tables just to extend the concepts. We'll join the EMPLOYEE, EMP_PAY and EMP_PAY_HIST tables for pay date February 15 as follows:

```
SELECT A.EMP_ID,
A.EMP_LAST_NAME,
B.EMP_REGULAR_PAY,
C.EMP_PAY_AMT
FROM HRSCHEMA.EMPLOYEE A
    INNER JOIN
       HRSCHEMA.EMP_PAY  B ON A.EMP_ID = B.EMP_ID
    INNER JOIN
       HRSCHEMA.EMP_PAY_HIST C ON B.EMP_ID = C.EMP_ID
   WHERE C.EMP_PAY_DATE = '2/15/2017'
---------+---------+---------+---------+---------+---------+-----
    EMP_ID  EMP_LAST_NAME       EMP_REGULAR_PAY  EMP_PAY_AMT
---------+---------+---------+---------+---------+---------+-----
      3217  JOHNSON                    65000.00     2708.33
      7459  STEWART                    85000.00     3541.66
      9134  FRANKLIN                   75000.00     3125.00
      4720  SCHULTZ                    80000.00     3333.33
      6288  WILLARD                    70000.00     2916.66
DSNE610I NUMBER OF ROWS DISPLAYED IS 5
```

Now let's move on to outer joins. There are three types of outer joins. A **left outer join** includes matching rows from both tables plus any rows from the first table (the LEFT table) that were missing from the other table but that otherwise satisfied the WHERE condition. A **right outer join** includes matching rows from both tables plus any rows from the second (the RIGHT) table that were missing from the join but that otherwise satisfied the WHERE condition. A **full outer join** includes matching rows from both tables, plus those in either table that were not matched but which otherwise satisfied the WHERE condition. We'll look at examples of all three types of outer joins.

Left Outer Join

Let's try a left outer join to include matching rows from the EMPLOYEE and EMP_PAY tables, plus any rows in the EMPLOYEE table that might not be in the EMP_PAY table. In this case we are not using a WHERE clause because the table is very small and we want to see all the results. But keep in mind that we could use a WHERE clause.

```
SELECT A.EMP_ID,
A.EMP_LAST_NAME,
A.EMP_FIRST_NAME,
B.EMP_REGULAR_PAY
FROM HRSCHEMA.EMPLOYEE A
LEFT OUTER JOIN
HRSCHEMA.EMP_PAY B
ON A.EMP_ID = B.EMP_ID
ORDER BY EMP_ID
```

```
---------+---------+---------+---------+---------+---------+---------+---
    EMP_ID  EMP_LAST_NAME    EMP_FIRST_NAME      EMP_REGULAR_PAY
---------+---------+---------+---------+---------+---------+---------+---
      3217  JOHNSON          EDWARD                     65000.00
      3333  FORD             JAMES                ---------------
      4720  SCHULTZ          TIM                        80000.00
      6288  WILLARD          JOE                        70000.00
      7459  STEWART          BETTY                      85000.00
      7777  HARRIS           ELISA                ---------------
      9134  FRANKLIN         BRIANNA                    75000.00
DSNE610I NUMBER OF ROWS DISPLAYED IS 7
```

As you can see, we've included two employees who have not been assigned an annual salary yet. James Ford and Elisa Harris have NULL as their regular pay. The LEFT JOIN says we want all records in the first (left) table that satisfy the query even if there is no matching record in the right table. That's why the query results included the two unmatched records.

Let's do another left join, and this time we'll join the EMPLOYEE table with the EMP_PAY_CHECK table. Like before, we want all records from the EMPLOYEE and EMP_PAY_CHECK tables that match on EMP_ID, plus any EMPLOYEE records that could not be matched to EMP_PAY_CHECK.

```
SELECT A.EMP_ID,
A.EMP_LAST_NAME,
A.EMP_FIRST_NAME,
B.EMP_SEMIMTH_PAY
FROM HRSCHEMA.EMPLOYEE A
LEFT OUTER JOIN
HRSCHEMA.EMP_PAY_CHECK B
ON A.EMP_ID = B.EMP_ID
```

391

```
    ORDER BY EMP_ID

    ---------+---------+---------+---------+---------+---------+---------+---
        EMP_ID  EMP_LAST_NAME      EMP_FIRST_NAME         EMP_SEMIMTH_PAY
    ---------+---------+---------+---------+---------+---------+---------+---
          3217  JOHNSON            EDWARD                         2708.33
          3333  FORD               JAMEs                  ---------------
          4720  SCHULTZ            TIM                            3333.33
          6288  WILLARD            JOE                            2916.66
          7459  STEWART            BETTY                          3541.66
          7777  HARRIS             ELISA                  ---------------
          9134  FRANKLIN           BRIANNA                        3125.00
    DSNE610I NUMBER OF ROWS DISPLAYED IS 7
```

Again we find two records in the EMPLOYEE table with no matching EMP_PAY_CHECK records. From a business standpoint that could be a problem unless the two are new hires who have not received their first pay check.

Right Outer Join

Meanwhile, now let us turn it around and do a right join. In this case we want all matching records in the EMPLOYEE and EMP_PAY_CHECK records plus any unmatched records in the EMP_PAY_CHECK table (the right hand table). We could also add a WHERE condition such that the EMP_SEMIMTH_PAY column has to be populated (cannot be NULL). Let's do that.

```
    SELECT B.EMP_ID,
    A.EMP_LAST_NAME,
    A.EMP_FIRST_NAME,
    B.EMP_SEMIMTH_PAY
    FROM HRSCHEMA.EMPLOYEE A
       RIGHT OUTER JOIN
          HRSCHEMA.EMP_PAY_CHECK B
             ON A.EMP_ID = B.EMP_ID
    WHERE EMP_SEMIMTH_PAY IS NOT NULL;


    ---------+---------+---------+---------+---------+---------+---------+---
        EMP_ID  EMP_LAST_NAME      EMP_FIRST_NAME         EMP_SEMIMTH_PAY
    ---------+---------+---------+---------+---------+---------+---------+---
          3217  JOHNSON            EDWARD                         2708.33
          4720  SCHULTZ            TIM                            3333.33
          6288  WILLARD            JOE                            2916.66
          7033  --------------------  --------------------       3208.00
          7459  STEWART            BETTY                          3541.66
          9134  FRANKLIN           BRIANNA                        3125.00
    DSNE610I NUMBER OF ROWS DISPLAYED IS 6
```

Now we have a case where there is a record in the EMP_PAY_CHECK table for employee 7033, but that same employee number is NOT in the EMPLOYEE table. That is absolutely something to research! It is important to find out why this condition exists (of course we know it exists because we intentionally added an unmatched record to set up the example).

392

But let's pause for a moment. You may be thinking that this is not a realistic example because any employee getting a paycheck would also **have** to be in the EMPLOYEE table, so this mismatch condition would never happen. I chose this example for a few reasons. One reason is to point out the importance of referential data integrity. The reason the above exception is even *possible* is because we haven't defined a referential integrity relationship between these two tables. For now just know that these things can and do happen when a system has not been designed with tight referential integrity in place.

A second reason I chose this example is to highlight outer joins as a useful tool in tracking down data discrepancies between tables (subqueries are another useful tool). Keep this example in mind when you are called on by your boss or your client to troubleshoot a data integrity problem in a high pressure, time sensitive situation. You need all the tools you can get.

The third reason for choosing this example is that it very clearly demonstrates what a right join is – it includes all records from both tables that can be matched and that satisfy the WHERE condition, plus any unmatched records in the "right" table that otherwise meet the WHERE condition (in this case that the EMP_SEMIMTH_PAY is populated).

Full Outer Join

Finally, let's do a full outer join to include both matched and unmatched records from both tables that meet the where condition. This will expose all the discrepancies we already uncovered, but now we'll do it with a single query.

```
SELECT A.EMP_ID,
  A.EMP_LAST_NAME,
  B.EMP_SEMIMTH_PAY
  FROM EMPLOYEE A
    FULL OUTER JOIN
      EMP_PAY_CHECK B
        ON A.EMP_ID = B.EMP_ID;
---------+---------+---------+---------+---------+--
  EMP_ID  EMP_LAST_NAME        EMP_SEMIMTH_PAY
---------+---------+---------+---------+---------+--
    3217  JOHNSON                      2708.33
    3333  FORD                 ---------------
    4720  SCHULTZ                      3333.33
    6288  WILLARD                      2916.66
----------  --------------------       3208.00
    7459  STEWART                      3541.66
    7777  HARRIS               ---------------
    9134  FRANKLIN                     3125.00
DSNE610I NUMBER OF ROWS DISPLAYED IS 8
```

So with the FULL OUTER join we have identified the missing EMPLOYEE record, as well as the two EMP_PAY_CHECK records that may be missing. Again these examples are intended both to explain the difference between the join types, and also to lend support to troubleshooting efforts where data integrity is involved.

One final comment. The outer join examples we've given so far point to potential issues with the data, and these joins are in fact helpful in diagnosing such problems. But there are many cases where an entry in one table does not necessarily imply an entry in another. For example, suppose we have an EMP_SPOUSE table that exists to administer company benefits. A person who is single has no spouse, so they would not have an entry in the EMP_SPOUSE table. When querying for all persons covered by company benefits, an inner join between EMPLOYEE and EMP_SPOUSE would incorrectly exclude any employee who doesn't have a spouse. So you'd need a LEFT JOIN using EMPLOYEE and EMP_SPOUSE to return all insured employees plus their spouses. Your data model will govern what type of joins are needed, so be familiar with it.

UNION and INTERSECT

Another way to combine the results from two or more tables (or in some complex cases, to combine different result sets from a single table) is to use the UNION and INTERSECT statements. In some cases this can be preferable to doing a join.

Union

The UNION predicate combines the result sets from sub-SELECT queries. To understand how this might be useful, let's look at three examples. First, let's say we have two companies that have merged to form a third company. We have two tables EMP_COMPA and EMP_COMPB that we have structured with an EMP_ID, EMP_LAST_NAME and EMP_FIRST_NAME. We are going to structure a third table which will create all new employee ids by generation using an identity column. The DDL for the new table looks like this:

```
CREATE TABLE HRSCHEMA.EMPLOYEE_NEW(
EMP_ID INT GENERATED ALWAYS AS IDENTITY,
EMP_OLD_ID INTEGER,
EMP_LAST_NAME VARCHAR(30) NOT NULL,
EMP_FIRST_NAME VARCHAR(20) NOT NULL)
IN TSHR;
```

Now we can load the table using a UNION as follows:

```
INSERT INTO
HRSCHEMA.EMPLOYEE_NEW

SELECT EMP_ID,
EMP_LAST_NAME,
EMP_FIRST_NAME
FROM HRSCHEMA.EMP_COMPA

UNION

SELECT EMP_ID,
EMP_LAST_NAME,
EMP_FIRST_NAME
FROM HRSCHEMA.EMP_COMPB;
```

This will load the new table with data from both the old tables, and the new employee numbers will be auto-generated. Notice that by design we keep the old employee numbers for cross reference if needed.

When using a UNION, the column list must be identical in terms of the number of columns and data types, but the column names need not be the same. The UNION operation looks at the columns by position in the subqueries, not by name.

Let's look at two other examples of UNION queries. First, recall that earlier we used a full outer join to return all employee ids, including those that exist in one table but not the other.

```
SELECT A.EMP_ID,
B.EMP_ID,
A.EMP_LAST_NAME,
B.EMP_SEMIMTH_PAY
FROM HRSCHEMA.EMPLOYEE A
    FULL OUTER JOIN
        HRSCHEMA.EMP_PAY_CHECK B
            ON A.EMP_ID = B.EMP_ID;
```

If we just needed a unique list of employee id numbers from the EMPLOYEE and EMP_PAY_CHECK tables, we could instead use this UNION SQL:

```
SELECT EMP_ID
FROM HRSCHEMA.EMPLOYEE
UNION
SELECT EMP_ID
FROM HRSCHEMA.EMP_PAY_CHECK
```

```
---------+---------+---------+---------+-
     EMP_ID
---------+---------+---------+---------+-
       3217
       3333
       4720
       6288
       7033
       7459
       7777
       9134
DSNE610I NUMBER OF ROWS DISPLAYED IS 8
```

If you are wondering why we didn't get duplicate employee numbers in our list, it is because the UNION statement automatically eliminates duplicates. If for some reason you need to retain the duplicates, you would need to specify UNION ALL.

One final example will show how handy the UNION predicate is. Suppose that you want to query the EMPLOYEE table to get a list of all employee names for an upcoming company party. But you also have a contractor who (by business rules) cannot be in the EMPLOYEE table. You still want to include the contractor's name in the result set for whom to invite to the party. Let's say you want to identify the contractor with a pseudo-employee-id of 9999, and the contractor's name is Janet Ko.

You could code the query as follows:

```
SELECT EMP_ID,
EMP_LAST_NAME,
EMP_FIRST_NAME
FROM HRSCHEMA.EMPLOYEE
UNION
SELECT 9999,
'KO',
'JANET'
FROM SYSIBM.SYSDUMMY1;
---------+---------+---------+---------+-----

---------+---------+---------+---------+-----
       3217  JOHNSON             EDWARD
       3333  FORD                JAMES
       4720  SCHULTZ             TIM
       6288  WILLARD             JOE
       7459  STEWART             BETTY
       7777  HARRIS              ELISA
       9134  FRANKLIN            BRIANNA
       9999  KO                  JANET
DSNE610I NUMBER OF ROWS DISPLAYED IS 8
```

Now you have listed all the employees plus your contractor friend Janet on your query results. This is a useful technique when you have a "mostly" table driven system that also has some exceptions to the business rules. Sometimes a system has one-off situations that simply don't justify full blown changes to the system design. UNION can help in these cases.

Intersect

The INTERSECT predicate returns a combined result set that consists of all of the matching rows (existing in **both** result sets). In one of the earlier UNION examples, we wanted all employee ids as long as they existed in either the EMPLOYEE table or the EMP_PAY_CHECK table.

```
SELECT EMP_ID
FROM HRSCHEMA.EMPLOYEE
UNION
SELECT EMP_ID
FROM HRSCHEMA.EMP_PAY_CHECK;

---------+---------+---------
     EMP_ID
---------+---------+---------
       3217
       4720
       6288
       7033
       7459
       9134
```

Now let's say we only want a list of employee ids that appear in both tables. The INTERSECT will accomplish that for us and we only need to change that one word in the query:

```
SELECT EMP_ID
FROM HRSCHEMA.EMPLOYEE
INTERSECT
SELECT EMP_ID
FROM HRSCHEMA.EMP_PAY_CHECK;

---------+---------+---------+---------+--------
     EMP_ID
---------+---------+---------+---------+--------
       3217
       4720
       6288
       7459
       9134
DSNE610I NUMBER OF ROWS DISPLAYED IS 5
```

Common Table Expression

A common table expression is a result set that you can create and then reference in a query as though it were a table. It sometimes makes coding easier. Take this as an example. Suppose we need to work with an aggregated year-to-date total pay for each employee. Recall that our table named EMPL_PAY_HIST includes these fields:

```
(EMP_ID INTEGER NOT NULL,
EMP_PAY_DATE DATE NOT NULL,
EMP_PAY_AMT DECIMAL (8,2) NOT NULL);
```

Assume further that we have created the following SQL that includes aggregated totals for the employees' pay:

```
WITH EMP_PAY_SUM (EMP_ID, EMP_PAY_TOTAL) AS
(SELECT EMP_ID,
SUM(EMP_PAY_AMT)
AS EMP_PAY_TOTAL
FROM EMP_PAY_HIST
GROUP BY EMP_ID)

SELECT B.EMP_ID,
A.EMP_LAST_NAME,
A.EMP_FIRST_NAME,
B.EMP_PAY_TOTAL
FROM EMPLOYEE A
INNER JOIN
EMP_PAY_SUM B
ON A.EMP_ID = B.EMP_ID;
```

What we've done is to create a temporary result set named EMP_PAY_SUM that can be queried by SQL as if it were a table. This helps break down the data requirement into two pieces, one of which summarizes the pay data and the other of which adds columns from other tables.

This example may not seem like much because you could have as easily combined the two SQLs into one. But as your data stores get more numerous, and your queries and joins grow more complex, you may find that common table expressions can simplify queries both for you and for the developer that follows you.

Here's the result of our common table expression and the query against it.

```
WITH EMP_PAY_SUM (EMP_ID, EMP_PAY_TOTAL) AS
(SELECT EMP_ID,
SUM(EMP_PAY_AMT)
AS EMP_PAY_TOTAL
FROM EMP_PAY_HIST
GROUP BY EMP_ID)

SELECT B.EMP_ID,
A.EMP_LAST_NAME,
A.EMP_FIRST_NAME,                                                    B.EMP_
PAY_TOTAL
FROM EMPLOYEE A
INNER JOIN
EMP_PAY_SUM B
ON A.EMP_ID = B.EMP_ID;


---------+---------+---------+---------+---------+---------+---------+---
    EMP_ID  EMP_LAST_NAME    EMP_FIRST_NAME         EMP_PAY_TOTAL
---------+---------+---------+---------+---------+---------+---------+---
      3217  JOHNSON          EDWARD                    10833.32
      4720  SCHULTZ          TIM                       13333.32
      6288  WILLARD          JOE                       11666.64
      7459  STEWART          BETTY                     14166.64
      9134  FRANKLIN         BRIANNA                   12500.00
DSNE610I NUMBER OF ROWS DISPLAYED IS 5
```

XML

XML is a highly used standard for exchanging self-describing data files or documents. Even if you work in a shop that does not use the DB2 XML data type or XML functions, it is good to know how to use these. A complete tutorial on XML is well beyond the scope of this book. We'll review some XML basics, but if you have little or no experience with XML, I strongly suggest that you purchase some books to acquire this knowledge. The following are a few that can help fill in the basics:

XML in a Nutshell, Third Edition 3rd Edition by Elliotte Rusty Harold
(ISBN 978-0596007645)

XSLT 2.0 and XPath 2.0 Programmer's Reference by Michael Kay
(ISBN: 978-0470192740)

XQuery: Search Across a Variety of XML Data by Priscilla Walmsley
(ISBN: ISBN-13: 978-1491915103)

Basic XML Concepts

You may know that XML stands for Extensible Markup Language. XML technology is cross-platform and independent of machine and software. It provides a structure that consists of both data and data element tags, and so it describes the data in both human readable and machine readable format. The tag names for the elements are defined by the developer/user of the data.

XML Structure

XML has a tree type structure that is required to begin with a root element and then it expands to the branches. To continue our discussion of the EMPLOYEE domain, let's take a simple XML example with an employee profile as the root. We'll include the employee id, the address and birth date. The XML document might look like this:

```
<?xml version="1.0" encoding="UTF-8"?>
<EMP_PROFILE>
      <EMP_ID>4175</EMP_ID>
      <EMP_ADDRESS>
<STREET>6161 MARGARET LANE</STREET>
<CITY>ERINDALE</CITY>
<STATE>AR</STATE>
<ZIP_CODE>72653</ZIP_CODE>
</EMP_ADDRESS>
<BIRTH_DATE>07/14/1991</BIRTH_DATE>
</EMP_PROFILE>
```

XML documents frequently begin with a declaration which includes the XML version and the encoding scheme of the document. In our example, we are using XML version 1.0 which is still very common. This declaration is optional but it's a best practice to include it.

Notice after the version specification that we continue with the tag name EMP_PROFILE enclosed by the <> symbols. The employee profile element ends with /EMP_PROFILE enclosed by the <> symbols. Similarly each sub-element is tagged and enclosed and the value (if any) appears between the opening and closing of the element.

XML documents must have a single root element, i.e., one element that is the root of all other elements. If you want more than one EMP_PROFILE in a document, then you would need a higher level element to contain the profiles. For example you could have a DEPARTMENT element that contains employee profiles, and a COMPANY element that contains DEPARTMENTS.

All elements must have a closing tag. Elements that are not populated can be represented by an opening and closing with nothing in between. For example, if an employee's birthday is not known, it can be represented by <BIRTH_DATE></BIRTH_DATE> or you can use the short hand form <BIRTH_DATE/>.

The example document includes elements such as the employee id, address and birth date. The address is broken down into a street name, city, state and zip code. Comments can be included in an XML document by following the following format:

```
<!--  This is a sample comment  -->
```

By default, white space is preserved in XML documents.

Ok, so we've given you a drive-thru version of XML. We have almost enough information to move on to how to manipulate XML data in DB2. Before we get to that, let's briefly look at two XML-related technologies that we will need.

XML Related Technologies

XPath

The extensible path language (XPath) is used to locate and extract information from an XML document using "path" expressions through the XML nodes. For example, in the case of the employee XML document we created earlier, you could locate and return a zip code value by specifying the path.

Recall this structure:

```
<EMP_PROFILE>
    <EMP_ID>4175</EMP_ID>
    <EMP_ADDRESS>
        <STREET>6161 MARGARET LANE</STREET>
        <CITY>ERINDALE</CITY>
        <STATE>AR</STATE>
        <ZIP_CODE>72653</ZIP_CODE>
    </EMP_ADDRESS>
    <BIRTH_DATE>07/14/1991</BIRTH_DATE>
</EMP_PROFILE>
```

In this example, the employee profile nodes with zip code 72653 can be identified using the following path:

```
/EMP_PROFILE/ADDRESS[ZIP_CODE=72653]
```

The XPath expression for all employees who live in Texas as follows:

```
/EMP_PROFILE/ADDRESS[STATE="TX"]
```

XQuery

XQuery enables us to query XML data using XPath expressions. It is similar to how we query relational data using SQL, but of course the syntax is different. Here's an example of pulling the employee id of every employee who lives at a zip code greater than 90000 from an XML document named **employees.xml**.

```
for $x in doc("employees.xml") employee/profile/address/zipcode
where $x/zipcode>90000
order by $x/zipcode
return $x/empid
```

In DB2 you run an XQuery using the built-in function **XMLQUERY**. We'll show you some examples using XMLQUERY shortly.

DB2 Support for XML

The **pureXML** technology provides support for XML under DB2 for z/OS. DB2 includes an XML data type and many built-in DB2 functions to validate, traverse and manipulate XML data. The DB2 XML data type can store well-formed XML documents in their hierarchical form and retrieve entire documents or portions of documents.

You can execute DML operations such as inserting, updating and deleting XML documents. You can index and create triggers on XML columns. Finally, you can extract data items from an XML document and then store those values in columns of relational tables using the SQL XMLTABLE built-in function.

XML Examples

XML for the EMPLOYEE table

Suppose that we need to implement a new interface with our employee benefits providers who use XML as the data exchange format. This could give us a reason to store our detailed employee information in an XML structure within the EMPLOYEE table. For our purposes, we will add a column named EMP_PROFILE to the EMPLOYEE table and make it an XML column. Here's the DDL:

```
ALTER TABLE HRSCHEMA.EMPLOYEE
ADD COLUMN EMP_PROFILE XML;
```

We could also establish an XML schema to validate our data structure, but for the moment we'll just deal with the basic SQL operations. As long as the XML is well formed, DB2 will accept it without a schema to validate against.

Let's assume we are going to add a record to the EMPLOYEE table for employee Fred Turnbull who has employee id 4175, has 1 year if service and was promoted on 12/1/2016. Here's a sample XML document structure we want for storing the employee profile:

```
<EMP_PROFILE>
   <EMP_ID>4175</EMP_ID>
   <EMP_ADDRESS>
      <STREET>6161 MARGARET LANE</STREET>
      <CITY>ERINDALE</CITY>
      <STATE>AR</STATE>
      <ZIP_CODE>72653</ZIP_CODE>
   </EMP_ADDRESS>
   <BIRTH_DATE>07/14/1991</BIRTH_DATE>
</EMP_PROFILE>
```

INSERT With XML

Now we can insert the new record as follows:

```
INSERT INTO HRSCHEMA.EMPLOYEE
(EMP_ID,
 EMP_LAST_NAME,
 EMP_FIRST_NAME,
 EMP_SERVICE_YEARS,
 EMP_PROMOTION_DATE,
 EMP_PROFILE)
VALUES (4175,
'TURNBULL',
'FRED',
1,
'12/01/2016',
'
<EMP_PROFILE>
  <EMP_ID>4175</EMP_ID>
  <EMP_ADDRESS>
<STREET>6161 MARGARET LANE</STREET>
<CITY>ERINDALE</CITY>
<STATE>AR</STATE>
<ZIP_CODE>72653</ZIP_CODE>
</EMP_ADDRESS>
<BIRTH_DATE>07/14/1991</BIRTH_DATE>
</EMP_PROFILE>
');
```

SELECT With XML

You can do a SELECT on an XML column and depending on what query tool you are using, you can display the content of the record in fairly readable form. Since the XML data is stored as one long string, it may be difficult to read in its entirety without reformatting. We'll look at some options for that later. Let's select the column we just added using SPUFI.

```
SELECT EMP_ID, EMP_PROFILE FROM HRSCHEMA.EMPLOYEE
WHERE EMP_ID = 4175;
------+--------+---------+---------+---------+---------+---------+-----
    EMP_ID   EMP_PROFILE
------+--------+---------+---------+---------+---------+---------+-----
      4175   <?xml version="1.0" encoding="IBM037"?><EMP_PROFILE><EMP_ID>41
```

In SPUFI, you would need to scroll to the right to see the rest of the column contents.

UPDATE With XML

To update an XML column you can use standard SQL if you want to update the entire content of the column. Suppose we want to change the address. This SQL will do it:

```
UPDATE HRSCHEMA.EMPLOYEE
SET EMP_PROFILE
 = '<EMP_PROFILE>
         <EMP_ID>3217</EMP_ID>
         <EMP_ADDRESS>
               <STREET>2913 PATE DR</STREET>
               <CITY>FORT WORTH</CITY>
               <STATE>TX</STATE>
               <ZIP_CODE>76105</ZIP_CODE>
         </EMP_ADDRESS>
         <BIRTH_DATE>03/15/1952</BIRTH_DATE>
    </EMP_PROFILE>
    '
WHERE EMP_ID = 3217;
```

DELETE With XML

If you wish to delete the entire EMP_PROFILE, you can set it to NULL as follows:

```
UPDATE HRSCHEMA.EMPLOYEE
SET EMP_PROFILE = NULL
WHERE EMP_ID = 3217;


SELECT EMP_ID, EMP_PROFILE FROM HRSCHEMA.EMPLOYEE
WHERE EMP_ID = 3217;
------+--------+---------+---------+---------+---------+---------+-----
    EMP_ID   EMP_PROFILE
------+--------+---------+---------+---------+---------+---------+-----
      3217   ------------------------------------------------------------
```

As you can see, the EMP_PROFILE column has been set to NULL. At this point, only one row in the EMPLOYEE table has the EMP_PROFILE populated.

```
SELECT EMP_ID, EMP_PROFILE FROM HRSCHEMA.EMPLOYEE;

-------+---------+---------+---------+---------+---------+---------+--------
  EMP_ID  EMP_PROFILE
-------+---------+---------+---------+---------+---------+---------+--------
    3217  -------------------------------------------------------------
    7459  -------------------------------------------------------------
    9134  -------------------------------------------------------------
    4175  <?xml version="1.0" encoding="IBM037"?><EMP_PROFILE><EMP_ID>4175<
```

Let's go ahead and add the XML data back to this record so we can use it later for other XML queries.

```
UPDATE HRSCHEMA.EMPLOYEE
SET EMP_PROFILE
 = '<EMP_PROFILE>
          <EMP_ID>3217</EMP_ID>
          <EMP_ADDRESS>
                  <STREET>2913 PATE DR</STREET>
                  <CITY>FORT WORTH</CITY>
                  <STATE>TX</STATE>
                  <ZIP_CODE>76105</ZIP_CODE>
          </EMP_ADDRESS>
          <BIRTH_DATE>03/15/1952</BIRTH_DATE>
     </EMP_PROFILE>
   '
WHERE EMP_ID = 3217;
```

Also, let's update one more record so we have a bit more data to work with.

```
UPDATE HRSCHEMA.EMPLOYEE
SET EMP_PROFILE
 = '<EMP_PROFILE>
  <EMP_ID>7459</EMP_ID>
  <EMP_ADDRESS>
      <STREET>6742 OAK ST</STREET>
      <CITY>DALLAS</CITY>
      <STATE>TX</STATE>
      <ZIP_CODE>75277</ZIP_CODE>
    </EMP_ADDRESS>
    <BIRTH_DATE>09/22/1963</BIRTH_DATE>
   </EMP_PROFILE>
   '
WHERE EMP_ID = 7459;
```

XML BUILTIN FUNCTIONS

XMLQUERY

XMLQUERY is the DB2 built-in function that enables you to run XQuery. Here is an example of using XMLQUERY with the XQuery **xmlcolumn** function to retrieve an XML element from the EMP_PROFILE element. In this case we will select the zip code for employee 4175.

```
SELECT XMLQUERY
('for $info
in db2-fn:xmlcolumn("HRSCHEMA.EMPLOYEE.EMP_PROFILE")/EMP_PROFILE
return $info/EMP_ADDRESS/ZIP_CODE') AS ZIPCODE
from HRSCHEMA.EMPLOYEE
where EMP_ID = 4175


ZIPCODE
--------------------------
<ZIP_CODE>72653</ZIP_CODE>
```

Notice that the data is returned in XML format. If you don't want the data returned with its XML structure, simply add the XQuery text() function at the end of the return string, as below:

```
SELECT XMLQUERY
('for $info
in db2-fn:xmlcolumn("HRSCHEMA.EMPLOYEE.EMP_PROFILE")/EMP_PROFILE
return $info/EMP_ADDRESS/ZIP_CODE/text()') AS ZIPCODE
FROM HRSCHEMA.EMPLOYEE
WHERE EMP_ID = 4175;
```

The result of this query will not include the XML format.

```
ZIPCODE
-------
 72653
```

XMLEXISTS

The XMLEXISTS predicate specifies an XQuery expression. If the XQuery expression returns an empty sequence, the value of the XMLEXISTS predicate is false. Otherwise, XMLEXISTS returns true and those rows matching the XMLEXISTS value of true are returned.

XMLEXISTS enables us to specify rows based on the XML content which is often what you want to do. Suppose you want to return the first and last names of all employees who live in the state of Texas? This query with XMLEXISTS would accomplish it:

```
SELECT EMP_LAST_NAME, EMP_FIRST_NAME
FROM HRSCHEMA.EMPLOYEE
WHERE
XMLEXISTS('$info/EMP_PROFILE[EMP_ADDRESS/STATE/text()="TX"]'
PASSING EMP_PROFILE AS "info");


---------+---------+---------+---------+---------+---------+---
EMP_LAST_NAME                        EMP_FIRST_NAME
---------+---------+---------+---------+---------+---------+---
JOHNSON                              EDWARD
STEWART                              BETTY
```

You can also use XMLEXISTS with update and delete functions.

XMLSERIALIZE

The XMLSERIALIZE function returns a serialized XML value of the specified data type that is generated from the first argument. You can use this function to generate an XML structure from relational data. Here's an example.

```
SELECT E.EMP_ID,
XMLSERIALIZE(XMLELEMENT ( NAME "EMP_FULL_NAME",
   E.EMP_FIRST_NAME || ' ' || E.EMP_LAST_NAME)
               AS CLOB(100)) AS "RESULT"
     FROM HRSCHEMA.EMPLOYEE E;
---------+---------+---------+---------+---------+---------+-
    EMP_ID   RESULT
---------+---------+---------+---------+---------+---------+-
      3217   <EMP_FULL_NAME>EDWARD JOHNSON</EMP_FULL_NAME>
      7459   <EMP_FULL_NAME>BETTY STEWART</EMP_FULL_NAME>
      9134   <EMP_FULL_NAME>BRIANNA FRANKLIN</EMP_FULL_NAME>
      4175   <EMP_FULL_NAME>FRED TURNBULL</EMP_FULL_NAME>
      4720   <EMP_FULL_NAME>TIM SCHULTZ</EMP_FULL_NAME>
      6288   <EMP_FULL_NAME>JOE WILLARD</EMP_FULL_NAME>
      3333   <EMP_FULL_NAME>JAMEs FORD</EMP_FULL_NAME>
      7777   <EMP_FULL_NAME>ELISA HARRIS</EMP_FULL_NAME>
DSNE610I NUMBER OF ROWS DISPLAYED IS 8
```

XMLTABLE

The XMLTABLE function can be used to convert XML data to relational data. You can then use it for traditional SQL such as in joins. To use XMLTABLE you must specify the relational column names you want to use. Then you point these column names to the XML content using path expressions. For this example we'll pull address information from the profile:

```
SELECT X.*
FROM HRSCHEMA.EMPLOYEE,
```

```
XMLTABLE ('$x/EMP_PROFILE'
          PASSING EMP_PROFILE as "x"

    COLUMNS
      STREET  VARCHAR(20)  PATH 'EMP_ADDRESS/STREET',
      CITY    VARCHAR(20)  PATH 'EMP_ADDRESS/CITY',
      STATE   VARCHAR(02)  PATH 'EMP_ADDRESS/STATE',
      ZIP     VARCHAR(10)  PATH 'EMP_ADDRESS/ZIP_CODE') ;
      AS X
---------+---------+---------+---------+---------+---------+-
STREET                 CITY                  STATE  ZIP
---------+---------+---------+---------+---------+---------+-
2913 PATE DR           FORT WORTH             TX     76105
6742 OAK ST            DALLAS                 TX     75277
6161 MARGARET LANE     ERINDALE               AR     72653
DSNE610I NUMBER OF ROWS DISPLAYED IS 3
```

XMLMODIFY

XMLMODIFY allows you to make changes within the XML document. There are three expressions available for XMLMODIFY: insert, delete and replace. Here is a sample of using the replace expression to change the ZIP_CODE element of the EMP_ADDRESS for employee 4175:

```
UPDATE HRSCHEMA.EMPLOYEE
SET EMP_PROFILE
= XMLMODIFY('replace value of node
HRSCHEMA.EMPLOYEE/EMP_PROFILE/EMP_ADDRESS/ZIP_CODE
with "72652" ')
WHERE EMP_ID = 4175;
```

Now let's verify that the statement worked successfully by finding the zip code on EMP_ID 4175.

```
SELECT XMLQUERY
('for $info
in db2-fn:xmlcolumn("HRSCHEMA.EMPLOYEE.EMP_PROFILE")/EMP_PROFILE
return $info/EMP_ADDRESS/ZIP_CODE/text()') AS ZIPCODE
from HRSCHEMA.EMPLOYEE
where EMP_ID = 4175;


-------------------------------------------
ZIPCODE
-------------------------------------------
 72652
```

Important: to use XMLMODIFY, you must have created the table in a universal table space (UTS). Otherwise you will receive this SQLCODE error when you try to use the XMLMODIFY function:

408

```
DSNT408I SQLCODE = -4730, ERROR:  INVALID SPECIFICATION OF XML COLUMN
         EMPLOYEE.EMP_PROFILE IS NOT DEFINED IN THE XML VERSIONING
         FORMAT,REASON 1
```

SPECIAL REGISTERS

Special registers allow you to access detailed information about the DB2 instance settings as well as certain session information. CURRENT DATE is an example of a special register that is often used in programming (see example below).

The following are SQL examples of some commonly used special registers. I suggest that you focus on these, and then browse the complete list of special registers and what they return.

CURRENT CLIENT_USERID

CURRENT CLIENT_USERID contains the value of the client user ID from the client information that is specified for the connection. In the following example, the TSO logon id of the user is HRSCHEMA.

```
SELECT CURRENT CLIENT_USERID
FROM SYSIBM.SYSDUMMY1;
---------+---------+---------

---------+---------+---------
HRSCHEMA
```

CURRENT DATE

CURRENT DATE specifies a date that is based on a reading of the time-of-day clock when the SQL statement is executed at the current server. This is often used in application programs to establish the processing date.

```
SELECT CURRENT DATE
FROM SYSIBM.SYSDUMMY1;
---------+---------+--

---------+---------+--
2017-01-13
```

CURRENT DEGREE

CURRENT DEGREE specifies the degree of parallelism for the execution of queries that are dynamically prepared by the application process. A value of "ANY" enables parallel processing. A value of 1 prohibits parallel processing. You can query for the value of the CURRENT DEGREE as follows:

```
SELECT CURRENT DEGREE
FROM SYSIBM.SYSDUMMY1;
---------+---------+-----

---------+---------+-----
1
```

CURRENT MEMBER

CURRENT MEMBER specifies the member name of a current DB2 data sharing member on which a statement is executing. The value of CURRENT MEMBER is a character string. More information on data sharing is provided later.

CURRENT OPTIMIZATION HINT

CURRENT OPTIMIZATION HINT specifies the user-defined optimization hint that DB2 should use to generate the access path for dynamic statements.

CURRENT RULES

CURRENT RULES specifies whether certain SQL statements are executed in accordance with DB2 rules or the rules of the SQL standard.

```
SELECT CURRENT RULES
FROM SYSIBM.SYSDUMMY1;
---------+---------+----

---------+---------+----
DB2
```

CURRENT SCHEMA

CURRENT SCHEMA specifies the schema name used to qualify unqualified database object references in dynamically prepared SQL statements.

```
SELECT CURRENT SCHEMA
FROM SYSIBM.SYSDUMMY1;

---------+---------+---
HRSCHEMA
```

CURRENT SERVER

CURRENT SERVER specifies the location name of the current server.

```
SELECT CURRENT SERVER
FROM SYSIBM.SYSDUMMY1;
---------+---------+--------

---------+---------+--------
LOCRGNA
```

CURRENT SQLID

CURRENT SQLID specifies the SQL authorization ID of the process.

```
SELECT CURRENT SQLID
FROM SYSIBM.SYSDUMMY1;
---------+---------+----

HRSCHEMA
```

CURRENT TEMPORAL BUSINESS_TIME

CURRENT TEMPORAL BUSINESS_TIME specifies a TIMESTAMP(12) value that is used in the default BUSINESS_TIME period specification for references to application-period temporal tables.

CURRENT TEMPORAL SYSTEM_TIME

CURRENT TEMPORAL SYSTEM_TIME specifies a TIMESTAMP(12) value that is used in the default SYSTEM_TIME period specification for references to system-period temporal tables.

CURRENT TIME

The CURRENT TIME special register specifies a time that is based on a reading of the time-of-day clock when the SQL statement is executed at the current server.

```
SELECT CURRENT TIME
FROM SYSIBM.SYSDUMMY1;


---------+---------+----

10.12.12
```

411

CURRENT TIMESTAMP

The CURRENT TIMESTAMP special register specifies a timestamp based on the time-of-day clock at the current server.

```
SELECT CURRENT TIMESTAMP
FROM SYSIBM.SYSDUMMY1;
---------+---------+--------

---------+---------+--------
2017-01-13-10.12.51.778225
```

SESSION_USER

SESSION_USER specifies the primary authorization ID of the process.

```
SELECT SESSION_USER
FROM SYSIBM.SYSDUMMY1;
---------+---------+-----

---------+---------+-----
HRSCHEMA
```

Complete List of Special Registers

The DB2 special registers are listed below and this information is from the DB2 product documentation: [18]

CURRENT APPLICATION COMPATIBILITY	Specifies the application compatibility level support for dynamic SQL statements in the package.
CURRENT APPLICATION ENCODING SCHEME	Specifies which encoding scheme is to be used for dynamic statements.
CURRENT CLIENT_ACCTNG	Contains the value of the accounting string from the client information that is specified for the connection.

---

18   https://www.ibm.com/support/knowledgecenter/SSEPEK_11.0.0/sqlref/src/tpc/db2z_specialregistersintro.html

CURRENT CLIENT_APPLNAME	Contains the value of the application name from the client information that is specified for the connection.
CURRENT CLIENT_CORR_TOKEN	Contains the value of the client correlation token from the client information that is specified for the connection.
CURRENT CLIENT_USERID	Contains the value of the client user ID from the client information that is specified for the connection.
CURRENT CLIENT_WRKSTNNAME	Contains the value of the workstation name from the client information that is specified for the connection.
CURRENT DATE	Specifies a date that is based on a reading of the time-of-day clock when the SQL statement is executed at the current server.
CURRENT DEBUG MODE	Specifies the default value for the DEBUG MODE option when certain routines are created. The DEBUG MODE option specifies whether the routine should be built with the ability to run in debugging mode.
CURRENT DECFLOAT ROUNDING MODE	Specifies the default rounding mode that is used for DECFLOAT values.
CURRENT DEGREE	Specifies the degree of parallelism for the execution of queries that are dynamically prepared by the application process.
CURRENT EXPLAIN MODE	Contains the values that control the EXPLAIN behavior in regards to eligible dynamic SQL statements.

CURRENT GET_ACCEL_ARCHIVE	Special register specifies whether a dynamic SQL query that references a table that is archived on an accelerator server uses the archived data. The special register does not apply to static SQL queries.
CURRENT LOCALE LC_CTYPE	Specifies the LC_CTYPE locale that will be used to execute SQL statements that use a built-in function that references a locale. Functions LCASE, UCASE, and TRANSLATE (with a single argument) refer to the locale when they are executed.
CURRENT MAINTAINED TABLE TYPES FOR OPTIMIZATION	Specifies a value that identifies the types of objects that can be considered to optimize the processing of dynamic SQL queries. This register contains a keyword representing table types.
CURRENT MEMBER	Specifies the member name of a current DB2 data sharing member on which a statement is executing. The value of CURRENT MEMBER is a character string.
CURRENT OPTIMIZATION HINT	Specifies the user-defined optimization hint that DB2 should use to generate the access path for dynamic statements.
CURRENT PACKAGE PATH	Specifies a value that identifies the path used to resolve references to packages that are used to execute SQL statements. This special register applies to both static and dynamic statements.
CURRENT PACKAGESET	Specifies an empty string, a string of blanks, or the collection ID of the package that will be used to execute SQL statements.

CURRENT PATH	Specifies the SQL path used to resolve unqualified data type names and function names in dynamically prepared SQL statements. It is also used to resolve unqualified procedure names that are specified as host variables in SQL CALL statements (CALL host-variable).
CURRENT PRECISION	Specifies the rules to be used when both operands in a decimal operation have precisions of 15 or less.
CURRENT QUERY ACCELERATION	Specifies a value that identifies when DB2 sends dynamic SQL queries to an accelerator server and what DB2 does if the accelerator server fails. The special register does not apply to static SQL queries.
CURRENT REFRESH AGE	Specifies a timestamp duration value. This duration is the maximum duration since a REFRESH TABLE statement has been processed on a system-maintained REFRESH DEFERRED materialized query table such that the materialized query table can be used to optimize the processing of a query.
CURRENT ROUTINE VERSION	Specifies the version identifier that is to be used when invoking a native SQL procedure. CURRENT ROUTINE VERSION is used for CALL statements that use a host variable to specify the procedure name.
CURRENT RULES	Specifies whether certain SQL statements are executed in accordance with DB2 rules or the rules of the SQL standard.
CURRENT SCHEMA	Specifies the schema name used to qualify unqualified database object references in dynamically prepared SQL statements.
CURRENT SERVER	Specifies the location name of the current server.

CURRENT SQLID	Specifies the SQL authorization ID of the process.
CURRENT TEMPORAL BUSINESS_TIME	Specifies a TIMESTAMP(12) value that is used in the default BUSINESS_TIME period specification for references to application-period temporal tables.
CURRENT TEMPORAL SYSTEM_TIME	Specifies a TIMESTAMP(12) value that is used in the default SYSTEM_TIME period specification for references to system-period temporal tables.
CURRENT TIME	Specifies a time that is based on a reading of the time-of-day clock when the SQL statement is executed at the current server.
CURRENT TIMESTAMP	Specifies a timestamp that is based on a reading of the time-of-day clock when the SQL statement is executed at the current server.
CURRENT TIME ZONE	Specifies a value that contains the difference between UTC and local time as defined by the current server, if the SESSION TIME ZONE special register has not been set.
ENCRYPTION PASSWORD	Specifies the encryption password and the password hint (if one exists) that are used by the encryption and decryption built-in functions.
SESSION_USER	Specifies the primary authorization ID of the process.
SESSION TIME ZONE	Specifies a value that identifies the time zone of the application process.

USER	Specifies the primary authorization ID of the process. Same as SESSION_USER.

NOTE: You can use all special registers in a user-defined function or a stored procedure. However, you can modify only some of the special registers. The following are the special registers that can be modified:

- CURRENT APPLICATION COMPATIBILITY
- CURRENT APPLICATION ENCODING SCHEME
- CURRENT DEBUG MODE
- CURRENT DECFLOAT ROUNDING MODE
- CURRENT DEGREE
- CURRENT EXPLAIN MODE
- CURRENT GET_ACCEL_ARCHIVE
- CURRENT LOCALE LC_CTYPE
- CURRENT MAINTAINED TABLE TYPES FOR OPTIMIZATION
- CURRENT OPTIMIZATION HINT
- CURRENT PACKAGE PATH
- CURRENT PACKAGESET
- CURRENT PATH
- CURRENT PRECISION
- CURRENT QUERY ACCELERATION
- CURRENT REFRESH AGE
- CURRENT ROUTINE VERSION
- CURRENT RULES
- CURRENT SCHEMA
- CURRENT SQLID1
- CURRENT TEMPORAL BUSINESS_TIME
- CURRENT TEMPORAL SYSTEM_TIME
- ENCRYPTION PASSWORD
- SESSION TIME ZONE

BUILT-IN FUNCTIONS

Built-in functions can be used in SQL statements to return a result based on an argument. These functions are great productivity tools because they can replace custom coded function-

417

ality in an application program. Whether your role is application developer, DBA or business services professional, the DB2 built-in functions can save you a great deal of time and effort if you know what they are and how to use them.

There are three types of builtin functions:

1. Aggregate
2. Scalar
3. Table

We'll look at examples of each of these types, and then we'll provide a complete list for your study.

AGGREGATE Functions

An aggregate function receives a set of values for each argument (such as the values of a column) and returns a single-value result for the set of input values. These are especially useful in data analytics. Here are some examples of commonly used aggregate functions.

AVERAGE

The AVERAGE function returns the average of a set of numbers. Using our EMP_PAY table, you could get the average EMP_REGULAR_PAY for your employees like this:

```
SELECT AVG(EMP_REGULAR_PAY)
FROM EMP_PAY;
---------+---------+---------+---------+--

---------+---------+---------+---------+--
     75000.000000000000000000000
DSNE610I NUMBER OF ROWS DISPLAYED IS 1
```

COUNT

The COUNT function returns the number of rows or values in a set of rows or values. Suppose you want to know how many employees you have. You could use this SQL to find out:

```
SELECT COUNT(*)
FROM EMPLOYEE;

8

DSNE610I NUMBER OF ROWS DISPLAYED IS 1
```

MAX

The `MAX` function returns the maximum value in a set of values.

MIN

The `MIN` function returns the minimum value in a set of values.

In the next two examples, we use the MAX and MIN functions to determine the highest and lowest paid employees:

```
SELECT MAX(EMP_REGULAR_PAY)
FROM EMP_PAY;
--------+--------+-------

--------+--------+-------
   85000.00
```

Now if we want know which both the maximum salary and the employee who earns it, it is a bit more complex, but not much:

```
SELECT EMP_ID, EMP_REGULAR_PAY
FROM EMP_PAY
WHERE EMP_REGULAR_PAY =
(SELECT MAX(EMP_REGULAR_PAY) FROM EMP_PAY);
--------+--------+--------+--------+------
     EMP_ID  EMP_REGULAR_PAY
--------+--------+--------+--------+------
 7459          85000.00
```

Similarly, we can find the minimum using the MIN function.

```
SELECT MIN(EMP_REGULAR_PAY)
FROM EMP_PAY;
--------+--------+--------+-

--------+--------+--------+-
   65000.00

SELECT EMP_ID, EMP_REGULAR_PAY
FROM EMP_PAY
WHERE EMP_REGULAR_PAY =
(SELECT MIN(EMP_REGULAR_PAY) FROM EMP_PAY);
--------+--------+--------+--------+---
     EMP_ID  EMP_REGULAR_PAY
--------+--------+--------+--------+---
     3217         65000.00
```

SUM

The SUM function returns the sum of a set of numbers. Suppose you need to know what your base payroll will be for the year. You could find out with this SQL:

```
SELECT SUM(EMP_REGULAR_PAY)
FROM EMP_PAY;
---------+---------+---------+---------

---------+---------+---------+---------
         375000.00
DSNE610I NUMBER OF ROWS DISPLAYED IS 1
```

Complete List of Aggregate Functions

The following table is a complete list of aggregate functions as described in the IBM product documentation:[19]

FUNCTION	DESCRIPTION
ARRAY_AGG	Returns an array in which each value of the input Set is assigned to an element of the array.
AVG	Returns the average of a set of numbers.
CORRELATION	Returns the coefficient of the correlation of a set of number pairs.
COUNT	Returns the number of rows or values in a set of rows or values.
COUNT_BIG	Returns the number of rows or values in a set of rows or values. It is similar to COUNT except that the result can be greater than the maximum value of an integer.
COVARIANCE	Returns the covariance of a set of number pairs.
GROUPING	When used in conjunction with grouping-sets and super-groups, the GROUPING function returns a value that indicates if a row returned in a GROUP BY result is a row generated by a grouping set that excludes the column represented by expression.
MAX	Returns the highest value in a set of values.
MEDIAN	Returns the median of a set of numbers.
MIN	Returns the minimum value in a set of values.

19 http://www.ibm.com/support/knowledgecenter/SSEPEK_11.0.0/sqlref/src/tpc/db2z_aggregatefunctionsintro.html

STDDEV	The STDDEV or STDDEV_SAMP function returns the standard deviation (/n), or the sample standard deviation (/n-1), of a set of numbers.
SUM	Returns the sum of a set of numbers.
VARIANCE	Returns the biased variance (/n) of a set of numbers.
XMLAGG	Returns an XML sequence that contains an item for each non-null value in a set of XML values.

SCALAR Functions

A scalar function can be used wherever an expression can be used. It is often used to calculate a value or to influence the result of a query. Again we'll provide some examples, and then a complete list of the scalar functions and what they do.

COALESCE

The COALESCE function returns the value of the first nonnull expression. It is normally used to assign some alternate value when a NULL value is encountered that would otherwise cause an entire record to be excluded from the results. For example, consider the EMP_PAY table with data as follows:

```
SELECT *
FROM EMP_PAY;
---------+---------+---------+---------+---------
    EMP_ID  EMP_REGULAR_PAY  EMP_BONUS_PAY
---------+---------+---------+---------+---------
      3217         65000.00          5500.00
      7459         85000.00          4500.00
      9134         75000.00          2500.00
      4720         80000.00          2500.00
      6288         70000.00          2000.00
DSNE610I NUMBER OF ROWS DISPLAYED IS 5
```

To demonstrate how COALESCE works, let's change the bonus pay for employee 9134 to NULL.

```
UPDATE EMP_PAY
SET EMP_BONUS_PAY = NULL
WHERE EMP_ID = 9134;
```

Now our data looks like this:

```
SELECT *
FROM EMP_PAY;
---------+---------+---------+---------+------
    EMP_ID  EMP_REGULAR_PAY  EMP_BONUS_PAY
---------+---------+---------+---------+------
      3217          65000.00          5500.00
      7459          85000.00          4500.00
      9134          75000.00      -------------
      4720          80000.00          2500.00
      6288          70000.00          2000.00
DSNE610I NUMBER OF ROWS DISPLAYED IS 5
```

Ok, here's the example. Let's find the average bonus pay in the EMP_PAY table.

```
SELECT AVG(EMP_BONUS_PAY)
AS AVERAGE_BONUS
FROM EMP_PAY;
---------+---------+---------+---------
                AVERAGE_BONUS
---------+---------+---------+---------
      3625.0000000000000000000000
```

There is a potential problem here! The problem is that the average bonus is not 3625, it is 2900 (total 14,500 divided by five employees). The problem here is that one of the employee records has NULL in the EMP_BONUS_PAY column. Consequently this record was excluded from the calculated average because NULL is not a numeric value and therefore cannot be included in a computation.

Assuming that you do want to include this record in your results to get the correct average, you will need to convert the NULL to numeric value zero. You can do this using the COALESCE function.

```
SELECT AVG(COALESCE(EMP_BONUS_PAY,0))
AS AVERAGE_BONUS
FROM EMP_PAY;
---------+---------+---------+---------+-----
                AVERAGE_BONUS
---------+---------+---------+---------+-----
        2900.00000000000000000
DSNE610I NUMBER OF ROWS DISPLAYED IS 1
```

The above says calculate the average EMP_BONUS_PAY using the first non-null value of EMP_BONUS_PAY or zero. Since employee 9134 has a NULL value in the EMP_BONUS_PAY field, DB2 substitutes a zero instead of the NULL. Zero is a numeric value, so this record can now be included in the computation of the average. This gives the correct average which is 2900.

Before we move on let's reset the bonus pay on our employee 9134 so that it can be used correctly for other queries later in the study guide.

```
UPDATE HRSCHEMA.EMP_PAY
SET EMP_BONUS_PAY = 2500.00
WHERE EMP_ID = 9134;
```

You can use COALESCE anytime you need to include a record that would otherwise be excluded due to a NULL value. Converting the NULL to a value will ensure the record can be included in the results.

CONCAT

The CONCAT function combines two or more strings. Suppose for example you want to list each employee's first and last names from the EMPLOYEE table. You could so it with this SQL:

```
SELECT
CONCAT(CONCAT(EMP_FIRST_NAME,' '),EMP_LAST_NAME)
AS EMP_FULL_NAME FROM HRSCHEMA.EMPLOYEE;

---------+---------+---------+---------+---------+---
EMP_FULL_NAME
---------+---------+---------+---------+---------+---
EDWARD JOHNSON
BETTY STEWART
BRIANNA FRANKLIN
FRED TURNBULL
TIM SCHULTZ
JOE WILLARD
JAMEs FORD
ELISA HARRIS
DSNE610I NUMBER OF ROWS DISPLAYED IS 8
```

LCASE

The LCASE function returns a string in which all the characters are converted to lowercase characters. I can't think of many good applications for this, but here is an example of formatting the last name of each employee to lower case. Note: this function does not change any value on the table, it is only formatting the value for presentation.

```
SELECT EMP_ID, LCASE(EMP_LAST_NAME)
 FROM HRSCHEMA.EMPLOYEE;
---------+---------+---------+---------+---
    EMP_ID
---------+---------+---------+---------+---
     3217  johnson
```

```
        7459   stewart
        9134   franklin
        4175   turnbull
        4720   schultz
        6288   willard
        3333   ford
        7777   harris
DSNE610I NUMBER OF ROWS DISPLAYED IS 8
```

LEFT

The LEFT function returns a string that consists of the specified number of leftmost bytes of the specified string units. Suppose you have an application that needs the first four letters of the last name (my pharmacy does this as part of the automated prescription filling process). You could accomplish that with this SQL:

```
SELECT EMP_ID, LEFT(EMP_LAST_NAME,4)
FROM HRSCHEMA.EMPLOYEE;
---------+---------+---------+---------+-----
    EMP_ID
---------+---------+---------+---------+-----
      3217   JOHN
      7459   STEW
      9134   FRAN
      4175   TURN
      4720   SCHU
      6288   WILL
      3333   FORD
      7777   HARR
DSNE610I NUMBER OF ROWS DISPLAYED IS 8
```

MAX

The MAX function returns the maximum value in a set of values. For example if we wanted to know the largest base pay for our EMP_PAY table, we could use this SQL:

```
SELECT MAX(EMP_REGULAR_PAY)
AS HIGHEST_PAY
FROM HRSCHEMA.EMP_PAY;
---------+---------+-------
HIGHEST_PAY
---------+---------+-------
   85000.00
```

MIN

The MIN scalar function returns the minimum value in a set of values. For example if we wanted to know the largest base pay for our EMP_PAY table, we could use this SQL:

```
SELECT MIN(EMP_REGULAR_PAY)
AS LOWEST_PAY
```

```
FROM HRSCHEMA.EMP_PAY

---------+---------+-------
LOWEST_PAY
---------+---------+-------
  65000.00
```

MONTH

The MONTH function returns the month part of a date value. We used this one earlier to compare the month of the employee's promotion to the current month.

```
SELECT
EMP_ID,
EMP_PROMOTION_DATE,
CURRENT DATE AS RQST_DATE
FROM HRSCHEMA.EMPLOYEE
WHERE MONTH(EMP_PROMOTION_DATE)
 = MONTH(CURRENT DATE);


---------+---------+---------+---------+---------+----
    EMP_ID  EMP_PROMOTION_DATE  RQST_DATE
---------+---------+---------+---------+---------+----
      3217  2017-01-01          2017-01-19
      7459  2016-01-01          2017-01-19
      4720  2017-01-01          2017-01-19
      6288  2016-01-01          2017-01-19
DSNE610I NUMBER OF ROWS DISPLAYED IS 4
```

REPEAT

The REPEAT function returns a character string that is composed of an argument that is repeated a specified number of times. Suppose for example that you wanted to display 10 asterisks as a literal field on a report. You could specify it this way:

```
SELECT
EMP_ID,
REPEAT('*',10) AS "FILLER LITERAL",
EMP_SERVICE_YEARS
FROM HRSCHEMA.EMPLOYEE;
---------+---------+---------+---------+---------+-
    EMP_ID  FILLER LITERAL  EMP_SERVICE_YEARS
---------+---------+---------+---------+---------+-
      3217  **********                      6
      7459  **********                      7
      9134  **********                      0
      4175  **********                      1
      4720  **********                      9
      6288  **********                      6
      3333  **********                      7
      7777  **********                      2
DSNE610I NUMBER OF ROWS DISPLAYED IS 8
```

SPACE

The SPACE function returns a character string that consists of the number of blanks that the argument specifies. You could use this in place of the quotation literals (especially when you want a lot of spaces). The example I'll give uses the SPACE function instead of having to concatenate an empty string using quotation marks.

```
SELECT
CONCAT(CONCAT(EMP_FIRST_NAME,SPACE(1)),
EMP_LAST_NAME)
AS EMP_FULL_NAME
FROM HRSCHEMA.EMPLOYEE;


---------+---------+---------+---------+----
EMP_FULL_NAME
---------+---------+---------+---------+----
EDWARD JOHNSON
BETTY STEWART
BRIANNA FRANKLIN
FRED TURNBULL
TIM SCHULTZ
JOE WILLARD
JAMEs FORD
ELISA HARRIS
DSNE610I NUMBER OF ROWS DISPLAYED IS 8
```

SUBSTR

The SUBSTR function returns a substring of a string. Let's use the earlier example of retrieving the first four letters of the last name via the LEFT function. You could also accomplish that with this SQL:

```
SELECT EMP_ID, SUBSTR(EMP_LAST_NAME,1,4)
FROM HRSCHEMA.EMPLOYEE;
---------+---------+---------+---------+---
    EMP_ID
---------+---------+---------+---------+---
      3217  JOHN
      7459  STEW
      9134  FRAN
      4175  TURN
      4720  SCHU
      6288  WILL
      3333  FORD
      7777  HARR
DSNE610I NUMBER OF ROWS DISPLAYED IS 8
```

426

The 1,4 means starting in position one for a length of four. Of course, you could use a different starting position. An example that might make more sense is reformatting the current date. For example:

```
SELECT CURRENT DATE,
SUBSTR(CHAR(CURRENT DATE),6,2)
|| '/'
||SUBSTR(CHAR(CURRENT DATE),9,2)
|| '/'
|| SUBSTR(CHAR(CURRENT DATE),1,4)
AS REFORMED_DATE FROM SYSIBM.SYSDUMMY1;
---------+---------+---------+---
            REFORMED_DATE
---------+---------+---------+---
2017-01-12  01/12/2017
```

UCASE

The UCASE function returns a string in which all the characters are converted to uppercase characters. Here is an example of changing the last name of each employee to upper case. First we will have to covert the uppercase EMP_LAST_NAME values to lowercase. We can do that using the LOWER function. Let's do this for a single row:

```
UPDATE HRSCHEMA.EMPLOYEE
SET EMP_LAST_NAME
= LOWER(EMP_LAST_NAME)
WHERE EMP_ID = 3217;
```

We can verify that the data did in fact get changed to lower case.

```
SELECT EMP_LAST_NAME
FROM HRSCHEMA.EMPLOYEE
WHERE EMP_ID = 3217;
---------+---------+---------+---------
EMP_LAST_NAME
---------+---------+---------+---------
johnson
DSNE610I NUMBER OF ROWS DISPLAYED IS 1
```

Now let's use the UCASE function to have the EMP_LAST_NAME display as upper case.

```
SELECT EMP_ID, UCASE(EMP_LAST_NAME)
FROM HRSCHEMA.EMPLOYEE
WHERE EMP_ID = 3217;
---------+---------+---------+---------+---------
     EMP_ID
---------+---------+---------+---------+---------
     3217  JOHNSON
DSNE610I NUMBER OF ROWS DISPLAYED IS 1
```

Note that the SELECT query did not change any data on the table. We have simply reformatted the data for presentation. Now let's actually convert the data on the record back to upper case:

```
UPDATE HRSCHEMA.EMPLOYEE
SET EMP_LAST_NAME = UPPER(EMP_LAST_NAME)
WHERE EMP_ID = 3217;
```

And we'll verify that it reverted back to uppercase:

```
SELECT EMP_LAST_NAME
FROM HRSCHEMA.EMPLOYEE
WHERE EMP_ID = 3217;
---------+---------+---------+---------+---
EMP_LAST_NAME
---------+---------+---------+---------+---
JOHNSON
DSNE610I NUMBER OF ROWS DISPLAYED IS 1
```

YEAR

The YEAR function returns the year part of a value that is a character or graphic string. The value must be a valid string representation of a date or timestamp.

```
SELECT CURRENT DATE AS TODAYS_DATE,
YEAR(CURRENT DATE) AS CURRENT_YEAR
FROM SYSIBM.SYSDUMMY1;

---------+---------+---------+--------
TODAYS_DATE  CURRENT_YEAR
---------+---------+---------+--------
2017-01-12          2017
```

Complete List of Scalar Functions

A complete list of scalar functions is available below and is from the IBM DB2 product documentation:[20]

FUNCTION	DESCRIPTION
ABS	Returns the absolute value of a number.
ACOS	Returns the arc cosine of the argument as an angle, expressed in radians.
ADD_MONTHS	Returns a date that represents expression plus a specified number of months.
ARRAY_DELETE	Deletes elements from an array.
ARRAY_FIRST	Returns the minimum index value of an array.
ARRAY_LAST	Returns the maximum index value of an array.
ARRAY_NEXT	Returns the next larger array index value for an array, relative to a specified array index argument.
ARRAY_PRIOR	Returns the next smaller array index value for an array, relative to a specified array index argument.
ASCII	Returns the leftmost character of the argument as an integer.
ASCII_CHR	Returns the character that has the ASCII code value that is specified by the argument.
ASCII_STR	Returns an ASCII version of the string in the system ASCII CCSID.
ASIN	Returns the arc sine of the argument as an angle, expressed in radians. The ASIN and SIN functions are inverse operations.
ATAN	Returns the arc tangent of the argument as an angle, expressed in radians. The ATAN and TAN functions are inverse operations.
ATANH	Returns the hyperbolic arc tangent of a number, expressed in radians. The ATANH and TANH functions are inverse operations.

20 https://www.ibm.com/support/knowledgecenter/SSEPEK_11.0.0/sqlref/src/tpc/db2z_scalarfunctionsintro.html

ATAN2	Returns the arc tangent of x and y coordinates as an angle, expressed in radians.
BIGINT	Returns a big integer representation of either a number or a character or graphic string representation of a number.
BINARY	Returns a BINARY (fixed-length binary string) representation of a string of any type or of a row ID type.
BITAND, BITANDNOT, BITOR, BITXOR, and BITNOT	The bit manipulation functions operate on the twos complement r representation of the integer value of the input arguments. The functions return the result as a corresponding base 10 integer value in a data type that is based on the data type of the input arguments.
BLOB	Returns a BLOB representation of a string of any type or of a row ID type.
CARDINALITY	Returns a value of type BIGINT that represents the number of elements of an array.
CCSID_ENCODING	Returns a string value that indicates the encoding scheme of a CCSID that is specified by the argument.
CEILING	Returns the smallest integer value that is greater than or equal to the argument.
CHAR	Returns a fixed-length character string representation of the argument.
CHAR9	Returns a fixed-length character string representation of the argument.
CHARACTER_LENGTH	Returns the length of the first argument in the specified string unit.
CLOB	Returns a CLOB representation of a string.
COALESCE	Returns the value of the first nonnull expression.
COLLATION_KEY	Returns a varying-length binary string that represents the collation key of the argument in the specified collation.

COMPARE_DECFLOAT	Returns a SMALLINT value that indicates whether the two arguments are equal or unordered, or whether one argument is greater than the other.
CONCAT	Combines two compatible string arguments.
CONTAINS	Searches a text search index using criteria that are specified in a search argument and returns a result about whether or not a match was found.
COS	Returns the cosine of the argument, where the argument is an angle, expressed in radians. The COS and ACOS functions are inverse operations.
COSH	Returns the hyperbolic cosine of the argument, where the argument is an angle, expressed in radians.
DATE	Returns a date that is derived from a value.
DAY	Returns the day part of a date value.
DAYOFMONTH	Returns the day part of a date value.
DAYOFWEEK	Returns an integer, in the range of 1 to 7, that represents the day of the week, where 1 is Sunday and 7 is Saturday.
DAYOFWEEK_ISO	Returns an integer, in the range of 1 to 7, that represents the day of the week, where 1 is Monday and 7 is Sunday.
DAYOFYEAR	Returns an integer, in the range of 1 to 366, that represents the day of the year, where 1 is January 1.
DAYS	Returns an integer representation of a date.
DBCLOB	Returns a DBCLOB representation of a character string value.
DECFLOAT	Returns a decimal floating-point representation of either a number or a character string representation of a number, a decimal number, an integer, a floating-point number, or a decimal floating-point number.

DECFLOAT_FORMAT	The DECFLOAT_FORMAT function returns a q DECFLOAT(34) value that is based on the interpretation of the input string using the specified format.
DECFLOAT_SORTKEY	Returns a binary value that can be used when sorting DECFLOAT values.
DECIMAL or DEC	Returns a decimal representation of either a number or a character-string or graphic-string representation of a number, an integer, or a decimal number.
DECODE	Compares each expression2 to expression1. If expression1 is equal to expression2, or both expression1 and expression2 are null, the value of the result-expression is returned. If no expression2 matches expression1, the value of else-expression is returned. Otherwise a null value is returned.
DECRYPT_BINARY, DECRYPT_BIT, DECRYPT_CHAR, and DE-CRYPT_DB	The decryption functions return a value that is the result of decrypting encrypted data. The decryption functions can decrypt only values that are encrypted by using the ENCRYPT_TDES function.
DEGREES	Returns the number of degrees of the argument, which is an angle, expressed in radians.
DIFFERENCE	Returns a value, from 0 to 4, that represents the difference between the sounds of two strings, based on applying the SOUNDEX function to the strings. A value of 4 is the best possible sound match.
DIGITS	Returns a character string representation of the absolute value of a number.
DOUBLE_PRECISION or DOUBLE	Returns a floating-point representation of either a number or a character-string or graphic-string representation of a number, an integer, a decimal number, or a floating-point number.

DSN_XMLVALIDATE	Returns an XML value that is the result of applying XML schema validation to the first argument of the function. DSN_XMLVALIDATE can validate XML data that has a maximum length of 2 GB - 1 byte.
EBCDIC_CHR	Returns the character that has the EBCDIC code value that is specified by the argument.
EBCDIC_STR	Returns a string, in the system EBCDIC CCSID, that is an EBCDIC version of the string.
ENCRYPT_TDES	Returns a value that is the result of encrypting the first argument by using the Triple DES encryption algorithm. The function can also set the password that is used for encryption.
EXP	Returns a value that is the base of the natural logarithm (e), raised to a power that is specified by the argument. The EXP and LN functions are inverse operations.
EXTRACT	Returns a portion of a date or timestamp, based on its arguments.
FLOAT	Returns a floating-point representation of either a number or a string representation of a number. FLOAT is a synonym for the DOUBLE function.
FLOOR	Returns the largest integer value that is less than or equal to the argument.
GENERATE_UNIQUE	Returns a bit data character string that is unique, compared to any other execution of the same function.
GETHINT	Returns a hint for the password if a hint was embedded in the encrypted data. A password hint is a phrase that helps you remember the password with which the data was encrypted. For example, 'Ocean' might be used as a hint to help remember the password 'Pacific'.

GETVARIABLE	Returns a varying-length character-string representation of the current value of the session variable that is identified by the argument.
GRAPHIC	Returns a fixed-length graphic-string representation of a character string or a graphic string value, depending on the type of the first argument.
HEX	Returns a hexadecimal representation of a value.
HOUR	Returns the hour part of a value.
IDENTITY_VAL_LOCAL	Returns the most recently assigned value for an identity column.
IFNULL	Returns the first non-null expression.
INSERT	Returns a string where, beginning at start in source-string, length characters have been deleted and insert-string has been inserted.
INTEGER or INT	Returns an integer representation of either a number or a character string or graphic string representation of an integer.
JULIAN_DAY	Returns an integer value that represents a number of days from January 1, 4713 B.C. (the start of the Julian date calendar) to the date that is specified in the argument.
LAST_DAY	Returns a date that represents the last day of the month of the date argument.
LCASE	Returns a string in which all the characters are

converted to lowercase characters. |
LEFT	Returns a string that consists of the specified number of leftmost bytes of the specified string units.
LENGTH	Returns the length of a value.
LN	Returns the natural logarithm of the argument.
LOCATE	Returns the position at which the first

occurrence of an argument starts within another argument. |

LOCATE_IN_STRING	Returns the position at which an argument starts within a specified string.
LOG10	Returns the common logarithm (base 10) of a number.
LOWER	Returns a string in which all the characters are converted to lowercase characters.
LPAD	Returns a string that is composed of string-expression that is padded on the left, with pad or blanks. The LPAD function treats leading or trailing blanks in string-expression as significant.
LTRIM	Removes bytes from the beginning of a string expression based on the content of a trim expression.
MAX	Returns the maximum value in a set of values.
MAX_CARDINALITY	Returns a value of type BIGINT that represents the maximum number of elements that an array can contain. This value is the cardinality that was specified in the CREATE TYPE statement for an ordinary array type.
MICROSECOND	Returns the microsecond part of a value.
MIDNIGHT_SECONDS	Returns an integer, in the range of 0 to 86400,that represents the number of seconds between midnight and the time that is specified in the argument.
MIN	Returns the minimum value in a set of values.
MINUTE	Returns the minute part of a value.
MOD	Divides the first argument by the second argument and returns the remainder.
MONTH	Returns the month part of a date value.
MONTHS_BETWEEN	Returns an estimate of the number of months between two arguments.

435

MQREAD	Returns a message from a specified MQSeries® location without removing the message from the queue.
MQREADCLOB	Returns a message from a specified MQSeries location without removing the message from the queue.
MQRECEIVE	Returns a message from a specified MQSeries location and removes the message from the queue.
MQRECEIVECLOB	Returns a message from a specified MQSeries location and removes the message from the queue.
MQSEND	Sends data to a specified MQSeries location, and returns a varying-length character string that indicates whether the function was successful or unsuccessful.
MULTIPLY_ALT	Returns the product of the two arguments. This function is an alternative to the multiplication operator and is especially useful when the sum of the precisions of the arguments exceeds 31.
NEXT_DAY	Returns a datetime value that represents the first weekday, named by string-expression, that is later than the date in expression.
NORMALIZE_DECFLOAT	Returns a DECFLOAT value that is the result of the argument, set to its simplest form. That is, a non-zero number that has any trailing zeros in the coefficient has those zeros removed by dividing the coefficient by the appropriate power of ten and adjusting the exponent accordingly. A zero has its exponent set to 0.
NORMALIZE_STRING	Takes a Unicode string argument and returns a normalized string that can be used for comparison.
NULLIF	Returns the null value if the two arguments are equal; otherwise, it returns the value of the first argument.
NVL	Returns the first argument that is not null.

OVERLAY	Returns a string that is composed of one argument that is inserted into another argument at the same position where some number of bytes have been deleted.
PACK	Returns a binary string value that contains a data type array and a packed representation of each non-null expression argument.
POSITION	Returns the position of the first occurrence of an argument within another argument, where the position is expressed in terms of the string units that are specified.
POSSTR	Returns the position of the first occurrence of an argument within another argument.
POWER	Returns the value of the first argument to the power of the second argument.
QUANTIZE	Returns a DECFLOAT value that is equal in value (except for any rounding) and sign to the first argument and that has an exponent that is set to equal the exponent of the second argument.
QUARTER	Returns an integer between 1 and 4 that represents the quarter of the year in which the date resides. For example, any dates in January, February, or March return the integer 1.
RADIANS	Returns the number of radians for an argument that is expressed in degrees.
RAISE_ERROR	Causes the statement that invokes the function to return an error with the specified SQLSTATE (along with SQLCODE -438) and error condition. The RAISE_ERROR function always returns the null value with an undefined data type.
RAND	Returns a random floating-point value between 0 and 1. An argument can be specified as an optional seed value.

REAL	Returns a single-precision floating-point representation of either a number or a string representation of a number.
REPEAT	Returns a character string that is composed of an argument that is repeated a specified number of times.
REPLACE	Replaces all occurrences of search-string in source-string with replace-string. If search-string is not found in source-string, source-string is returned unchanged.
RID	Returns the record ID (RID) of a row. The RID is used to uniquely identify a row.
RIGHT	Returns a string that consists of the specified number of rightmost bytes or specified string unit from a string.
ROUND	Returns a number that is rounded to the specified number of places to the right or left of the decimal place.
ROUND_TIMESTAMP	Returns a timestamp that is rounded to the unit that is specified by the timestamp format string.
ROWID	ROWID Returns a row ID representation of its argument.
RPAD	Returns a string that is padded on the right with blanks or a specified string.
RTRIM	Removes bytes from the end of a string expression based on the content of a trim expression.
SCORE	Searches a text search index using criteria that are specified in a search argument and returns a relevance score that measures how well a document matches the query.
SECOND	Returns the seconds part of a value with optional fractional seconds.
SIGN	Returns an indicator of the sign of the argument.
SIN	Returns the sine of the argument, where the argument is an angle, expressed in radians.

SINH	Returns the hyperbolic sine of the argument, where the argument is an angle, expressed in radians.
SMALLINT	Returns a small integer representation either of a number or of a string representation of a number.
SOUNDEX	Returns a 4-character code that represents the sound of the words in the argument. The result can be compared to the results of the SOUNDEX function of other strings.
SOAPHTTPC and SOAPHTTPV	Returns a CLOB representation of XML data that results from a SOAP request to the web service that is specified by the first argument. The SOAPHTTPV function returns a VARCHAR representation of XML data that results from a SOAP request to the web service that is specified by the first argument.
SOAPHTTPNC and SOAPHTTPNV	The SOAPHTTPNC and SOAPHTTPNV functions allow you to specify a complete SOAP message as input and to return complete SOAP messages from the specified web service. The returned SOAP messages are CLOB or VARCHAR representations of the returned XML data.
SPACE	Returns a character string that consists of the number of SBCS blanks that the argument specifies.
SQRT	Returns the square root of the argument.
STRIP	Removes blanks or another specified character from the end, the beginning, or both ends of a string expression.
SUBSTR or SUBSTRING	Returns a substring of a string.
TAN	Returns the tangent of the argument, where the argument is an angle, expressed in radians.
TANH	Returns the hyperbolic tangent of the argument, where the argument is an angle, expressed in radians.

TIME	Returns a time that is derived from a value.
TIMESTAMP	Returns a TIMESTAMP WITHOUT TIME ZONE value from its argument or arguments.
TIMESTAMPADD	Returns the result of adding the specified number of the designated interval to the timestamp value.
TIMESTAMPDIFF	Returns an estimated number of intervals of the type that is defined by the first argument, based on the difference between two timestamps.
TIMESTAMP_FORMAT	Returns a TIMESTAMP WITHOUT TIME ZONE value that is based on the interpretation of the input string using the specified format.
TIMESTAMP_ISO	Returns a timestamp value that is based on a date, a time, or a timestamp argument.
TIMESTAMP_TZ	Returns a TIMESTAMP WITH TIME ZONE value from the input arguments.
TO_CHAR	Returns a character string representation of a timestamp value that has been formatted using a specified character template.
TO_DATE	Returns a timestamp value that is based on the interpretation of the input string using the specified format.
TO_NUMBER	Returns a DECFLOAT(34) value that is based on the interpretation of the input string using the specified format.
TOTALORDER	Returns an ordering for DECFLOAT values. The TOTALORDER function returns a small integer value that indicates how expression1 compares with expression2.
TRANSLATE	Returns a value in which one or more characters of the first argument might have been converted to other characters.
TRIM	Removes bytes from the beginning, from the end, or from both the beginning and end of a string expression.

TRIM_ARRAY	Deletes elements from the end of an ordinary array.
TRUNCATE or TRUNC	Returns the first argument, truncated as specified. Truncation is to the number of places to the right or left of the decimal point this is specified by the second argument.
TRUNC_TIMESTAMP	Returns a TIMESTAMP WITHOUT TIME ZONE value that is the expression, truncated to the unit that is specified by the format-string.
UCASE	Returns a string in which all the characters have been converted to uppercase characters, based on the CCSID of the argument. The UCASE function is identical to the UPPER function.
UNICODE	Returns the Unicode UTF-16 code value of the leftmost character of the argument as an integer.
UNICODE_STR	Returns a string in Unicode UTF-8 or UTF-16, depending on the specified option. The string represents a Unicode encoding of the input string.
UPPER	Returns a string in which all the characters have been converted to uppercase characters.
VALUE	Returns the value of the first non-null expression.
VARBINARY	Returns a VARBINARY (varying-length binary string) representation of a string of any type.
VARCHAR	Returns a varying-length character string representation of the value specified by the first argument.

VARCHAR_BIT_FORMAT	Returns a bit data string representation of a character string that has been formatted using a format-string.
VARCHAR9	Returns a fixed-length character string representation of the argument. The VARCHAR9 function is intended for compatibility with previous releases of DB2 for z/OS that depend on the result format that is returned for decimal input values in Version 9 and earlier.
VARCHAR_FORMAT	Returns a character string representation of the first argument in the format indicated by format string.
VARGRAPHIC	Returns a varying-length graphic string representation of a the first argument. The first argument can be a character string value or a graphic string value.
VERIFY_GROUP_FOR_USER	Returns a value that indicates whether the primary authorization ID and the secondary authorization IDs that are associated with the first argument are in the authorization names that are specified in the list of the second argument.
VERIFY_ROLE_FOR_USER	Returns a value that indicates whether the roles that are associated with the authorization ID that is specified in the first argument are included in the role names that are specified in the list of the second argument.
VERIFY_TRUSTED_CONTEXT_ ROLE_FOR_USER	Returns a value that indicates whether the authorization ID that is associated with first argument has acquired a role in a trusted connection and whether that acquired role is included in the role names that are specified in the list of the second argument.

442

WEEK	Returns an integer in the range of 1 to 54 that represents the week of the year. The week starts with Sunday, and January 1 is always in the first week.
WEEK_ISO	Returns an integer in the range of 1 to 53 that represents the week of the year. The week starts with Monday and includes seven days. Week 1 is the first week of the year that contains a Thursday, which is equivalent to the first week that contains January 4.
XMLATTRIBUTES	Constructs XML attributes from the arguments. This function can be used as an argument only for the XMLELEMENT function.
XMLCOMMENT	Returns an XML value with a single comment node from a string expression. The content of the comment node is the value of the input string expression, mapped to Unicode (UTF-8).
XMLCONCAT	Returns an XML sequence that contains the concatenation of a variable number of XML input arguments.
XMLDOCUMENT	Returns an XML value with a single document node and zero or more nodes as its children. The content of the generated XML document node is specified by a list of expressions.
XMLELEMENT	Returns an XML value that is an XML element node.
XMLFOREST	Returns an XML value that is a sequence of XMLElement nodes.
XMLMODIFY	Returns an XML value that might have been Modified by the evaluation of an XQuery updating Expression and XQuery variables that are specified as input arguments.

XMLNAMESPACES	Constructs namespace declarations from the

arguments. This function can be used as an argument only for specific functions, such as the XMLELEMENT function and the XMLFOREST function. |
XMLPARSE	Parses the argument as an XML document and returns an XML value.
XMLPI	Returns an XML value with a single processing instruction node.
XMLQUERY	Returns an XML value from the evaluation of an XQuery expression, by using specified input arguments, a context item, and XQuery variables.
XMLSERIALIZE	Returns a serialized XML value of the specified data type that is generated from the first argument.
XMLTEXT	Returns an XML value with a single text node that contains the value of the argument.
XMLXSROBJECTID	Returns the XSR object identifier of the XML schema that is used to validate the XML document specified in the argument.
XSLTRANSFORM	Transforms an XML document into a different data format. The output can be any form possible for the XSLT processor, including but not limited to XML, HTML, and plain text.
YEAR	Returns the year part of a value that is a character or graphic string. The value must be a valid string representation of a date or timestamp.

Application Programming with DB2

CURSORS

Types of Cursors

Cursors are scrollable or nonscrollable, sensitive or insensitive, static or dynamic. A non-scrollable cursor moves sequentially through a result set. A scrollable cursor can move where you want it to move within the result set. Scrollable cursors can be sensitive or insensitive. A sensitive cursor can be static or dynamic.

To declare a cursor as scrollable, you use the SCROLL keyword. In addition, a scrollable cursor is either sensitive or insensitive, and you specify this with the SENSITIVE and INSENSITIVE keywords. Finally to specify a sensitive cursors as static or dynamic, use the STATIC or DYNAMIC keyword.

INSENSITIVE SCROLLABLE

If you declare a cursor as INSENSITIVE, it means that:

- The size and row order in the result set are static. Also the row values cannot change, which means you cannot do updates using the cursor.

- Any changes to the underlying table after the cursor is opened are not visible to the cursor and will not be reflected in the result set.

SENSITIVE STATIC SCROLLABLE

If you declare a cursor as SENSITIVE STATIC SCROLLABLE, it means that::

- The size of the result set and ordering of the rows is static. Neither can be changed.

- Any changes to the underlying table after the cursor is opened will not be reflected in the cursor's result set.

- Rows in the cursor's result set can be changed and the changes are reflected in the result set.

- If a row is changed and its value no longer satisfies the original SELECT query, then the row disappears from the result set.

- Changes made to the underlying table will be visible in the cursor's result set if you specify SENSITIVE on the FETCH statement.

- If a row in a result set is deleted from the underlying table, the row disappears from the result set.

SENSITIVE DYNAMIC SCROLLABLE

If you declare a cursor as SENSITIVE DYNAMIC SCROLLABLE, it means that:

- The size of the result set and the ordering can vary each time you do a fetch.

- Any changes to the underlying table by other processes after the cursor is opened will be reflected in the cursor's result set once the changes are committed.

- Rows in the cursor's result set can be changed and deleted, and those changes/deletions are reflected in the result set.

- If a row is changed and its value no longer satisfies the original SELECT query, then the row disappears from the result set.

- Changes made to the underlying table will be visible in the cursor's result set if you specify SENSITIVE on the FETCH statement.

- If a row in a result set is deleted from the underlying table, the row disappears from the result set.

Additional Cursor Options

A cursor can specify WITHOUT HOLD or WITH HOLD, the main difference being whether or not the cursor is closed on a COMMIT. Specifying WITHOUT HOLD allows a cursor to be closed when a COMMIT operation occurs. Specifying WITH HOLD prevents the cursor from being closed when a COMMIT takes place.

A cursor can specify WITHOUT RETURN or WITH RETURN, the difference being whether the result set is intended to be returned to a calling program or procedure. Specifying WITH RE-TURN means that the result set is meant to be returned from the procedure it is generated in. Specifying WITHOUT RETURN means that the cursor's result set is not intended to be returned from the procedure it is generated in.

A cursor can also specify WITH ROWSET POSITIONING or WITHOUT ROWSET POSITIONING.

If you specify WITH ROWSET POSITIONING, then your cursor can return either a single row or rowset (multiple rows) with a single FETCH statement. If WITHOUT ROWSET POSITIONING is specified, it means the cursor can only return a single row with a FETCH statement.

Sample Program

To use cursors in a program, you must:

1. Declare the cursor

2. Open the Cursor

3. Fetch the cursor (one or more times)

4. Close the cursor

I suggest you memorize the sequence above.

Here is a basic program that uses a cursor to retrieve and update records. We showed this program earlier in the DML section to demonstrate the positioned UPDATE operation. If you haven't used cursors much, I suggest getting very familiar with the structure of this program.

Let's say that we want to check all records in the EMPLOYEE table and if the last name is in lower case, we want to change it to upper case and display the employee number of the corrected record. Let's first set up some test data:

```
UPDATE HRSCHEMA.EMPLOYEE
SET EMP_LAST_NAME = LOWER(EMP_LAST_NAME)
WHERE EMP_LAST_NAME IN ('JOHNSON', 'STEWART', 'FRANKLIN');
```

After you execute this SQL, here's the current content of the EMPLOYEE table:

```
    SELECT EMP_ID, EMP_LAST_NAME, EMP_FIRST_NAME
    FROM HRSCHEMA.EMPLOYEE;
---------+---------+---------+---------+---------+-----
    EMP_ID  EMP_LAST_NAME          EMP_FIRST_NAME
---------+---------+---------+---------+---------+-----
      3217  johnson                EDWARD
      7459  stewart                BETTY
      9134  franklin               BRIANNA
      4720  SCHULTZ                TIM
      6288  WILLARD                JOE
      1122  JENKINS                DEBBIE
      4175  TURNBULL               FREDERICK
      1001  HENDERSON              JOHN
DSNE610I NUMBER OF ROWS DISPLAYED IS 8
```

To accomplish our objective we'll define and open a cursor on the EMPLOYEE table. We can specify a WHERE clause that limits the result set to only those records that contain lower case characters. After we find them, we will change the case to upper and replace the records.

First we need to identify the rows that include lower case letters in column EMP_LAST_NAME. We can do this using the UPPER function. We'll compare the current contents of the EMP_LAST_NAME to the value of UPPER(EMP_LAST_NAME) and if the results are not identical, the row in question has lower case and needs to be changed. Our result set should include all rows where these two values are not identical. So our SQL would be:

```
SELECT EMP_ID, EMP_LAST_NAME
FROM EMPLOYEE
WHERE EMP_LAST_NAME <> UPPER(EMP_LAST_NAME)
```

Once we've placed the last name value in the host variable EMP-LAST-NAME, we can use the COBOL Upper-case function to convert lowercase to uppercase.

```
MOVE FUNCTION UPPER-CASE (EMP-LAST-NAME) TO EMP-LAST-NAME
```

Now we are ready to write the program. So we define and open the cursor, cycle through the result set using FETCH, modify the data and then do the UPDATE action specifying the current record of the cursor. That is what is meant by a positioned update – the cursor is positioned on the record to be changed, hence you do not need to specify a more elaborate WHERE clause in the UPDATE. Only the **WHERE CURRENT OF <cursor name>** clause need be specified. Also we will include the **FOR UPDATE** clause in our cursor definition to ensure DB2 knows our intent is to update the data we retrieve.

The program code follows:

```
    IDENTIFICATION DIVISION.
    PROGRAM-ID. COBEMP2.

 *****************************************************
 *        PROGRAM USING DB2 CURSOR HANDLING          *
 *****************************************************

    ENVIRONMENT DIVISION.
    DATA DIVISION.
    WORKING-STORAGE SECTION.

        EXEC SQL
          INCLUDE SQLCA
```

```
          END-EXEC.


       EXEC SQL
          INCLUDE EMPLOYEE
       END-EXEC.

       EXEC SQL
            DECLARE EMP-CURSOR CURSOR FOR
            SELECT EMP_ID, EMP_LAST_NAME
            FROM EMPLOYEE
            WHERE EMP_LAST_NAME <> UPPER(EMP_LAST_NAME)
            FOR UPDATE OF EMP_LAST_NAME
       END-EXEC.


   PROCEDURE DIVISION.

   MAIN-PARA.
       DISPLAY "SAMPLE COBOL PROGRAM: UPDATE USING CURSOR".

       EXEC SQL
            OPEN EMP-CURSOR
       END-EXEC.

       DISPLAY 'OPEN CURSOR SQLCODE: ' SQLCODE.

       PERFORM FETCH-CURSOR
          UNTIL SQLCODE NOT EQUAL 0.

       EXEC SQL
            CLOSE EMP-CURSOR
       END-EXEC.

       DISPLAY 'CLOSE CURSOR SQLCODE: ' SQLCODE.

       STOP RUN.

   FETCH-CURSOR.

       EXEC SQL
            FETCH EMP-CURSOR INTO :EMP-ID, :EMP-LAST-NAME
       END-EXEC.

       IF SQLCODE = 0
          DISPLAY 'BEFORE CHANGE  ', EMP-LAST-NAME
          MOVE FUNCTION UPPER-CASE (EMP-LAST-NAME)
             TO EMP-LAST-NAME
          EXEC SQL
             UPDATE EMPLOYEE
             SET EMP_LAST_NAME = :EMP-LAST-NAME
```

```
            WHERE CURRENT OF EMP-CURSOR
        END-EXEC

    END-IF.

    IF SQLCODE = 0
        DISPLAY 'AFTER CHANGE    ', EMP-LAST-NAME
    END-IF.
```

Here is the output from running the program:

```
SAMPLE COBOL PROGRAM: UPDATE USING CURSOR
OPEN CURSOR SQLCODE: 0000000000
BEFORE CHANGE    johnson
AFTER CHANGE     JOHNSON
BEFORE CHANGE    stewart
AFTER CHANGE     STEWART
BEFORE CHANGE    franklin
AFTER CHANGE     FRANKLIN
CLOSE CURSOR SQLCODE: 0000000000
```

And here is the modified table:

```
    SELECT EMP_ID,
    EMP_LAST_NAME,
    EMP_FIRST_NAME
    FROM HRSCHEMA.EMPLOYEE;
    ---------+---------+---------+---------+---------+-----
        EMP_ID  EMP_LAST_NAME     EMP_FIRST_NAME
    ---------+---------+---------+---------+---------+-----
        3217  JOHNSON           EDWARD
        7459  STEWART           BETTY
        9134  FRANKLIN          BRIANNA
        4720  SCHULTZ           TIM
        6288  WILLARD           JOE
        1122  JENKINS           DEBBIE
        4175  TURNBULL          FREDERICK
        1001  HENDERSON         JOHN
DSNE610I NUMBER OF ROWS DISPLAYED IS 8
```

This method of using a positioned cursor update is something you will use often, particularly when you do not know your result set beforehand, or anytime you need to examine the content of the record before you perform the update.

Error Handling

In over three decades of experience with DB2, I believe one of the most neglected areas in programmer training is problem resolution. I'm not sure why this is, but I'd like to provide some standards that may help save time and make programmers more effective. First, let's look at SQLCODE processing, and then we'll look at standardizing an error reporting routine.

450

SQLCODES

When using embedded SQL with DB2 you include a SQLCA structure which includes an SQLCODE variable. DB2 sets the SQLCODE after each SQL statement. The SQLCODE should be interrogated to determine the success or failure of the SQL statement.

The value of the SQLCODE can be interpreted generally as follows:

> If SQLCODE = 0, execution was successful.
> If SQLCODE > 0, execution was successful with a warning.
> If SQLCODE < 0, execution was not successful.
> SQLCODE = 100, "no data" was found.

Here's an example of an SQL error message when a query is executed via SPUFI. In this case, the last name column is incorrectly spelled (it should be EMP_LAST_NAME) so DB2 does not recognize it. The -206 is accompanied by an explanation. A more complete explanation and recommendations for action to take is available if you look up the SQLCODE on the IBM product documentation web site.

```
   SELECT EMP_ID, EMP_LASTNAME
   FROM HRSCHEMA.EMPLOYEE;
---------+---------+---------+---------+---------+---------+---------+---
DSNT408I SQLCODE = -206, ERROR:  EMP_LASTNAME IS NOT VALID IN THE CONTEXT
         WHERE IT IS USED
DSNT418I SQLSTATE    = 42703 SQLSTATE RETURN CODE
DSNT415I SQLERRP     = DSNXORSO SQL PROCEDURE DETECTING ERROR
DSNT416I SQLERRD     = -100 0  0  -1  0  0 SQL DIAGNOSTIC INFORMATION
DSNT416I SQLERRD     = X'FFFFFF9C'  X'00000000'  X'00000000'  X'FFFFFFFF'
         X'00000000'  X'00000000' SQL DIAGNOSTIC INFORMATION
```

There are far too many SQL codes to memorize! I suggest concentrating on the codes listed below. Most developers have run into these at one time or another. They tend to be pretty common.

Common Error SQLCODES

Code	Explanation
-117	THE NUMBER OF VALUES ASSIGNED IS NOT THE SAME AS THE NUMBER OF SPECIFIED OR IMPLIED COLUMNS
-180	THE DATE, TIME, OR TIMESTAMP VALUE value IS INVALID
-181	THE STRING REPRESENTATION OF A DATETIME VALUE IS NOT A VALID DATETIME VALUE
-203	A REFERENCE TO COLUMN column-name IS AMBIGUOUS
-206	Object-name IS NOT VALID IN THE CONTEXT WHERE IT IS USED
-305	THE NULL VALUE CANNOT BE ASSIGNED TO OUTPUT HOST VARIABLE NUMBER position-number BECAUSE NO INDICATOR VARIABLE IS SPECIFIED
-501	THE CURSOR IDENTIFIED IN A FETCH OR CLOSE STATEMENT IS NOT OPEN
-502	THE CURSOR IDENTIFIED IN AN OPEN STATEMENT IS ALREADY OPEN
-803	AN INSERTED OR UPDATED VALUE IS INVALID BECAUSE THE INDEX IN INDEX SPACE indexspace-name CONSTRAINS COLUMNS OF THE TABLE SO NO TWO ROWS CAN CONTAIN DUPLICATE VALUES IN THOSE COLUMNS. RID OF EXISTING ROW IS X record-id
-805	DBRM OR PACKAGE NAME location-name.collection-id.dbrm-name.consistency-token NOT FOUND IN PLAN plan-name. REASON reason-code
-811	THE RESULT OF AN EMBEDDED SELECT STATEMENT OR A SUBSELECT IN THE SET CLAUSE OF AN UPDATE STATEMENT IS A TABLE OF MORE THAN ONE ROW, OR THE RESULT OF A SUBQUERY OF A BASIC PREDICATE IS MORE THAN ONE VALUE
-818	THE PRECOMPILER-GENERATED TIMESTAMP x IN THE LOAD MODULE IS DIFFERENT FROM THE BIND TIMESTAMP y BUILT FROM THE DBRM z
-904	UNSUCCESSFUL EXECUTION CAUSED BY AN UNAVAILABLE RESOURCE. REASON reason-code, TYPE OF RESOURCE resource-type, AND RESOURCE NAME resource-name.
-911	THE CURRENT UNIT OF WORK HAS BEEN ROLLED BACK DUE TO DEADLOCK OR TIMEOUT. REASON reason-code, TYPE OF RESOURCE resource-type, AND RESOURCE NAME resource-name
-913	UNSUCCESSFUL EXECUTION CAUSED BY DEADLOCK OR TIMEOUT. REASON CODE reason-code, TYPE OF RESOURCE resource-type, AND RESOURCE NAME resource-name.
-922	AUTHORIZATION FAILURE: error-type ERROR. REASON reason-code.

Standardizing an Error Routine:

For optimal use of the SQLCODEs returned from DB2, I suggest you create and use a standard error routine in all your embedded SQL programs. I'll provide a model you can use, but first let's create a simple program and force an error to demonstrate the kind of problem resolution information that would be useful.

In this case, let's select a value into a host variable using a fullselect query. But we will make sure the fullselect query encounters more than one row. That will cause a -811 SQLCODE error which we will trap.

```
IDENTIFICATION DIVISION.
PROGRAM-ID. COBEMP5.
*****************************************************
*      PROGRAM USING DB2 SELECT WITH ERROR TO      *
*      DEMONSTRATE COMMON ERROR ROUTINE            *
*****************************************************
ENVIRONMENT DIVISION.
DATA DIVISION.
WORKING-STORAGE SECTION.

01 HV-EMP-VARIABLES.
   10   HV-ID             PIC S9(9) USAGE COMP.
   10   HV-LAST-NAME      PIC X(30).
   10   HV-FIRST-NAME     PIC X(20).
   10   HV-SERVICE-YEARS  PIC S9(9) USAGE COMP.
   10   HV-PROMOTION-DATE PIC X(10).

77 ERR-CNT               PIC S9(9) USAGE COMP.
77 RET-SQL-CODE          PIC -9(4).

   EXEC SQL
     INCLUDE SQLCA
   END-EXEC.

   EXEC SQL
     INCLUDE EMPLOYEE
   END-EXEC.

PROCEDURE DIVISION.

MAIN-PARA.
    DISPLAY "SAMPLE COBOL PROGRAM: COMMON ERROR ROUTINE".

*  LOAD THE EMPLOYEE ARRAY

   EXEC SQL
     SELECT EMP_ID
     INTO :HV-ID
```

453

```
        FROM HRSCHEMA.EMPLOYEE
        WHERE EMP_ID >= 3217

    END-EXEC.


    IF SQLCODE NOT EQUAL 0

        MOVE SQLCODE TO RET-SQL-CODE
        DISPLAY 'ERROR - SQL CODE = ' RET-SQL-CODE.

    STOP RUN.
```

This would generate the following output to SYSPRINT:

```
    SAMPLE COBOL PROGRAM: COMMON ERROR ROUTINE
    ERROR - SQL CODE = -0811
```

Ok, it's good that we trapped the error SQLCODE. But it would be better if DB2 returned the full error description. We can do that if we call a utility program named DSNTIAR. So let's create a common DB2 error subroutine with DSNTIAR so that we can use it in all our DB2 programs.

First, create these working storage variables.

```
    77 ERR-TXT-LGTH          PIC S9(9) USAGE COMP VALUE +72.

    01 ERR-MSG.
        05 ERR-MSG-LGTH      PIC S9(4) COMP VALUE +960.
        05 ERR-MSG-TXT       PIC X(72) OCCURS 12 TIMES
                                       INDEXED BY ERR-NDX.
```

The message and text length variables as well as the SQLCA structure will be passed to DSNTIAR.

Next, create two subroutines as follows. Combined, these subroutines will call DSNTIAR and display the returned output from that utility.

```
    P9999-SQL-ERROR.

        DISPLAY ERR-REC.

        CALL 'DSNTIAR' USING SQLCA,
                             ERR-MSG,
                             ERR-TXT-LGTH.
```

```
        IF RETURN-CODE IS EQUAL TO ZERO

            PERFORM P9999-DISP-ERR
                VARYING ERR-NDX FROM 1 BY 1
                UNTIL ERR-NDX > 12

        ELSE
            DISPLAY 'DSNTIAR ERROR CODE = ' RETURN-CODE
            STOP RUN.


    P9999-DISP-ERR.

        DISPLAY ERR-MSG-TXT(ERR-NDX).

    P9999-DISP-ERR-EXIT.
```

Finally we can modify our code to call the error routine anytime a bad SQL code is returned.

```
EXEC SQL
    SELECT EMP_ID
    INTO :HV-ID
    FROM HRSCHEMA.EMPLOYEE
    WHERE EMP_ID >= 3217

END-EXEC.

IF SQLCODE IS NOT EQUAL TO ZERO
    MOVE SQLCODE TO SQLCODE-VIEW
    MOVE 'EMPLOYEE' TO ERR-TAB
    MOVE 'MAIN'      TO ERR-PARA
    MOVE EMP-ID      TO ERR-DETAIL
    PERFORM P9999-SQL-ERROR.
```

Now when we recompile, bind and rerun the program, the outlook looks like this:

```
SAMPLE COBOL PROGRAM: COMMON ERROR ROUTINE
SQLCODE = -811     EMPLOYEE        MAIN             000000000
 DSNT408I SQLCODE = -811, ERROR:  THE RESULT OF AN EMBEDDED SELECT
         STATEMENT OR A SUBSELECT IN THE SET CLAUSE OF AN UPDATE
         STATEMENT IS A TABLE OF MORE THAN ONE ROW, OR THE RESULT OF A
         SUBQUERY OF A BASIC PREDICATE IS MORE THAN ONE VALUE
 DSNT418I SQLSTATE   = 21000 SQLSTATE RETURN CODE
 DSNT415I SQLERRP    = DSNXREMS SQL PROCEDURE DETECTING ERROR
 DSNT416I SQLERRD    = -140  0  0  -1  0  0 SQL DIAGNOSTIC INFORMATION
 DSNT416I SQLERRD    = X'FFFFFF74'  X'00000000'  X'00000000'
         X'FFFFFFFF'  X'00000000'  X'00000000' SQL DIAGNOSTIC
         INFORMATION
```

455

The SQLCODE details, along with the other information we printed in the ERR-REC (such as the table name and paragraph name) are now displayed to SYSPRINT. This information is more helpful for debugging than just having the SQL code. Moreover, you can create these declarations and routines as copybooks and include them in all your DB2 programs. When used, they ensure standardization throughout your shop. They also will save a lot of time by making useful error-related information available on a consistent basis. I strongly recommend that you implement this standard in your shop!

Here is the entire program listing:

```
       IDENTIFICATION DIVISION.
       PROGRAM-ID. COBEMP5.
      *******************************************************
      *       PROGRAM USING DB2 SELECT WITH ERROR TO        *
      *       DEMONSTRATE COMMON ERROR ROUTINE              *
      *******************************************************

       ENVIRONMENT DIVISION.
       DATA DIVISION.
       WORKING-STORAGE SECTION.

       01 HV-EMP-VARIABLES.
          10   HV-ID               PIC S9(9) USAGE COMP.
          10   HV-LAST-NAME        PIC X(30).
          10   HV-FIRST-NAME       PIC X(20).
          10   HV-SERVICE-YEARS    PIC S9(9) USAGE COMP.
          10   HV-PROMOTION-DATE   PIC X(10).

       01 ERR-REC.
          05 FILLER                PIC X(10) VALUE 'SQLCODE = '.
          05 SQLCODE-VIEW           PIC -999.
          05 FILLER                PIC X(005) VALUE SPACES.
          05 ERR-TAB               PIC X(016).
          05 ERR-PARA              PIC X(015).
          05 ERR-DETAIL            PIC X(040).

       77 ERR-TXT-LGTH             PIC S9(9) USAGE COMP VALUE +72.

       01 ERR-MSG.
          05 ERR-MSG-LGTH          PIC S9(04) COMP VALUE +864.
          05 ERR-MSG-TXT           PIC X(072) OCCURS 12 TIMES
                                             INDEXED BY ERR-NDX.

          EXEC SQL
            INCLUDE SQLCA
          END-EXEC.

          EXEC SQL
```

456

```
        INCLUDE EMPLOYEE
     END-EXEC.

 PROCEDURE DIVISION.

 MAIN-PARA.
     DISPLAY "SAMPLE COBOL PROGRAM: COMMON ERROR ROUTINE".

*   SELECT AN EMPLOYEE

     EXEC SQL
        SELECT EMP_ID
        INTO :HV-ID
        FROM HRSCHEMA.EMPLOYEE
        WHERE EMP_ID >= 3217

     END-EXEC.

     IF SQLCODE IS NOT EQUAL TO ZERO

        MOVE SQLCODE TO SQLCODE-VIEW
        MOVE 'EMPLOYEE' TO ERR-TAB
        MOVE 'MAIN'     TO ERR-PARA
        MOVE EMP-ID     TO ERR-DETAIL
        PERFORM P9999-SQL-ERROR.

     STOP RUN.

 P9999-SQL-ERROR.

     DISPLAY ERR-REC.

     CALL 'DSNTIAR' USING SQLCA,
                   ERR-MSG,
                   ERR-TXT-LGTH.

     IF RETURN-CODE IS EQUAL TO ZERO

        PERFORM P9999-DISP-ERR
           VARYING ERR-NDX FROM 1 BY 1
           UNTIL ERR-NDX > 12

     ELSE
        DISPLAY 'DSNTIAR ERROR CODE = ' RETURN-CODE
        STOP RUN.

 P9999-DISP-ERR.

     DISPLAY ERR-MSG-TXT(ERR-NDX).

 P9999-DISP-ERR-EXIT.
```

Dynamic versus Static SQL

Static SQL

Static SQL statements are embedded within an application program that is written in a traditional programming language such as COBOL or PL/I. The statement is prepared before the program is executed, and the executable statement persists after the program ends. You can use static SQL when you know before run time what SQL statements your application needs to use.

As a practical matter, when you use static SQL you cannot change the form of SQL statements unless you make changes to the program and recompile and bind it. However, you can increase the flexibility of those statements by using host variables. So for example you could write an SQL that retrieves employee information for all employees with X years of service where the X becomes a host variable that you load at run time. Using static SQL and host variables is more secure than using dynamic SQL.

Dynamic SQL

Unlike static SQL which is prepared before the program runs, with dynamic SQL DB2 prepares and executes the SQL statements at run time as part of the program's execution. Dynamic SQL is a good choice when you do not know the format of an SQL statement before you write or run a program. An example might be a user interface that allows a web application to submit SQL statements to a background COBOL program for execution. In this case, you wouldn't know the structure of the statement the client submits until run time.

Applications that use dynamic SQL create an SQL statement in the form of a character string. A typical dynamic SQL application takes the following steps:

- Translates the input data into an SQL statement.

- Prepares the SQL statement to execute and acquires a description of the result table (if any).

- Obtains, for SELECT statements, enough main storage to contain retrieved data.

- Executes the statement or fetches the rows of data.

- Processes the returned information.

- Handles SQL return codes.

Performance Comparison of Static versus Dynamic SQL

Ordinarily static SQL is more efficient than dynamic because the former is prepared and optimized before the program executes. For static SQL statements DB2 typically determines the access path when you bind the plan or package - the exception being if you code REOPT (AL-WAYS) in your bind statement. If you code REOPT (ALWAYS) on a package that has static SQL, DB2 will determine the access path when you bind the plan or package and again at run time using the values of host variables and parameter markers (if included).

For dynamic SQL statements, DB2 determines the access path at run time, when the statement is prepared. The cost of preparing a dynamic statement many times can lead to a performance that is worse than with static SQL. However you can consider these options to improve your performance with dynamic SQL:

1. You can improve performance by caching dynamic statements. To do this, set subsystem parameter CACHEDYN=YES.

2. With dynamic SQL you can also re-optimize your query by using the REOPT bind options. If you are not using the CACHEDYN=YES, you can use the REOPT (ALWAYS) bind option to ensure the best access path. But keep in mind this may slow performance for frequently used dynamic statements.

3. If you are using the CACHEDYN=YES subsystem parameter setting, you can use bind option REOPT (ONCE) and DB2 will only determine the optimal access path the first time the statement is executed. It saves that access path in the dynamic statement cache.

4. If you specify REOPT (AUTO), DB2 will look at any statements with parameter markers and determine whether a new access path might improve performance. If it determines that it would, DB2 will generate a new access path.

To conclude this section, you generally want to use static SQL when you know the structure of your SQL statement and when performance is a significant goal. Use dynamic SQL when you need the flexibility of not knowing the structure of your SQL until run time.

Program Preparation

Before a DB2 program can be run, it must be prepared. Depending on what type of application it is, the programs may need to be pre-compiled, compiled, link-edited and bound. Let's consider each of these steps.

Precompile

Embedded SQL programs (those for which the SQL is embedded in an application program such as COBOL or PL/I) must be precompiled using either the DB2 precompiler or the DB2 coprocessor. The reason is the language compilers such as COBOL do not recognize SQL statements. The precompiler does two things:

- It translates the SQL statements into something that can be compiled.

- It outputs a DBRM (database request module) which is a file that includes all the SQL statements and is used to communicate with DB2.

Compile, link-edit

The program must also be compiled and link-edited to produce an executable load module. DB2 keeps track of the timestamp on the executable module and the timestamp on the DBRM module and these must match or you will receive a -805 SQL error.

Bind

After the precompile, the DBRM must be bound to a package. A package is a compiled version of a DBRM and so it includes the executable versions of SQL statements. You can also specify a collection name when you bind a package. A collection is a group of related packages. Here is a sample BIND PACKAGE statement:

```
BIND PACKAGE(HRSCHEMA) -
MEMBER(COBEMP6)         -
OWNER(HRSCHEMA)          -
QUALIFIER(HRSCHEMA)     -
ACTION(REPLACE)      -
CURRENTDATA(NO)      -
EXPLAIN(NO)          -
ISOLATION(CS)        -
VALIDATE  (BIND)     -
RELEASE   (COMMIT)
```

Packages themselves are not executable without being added to a DB2 plan. Here is a sample BIND PLAN statement:

```
BIND   PLAN     (COBEMP6) -
       PKLIST   (HRSCHEMA.COBEMP6) -
       ACTION   (REP)      -
       ISOLATION (CS)       -
       EXPLAIN  (YES)       -
       VALIDATE (BIND)      -
       RELEASE  (COMMIT)    -
       OWNER    (HRSCHEMA)    -
       QUALIFIER (HRSCHEMA)
```

Non-Embedded SQL Applications

Some application types do not require the precompile, compile/link-edit and bind steps.

- REXX procedures are interpreted and not compiled, so they do not need to be precompiled, compiled/link-edited and bound.

- ODBC applications use dynamic SQL only, so they do not require precompile.

- Java applications containing only JDBC do not need precompile or binding. However Java applications using the SQLJ interface are embedded SQL and they need precompile and bind steps.

Data Concurrency

Isolation Levels & Bind Release Options

Isolation level means the degree to which a DB2 application's activities are isolated from the operations of other DB2 applications. The isolation level for a package is specified when the package is bound, although you can override the package isolation level in an SQL statement. There are four isolation levels: Repeatable Read, Read Stability, Cursor Stability and Uncommitted Read.

ISOLATIONS LEVELS

Repeatable Read (RR)

Repeatable Read ensures that a query issued multiple times within the same unit of work will produce the exact same results. It does this by locking all rows that could affect the result. It does not permit any adds/changes/deletes to the table that could affect the result.

461

Read Stability (RS)

Read Stability locks for the duration of the transaction those rows that are returned by a query, but it allows additional rows to be added to the table.

Cursor Stability (CS)

Cursor Stability only locks the row that the cursor is placed on (and any rows it has updated during the unit of work). This is the default isolation level if no other is specified.

Uncommitted Read (UR)

Uncommitted Read permits reading of uncommitted changes which may never be applied to the database. It does not lock any rows at all unless the row(s) is updated during the unit of work.

An IBM recommended best practice prefers isolation levels in this order:

1. Cursor stability (CS)

2. Uncommitted read (UR)

3. Read stability (RS)

4. Repeatable read (RR)

Of course the chosen isolation level depends on the scenario. We'll look at specific scenarios now.

Isolation Levels for Specific Situations

When your environment is basically read-only (such as with data warehouse environments), use UR **(UNCOMMITTED READ)** because it incurs the least overhead.

If you want to maximize data concurrency without seeing uncommitted data, use the CS **(CURSOR STABILITY)** isolation level. CS only locks the row where the cursor is placed (and any other rows which have been changed since the last commit point), thus maximizing concurrency compared to RR or RS.

If you want no existing rows that were retrieved to be changed by other processes during your unit of work, but you don't mind if new rows are inserted, use RS (**READ STABILITY**).

Finally if you must lock all rows that satisfy the query and also not permit any new rows to be added that could change the result of the query, use RR (**REPEATABLE READ**).

Based on the above, if we wanted to order the isolation levels from most to least impact on performance, the order would be:

1. REPEATABLE READ (RR)

2. READ STABILITY (RS)

3. CURSOR STABILITY (CS)

4. UNCOMMITTED READ (UR)

Finally, in DB2 11 there is a `SKIP LOCKED DATA` clause for the SELECT statement that allows it to bypass any rows that are current locked by other applications. For example:

```
SELECT *
FROM HRSCHEMA.EMP_PAY
SKIP LOCKED DATA;
```

To use `SKIP LOCKED DATA` the application must use either cursor stability (CS) or read stability (RS) isolation level. The `SKIP LOCKED DATA` clause is ignored if the isolation level is uncommitted read (UR) or repeatable read (RR).

How to Specify/Override Isolation Level

To specify an isolation level at bind time, use the ISOLATION keyword with the abbreviated form of the isolation level you want. For example:

```
ISOLATION(CS)
```

If you want to override an isolation level in a query, specify the override at the end of the query by using the WITH <isolation level abbreviation> clause. For example, to override the default isolation level of CS to use UR instead on a query, code the following and notice we've used `WITH UR` at the end of the query:

```
SELECT EMP_ID,
EMP_LAST_NAME,
EMP_FIRST_NAME
FROM EMPLOYEE
ORDER BY EMP_ID
WITH UR;
```

Bind Release Options

The RELEASE bind option determines when any acquired locks are released. The two options are DEALLOCATE and COMMIT. Specifying RELEASE(DEALLOCATE) means the acquired locks will be released when the application session ends. Specifying RELEASE(COMMIT) means locks are released at a commit point. Under TSO this means when a DB2 COMMIT statement is issued. Under IMS a commit occurs when a CKPT or SYNC IMS call is issued. Under CICS a commit occurs when a SYNCPOINT is issued.

As a practical matter, the best concurrency is achieved by using RELEASE(COMMIT) because locks are generally released sooner than the end of the application. However, assuming the program commits frequently, this will result in more processing time than if using RELEASE(DEALLOCATE). So you must weigh your objectives and decide accordingly.

The RELEASE option is only applicable to static SQL statements, i.e., those bound before your program runs. Dynamic SQL statements release the locks at the next commit point.

COMMIT, ROLLBACK, and SAVEPOINTS

Central to understanding transaction management is the concept of a unit of work. A unit of work begins when a program is initiated. Multiple adds, changes and deletes may then take place during the same unit of work. The changes are not made permanent until a commit point is reached. A unit of work ends in one of three ways:

1. When a commit is issued.

2. When a rollback is issued.

3. When the program ends.

Let's look at each of these.

COMMIT

The COMMIT statement ends a transaction and makes the changes permanent and visible to other processes. Also, when a program ends, there is an implicit COMMIT. This is important to know; however an IBM recommended best practice is to do an explicit COMMIT at the end of the program.

Here are some other points about COMMIT to know and remember:

- For an IMS/DB2 program, an IMS **CKPT** call causes a commit of both DB2 and IMS changes made during the unit of work. For CICS, an **EXEC CICS SYNCPOINT** call is made to commit DB2 data.

- The DB2 COMMIT statement does not work in an IMS/DB2 or CICS/DB2 program because in those cases transaction management is performed by the IMS and CICS transaction managers. You won't receive an error for issuing the COMMIT statement, it simply will not work.

- Autonomous procedures were introduced in DB2 11; these procedures run with their own units of work, separate from the calling program.

ROLLBACK

A ROLLBACK statement ends a transaction without making changes permanent – the changes are simply discarded. This is done either intentionally by the application when it determines there is a reason to ROLLBACK the changes and it issues a ROLLBACK explicitly, or because the system traps an error that requires it to do a ROLLBACK of changes. In both cases, the rolled back changes are those that have been made since the last COMMIT point. If no COMMITs have been issued, then all changes made in the session are rolled back.

You can also issue a ROLLBACK TO <savepoint> if you are using SAVEPOINTS. We'll take a look at that shortly.

Here are some other points about ROLLBACK to know and remember:

- The abend of a process causes an implicit ROLLBACK.

- Global variable contents are not affected by ROLLBACK.

SAVEPOINT

The SAVEPOINT statement creates a point within a unit of recovery to which you can roll back changes. This is similar to using ROLLBACK to back out changes since the last COMMIT point, except a SAVEPOINT gives you even more control because it allows a partial ROLLBACK **between** COMMIT points.

You might wonder what the point is of using a SAVEPOINT. Let's take an example. Suppose you have a program that does INSERT statements and you program logic to COMMIT every 500 inserts. If you issue a ROLLBACK, then all updates since the last COMMIT will be backed out. That's pretty straightforward.

But suppose you are updating information for vendors from a file of updates that is sorted by vendor, and if there is an error you want to roll back to where you started updating records for that vendor. And you want all other updates since the last COMMIT point to be applied to the database. This is different than rolling back to the last COMMIT point, and you can do it by setting a new SAVEPOINT each time the vendor changes. Issuing a SAVEPOINT enables you to execute several SQL statements as a single executable block between COMMIT statements. You can then undo changes back out to that savepoint by issuing a ROLLBACK TO SAVEPOINT statement.

Example

Let's do a simple example. First, create a new table and then add some records to the table. We'll create a copy of EMP_PAY.

```
CREATE TABLE HRSCHEMA.EMP_PAY_X
LIKE HRSCHEMA.EMP_PAY
IN TSHR;
```

Now let's add some records. We'll add one record, then create a SAVEPOINT, add another record and then roll back to the SAVEPOINT. This should leave us with only the first record in the table.

```
INSERT INTO EMP_PAY_X
VALUES(1111,
45000.00,
1200.00);

SAVEPOINT A ON ROLLBACK RETAIN CURSORS;

INSERT INTO EMP_PAY_X
VALUES(2222,
55000.00,
```

```
1500.00);

ROLLBACK TO SAVEPOINT A;
```

We can verify that only the first record was added to the table:

```
SELECT * FROM HRSCHEMA.EMP_PAY_X;
---------+---------+---------+---------+---------+----
    EMP_ID  EMP_REGULAR_PAY  EMP_BONUS_PAY
---------+---------+---------+---------+---------+----
      1111         45000.00        1200.00
```

If you have multiple SAVEPOINT S and you ROLLBACK to one of them, then the ROLLBACK will include updates made after any later SAVEPOINT S. Let's illustrate this with an example. We'll set three SAVEPOINT s: A, B and C. We'll add a record, then issue savepoint and we'll do this three times. Then we'll ROLLBACK to the first SAVEPOINT which is A. What we're saying is that any updates made after A will be backed out, which includes the INSERTs made after SAVEPOINTs B and C. Let's try this:

```
INSERT INTO EMP_PAY_X
VALUES(2222,
55000.00,
1500.00);

SAVEPOINT A ON ROLLBACK RETAIN CURSORS;

INSERT INTO EMP_PAY_X
VALUES(3333,
65000.00,
2500.00);

SAVEPOINT B ON ROLLBACK RETAIN CURSORS;

INSERT INTO EMP_PAY_X
VALUES(4444,
75000.00,
2000.00);

SAVEPOINT C ON ROLLBACK RETAIN CURSORS;

ROLLBACK TO SAVEPOINT A;

  SELECT * FROM HRSCHEMA.EMP_PAY_X;
```

```
---------+---------+---------+---------+---------
    EMP_ID  EMP_REGULAR_PAY  EMP_BONUS_PAY
---------+---------+---------+---------+---------
      1111          45000.00         1200.00
      2222          55000.00         1500.00
DSNE610I NUMBER OF ROWS DISPLAYED IS 2
```

Now as you can see, only the first record (2222) was inserted because we specified ROLL-BACK all the way to SAVEPOINT A. Note that the 1111 record was already in the table.

Things to Remember about SAVEPOINT

- If you specify UNIQUE in the SAVEPOINT declaration, you cannot reuse the SAVEPOINT name in the same unit of work.

- If you specify ON ROLLBACK RETAIN CURSORS it means cursors are not closed after a rollback to SAVEPOINT.

- If you specify ON ROLLBACK RETAIN LOCKS this means that any locks acquired after the SAVEPOINT are not released. This is also the default.

- If the SAVEPOINT name is not specified on a ROLLBACK, then all updates back to the last COMMIT point are backed out and all SAVEPOINTs are erased.

Units of Work

A unit of work is a set of database operations in an application that is ended by a commit, a rollback or the end of the application process. A commit or rollback operation applies only to the set of changes made within that unit of work. An application process can involve one or many units of work.

Once a commit action occurs, the database changes are permanent and visible to other application processes. Any locks obtained by the application process are held until the end of the unit of work. So if you update 10 records within one unit of work, the records are all locked until a commit point.

As explained elsewhere, in distributed environments where you update data stores on more than one system, a two-phase commit is performed. The two phase commit ensures that data is consistent between the two systems by either fully commiting or fully rolling back the unit of work. The two phase commit consists of a commit-request phase and an actual commit phase.

Autonomous Transactions

Autonomous Transactions Basics

Autonomous procedures were introduced in DB2 11, so some questions about these transactions are very likely to appear on the exam. Autonomous transactions are native SQL procedures which run with their own units of work, separate from the calling program. If a calling program issues a ROLLBACK to back out its changes, the committed changes of the autonomous procedure are not affected.

Autonomous procedures can be called by normal application programs, other stored procedures, user-defined functions or triggers. Autonomous procedures can also invoke triggers, perform SQL statements, and execute commit and rollback statements.

Restrictions

Be sure to be familiar with these restrictions and limitations on using autonomous procedures:

- Only native SQL procedures can be defined as autonomous.

- Parallelism is disabled for autonomous procedures.

- An autonomous procedure cannot call another autonomous procedure.

- Autonomous procedures cannot see uncommitted changes from the calling application.

- DYNAMIC RESULT SETS 0 must be specified when autonomous procedures are used.

- Stored procedure parameters must not be defined as a LOB data type, or any distinct data type that is based on a LOB or XML value.

- Autonomous procedures do not share locks with the calling application, meaning that the autonomous procedure might timeouts because of lock contention with the calling application.

Autonomous procedures are useful for logging information about error conditions encountered by an application program. Similarly they can be used for creating an audit trail of activity for transactions.

Checkpoint/Restart processing

This section concerns the commit, rollback and recovery of an application or application program. We already covered the use of COMMIT, ROLLBACK and SAVEPOINTs in prior sub-sections. Here we'll apply the COMMIT and ROLLBACK in a DB2 program.

DB2 Program

For the DB2 program, we use the COMMIT statement at appropriate intervals. Let's use our update COBOL program and employ both the COMMIT and the ROLLBACK options. Let's say that we'll commit every 5 records. So we set up a commit counter called COMMIT-CTR and we'll increment it each time we update a record. Once the counter reaches 5 updates, we'll issue a COMMIT statement and reset our record counter to zero. If we perform an update that fails, we'll issue a ROLLBACK.

Note that we also added our generic SQL error handling routine. This will simplify our problem determination in case we encounter an error. Note that you must define the cursor WITH HOLD in order to keep it open when using COMMIT. Otherwise the COMMIT will close the cursor.

```
    IDENTIFICATION DIVISION.
    PROGRAM-ID. COBEMPC.
    ****************************************************
    *       PROGRAM DEMONSTRATING USE OF COMMIT AND      *
    *       ROLLBACK PROCESSING.                         *
    ****************************************************
    ENVIRONMENT DIVISION.
    DATA DIVISION.
    WORKING-STORAGE SECTION.

        EXEC SQL
          INCLUDE SQLCA
        END-EXEC.

        EXEC SQL
          INCLUDE EMPLOYEE
        END-EXEC.

        EXEC SQL
            DECLARE EMP-CURSOR CURSOR WITH HOLD FOR
            SELECT EMP_ID, EMP_LAST_NAME
            FROM HRSCHEMA.EMPLOYEE
            WHERE EMP_LAST_NAME <> UPPER(EMP_LAST_NAME)
            FOR UPDATE OF EMP_LAST_NAME
        END-EXEC.

    01 COMMIT-CTR    PIC S9(9) USAGE COMP  VALUE 0.

    01 ERR-REC.
```

470

```
    05  FILLER              PIC X(10) VALUE 'SQLCODE = '.
    05  SQLCODE-VIEW         PIC -999.
    05  FILLER              PIC X(005) VALUE SPACES.
    05  ERR-TAB             PIC X(016).
    05  ERR-PARA            PIC X(015).
    05  ERR-DETAIL          PIC X(040).

77 ERR-TXT-LGTH            PIC S9(9) USAGE COMP VALUE +72.

01 ERR-MSG.
    05  ERR-MSG-LGTH        PIC S9(04) COMP VALUE +864.
    05  ERR-MSG-TXT         PIC X(072) OCCURS 12 TIMES
                                        INDEXED BY ERR-NDX.

PROCEDURE DIVISION.

MAIN-PARA.
    DISPLAY "SAMPLE COBOL PROGRAM: UPDATE USING CURSOR".

    EXEC SQL
        OPEN EMP-CURSOR
    END-EXEC.

    IF SQLCODE NOT EQUAL 0
       PERFORM P9999-SQL-ERROR

    DISPLAY 'OPEN CURSOR SQLCODE: ' SQLCODE.

    PERFORM FETCH-CURSOR
      UNTIL SQLCODE NOT EQUAL 0.

    EXEC SQL
        CLOSE EMP-CURSOR
    END-EXEC.

    IF SQLCODE NOT EQUAL 0
       PERFORM P9999-SQL-ERROR

    DISPLAY 'CLOSE CURSOR SQLCODE: ' SQLCODE.

    STOP RUN.

FETCH-CURSOR.

    EXEC SQL
        FETCH EMP-CURSOR INTO :EMP-ID, :EMP-LAST-NAME
    END-EXEC.

    IF SQLCODE = 0
       DISPLAY 'BEFORE CHANGE  ', EMP-LAST-NAME
       MOVE FUNCTION UPPER-CASE (EMP-LAST-NAME)
```

471

```
                TO EMP-LAST-NAME
        EXEC SQL
            UPDATE HRSCHEMA.EMPLOYEE
            SET EMP_LAST_NAME = :EMP-LAST-NAME
            WHERE CURRENT OF EMP-CURSOR
        END-EXEC

    END-IF.

    IF SQLCODE = 0
        DISPLAY 'AFTER CHANGE    ', EMP-LAST-NAME
        ADD +1 TO COMMIT-CTR
        IF COMMIT-CTR >= 5
            EXEC SQL
                COMMIT
            END-EXEC
            MOVE ZERO TO COMMIT-CTR
        ELSE
            NEXT SENTENCE
        END-IF
    ELSE
        PERFORM P9999-SQL-ERROR
        EXEC SQL
            ROLLBACK
        END-EXEC
        GOBACK
    END-IF.

P9999-SQL-ERROR.

    DISPLAY ERR-REC.

    CALL 'DSNTIAR' USING SQLCA,
                   ERR-MSG,
                   ERR-TXT-LGTH.

    IF RETURN-CODE IS EQUAL TO ZERO

        PERFORM P9999-DISP-ERR
            VARYING ERR-NDX FROM 1 BY 1
            UNTIL ERR-NDX > 12

    ELSE
        DISPLAY 'DSNTIAR ERROR CODE = ' RETURN-CODE
        STOP RUN.

P9999-DISP-ERR.

    DISPLAY ERR-MSG-TXT(ERR-NDX).

P9999-DISP-ERR-EXIT.
```

IMS Programs

For an IMS/DB2 program, you must issue the IMS CKPT call to commit DB2 data (as well as IMS data if you are updating IMS databases). Here are some important things to know about DB2 updates in an IMS program:

1. In an IMS program, the DB2 COMMIT statement will **not** commit DB2 changes.

2. IMS/DB2 will not tell you that your DB2 COMMIT statement didn't work – it will not generate an error. The COMMIT statement will simply have no effect.

3. You must use the IMS CKPT statement to commit both IMS and DB2 data.

4. Similarly, if you want to back out uncommitted DB2 changes, the DB2 ROLLBACK statement will not work. You must use the IMS ROLL or ROLB statements.

5. ROLB means that any changes are backed out to the last checkpoint, and then control is returned to the calling program which can continue processing. ROLL means that any changes are backed out to the last checkpoint, and then the program is terminated with abend code U0778.

Testing & Validating Results

You obviously need to test and validate results for new and modified applications. This section is both a reminder to perform structured testing and some hints for how to go about it. If you have been an application developer for very long, most or all of this will not be new to you. Still, this is basic training and these are vitally important principles that must be followed to ensure high quality testing and validation.

Test Structures

You normally have a test environment which is typically a separate instance of DB2 and is used primarily or exclusively for testing. You or your DBA will create test objects (tables, indexes, views) in the test environment. The DDL is usually saved and then modified as necessary to recreate the same object in another DB2 instance, or to drop and create a new version in the same instance.

If you do both production support and new development activities in the same test environment (not recommended but it's the case in many shops), it is important to have a strategy for dealing with these work flows so they don't impact each other. For example if you add a column to a table in your test system and change a copybook, that may be fine for the development work. But if someone else is changing the same program (that obviously uses the same copybook) to resolve a production problem, they may inadvertently move something incorrect to production. So coordination is necessary and essential when sharing a test environment.

Test Data

It is vital to know the business rules of the application so that you select a robust set of test data for your application. To test successfully, all branches of your program or package must be tested, and then all components must be tested together (integration testing). It is especially important that in addition to testing new code and SQL, you also perform regression testing on existing code. Take the time to develop a good structured, comprehensive test plan that can be reused many times.

You can create brand new test data, or you can extract data from production and load it to test, or you can do both. Here are a few basic methods for loading test data.

1. Create the data in a flat file and write an application program to load it using `INSERT` statements.

2. Extract the data from production using a utility such as `DSNTIAUL`, then load it using an application program as in method 1.

3. Use the DB2 UNLOAD and LOAD utilities to extract data from production and load to test. We'll go over details for this later in this section.

Testing SQL Statements

Before coding SQL statements in an application program, you should test them using SPUFI. If you have been using DB2 for z/OS for any length of time you are probably familiar with SPUFI. If you are not familiar, here is a sample query executed from SPUFI.

From the ISPF main menu select the DB2 option.

```
   Menu  Utilities  Compilers  Options  Status  Help

                           ISPF Primary Option Menu
 Option ===>

 0  Settings      Terminal and user parameters      User ID . : HRDEV1
 1  View          Display source data or listings   Time. . . : 21:19
 2  Edit          Create or change source data      Terminal. : 3278
 3  Utilities     Perform utility functions         Screen. . : 1
 4  Foreground    Interactive language processing   Language. : ENGLISH
 5  Batch         Submit job for language processing Appl ID . : ISR
 6  Command       Enter TSO or Workstation commands  TSO logon : MATPROC
 7  Dialog Test   Perform dialog testing            TSO prefix: HRDEV1
 10 SCLM          SW Configuration Library Manager   System ID : MATE
 11 Workplace     ISPF Object/Action Workplace       MVS acct. : MT529
 12 DITTO         DITTO/ESA for MVS                  Release . : ISPF 6.0
 13 FMN           File Manager
 15 DB2           DB2 Primary Menu
 17 QMF           DB2 Query Management Facility
 S  SDSF          Spool Search and Display Facility

        Enter X to Terminate using log/list defaults
```

From the DB2 Interactive Primary Option menu, select option 1 (SPUFI).

```
                            DB2I PRIMARY OPTION MENU           SSID: DB2X
COMMAND ===>

Select one of the following DB2 functions and press ENTER.

    1  SPUFI                 (Process SQL statements)
    2  DCLGEN                (Generate SQL and source language declarations)
    3  PROGRAM PREPARATION   (Prepare a DB2 application program to run)
    4  PRECOMPILE            (Invoke DB2 precompiler)
    5  BIND/REBIND/FREE      (BIND, REBIND, or FREE plans or packages)
    6  RUN                   (RUN an SQL program)
    7  DB2 COMMANDS          (Issue DB2 commands)
    8  UTILITIES             (Invoke DB2 utilities)
    D  DB2I DEFAULTS         (Set global parameters)
    Q  QMF                   (Query Management Facility
    X  EXIT                  (Leave DB2I)




PRESS:                      END to exit      HELP for more information
```

Now enter the data set name where you will code and save your commands (DDL, DML or DCL), as well as the dataset to capture your output.

```
                        SPUFI                          SSID: DB2X
  ===>

Enter the input data set name:       (Can be sequential or partitioned)
  1  DATA SET NAME ... ===> 'HRDEV1.SPUFI.CNTL(EXECSQL)'
  2  VOLUME SERIAL ... ===>         (Enter if not cataloged)
  3  DATA SET PASSWORD ===>         (Enter if password protected)

Enter the output data set name:      (Must be a sequential data set)
  4  DATA SET NAME ... ===> 'HRDEV1.SPUFI.OUT'

Specify processing options:
  5  CHANGE DEFAULTS   ===> NO      (Y/N - Display SPUFI defaults panel?)
  6  EDIT INPUT ...... ===> YES     (Y/N - Enter SQL statements?)
  7  EXECUTE ......... ===> YES     (Y/N - Execute SQL statements?)
  8  AUTOCOMMIT ...... ===> YES     (Y/N - Commit after successful run?)
  9  BROWSE OUTPUT ... ===> YES     (Y/N - Browse output data set?)

For remote SQL processing:
 10  CONNECT LOCATION  ===>


PRESS:  ENTER to process    END to exit            HELP for more information
```

When you press enter, you will be in EDIT mode with the input dataset you specified. In this case we will code a query to select all the EMPLOYEE **records:**

```
  File  Edit  Edit_Settings  Menu  Utilities  Compilers  Test  Help


EDIT       HRDEV1.SPUFI.CNTL(EXECSQL) - 01.00          Columns 00001 00072
Command ===>                                            Scroll ===> PAGE
****** *************************** Top of Data ***************************
000001 SELECT * FROM EMPLOYEES;
****** *************************** Bottom of Data ************************
```

Press PF3 to save your code, then press ENTER to execute the query. Here are our results:

```
  Menu  Utilities  Compilers  Help
SSSSSSSSSSSSSSSSSSSSSSSSSSSSSSSSSSSSSSSSSSSSSSSSSSSSSSSSSSSSSSSSSSSSSSSSSS
BROWSE    HRDEV1.SPUFI.OUT                        Line 00000000 Col 001 080
Command ===>                                            Scroll ===> PAGE
***************************** Top of Data ***************************
--------+---------+---------+---------+---------+---------+---------+-----
SELECT * FROM EMPLOYEE;
--------+---------+---------+---------+---------+---------+---------+-----
    EMPNO  NAME
--------+---------+---------+---------+---------+---------+---------+-----
      100  SMITH
      200  JONES
DSNE610I NUMBER OF ROWS DISPLAYED IS 2
DSNE616I STATEMENT EXECUTION WAS SUCCESSFUL, SQLCODE IS 100
--------+---------+---------+---------+---------+---------+---------+---------+
--------+---------+---------+---------+---------+---------+---------+---------+
DSNE617I COMMIT PERFORMED, SQLCODE IS 0
DSNE616I STATEMENT EXECUTION WAS SUCCESSFUL, SQLCODE IS 0
--------+---------+---------+---------+---------+---------+---------+---------+
DSNE601I SQL STATEMENTS ASSUMED TO BE BETWEEN COLUMNS 1 AND 72
DSNE620I NUMBER OF SQL STATEMENTS PROCESSED IS 1
DSNE621I NUMBER OF INPUT RECORDS READ IS 2
DSNE622I NUMBER OF OUTPUT RECORDS WRITTEN IS 18
***************************** Bottom of Data ***************************
```

Debugging Programs

When you are getting unexpected results from your program, make sure to review these items from your compiler listing. I'd had many problems referred to me where the problem could clearly be found in the output, but the developer did not look closely enough.

- **Output from the precompiler** – check for errors and warnings.   Resolve any that appear. Sometimes a host variable will be undefined or improperly defined for the

DB2 column you are trying to select into it. Suppose I specify host variable HV-ID in a select query but I forgot to actually define it? When that happens you will see something like this in the precompiler output:

```
DB2 SQL PRECOMPILER       MESSAGES
DSNH312I E      DSNHSMUD LINE 87 COL 21  UNDEFINED OR UNUSABLE HOST
VARIABLE "HV-ID"
```

• **Output from the language compiler.** If you receive any COBOL, PL/I or assembler errors or warnings, make sure to resolve them. These are typically the easiest to understand and correct because the compiler tells you the exact statement number and the error.

Suppose for example that I inadvertently defined a record counter variable REC-COUNT as PIC X(3). Then later in the program I try to increment it with the statement:

```
    ADD +1 TO REC-COUNT.
```

This would cause the following compiler error:

```
PP 5655-S71 IBM Enterprise COBOL for z/OS  4.2.0              COBEMPZ
Date 0
LineID  Message code  Message text
  186  IGYPA3074-S   "REC-COUNT (ALPHANUMERIC)" was not numeric, but was
a sender in an arithmetic expression.  The statement was discarded.
```

• **Output from the linkage editor.** If you are using standard compile JCL for your shop, linkage editor errors will be rare. However, do make sure you don't have any unresolved references. Suppose for example you are compiling the DB2 connection program DSNULI but in the linkage editor step it is misspelled as DSNULX. In this case you will get a linkage editor error as follows:

```
    IEW2278I B352 INVOCATION PARAMETERS - MAP,XREF
    IEW2322I 1220  1     INCLUDE SYSLIB(DSNULX)
    IEW2303E 1030 MEMBER DSNULX OF THE DATA SET SPECIFIED BY SYSLIB COULD NOT BE
    FOUND.
```

• **Output from the bind process**. Did you have any error messages? How about warning messages? Bind errors must be resolved or you will typically receive a -805 SQL code when you try to run the program. Here is a case where a bind is being attempted for a package that does not exist.

```
BIND   MEMBER     (COBEMPY)       PLAN      (COBEMPY)      ACTION     (RE-
PLACE)
       VALIDATE   (BIND)          RELEASE    (COMMIT)        OWNER
(HRSCHEMA)
DSNT230I  -DB2X BIND DBRM-MEMBER-NAME ERROR
          USING HRSCHEMA AUTHORITY
          PLAN=COBEMPY
          MEMBER COBEMPY NOT FOUND IN PDS SEARCH ORDER
DSNT201I  -DB2X  BIND FOR PLAN COBEMPY  NOT SUCCESSFUL
```

In summary, when you encounter an error running a program, make sure you have thoroughly checked your outputs from the pre-compile, language compile, link-edit and bind steps. These almost always provide you with the information you need to diagnose the problem. Even if you end up asking for help, you should gather the available diagnostic information to show your colleague what you have looked at so far to resolve the problem.

Testing Connections and Stored Procedures

How to Set up a Data Source
Here are the basic instructions for how to set up a data source to use for testing.

.NET
For .NET you can set up a data source using any of three providers:

1. OLE DB
2. ODBC
3. IBM Data Server Provider for .NET

While you can use any of the three .NET providers above, IBM recommends using the IBM Data Server Provider for .NET.

OLE DB
A connection string sample for OLE DB written in C# for the HR database is as follows, as assume that :

```
OleDbConnection con = new OleDbConnection("Provider=IBMDADB2;" +
              "Data Source=DBHR;UID=HRUSER01;PWD=<your password>;" );
con.Open()
```

ODBC

The ODBC .NET provider uses the same connection parameters as the CLI interface (see example below under ODBC). Here is a sample connection in C# language for the HR database:

```
OdbcConnection con = new OdbcConnection("DSN=hrDSN;UID=HRUS-
ER01;PWD=<your password>;");
con.Open()
```

IBM Data Server Provider for .NET

To use the IBM Data Server Provider for .NET, you will need to create a connection string that includes the database and login credentials. Here is a sample connection string for the HR database.

```
String cs = "Database=DBHR;UID=HRUSER01;PWD=<your password>";
DB2Connection conn = new DB2Connection(connectString);
conn.Open();
return conn;
```

JAVA

For Java, you load the DB2 driver and then establish a connection using the get Connection method. Here is sample code to load the driver:

```
try {
  // Load the IBM Data Server Driver for JDBC and SQLJ with
  // DriverManager

  Class.forName("com.ibm.db2.jcc.DB2Driver");
} catch (ClassNotFoundException e) {
    e.printStackTrace();
}
```

And this code will establish the connection:

```
String url = "jdbc:db2://HOUSTON1:5021/DBHR:" +
"user=HRUSER01;password=PASS3454;";

Connection con = DriverManager.getConnection(url);
```

ODBC

You will need to install the IBM Data Server Driver for ODBC and CLI. From there you can specify configuration information in one of three methods:

1. Specify in the connect string when using the SQLDriverConnect function.
2. Store configuration parameters in the db2cli.ini file.
3. Store configuration parameters in the db2dsdriver.cfg file.

If you store the connection parameters in the db2dsdriver.cfg file, create an XML structure like this:

```
<configuration>
  <dsncollection>
    <dsn alias="hrDSN" name="hrConnect" host="HRserver.domain.com"
port="446">
    </dsn>
  </dsncollection>
  <databases>
    <database name="DBHR" host=" HRserver.domain.com " port="446">
      <parameter name="CommProtocol" value="TCPIP"/>
      <parameter name="UID" value="HRUSER01"/>
    </database>
  </databases>
</configuration>
```

Now you can set pass the connection string in SQLDriverConnect() as:

```
DSN=hrDSN;PWD=<your password>:
```

How to Test a Connection

The CONNECT statement is used to connect a user or application to a DB2 database server. The basic syntax for the CONNECT statement is:

```
CONNECT TO <server-name> USER <user id> USING <password>
```

The server or the local DB2 subsystem checks the authorization ID and password to verify that the user is authorized to connect to the server.

Local Access

Local DB2 access (meaning you are logged onto the same machine where you bind your packages) is typically governed by local security. If you are logged onto TSO then you have already been authenticated by the z/OS security subsystem. If your TSO id is used as your primary DB2 authorization id (typically it is) then RACF checks to make sure you are authorized to access DB2. If you are, then a connection occurs.

Remote Access

A remote server can be either another instance of DB2 for z/OS or it can be an instance of another product. The server name location must be in the SYSIBM.LOCATIONS table on the local system. The LINKNAME in the SYSIBM.LOCATIONS table corresponds to LINKNAME in the SYSIBM.IPNAMES table which includes the IPADDR (IP address) of the remote system.

There are two ways to connect to a remote server:

1. Using an explicit CONNECT statement.
2. Using three part naming.

We'll look at examples of each.

Connecting Via Explicit CONNECT Statement

The application can connect to a server based on the location name in the CONNECT statement. For example, suppose we want to run a select statetment against an HREMP table owned by schema HRGROUP which resides on a server whose LOCATION value in the locations table is HOUSTON1. We also have a user id and password (HR001 and ACDVXZ84) that has authority to access the remote system. If our local id does not have access to the remote system, we can expicitly connect using these credentials. Our first example will do that.

To accomplish our task in an embedded SQL program, we would code the following:

```
EXEC SQL
   CONNECT TO HOUSTON1 USER HR001 USING ACDVXZ84;
   SELECT * FROM HRGROUP.HREMP
   WHERE EMP_ID = 3217;
```

Similarly you can call a stored procedure GET_EMPLOYEE (provided it is bound on the remote server) by coding the following:

```
EXEC SQL
   CONNECT TO HOUSTON1 USER HR001 USING ACDVXZ84;

EXEC SQL
   CALL GET_EMPLOYEE (3217);
```

Connection Via Three Part Names

A three-part name can be used consisting of a location that uniquely identifies the remote server, an AUTHORIZATION ID that identifies the owner of the object, and the OBJECT name that identifies the object at the location that you want to access. When using three part naming an implicit CONNECT takes place provided the remote server is defined in the SYS-IBM.LOCATIONS table, and that the user has security to connect to that remote server. Int his case we will assume the user does have the connect privilege on the remote server.

Let's use the same example of running a select statement against an HREMP table owned by schema HRGROUP which resides on a server whose LOCATION is HOUSTON1.

```
EXEC SQL
   SELECT * FROM HOUSTON1.HRGROUP.HREMP
   WHERE EMP_ID = 3217;
```

As you can see, you only need to add the location name in front of the schema/table name. The connect takes place implicitly.

Finally, you can create an ALIAS for the object using the three part name, and your SQL can then reference the ALIAS instead of the three part name. For example:

```
CREATE ALIAS EMPTBLH FOR HOUSTON1.HRGROUP.HREMP;
```

Now you can write your select as:

```
SELECT * FROM EMPTBLH;
```

Things to remember:

- When you connect from your local DB2 subsystem to a remote DB2 subsystem, the remote server location name **mus**t reside in SYSIBM.LOCATIONS table on your local DB2 subsystem.

- When you connect successfully to a remote server, the location name of the remote server is placed in the CURRENT SERVER special register.

- Before you can execute a package on a remote server, you must bind the package on the remote server, and connect to the remote server (either explicitly with a connect statement or using three part naming).

- Issuing a CONNECT statement with no operand is a special form of CONNECT. It returns information about the current server in the SQLERRP field of the SQLCA. Note: SQLERRP returns blanks if the application process is in an unconnected state.

Performance Guidelines for Remote Data Access

The following are some IBM recommended guidelines to improve performance when accessing distributed data:

- For read-only queries, bind package with DBPROTOCOL(DRDACBF).

- For read-only queries, specify FOR READ ONLY (or FOR FETCH ONLY) in SQL.

- Exclude any unnecessary columns from the query.

- Use FETCH FIRST X ROWS ONLY if you only need a subset of the rows that would otherwise be returned by a query.

- FOR ODBC and JDBC, specify the number or rows you want to retrieve in the rowset parameter.

- When retrieving LOB data, set the CURRENT RULES special register to DB2.

How to Test a Stored Procedure

There are several ways you can test a stored procedure. Unfortunately SPUFI is not one of them. Here are some choices: Application Program, Rexx, QMF and Data Studio.

Application Program

You can put together a program to call the stored procedure. We did this in previous examples. Refer to the program listings above for COBEMP7 and COBEMP8 as examples.

Rexx

You can create a Rexx procedure such as the below to test a stored procedure. The load library names and subsystem name below must be changed to whatever the correct library names and subsystem name for your system.

```
/* REXX */
SUBSYS = "DB2X"
address TSO
  "FREE  FI(STEPLIBX) DA('DSNTST.SDSNLOAD')"
  "ALLOC FI(STEPLIBX) DA('DSNTST.SDSNLOAD') SHR REUSE"
  if rc <> 0
  then do
       say "DB2 SDSNLOAD library for SSID="ssid" is not available!"
       say "Check z/OS LINKLIST or allocate to STEPLIB in advance!"
       signal error
```

```
    end
ADDRESS TSO "SUBCOM DSNREXX" /* HOST CMD ENV AVAILABLE ? */
IF RC <> 0 THEN S_RC = RXSUBCOM('ADD','DSNREXX','DSNREXX')
say 'About to connect...'
ADDRESS DSNREXX "CONNECT" SUBSYS
IF SQLCODE <> 0 THEN CALL SQLCA
say 'About to call SP...'
ADDRESS DSNREXX
/* Identify Stored procedure, define host variables */
STOPRO = "HRSCHEMA.GETEMP"
EMP_ID = +3217
EMP_LAST_NAME = ''
EMP_FIRST_NAME = ''
/* call the stored procedure  */
ADDRESS DSNREXX
"EXECSQL CALL :STOPRO(:EMP_ID,:EMP_LAST_NAME,:EMP_FIRST_NAME)"
IF SQLCODE <> 0 THEN CALL SQLCA
say SQLCODE
IF SQLCODE = 0 THEN
SAY "Stored Procedure: " GETEMP " was successful"
SAY "EMP_LAST_NAME= " EMP_LAST_NAME
SAY "EMP_FIRST_NAME= " EMP_FIRST_NAME
```

The output is displayed as follows:

```
Connected...
About to call SP...
0
Stored Procedure:  GETEMP  was successful
EMP_ID       =   3217
EMP_LAST_NAME=   JOHNSON
EMP_FIRST_NAME=  EDWARD
***
```

485

QMF

You can use QMF to run a stored procedure but it's use is limited as you cannot get any OUT variables displayed.

```
SQL QUERY                                          MODIFIED  LINE    1

CALL HRSCHEMA.GETEMP (&A01, &B01, &C01)




*** END ***


1=Help        2=Run         3=End          4=Print     5=Chart      6=Draw
7=Backward    8=Forward     9=Form         10=Insert   11=Delete    12=Report
OK, cursor positioned.
COMMAND ===>                                               SCROLL ===> PAGE
```

When you press the PF2 key to run the stored procedure you will be prompted for the parameter values. Enter the values and press enter.

```
SQL QUERY                                           MODIFIED  LINE    1
+---------------------------------------------------------------------------+
|                 RUN Command Prompt - Values of Variables                  |
|                                                                           |
| Your RUN command runs a query or procedure with variables that need       |
| values. Fill in a value for each variable named below:                    |
|                                                             1  to 10 of 10 |
| &A01             3217                                                      |
| &B01             NULL                                                      |
| &C01             NULL                                                      |
|                                                                           |
|                                                                           |
|                                                                           |
|                                                                           |
|                                                                           |
+---------------------------------------------------------------------------+
| F1=Help  F3=End  F7=Backward  F8=Forward                                  |
+---------------------------------------------------------------------------+


    Please give a value for each variable name.
```

486

Next we get a panel that says our stored procedure is successful. However, QMF cannot return the out parameters, so we have no output. This somewhat limits the usefulness of running stored procedures under QMF. You can however usefully execute queries under QMF where the output from the query is not needed, such as in update queries. We'll do an example of one on the next page.

```
QMF HOME PANEL                   Query      Management      Facility
Version 10 Release 1
                                 ******    **    **       ********    ____
Authorization ID                  **    **    ***   ***      **       ____
  HRSCHEMA                          **    **  ****  ****      *******  ____
                            **    **    **  **  **  **         **
Connected to                **   * **    **  ****  **    **
  LOCRGNA                    ******    **    ***    **   **            ____
                               **                                    _____
                                     http://www.ibm.com/qmf
Enter a command on the command line or press a function key.
For help, press the Help function key or enter the command HELP.

1=Help        2=List      3=End       4=Show      5=Chart      6=Query
7=Retrieve    8=Edit Table 9=Form     10=Proc     11=Profile   12=Report
OK, Your Stored Procedure has successfully completed.
COMMAND ===>
```

487

Let's create a stored procedure that updates the employee's years of service. Our IN parameters will be the employee id and the years of service:

```
CREATE PROCEDURE HRSCHEMA.UPDEMP
(IN EMP_NO INT, IN YRSSRV INT)

LANGUAGE SQL
MODIFIES SQL DATA

 BEGIN
    UPDATE HRSCHEMA.EMPLOYEE
    SET EMP_SERVICE_YEARS = YRSSRV
    WHERE EMP_ID = EMP_NO;
 END


---------+---------+---------+---------+---------+---------+-
DSNE616I STATEMENT EXECUTION WAS SUCCESSFUL, SQLCODE IS 0
```

Now let's move to QMF and run a query to see the current value of years of service for employee. And then we'll press PF2 to run the query.

```
SQL QUERY                                    MODIFIED   LINE    1

SELECT EMP_ID, EMP_SERVICE_YEARS
FROM HRSCHEMA.EMPLOYEE
WHERE EMP_ID = 3217;
```

```
*** END ***




1=Help       2=Run       3=End       4=Print    5=Chart      6=Draw
7=Backward   8=Forward   9=Form      10=Insert  11=Delete    12=Report
OK, cursor positioned.
COMMAND ===>                                         SCROLL ===> PAGE
```

Our query output will look like this:

```
REPORT                                      LINE 1       POS 1        79


                       EMP
            EMP      SERVICE
            ID        YEARS
        -----------  -----------
            3217           4


  *** END ***




 1=Help        2=            3=End        4=Print       5=Chart         6=Query
 7=Backward    8=Forward     9=Form      10=Left       11=Right        12=
OK, this is the REPORT from your RUN command.
 COMMAND ===>                                           SCROLL ===> PAGE
```

Then, let's press PF6 again to create a new query to call the UPDEMP stored procedure and we'll specify 5 years as the value we want to update for employee 3217. Since we are only using IN parameters, we can simply specify these values in parentheses.

Code the following and then press PF2 to execute.

```
SQL QUERY                                    MODIFIED  LINE    1

CALL HRSCHEMA.UPDEMP (3217, 5)







  *** END ***




 1=Help        2=Run         3=End        4=Print      5=Chart        6=Draw
 7=Backward    8=Forward     9=Form      10=Insert    11=Delete      12=Report
OK, cursor positioned.
 COMMAND ===>                                           SCROLL ===> PAGE
```

You'll then see the main menu with the message that the procedure has successfully executed.

```
QMF HOME PANEL                     Query      Management     Facility
Version 10 Release 1
                                   ******    **    **      ********      ____
Authorization ID                   **     **   ***   ***      **         ____
  HRSCHEMA                          **    **   ****  ****   *******       ____
                                 **    **   ** ** ** **    **             ____
Connected to                     **   *  **   **  ****  **    **          ____
  LOCRGNA                         ******    **    ***   **  **            _____
                                      **
                                           http://www.ibm.com/qmf
Enter a command on the command line or press a function key.
For help, press the Help function key or enter the command HELP.
```

```
1=Help        2=List       3=End       4=Show      5=Chart      6=Query
7=Retrieve    8=Edit Table 9=Form      10=Proc     11=Profile   12=Report
```
OK, Your Stored Procedure has successfully completed.
```
COMMAND ===>
```

Finally, let's confirm that the years of service value is now 5 instead of 4. Press PF6 to get back to the query screen and enter this query:

```
SQL QUERY                                    MODIFIED  LINE   1

SELECT EMP_ID, EMP_SERVICE_YEARS
FROM HRSCHEMA.EMPLOYEE
WHERE EMP_ID = 3217;




*** END ***



1=Help        2=Run        3=End       4=Print     5=Chart      6=Draw
7=Backward    8=Forward    9=Form      10=Insert   11=Delete    12=Report
OK, cursor positioned.
COMMAND ===>                                        SCROLL ===> PAGE
```

Now press PF2 to get the query result.

And our result shows that the stored procedure did in fact change the value of the years of service from 4 to 5.

```
REPORT                                          LINE 1      POS 1      79


                        EMP
            EMP       SERVICE
            ID         YEARS
        -----------  -----------
            3217              5


*** END ***




1=Help          2=              3=End        4=Print       5=Chart        6=Query
7=Backward      8=Forward       9=Form       10=Left       11=Right       12=
OK, this is the REPORT from your RUN command.
COMMAND ===>                                         SCROLL ===> PAGE
```

So in some cases it is convenient to test stored procedures using QMF. Generally, as long as you don't use or need output parameters, you can test successfully with QMF.

Data Studio

Using data studio you can run the stored procedure by clicking Stored Procedures in the object tree, right clicking on the name of the stored procedure – in our case GETEMP – and then clicking on **RUN**.

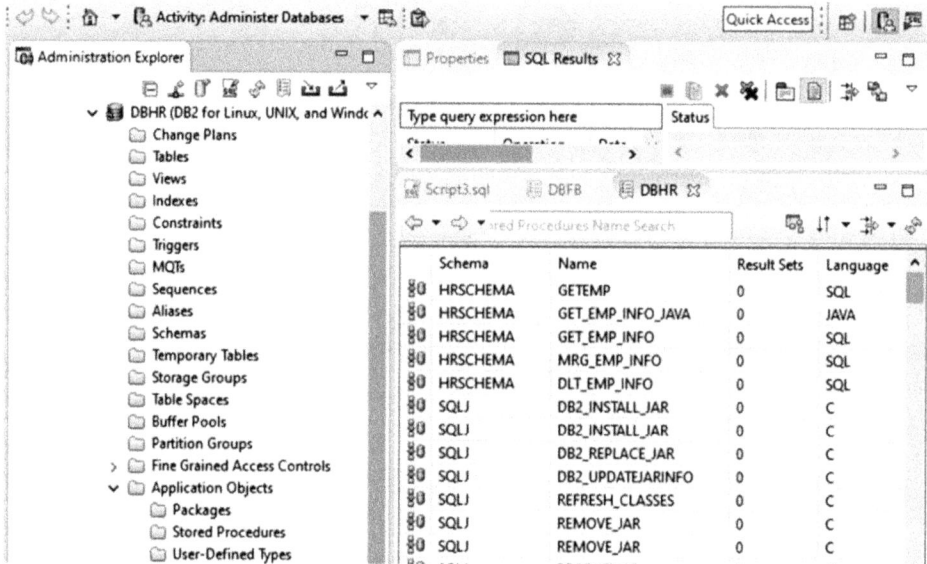

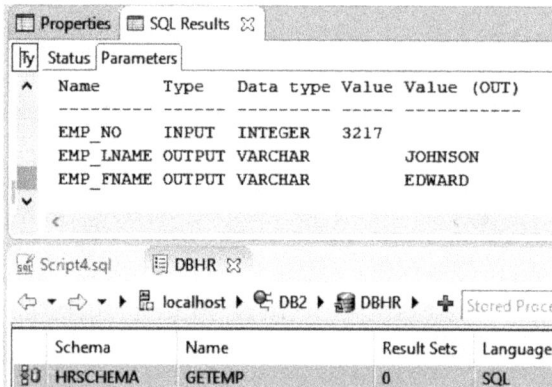

DB2 LOAD and UNLOAD Utilities

UNLOAD

Use the UNLOAD utility to copy the contents of a table into a flat file. This example copies the EMP_PAY table to file HRSCHEMA.EMPPAY.UNLOAD.

```
UNLOAD DATA FROM TABLE HRSCHEMA.EMP_PAY
```

If you created the JCL from the online panels, you will see this DD for the output. If you are reusing JCL that does not include it, you must add the below DD to your JCL.

```
//DSNUPROC.SYSREC DD DSN=HRSCHEMA.EMPPAY.UNLOAD,
//      DISP=(MOD,CATLG),
//      SPACE=(16384,(20,20),,,ROUND),
```

You can keep the unload output in internal DB2 format by specifying FORMAT INTERNAL. If you want the output to be in another format you can specify EBCDIC, ASCII or UNICODE. Also if you want the output file to be delimited, you can specify DELIMIT.

If you want a subset of the data in the table, you can specify this using either the SAMPLE or WHEN clauses. The first example below unloads the first 100 records from EMP_PAY. The second example pulls all records for which the EMP_ID is greater than 3217.

```
UNLOAD DATA
FROM TABLE HRSCHEMA.EMP_PAY
SAMPLE 100

UNLOAD DATA
FROM TABLE HRSCHEMA.EMP_PAY
WHEN (EMP_ID > 3217)
```

The following are the phases of the unload utility.

- UTILINIT   Performs initialization.
- UNLOAD   Unloads records to sequential data sets.
- UTILTERM   Performs cleanup.

LOAD

Use the LOAD utility when you want to load a DB2 table from an input file. This is useful when you are working in a test system and need to restore a baseline set of data. The input file can be in DB2 format or one of the other supported formats.

The load command can be taken from the SYSPUNCH file that was created when the table was unloaded. To take a simple example, let's use the file we unloaded from EMP_PAY and reload the EMP_PAY table with it. The content of our SYSPUNCH file is:

```
LOAD DATA INDDN SYSREC   LOG NO  RESUME YES
 UNICODE CCSID(00367,01208,01200)
 INTO TABLE
 "HRSCHEMA".
 "EMP_PAY"
 WHEN(00001:00002) = X'012E'
 NUMRECS                    5
 ( "EMP_ID"                           .
  POSITION(  00003:00006) INTEGER
 , "EMP_REGULAR_PAY"
  POSITION(  00007:00011) DECIMAL
 , "EMP_BONUS_PAY"
  POSITION(  00013:00017) DECIMAL
                       NULLIF(00012)=X'FF'
 )
```

Now we can substitute the SYSIN DD * in the load JCL with the name of the SYSPUNCH file and we are ready to run. Your output should look something like this:

```
DSNUGUTC - OUTPUT START FOR UTILITY, UTILID = TEMP
DSNUGTIS - PROCESSING SYSIN AS EBCDIC
DSNUGUTC -  LOAD DATA INDDN SYSREC LOG NO RESUME YES UNICODE CCSID(367, 1208, 12
.31 DSNURWI -  INTO TABLE "HRSCHEMA". "EMP_PAY" WHEN(1:2)=X'012E' NUMRECS 5
.31 DSNURWI -   ("EMP_ID" POSITION(3:6) INTEGER,
.31 DSNURWI -    "EMP_REGULAR_PAY" POSITION(7:11) DECIMAL,
.31 DSNURWI -    "EMP_BONUS_PAY" POSITION(13:17) DECIMAL NULLIF(12)=X'FF')
.73 DSNURWT - (RE)LOAD PHASE STATISTICS - NUMBER OF RECORDS=5 FOR TABLE HRSCHE-
MA.E
.73 DSNURWT - (RE)LOAD PHASE STATISTICS - TOTAL NUMBER OF RECORDS LOADED=5 FOR T

DSNURILD - (RE)LOAD PHASE STATISTICS - NUMBER OF INPUT RECORDS PROCESSED=5
DSNURILD - (RE)LOAD PHASE COMPLETE, ELAPSED TIME=00:00:00
.74 DSNUGSRX - TABLESPACE DBHR.TSHR IS IN COPY PENDING
```

Note that your tablespace is in COPY PENDING state. You can take a backup to clear this status, or you can use the REPAIR utility as follows:

```
REPAIR SET TABLESPACE DBHR.TSHR NOCHECKPEND
```

Here's a summary of factors that result in optimal DB2 application performance. I suggest you use this as a checklist for designing components and for resolving performance issues.

1. Make sure data concurrency is optimal.

   • Commit your work regularly to reduce contention due to locking. In some

494

cases this means committing after a certain amount of time as opposed to a certain number of transactions.

- Bind applications with the ISOLATION(CS) and CURRENTDATA(NO) options. These options are considered optimal to prevent unnecessary locking and to release locks as soon as possible.

- Include logic in your application program to retry after a deadlock or timeout to attempt recovery from the contention situation without assistance.

2. Use stored procedures to improve performance– they are compiled and run on the server for optimum performance.

3. Optimize your queries for small results sets whenever possible.

4. Use the latest statistics. Execute RUNSTATS followed by REBIND to your packages, especially when table data has grown significantly.

5. Check with the database administrator to see if the table needs to be reorganized.

6. Create/review EXPLAIN data for long running queries.

7. Eliminate unnecessary sorts.

8. Evaluate your SQL, checking to see if you are using stage 1 predicates versus stage 2 predicates.

9. Perform a Trace - refer to the subsection on the various traces available.

Additional Resources

DB2 11 for z/OS Application Programming and SQL Guide

DB2 11 for z/OS SQL Reference

Chapter 6 Review Questions

1. Which of the following is NOT a valid data type for use as an identity column?

   a. INTEGER
   b. REAL
   c. DECIMAL
   d. SMALLINT

2. You need to store numeric integer values of up to 5,000,000,000. What data type is appropriate for this?

   a. INTEGER
   b. BIGINT
   c. LARGEINT
   d. DOUBLE

3. Which of the following is NOT a LOB (Large Object) data type?
   a. CLOB
   b. BLOB
   c. DBCLOB
   d. DBBLOB

4. If you want to add an XML column VAR1 to table TBL1, which of the following would accomplish that?

   a. ALTER TABLE TBL1 ADD VAR1 XML
   b. ALTER TABLE TBL1 ADD COLUMN VAR1 XML
   c. ALTER TABLE TBL1 ADD COLUMN VAR1 (XML)
   d. ALTER TABLE TBL1 ADD XML COLUMN VAR1

5. If you want rows that have similar key values to be stored physically close to each other, what keyword should you specify when you create an index?

   a. UNIQUE
   b. ASC
   c. INCLUDE
   d. CLUSTER

6. Assume a table where certain columns contain sensitive data and you don't want all users to see these columns. Some other columns in the table must be made accessible to all users. What type of object could you create to solve this problem?

    a. INDEX
    b. SEQUENCE
    c. VIEW
    d. TRIGGER

7. To grant a privilege to all users of the database, grant the privilege to whom?

    a. ALL
    b. PUBLIC
    c. ANY
    d. DOMAIN

8. Tara wants to grant CONTROL of table TBL1 to Bill, and also allow Bill to grant the same privilege to other users. What clause should Tara use on the GRANT statement?

    a. WITH CONTROL OPTION
    b. WITH GRANT OPTION
    c. WITH USE OPTION
    d. WITH REVOKE OPTION

9. Which of the following will generate DB2 SQL data structures for a table or view that can be used in a PLI or COBOL program?

    a. DECLARE
    b. INCLUDE
    c. DCLGEN
    d. None of the above.

10. Assuming you are using a DB2 precompiler, which of the following orders the DB2 program preparation steps correctly?

    a. Precompile SQL, Bind Package, Bind Plan.
    b. Precompile SQL, Bind Plan, Bind Package.
    c. Bind Package, Precompile SQL, Bind Plan.
    d. Bind Plan, Precompile SQL, Bind Package.

11. To end a transaction without making the changes permanent, which DB2 statement should be issued?

    a. COMMIT
    b. BACKOUT
    c. ROLLBACK
    d. NO CHANGE

12. If you want to maximize data concurrency without seeing uncommitted data, which isolation level should you use?

    a. RR
    b. UR
    c. RS
    d. CS

13. To end a transaction and make the changes visible to other processes, which statement should be issued?

    a. ROLLBACK
    b. COMMIT
    c. APPLY
    d. CALL

14. Order the isolation levels, from greatest to least impact on performance.

    a. RR, RS, CS, UR
    b. UR, RR, RS, CS
    c. CS, UR, RR, RS
    d. RS, CS, UR, RR

15. Suppose you have created a test version of a production table, and you want to to use the UNLOAD utility to extract the first 1,000 rows from the production table to load to the test version. Which keyword would you use in the UNLOAD statement?

    a. WHEN
    b. SELECT
    c. SAMPLE
    d. SUBSET

16. Which of the following is NOT a way you could test a DB2 SQL statement?

    a. Running the statement from the DB2 command line processor.
    b. Running the statement from the SPUFI utility.
    c. Running the statement from IBM Data Studio.
    d. All of the above are valid ways to test an SQL statement.

Chapter 7: Customer Information Control System (CICS)

Introduction

CICS is an acronym for Customer Information Control System. This technology was developed by IBM in the 1960's. CICS is a transaction processing system as well as a telecommunication system that can support many hundreds of terminals. For example, CICS was the enabling technology that supported the early automated teller machines (ATM).

CICS Elements

CICS applications are comprised of several elements, including a screen mapset, a program and a transaction. We'll discuss each of these broadly, and then we'll get into more detail in later sections.

Screen Map

CICS applications consist of formatted screens that are an interface to CICS programs that provide some sort of processing, typically to display and update data in a data store. The screen component of CICS applications is usually developed using a CICS product called Basic Mapping Support (BMS).

To use BMS, you code instructions for how the screen is to be displayed, both the literal fields as well as user enterable fields. The instructions you provide to BMS will include the screen position and attributes of each field. Then you compile the instructions using a JCL (job control language) that produces what's called a mapset. A mapset contains one or more screen maps, and the map is what defines the screen that gets displayed.

As part of the assembly process, BMS will create a symbolic map which is a copybook that includes the layout of the screen including any enterable fields. The layout is used by the CICS program to receive data from the screen, as well as to send data back to the screen.

Here are some BMS terms you will need to know.

Map

A map is an individual screen format which you can create using BMS. The name of the mapset can be between 1 and 7 characters.

Mapset

A mapset is a set of maps which are defined and stored together to formulate a single load module. The name of the mapset can be between 1 and 7 characters.

BMS Macros

BMS maps are written in Assembler language. Fortunately you don't need to be concerned with learning Assembler. Instead three macros are used to define a screen mapset. These macros are DFHMSD, DFHMDI, and DFHMDF. We'll give you the high level about these macros and then discuss them in more detail when we create some screens.

DFHMSD

DFHMSD defines a mapset. In the above case, we've named our mapset EMPMMNU.

You can have more than one map in a mapset, but most applications I've worked with use one map per mapset. When defining the mapset, you have these parameters:

TYPE

TYPE indicates the kind of map to be generated. There is both the physical map which defines how the screen looks and behaves. There is also a symbolic map which is a copybook that is used to pass data back and forth from the screen to the program. We'll get into that momentarily. Meanwhile, here are the possible values for TYPE:

```
MAP             define the physical map
DSECT           define the symbolic map
&&SYSPARM       define both the physical and symbolic map
```

Typically we will specify &&SYSPARM in our mapset definition because we want both the physical map and the symbolic map to be generated.

MODE

MODE indicates whether input and/or output operations are allowed. You can specify IN, OUT or INOUT. INOUT is usually specified.

CTRL

CTRL defines device control requests. You can specify the following:

```
FREEKB       Unlocks the keyboard.
FRSET        Resets the MDT to zero status.
ALARM        Sets an audible alarm at screen display time (if the
             device supports it).
PRINT        Causes the mapset to be printed out on a printer.
```

LANG

LANG indicates whether the program that uses the mapset is Assembler, COBOL, PLI or C language. The corresponding values are:

```
ASM
COBOL
PLI
C
```

TIOAPFX

`TIOAPFX` determines whether a filler space for BMS commands will be included in the symbolic map. The values are `YES` or `NO`.

DSATTS

`DSATTS` indicates which attributes to include in the symbolic description map. The valid values are:

```
COLOR
HILIGHT
OUTLINE
PS
SOSI
TRANSP
VALIDN
```

In our screen maps we will specify `COLOR` and `HILITE`.

MAPATTS

`MAPATTS` indicates which attributes to include in the physical map. The valid values are:

```
COLOR
HILIGHT
OUTLINE
PS
SOSI
TRANSP
VALIDN
```

In our screen maps we will specify `COLOR` and `HILITE`.

STORAGE

`STORAGE` indicates whether the symbolic maps in the mapset are to be defined with separate storage areas, or whether they will use redefined storage space. Specifying `STORAGE=AUTO` means these maps will occupy separate storage areas.

Now let's actually look at a mapset definition. Here is the one we will use later for the EMP-MMNU mapset. We've indicated that the mapset name is EMPMMNU. We want it to generate both a physical and a symbolic map. The mapset is for a COBOL program to use. We want to be able to choose the COLOR of our fields and to set them to HILITE if desired.

```
----+----1----+----2----+----3----+----4----+----5----+----6----+----7----+----8
EMPMMNU   DFHMSD TYPE=&SYSPARM,MODE=INOUT,CTRL=(FREEKB,FRSET),        X
                 LANG=COBOL,TIOAPFX=YES,                              X
                 DSATTS=(COLOR,HILIGHT),                              X
                 MAPATTS=(COLOR,HILIGHT),                             X
                 STORAGE=AUTO
```

DFHMDI

The DFHMDI macro defines a screen map, and establishes that this is the beginning of the map. The map name can be up to 7 characters. Besides the map name you usually specify the number of lines and columns in the map. Typically this will be 24 lines and 80 columns. When defining the map, you have these parameters:

SIZE   The size of the map in lines and columns.

LINE   Specifies the starting line number of the map.

COLUMN   Specifies the starting column of the map.

JUSTIFY   Indicates the position of the map on the page. Valid values for this
                parameter are: LEFT, RIGHT, FIRST, LAST and BOTTOM

You can also specify CTRL and TIOAPFX at the map level, but we will not do that in our example. In fact, we'll simply specify the name and size for our EMPMNU map:

```
EMPMNU    DFHMDI SIZE=(24,80)
```

DFHMDF

The DFHMDF macro defines fields and their attributes. Literal fields will not usually have tag names, but any field you want to reference in your program must have a tag name specified in positions 1-7. The name can be from 1 to 7 characters. I strongly suggest you create meaningful names, as these will be the variable names generated in the symbolic map (that your program will use).

When defining the map, you have these parameters:

POS

POS establishes the position of the new field in terms of the line and column number.

LENGTH

LENGTH specifies the length of the field.

INITIAL

INITIAL specifies the initial value for the field, if any.

JUSTIFY

JUSTIFY specifies whether the field is to be left or right justified.

ATTRB

ATTRB describes the attribute(s) for this field. Valid ATTRB values are as follows:

ASKIP
Autoskip, the cursor skips to the next field.

PROT
The field is protected. You cannot enter data here.

UNPROT
The field is unprotected. You can enter data here.

NUM
Only numeric data can be entered in this field.

BRT
Field is bright (highlighted).

NORM
Normal display (this is the default).

DRK
Dark display.

IC
Insert cursor (place the cursor on this field).

FSET

Field set, ensures data is sent from terminal to program even if nothing changes on the screen.

`PICIN`
Picture in – this describes the format of numeric data for an input field. For example `PICIN` = `9(8)` describes an 8 digit numeric input.

`PICOUT`
Picture out – this describes the format of numeric data for an output field. For example `PI-COUT` = `9(8)` describes an 8 digit numeric output.

Sample Screen Map
Here's a sample of a screen we want to build.

```
EMPMMNU                    EMPLOYEE SUPPORT MENU                    EMNU
            ENTER THE NUMBER OF YOUR SELECTION,  THEN PRESS ENTER.


                    _      1. EMPLOYEE INQUIRY

                           2. EMPLOYEE ADD

                           3. EMPLOYEE CHANGE

                           4. EMPLOYEE DELETE






   F3 EXIT
```

Here is the BMS code to build the screen. I'm sure it looks somewhat cryptic. We'll explain each of the commands, tag names, etc.

```
----+----1----+----2----+----3----+----4----+----5----+----6----+----7----+----8
EMPMMNU   DFHMSD TYPE=&SYSPARM,MODE=INOUT,CTRL=(FREEKB,FRSET),          X
                LANG=COBOL,TIOAPFX=YES,                                 X
                DSATTS=(COLOR,HILIGHT),                                 X
                MAPATTS=(COLOR,HILIGHT),                                X
                STORAGE=AUTO
EMPMNU    DFHMDI SIZE=(24,80)
          DFHMDF POS=(01,1),LENGTH=07,COLOR=BLUE,                       X
                INITIAL='EMPMMNU'
          DFHMDF POS=(01,31),LENGTH=21,COLOR=BLUE,                      X
                INITIAL='EMPLOYEE SUPPORT MENU'
TRANID    DFHMDF POS=(01,76),LENGTH=04,INITIAL='EMNU',COLOR=BLUE
          DFHMDF POS=(03,15),LENGTH=35,COLOR=BLUE,                      X
                INITIAL='ENTER THE NUMBER OF YOUR SELECTION,'
          DFHMDF POS=(03,52),LENGTH=17,COLOR=BLUE,                      X
                INITIAL='THEN PRESS ENTER.'
ACTION    DFHMDF POS=(06,28),LENGTH=01,ATTRB=(IC,UNPROT),COLOR=GREEN,   X
                HILIGHT=UNDERLINE
          DFHMDF POS=(06,30),LENGTH=01,ATTRB=ASKIP
          DFHMDF POS=(06,32),LENGTH=19,COLOR=BLUE,                      X
                INITIAL='1. EMPLOYEE INQUIRY'
          DFHMDF POS=(08,32),LENGTH=15,COLOR=BLUE,                      X
                INITIAL='2. EMPLOYEE ADD'
          DFHMDF POS=(10,32),LENGTH=18,COLOR=BLUE,                      X
                INITIAL='3. EMPLOYEE CHANGE'
          DFHMDF POS=(12,32),LENGTH=18,COLOR=BLUE,                      X
                INITIAL='4. EMPLOYEE DELETE'
MESSAGE   DFHMDF POS=(23,02),LENGTH=67,COLOR=YELLOW
          DFHMDF POS=(24,02),LENGTH=07,ATTRB=PROT,COLOR=BLUE,           X
                INITIAL='F3 EXIT'
          DFHMSD TYPE=FINAL
          END
```

As you can see, we define the mapset using the **DFHMSD** macro and we name the mapset **EMPMMNU**. We specified that the mapset will be invoked by a program written in the COBOL language, and we provided a few other parameters.

Next we defined a map using the **DFHMDI** macro, and we named the map **EMPMNU**. We also specified the standard screen size of 24 lines of 80 bytes each.

Then we mapped how the screen will appear (both literal values and modifiable fields) with the DFHMDF macro. Note that tag names may be used for a field and if so the tag names begin in column 1.

As mentioned earlier, literal values such as the screen title do not require tag names, but input/

output variables such as the action code do. For example we named the action code field AC-TION. Then we specified by line number and column number exactly where the field is to be displayed on the screen (line 6, column 28).

We also specified each field's attributes. In this case we specified that the ACTION field is un-protected and the initial cursor position will be on this field. We further specified the color to be green, and that the field will be underlined to make it easy for the user to see.

Take a look at the rest of the fields that we defined, and reference their attributes back to the attribute list we provided earlier. This should give you a good idea what we are doing with the physical map.

Now let's take a look at the symbolic map that BMS generates for use in the application program..

Symbolic Map

The symbolic map is a copybook that includes 5 different fields for each input/output field. The variable names are the actual field name defined in the map, plus a suffix. For example the length of the field named ACTION is contained in a variable named ACTIONL. Here are the five suffixes and their usage:

L	Length variable that specifies the length of the entered data
F	Field variable indicates whether the field value has changed
I	Indicates an input variable - this is where the input data is stored
O	Indicates an output variable - this is where the output data is stored
A	Attribute variable that indicates the attributes for the field

In the screen map, we have a field named ACTION. We see below the corresponding variables are: **ACTIONL, ACTIONF, ACTIONA, ACTIONI, ACTIONO**

```
01  EMPMNUI.
    02  FILLER PIC X(12).
    02  TRANIDL    COMP  PIC  S9(4).
    02  TRANIDF    PICTURE X.
    02  FILLER REDEFINES TRANIDF.
      03 TRANIDA    PICTURE X.
    02  FILLER    PICTURE X(2).
    02  TRANIDI PIC X(4).
    02  ACTIONL    COMP  PIC  S9(4).
    02  ACTIONF    PICTURE X.
    02  FILLER REDEFINES ACTIONF.
      03 ACTIONA    PICTURE X.
    02  FILLER    PICTURE X(2).
```

```
02  ACTIONI  PIC X(1).
02  MESSAGEL    COMP  PIC  S9(4).
02  MESSAGEF    PICTURE X.
02  FILLER REDEFINES MESSAGEF.
  03 MESSAGEA    PICTURE X.
02  FILLER   PICTURE X(2).
02  MESSAGEI  PIC X(67).
01  EMPMNUO REDEFINES EMPMNUI.
02  FILLER PIC X(12).
02  FILLER PICTURE X(3).
02  TRANIDC    PICTURE X.
02  TRANIDH    PICTURE X.
02  TRANIDO  PIC X(4).
02  FILLER PICTURE X(3).
02  ACTIONC    PICTURE X.
02  ACTIONH    PICTURE X.
02  ACTIONO  PIC X(1).
02  FILLER PICTURE X(3).
02  MESSAGEC    PICTURE X.
02  MESSAGEH    PICTURE X.
02  MESSAGEO  PIC X(67).
```

Of course a screen map doesn't do anything without a program to do some processing. So let's move on to the CICS program.

Sample CICS Program

Our program will be named EMPPMNU, and the identification and environment divisions are simple.

```
IDENTIFICATION DIVISION.
PROGRAM-ID. EMPPGMNU.

***************************************************
*  MENU PROGRAM FOR EMPLOYEE APPLICATION        *
*                                               *
*  AUTHOR       : ROBERT WINGATE                *
*  DATE-WRITTEN : 2018-07-26                    *
***************************************************

ENVIRONMENT DIVISION.
```

Next, let's add the data division which will also be fairly simple.

```
DATA DIVISION.

WORKING-STORAGE SECTION.

01 WS-FLAGS.
   05 SW-VALID-SELECTION      PIC X(1) VALUE 'N'.
      88  VALID-SELECTION           VALUE 'Y'.
      88  NOT-VALID-SELECTION       VALUE 'N'.
01 WS-VARS.
   05 COMM-AREA               PIC X(20) VALUE SPACE.
```

```
    05  PROGRAM-NAME            PIC X(08) VALUE SPACES.
    05  INVALID-ACTION-MSG      PIC X(34)
      VALUE 'ENTER A VALID ACTION: 1, 2, 3 OR 4'.

    COPY EMPMMNU.
    COPY DFHAID.
    COPY DFHBMSCA.

LINKAGE SECTION.
01  DFHCOMMAREA          PIC X(20).
```

We've coded a flag to indicate if the user entered an invalid selection. We also define a communication area for data to be passed to other programs. We won't be passing data from this program but for consistency with the other programs we'll define a 20 byte are called COMM-AREA. Next, we've defined an 8 byte variable which will contain the name of the program we will transfer to if the user requests it. We'll also define a message literal that will be used when the user enters an invalid selection.

We're also including three copybooks as follows:

1. **EMPMMNU** is the symbolic map generated when we compile the mapset.

2. **DFHAID** contains the standard attention identifier list which is a set of literal names for the various key presses that are captured. For example DFHPF3 indicates that the PF3 key was pressed. DFHENTER means that the ENTER key was pressed. The complete list is on the IBM product support web site.

https://www.ibm.com/support/knowledgecenter/en/SSAL2T_8.1.0/com.ibm.cics.tx.doc/reference/r_attn_idntfr_consts_lst.html

3. **DFHBMSCA** contains constants for setting various values such as attribute characters. For example DFHBMFSE is a constant that means the value of a field has changed. For example, in our symbolic map the field ACTIONF is the attribute byte for the ACTIONI field. If the value of ACTIONF is equal to DFHBMFSE, it means the field value has changed. The complete list is on the IBM product support web site.

https://www.ibm.com/support/knowledgecenter/en/SSGMCP_5.3.0/com.ibm.cics.ts.applicationprogramming.doc/topics/dfhp4_bmsconstants.html

Finally, in the linkage section we must define a variable named DFHCOMMAREA. This is where another program can pass data to this program. We won't be doing that in the menu program, but again we want consistency to the other programs so we will define DFHCOMMAREA as character 20 bytes.

For the procedure division, let's divide our discussion into three parts:

1. The sending and receiving of data using the symbolic map.
2. The checking for keys pressed by the user.
3. The actual processing of the user's request.

In any transactional system there must be commands to retrieve the user's screen input, and to send output back to the screen. In CICS the commands are SEND and RECEIVE used with various parameters. To erase the screen and send the menu screen map, you would use the following command:

```
EXEC CICS SEND
    MAP    ('EMPMNU')
    MAPSET ('EMPMMNU')
    FROM   (EMPMNUO)
    ERASE
END-EXEC.
```

This directs CICS to send the EMPMNU map located in the EMPMMNU mapset to the user's terminal, and to populate the variables with data from the symbolic map EMPMNUO. The latter is the output map from the copybook EMPMMNU that was generated when we compiled the screen. We'll go through the steps in more detail in the coming sections. Also notice that we specified the ERASE parameter to remove any residual screen data from a previous screen.

Once the user has entered a selection and pressed the Enter key, we will receive the data into the program as follows:

```
EXEC CICS RECEIVE
    MAP    ('EMPMNU')
    MAPSET ('EMPMMNU')
    INTO   (EMPMNUI)
END-EXEC.
```

Here we command CICS to receive screen data using the specified map and mapset, and we indicate the symbolic input map as a container for the data.

Finally, once the program has processed the user's input, the program will send the map back to the screen where it will be displayed. Typically in this case, you only send the modified data back (not the entire screen) by specifying the DATAONLY parameter.

```
EXEC CICS SEND
    MAP    ('EMPMNU')
    MAPSET ('EMPMMNU')
    FROM   (EMPMNUO)
    DATAONLY
END-EXEC.
```

So let's program three procedures that use the SEND and RECEIVE commands as follows:

```
SEND-MAP.
    EXEC CICS SEND
        MAP     ('EMPMNU')
        MAPSET  ('EMPMMNU')
        FROM    (EMPMNUO)
        ERASE
    END-EXEC.

SEND-MAP-DATAONLY.
    EXEC CICS SEND
        MAP     ('EMPMNU')
        MAPSET  ('EMPMMNU')
        FROM    (EMPMNUO)
        DATAONLY
    END-EXEC.


RECEIVE-MAP.
    EXEC CICS RECEIVE
        MAP     ('EMPMNU')
        MAPSET  ('EMPMMNU')
        INTO    (EMPMNUI)
    END-EXEC.
```

Ok, next we want to code the program mainline to intercept and handle whatever conditions or keyed input occur. Here is our code below, and we'll explain each element.

There is a variable named EIBCALEN that indicates how many times this session has been through the program. If the value is zero, that means this is the first time through. If it is the first time through, we simply want to display the screen for the user. So in the code below, we initialize the symbolic output map, and then we call the send map procedure.

Next, we check to see if the user pressed the clear key, the PA keys, or the PF3 key and handle those as indicated. Next, if the user pressed the ENTER key, we call the procedure to process the user's selection. Finally, if the user pressed any other key, we set up the "invalid key pressed" error message and then send the map.

```
IF EIBCALEN > ZERO
    MOVE DFHCOMMAREA TO COMM-AREA
END-IF.

EVALUATE TRUE

    WHEN EIBCALEN = ZERO
        MOVE LOW-VALUES    TO  EMPMNUO
        PERFORM SEND-MAP

    WHEN EIBAID = DFHCLEAR
```

```
                MOVE LOW-VALUES   TO  EMPMNUO
                PERFORM SEND-MAP

            WHEN EIBAID = DFHPA1 OR DFHPA2 OR DFHPA3
               CONTINUE

            WHEN EIBAID = DFHPF3
               MOVE LOW-VALUES TO  EMPMNUO
               MOVE "BYE, PRESS CLEAR KEY TO ENTER A TRANSACTION ID"
                   TO MESSAGEO
               PERFORM SEND-MAP-DATAONLY

               EXEC CICS
                 RETURN
               END-EXEC

            WHEN EIBAID = DFHENTER
               PERFORM MAIN-PROCESS-PARA

            WHEN OTHER
               MOVE LOW-VALUES TO EMPMNUO
               MOVE "INVALID KEY PRESSED" TO MESSAGEO
               PERFORM SEND-MAP-DATAONLY

        END-EVALUATE.

        EXEC CICS
           RETURN TRANSID('EMNU')
           COMMAREA (COMM-AREA)
        END-EXEC.
```

Ok, at this point we've handled everything except processing the user's request. To do that, we will need two procedures. One is the MAIN-PROCESS-PARA which will evaluate the user's request and call the appropriate program. The other is the CICS routine to transfer control to the requested program. Here's the main process routine. Basically what we're doing is to:

1. Receive the screen map.
2. Check for a valid action number.
3. If valid, load the program name and call the branch-to procedure
4. If invalid, load the error message
5. Send the map with data only

Here's the main processing procedure:

```
MAIN-PROCESS-PARA.

    PERFORM RECEIVE-MAP.

    IF ACTIONI NOT = '1' AND '2' AND '3' AND '4'
       MOVE DFHREVRS TO ACTIONH
       MOVE INVALID-ACTION-MSG TO MESSAGEO
```

513

```
            SET NOT-VALID-SELECTION TO TRUE
        ELSE
            SET VALID-SELECTION TO TRUE
        END-IF.

        IF VALID-SELECTION
            EVALUATE ACTIONI
                WHEN '1'
                    MOVE 'EMPPGINQ' TO PROGRAM-NAME
                WHEN '2'
                    MOVE 'EMPPGADD' TO PROGRAM-NAME
                WHEN '3'
                    MOVE 'EMPPGCHG' TO PROGRAM-NAME
                WHEN '4'
                    MOVE 'EMPPGDEL' TO PROGRAM-NAME
            END-EVALUATE

            PERFORM BRANCH-TO-PROGRAM

        END-IF.

        PERFORM SEND-MAP-DATAONLY.
```

In our BRANCH-TO-PROGRAM paragraph we use the CICS XCTL command to transfer control to the specified program. We just need to provide the program name which we do using the variable PROGRAM-NAME.

```
        BRANCH-TO-PROGRAM.

            EXEC CICS
                XCTL PROGRAM(PROGRAM-NAME)
            END-EXEC

            MOVE 'PROGRAM NOT AVAILABLE' TO MESSAGEO.
```

Ok, that's it. Here is the complete program code. If some of this doesn't make 100% sense, don't worry. We'll explain more later. This program example is just for illustration now so you get a taste of CICS programming.

```
        IDENTIFICATION DIVISION.
        PROGRAM-ID. EMPPGMNU.
        ***********************************************
        *   MENU PROGRAM FOR EMPLOYEE APPLICATION       *
        *                                               *
        *   AUTHOR        : ROBERT WINGATE              *
        *   DATE-WRITTEN  : 2018-07-26                  *
        ***********************************************
        ENVIRONMENT DIVISION.
        DATA DIVISION.
        WORKING-STORAGE SECTION.
```

```
01 WS-FLAGS.
   05 SW-VALID-SELECTION      PIC X(1) VALUE 'N'.
      88  VALID-SELECTION              VALUE 'Y'.
      88  NOT-VALID-SELECTION          VALUE 'N'.

01 WS-VARS.
   05 COMM-AREA              PIC X(20) VALUE SPACE.
   05 PROGRAM-NAME           PIC X(08) VALUE SPACES.
   05 INVALID-ACTION-MSG     PIC X(34)
      VALUE 'ENTER A VALID ACTION: 1, 2, 3 OR 4'.

   COPY EMPMMNU.
   COPY DFHAID.
   COPY DFHBMSCA.

LINKAGE SECTION.

01 DFHCOMMAREA         PIC X(20).

PROCEDURE DIVISION.

   IF EIBCALEN > ZERO
     MOVE DFHCOMMAREA  TO COMM-AREA
   END-IF.

   EVALUATE TRUE

     WHEN EIBCALEN = ZERO
       MOVE LOW-VALUES    TO  EMPMNUO
       PERFORM SEND-MAP

     WHEN EIBAID = DFHCLEAR
       MOVE LOW-VALUES    TO  EMPMNUO
       PERFORM SEND-MAP

     WHEN EIBAID = DFHPA1 OR DFHPA2 OR DFHPA3
       CONTINUE

     WHEN EIBAID = DFHPF3
       MOVE LOW-VALUES TO  EMPMNUO
       MOVE "BYE, PRESS CLEAR KEY TO ENTER A TRANSACTION ID"
            TO MESSAGEO
       PERFORM SEND-MAP-DATAONLY

       EXEC CICS
          RETURN
       END-EXEC

     WHEN EIBAID = DFHENTER
       PERFORM MAIN-PROCESS-PARA

     WHEN OTHER
```

515

```
              MOVE LOW-VALUES TO EMPMNUO
              MOVE "INVALID KEY PRESSED" TO MESSAGEO
              PERFORM SEND-MAP-DATAONLY

       END-EVALUATE.

       EXEC CICS
          RETURN TRANSID('EMNU')
          COMMAREA (COMM-AREA)
       END-EXEC.

MAIN-PROCESS-PARA.

       PERFORM RECEIVE-MAP.

       IF ACTIONI NOT = '1' AND '2' AND '3' AND '4'
          MOVE DFHREVRS TO ACTIONH
          MOVE INVALID-ACTION-MSG TO MESSAGEO
          SET NOT-VALID-SELECTION TO TRUE
       ELSE
          SET VALID-SELECTION TO TRUE
       END-IF.

       IF VALID-SELECTION
          EVALUATE ACTIONI
             WHEN '1'
                MOVE 'EMPPGINQ' TO PROGRAM-NAME
             WHEN '2'
                MOVE 'EMPPGADD' TO PROGRAM-NAME
             WHEN '3'
                MOVE 'EMPPGCHG' TO PROGRAM-NAME
             WHEN '4'
                MOVE 'EMPPGDEL' TO PROGRAM-NAME
          END-EVALUATE

          PERFORM BRANCH-TO-PROGRAM

       END-IF.

       PERFORM SEND-MAP-DATAONLY.

BRANCH-TO-PROGRAM.

       EXEC CICS
          XCTL PROGRAM(PROGRAM-NAME)
       END-EXEC

       MOVE 'PROGRAM NOT AVAILABLE' TO MESSAGEO.

SEND-MAP.
       EXEC CICS SEND
          MAP    ('EMPMNU')
```

```
              MAPSET  ('EMPMMNU')
              FROM    (EMPMNUO)
              ERASE
          END-EXEC.

      SEND-MAP-DATAONLY.
          EXEC CICS SEND
              MAP     ('EMPMNU')
              MAPSET  ('EMPMMNU')
              FROM    (EMPMNUO)
              DATAONLY
          END-EXEC.


      RECEIVE-MAP.
          EXEC CICS RECEIVE
              MAP     ('EMPMNU')
              MAPSET  ('EMPMMNU')
              INTO    (EMPMNUI)
          END-EXEC.
```

Sample Transaction

You invoke a CICS program through a transaction. Transaction identifiers can be between 1 and 4 characters long. All shops I've ever worked in use 4 characters by convention. So we could define transaction EMNU for our menu program, and that is how the transaction is invoked (by typing EMNU on a blank CICS screen and pressing ENTER)

In order to make these components active, you must define your mapset, program and transaction to CICS. This is typically done using the CEDA utility. We will do that later in another section. For now, we've seen examples of a map, mapset and program, and we've discussed at a high level that these are linked using a transaction.

In the next section, we will provide specifications for a simple employee support application to be used by a fictitious HR department. This will give us something structured to work with as we explain more complex CICS concepts and start to do actual programming.

Employee Support Application Design

Purpose

This design is for a Human Resource application called Employee Support. It consists of a data store containing employee information such as employee number, name, years of service and social security number. There will be several CICS screen that allow the user to:

1. Select from a menu of options (Main Menu)
2. Display employee detail for a specified employee
3. Add detail for a new employee
4. Change detail for an existing employee
5. Delete detail for an employee

General Design

The following provides a high level design of the application in terms of data store, screen, program and transaction components.

Data Design

The following table specification is provided:

EMPLOYEE (key is EMP_ID).

Field Name	Type
EMP_ID	INTEGER
EMP_LAST_NAME	VARCHAR(30)
EMP_FIRST_NAME	VARCHAR(20)
EMP_SERVICE_YEARS	INTEGER
EMP_PROMOTION_DATE	DATE
EMP_SSN	VARCHAR(09)

The fields are self explanatory except for EMP_SSN which is the employee's social security number.

Application Elements

Screen Components

Here we specify the screen mapset names and the transactions that will be associated with them.

Screen Name	Trans ID	Mapset Name
Employee Support Menu	EMNU	EMPMMNU
Employee Inquiry	EMIN	EMPMINQ
Employee Add	EMAD	EMPMADD
Employee Change	EMCH	EMPMCHG
Employee Delete	EMDE	EMPMDEL

Hierarchy Chart

This is a simple functional hierarchy chart. It indicates that the menu program will call one of four different programs: the Employee Inquiry, Employee Add, Employee Change and Employee Delete.

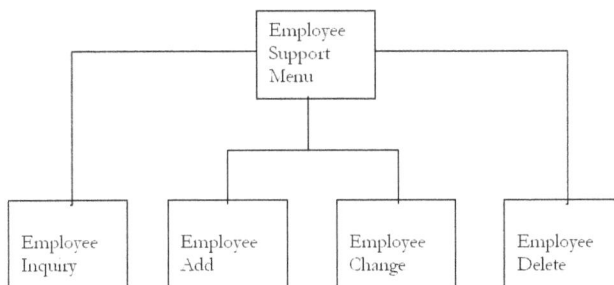

Program Elements (COBOL)

The following table summarizes the program components required for our application and the transactions that will invoke the programs.

Screen Name	Trans ID	Program Name
Employee Support Menu	EMNU	EMPPGMNU
Employee Inquiry	EMIN	EMPPGINQ
Employee Add	EMAD	EMPPGADD
Employee Change	EMCH	EMPPGCHG
Employee Delete	EMDE	EMPPGDEL

Detail Design
Detailed Data Design

In this section we will design and create the appropriate DB2 table for our application. You'll need to either have the DBA assign you a database to work with or create one yourself. If you have authority, I suggest you create a Human Resource database named DBHR. Also you can create a tablespace called TSHR, and a schema called HRSCHEMA.

The following is just sample DDL. You must know your system in order to supply the correct values.

```
CREATE DATABASE DBHR
STOGROUP SGHR
BUFFERPOOL BPHR
INDEXBP IBPHR
CCSID UNICODE;

CREATE TABLESPACE TSHR
IN DBHR
USING STOGROUP SGHR
PRIQTY 50
SECQTY 20
LOCKSIZE PAGE
BUFFERPOOL BPHR2;

CREATE SCHEMA HRSCHEMA
AUTHORIZATION USER01;      ← This should be your DB2 id, whatever it is.
```

Now you can create the EMPLOYEE table. Here is the DDL:

```
CREATE TABLE HRSCHEMA.EMPLOYEE(
EMP_ID INT NOT NULL,
EMP_LAST_NAME VARCHAR(30) NOT NULL,
EMP_FIRST_NAME VARCHAR(20) NOT NULL,
EMP_SERVICE_YEARS INT NOT NULL WITH DEFAULT 0,
EMP_PROMOTION_DATE DATE,
EMP_SSN  CHAR(09),
PRIMARY KEY(EMP_ID));
```

We also need to create a unique index to support the primary key:

```
CREATE UNIQUE INDEX NDX_EMPLOYEE
ON EMPLOYEE (EMP_ID);
```

Now let's insert some data.

```
INSERT INTO HRSCHEMA.EMPLOYEE
VALUES (3217,
'JOHNSON',
'EDWARD',
4,
'01/01/2017',
'397342007');

INSERT INTO HRSCHEMA.EMPLOYEE
VALUES (7459,
'STEWART',
'BETTY',
7,
'07/31/2016',
' 019572830');


INSERT INTO HRSCHEMA.EMPLOYEE
VALUES (9134,
'FRANKLIN',
'BRIANNA',
DEFAULT,
NULL,
' 937293598');


INSERT INTO HRSCHEMA.EMPLOYEE
VALUES (4720,
'SCHULTZ',
'TIM',
9,
'01/01/2017',
' 650450254');


INSERT INTO HRSCHEMA.EMPLOYEE
VALUES (6288,
'WILLARD',
'JOE',
6,
'01/01/2016',
' 209883920');
```

Ok that's enough data to start with. We'll use it when we get to the actual programming.

CICS Screen Designs

Employee Support Menu

This screen allows the user to select one of four options:

1. Employee Inquiry
2. Employee Add
3. Employee Change
4. Employee Delete

The following are the transaction, mapset and map names we will use for this program.

```
Transaction Name:  EMNU
Screen Mapset:     EMPMMNU
Program Name:      EMPPGMNU
Data Stores:       None
```

The display should be as follows. We'll see later that there is a message field just above the F3 EXIT display. For now this field is empty, so it does not show any message. It will be used when the program needs to display an error message, or in some cases to prompt the user for an additional action.

```
----+----1----+----2----+----3----+----4----+----5----+----6----+----7----+----8
EMPMMNU                       EMPLOYEE SUPPORT MENU                         EMNU

         ENTER THE NUMBER OF YOUR SELECTION,   THEN PRESS ENTER.


                    _    1. EMPLOYEE INQUIRY

                         2. EMPLOYEE ADD

                         3. EMPLOYEE CHANGE

                         4. EMPLOYEE DELETE




F3 EXIT
```

Note: The numbers on the top line of the display are just for reference – they are not part of the actual screen image.

The user types the number of the desired selection in the input field which is the underscore to the left of the first choice (item 1). So the user enters 1, 2 3, or 4 and then presses ENTER. The result is that the selected screen will display.

Employee Inquiry

This screen allows a user to enter an employee id. After doing so and pressing ENTER, the detailed information for the employee will be displayed. If the requested employee id is not in the EMPLOYEE table, then an error message will be returned. The user will also be able to switch to a different option such as the add, change or delete screen.

```
Transaction Name: EMIN
Screen Mapset:    EMPMINQ
Program Name:     EMPPGINQ
Data Stores:      EMPLOYEE
```

The display should be as follows. The user enters a valid employee id and presses ENTER. If the employee is found in the table, the detailed employee information is displayed on the screen. If the employee is not found, an error message is displayed.

```
----+----1----+----2----+----3----+----4----+----5----+----6----+----7----+----8
EMPMINQ                     EMPLOYEE INQUIRY                            EMIN


        EMPLOYEE ->     ____        ENTER EMPLOYEE ID, THEN PRESS ENTER

        EMPLOYEE ID

        EMP LAST NAME

        EMP FIRST NAME

        EMP SOCIAL SEC

        EMP YEARS SRVC

        EMP LAST PROM




  F2 INQ   F3 EXIT   F4 ADD   F5 CHG   F6 DEL
```

For example, if the user enters employee 3217, and presses the ENTER key, the following should result:

```
----+----1----+----2----+----3----+----4----+----5----+----6----+----7----+----8
EMPMINQ                    EMPLOYEE INQUIRY                              EMIN

        EMPLOYEE ->    3217        ENTER EMPLOYEE ID, THEN PRESS ENTER

        EMPLOYEE ID    3217

        EMP LAST NAME  JOHNSON

        EMP FIRST NAME EDWARD

        EMP SOCIAL SEC 397342007

        EMP YEARS SRVC 04

        EMP LAST PROM  2017-01-01




    F2 INQ   F3 EXIT   F4 ADD   F5 CHG   F6 DEL
```

If the user enters an invalid employee id, an error message should result.

```
----+----1----+----2----+----3----+----4----+----5----+----6----+----7----+----8
EMPMINQ                    EMPLOYEE INQUIRY                              EMIN

        EMPLOYEE ->    3188        ENTER EMPLOYEE ID, THEN PRESS ENTER

        EMPLOYEE ID

        EMP LAST NAME

        EMP FIRST NAME

        EMP SOCIAL SEC

        EMP YEARS SRVC

        EMP LAST PROM




EMPLOYEE ID 3188 NOT FOUND
    F2 INQ   F3 EXIT   F4 ADD   F5 CHG   F6 DEL
```

Employee Add

This screen allows a user to enter a new employee. The detailed information for the employee will be added to the EMPLOYEE table. The user will also be able to switch to a different option such as the inquiry, change or delete screen.

```
Transaction Name: EMAD
Screen Mapset:    EMPMADD
Program Name:     EMPPGADD
Data Stores:      EMPLOYEE
```

The initial display should be as follows. The user enters an employee id, last and first names, years of service, last promotion date and social security number. If the record is successfully added, a message will indicate this.

```
----+----1----+----2----+----3----+----4----+----5----+----6----+----7----+----8
EMPMADD                     EMPLOYEE ADD                              EMAD


                    ENTER EMPLOYEE INFO, THEN PRESS PF4

        EMPLOYEE ID      ____

        EMP LAST NAME    _____

        EMP FIRST NAME   _____

        EMP SOCIAL SEC   _____

        EMP YEARS SRVC   __

        EMP LAST PROM    _____




ENTER DATA FOR NEW EMPLOYEE, THEN PRESS PF4 TO ADD
F2 INQ   F3 EXIT   F4 ADD   F5 CHG   F6 DEL
```

Suppose we add employee 8888 with name Joan Sanders with social security number 432993928, 5 years of service and last promotion date 2018-07-01. Here is the screen filled in:

```
EMPMADD                          EMPLOYEE ADD                          EMAD


        EMPLOYEE ->    ENTER EMPLOYEE INFO, THEN PRESS PF4PRESS ENTER

        EMPLOYEE ID    8888

        EMP LAST NAME  Sanders

        EMP FIRST NAME Joan

        EMP SOCIAL SEC 432993928

        EMP YEARS SRVC 05

        EMP LAST PROM  2018-07-01




     ENTER DATA FOR NEW EMPLOYEE, THEN PRESS PF4 TO ADD
     F2 INQ   F3 EXIT   F4 ADD   F5 CHG   F6 DEL
```

Now press PF4. If successful you should see this screen with message indicating successfully added.

```
EMPMADD                          EMPLOYEE ADD                          EMAD


        EMPLOYEE ->    ENTER EMPLOYEE INFO, THEN PRESS PF4PRESS ENTER

        EMPLOYEE ID    8888

        EMP LAST NAME  SANDERS

        EMP FIRST NAME JOAN

        EMP SOCIAL SEC 432993928

        EMP YEARS SRVC 05

        EMP LAST PROM  2018-07-01




     EMPLOYEE ADDED SUCCESSFULLY
     F2 INQ   F3 EXIT   F4 ADD   F5 CHG   F6 DEL
```

Edits and Validations

We haven't yet specified the edits and validations for our business design. The fields must be checked for conformance with the following rules:

- Employee number is required and must be numeric and greater than zero.
- Last name is required.
- First name is required.
- Social Security Number is required and must be numeric and greater than zero.
- Year of Service is required and must be numeric and greater than zero.
- Last promotion date must be a valid date value in format YYYY-MM-DD.

Employee Change

This screen allows a user to enter an employee id, press ENTER and the detailed information for the employee will be returned. The user is prompted to make changes and then press PF5 to apply the changes to the table.

```
Transaction Name: EMCH
Screen Mapset:    EMPMCHG
Program Name:     EMPPGCHG
Data Stores:      EMPLOYEE


----+----1----+----2----+----3----+----4----+----5----+----6----+----7----+----8
EMPMCHG                    EMPLOYEE CHANGE                              EMCH


        EMPLOYEE ->     ____        ENTER EMPLOYEE ID, THEN PRESS ENTER

        EMPLOYEE ID

        EMP LAST NAME

        EMP FIRST NAME

        EMP SOCIAL SEC

        EMP YEARS SRVC

        EMP LAST PROM




  F2 INQ   F3 EXIT   F4 ADD   F5 CHG   F6 DEL
```

For example, suppose you want to change the years of service for employee 3217 from 4 year to 5 years. Here's the screen with the data pulled up:

```
EMPMCHG                    EMPLOYEE CHANGE                          EMCH


        EMPLOYEE ->    3217      ENTER EMPLOYEE ID, THEN PRESS ENTER

        EMPLOYEE ID    3217

        EMP LAST NAME  JOHNSON

        EMP FIRST NAME EDWARD

        EMP SOCIAL SEC 397342007

        EMP YEARS SRVC 04

        EMP LAST PROM  2017-01-01




    MAKE CHANGES AND THEN PRESS PF5
    F2 INQ   F3 EXIT   F4 ADD   F5 CHG   F6 DEL
```

If you make the change and press PF5, and the change is successful, you will receive a confirmation message that the change was successful.

```
EMPMCHG                    EMPLOYEE CHANGE                          EMCH


        EMPLOYEE ->    3217      ENTER EMPLOYEE ID, THEN PRESS ENTER

        EMPLOYEE ID    3217

        EMP LAST NAME  JOHNSON

        EMP FIRST NAME EDWARD

        EMP SOCIAL SEC 397342007

        EMP YEARS SRVC 05

        EMP LAST PROM  2017-01-01




    EMPLOYEE MODIFIED SUCCESSFULLY
    F2 INQ   F3 EXIT   F4 ADD   F5 CHG   F6 DEL
```

If the user presses ENTER instead of the PF5 key, they will continue to receive the confirmation message to change the record. Of course the user can also press another PF key to take another action such as switching to another screen or exiting.

Employee Delete

This screen allows a user to enter an employee id, press ENTER and the detailed information for the employee will be returned. The user will then be prompted to delete the employee from the table by pressing PF5.

```
Transaction Name:  EMDE
Screen Mapset:     EMPMDEL
Program Name:      EMPPGDEL
Data Stores:       EMPLOYEE
```

The initial display should be as follows.

```
----+----1----+----2----+----3----+----4----+----5----+----6----+----7----+----8
EMPMDET                          EMPLOYEE DETAIL                          EMDT


        EMPLOYEE ->      ____        ENTER EMPLOYEE ID, THEN PRESS ENTER

        EMPLOYEE ID

        EMP LAST NAME

        EMP FIRST NAME

        EMP SOCIAL SEC

        EMP YEARS SRVC

        EMP LAST PROM




   F2 INQ   F3 EXIT   F4 ADD   F5 CHG   F6 DEL
```

Once the employee information is returned to the screen the user will be prompted by a message to press PF6 to delete the record. If the user presses ENTER instead of the PF6 key, they will continue to receive the confirmation message to delete the record. Of course the user can also press another PF key to take another action such as switching to another screen or exiting.

```
EMPMDEL                        EMPLOYEE DELETE                              EMDE


        EMPLOYEE ->    8888       ENTER EMPLOYEE ID, THEN PRESS ENTER

        EMPLOYEE ID    8888

        EMP LAST NAME  WINGATE

        EMP FIRST NAME ROBERT

        EMP SOCIAL SEC 454084595

        EMP YEARS SRVC 05

        EMP LAST PROM  2016-01-01




PRESS PF6 TO DELETE EMPLOYEE
F2 INQ   F3 EXIT   F4 ADD   F5 CHG   F6 DEL
```

If they do press PF6 they will receive a confirmation message that the record was deleted.

```
EMPMDEL                        EMPLOYEE DELETE                              EMDE


        EMPLOYEE ->    8888       ENTER EMPLOYEE ID, THEN PRESS ENTER

        EMPLOYEE ID    8888

        EMP LAST NAME  WINGATE

        EMP FIRST NAME ROBERT

        EMP SOCIAL SEC 454084595

        EMP YEARS SRVC 05

        EMP LAST PROM  2016-01-01




EMPLOYEE DELETED SUCCESSFULLY
F2 INQ   F3 EXIT   F4 ADD   F5 CHG   F6 DEL
```

Ok this completes the business design of our CICS screens. In the next section we will write and compile the code to create the screens.

Employee Support Construction

BMS Coding Specifications and Assembly

BMS coding can be rather cryptic until you've done a few screens. We'll start out with the menu screen and explain how we create it. Then we'll jump into CICS to display it to make sure we have it correct.

Coding the EMPMMNU Screen

The BMS code for our menu screen is as follows. Here are a couple of general things to keep in mind. There are various types of entries, but almost all of them describe what goes on the screen and the attributes associated with those fields. Take a good look at the code and then we'll go through the basics of compiling, defining and installing the screen mapset.

```
----+----1----+----2----+----3----+----4----+----5----+----6----+----7----+----8
EMPMMNU  DFHMSD TYPE=&SYSPARM,MODE=INOUT,CTRL=(FREEKB,FRSET),        X
               LANG=COBOL,TIOAPFX=YES,                              X
               DSATTS=(COLOR,HILIGHT),                             X
               MAPATTS=(COLOR,HILIGHT),                            X
               STORAGE=AUTO
EMPMNU   DFHMDI SIZE=(24,80)
         DFHMDF POS=(01,1),LENGTH=07,COLOR=BLUE,                  X
               INITIAL='EMPMMNU'
         DFHMDF POS=(01,31),LENGTH=21,COLOR=BLUE,                 X
               INITIAL='EMPLOYEE SUPPORT MENU'
TRANID   DFHMDF POS=(01,76),LENGTH=04,INITIAL='EMNU',COLOR=BLUE
         DFHMDF POS=(03,15),LENGTH=35,COLOR=BLUE,                 X
               INITIAL='ENTER THE NUMBER OF YOUR SELECTION,'
         DFHMDF POS=(03,52),LENGTH=17,COLOR=BLUE,                 X
               INITIAL='THEN PRESS ENTER.'
ACTION   DFHMDF POS=(06,28),LENGTH=01,ATTRB=(IC,UNPROT),COLOR=GREEN, X
               HILIGHT=UNDERLINE
         DFHMDF POS=(06,30),LENGTH=01,ATTRB=ASKIP
         DFHMDF POS=(06,32),LENGTH=19,COLOR=BLUE,                 X
               INITIAL='1. EMPLOYEE INQUIRY'
         DFHMDF POS=(08,32),LENGTH=15,COLOR=BLUE,                 X
               INITIAL='2. EMPLOYEE ADD'
         DFHMDF POS=(10,32),LENGTH=18,COLOR=BLUE,                 X
               INITIAL='3. EMPLOYEE CHANGE'
         DFHMDF POS-(12,32),LENGTH-18,COLOR=BLUE,                 X
               INITIAL='4. EMPLOYEE DELETE'
MESSAGE  DFHMDF POS=(23,02),LENGTH=67,COLOR=YELLOW
         DFHMDF POS=(24,02),LENGTH=07,ATTRB=PROT,COLOR=BLUE,      X
               INITIAL='F3 EXIT'
         DFHMSD TYPE=FINAL
         END
```

Columns 1 – 8 are tag names that you can refer to in your program, including the mapset name, the map name, and any enterable fields you include to allow input and output. We'll talk about those momentarily.

Column 10 includes a keyword that begins an entry. These are the keywords that are normally required to define a map.

DFHMSD – this is the mapset name.

DFHMDI – this is the map name. You can include more than one map in a mapset. I have never worked in a shop that did this, but you can do it.

DFHMDF – this is where you actually define what goes on the screen.

```
DFHMSD TYPE=FINAL
END
```

The latter terminates the mapset.

Column 17 specifies detail about the entry just given. If the specification requires more than one line, a continuation character X must be placed in column 72. You can then continue the specification beginning in column 16 on the next line. Here's an example from the map:

```
ACTION    DFHMDF POS=(06,28),LENGTH=01,ATTRB=(IC,UNPROT),COLOR=GREEN,    X
               HILIGHT=UNDERLINE
```

The above created the input/output field called ACTION on line 6, column 28.

Compiling the EMPMMNU Screen

To generate the screen map, you must assemble it using the BMS assembler program. Henceforth in this book, we'll refer to this as compiling the mapset or screen. While technically we are "assembling it", compile is a more commonly used term for making a program executable, so we'll use it.

You'll need to locate the appropriate compile JCL used in your shop. I will show you the one I use, but it will undoubtedly be different that yours. Check with your supervisor or technical leader for your shop's JCL.

```
//USER01D JOB 'ASSEMBLE MAP',CLASS=A,
//           MSGLEVEL=(1,1),NOTIFY=&SYSUID
//*
//* TO ASSEMBLE CICS BMS MAP
//*
//CICSMAP  EXEC DFHMAPS,
//         COPYLIB=USER01.COPYLIB,       <= SYMBOLIC MAP COPY LIBRARY
//         SRCLIB=USER01.CICS.MAPLIB,    <= SOURCE LIBRARY
//         MEMBER=EMPMMNU                 <= MAP MEMBER NAME
//*
```

Always check the output for errors. The most common BMS error I've seen is overlapping fields. A good compile will have these words near the end of the output:

> **No Statements Flagged in this Assembly**

Using CEDA to Define and Install Components

Before you can use a CICS resource you must define and install it in CICS. CEDA is a CICS transaction that allows a developer or administrator to add and maintain resource definitions. To bring up CEDA, logon to CICS and enter transaction CEDA. You will see the following:

```
ENTER ONE OF THE FOLLOWING

ADd
ALter
APpend
CHeck
COpy
DEFine
DELete
DIsplay
Expand
Install
Lock
Move
REMove
REName
UNlock
USerdefine
View
                                           SYSID=CICS APPLID=CICSTS42


PF 1 HELP        3 END         6 CRSR         9 MSG          12 CNCL
```

Defining a Mapset

Type DEF or DEFINE and press enter. You will see this screen.

```
DEF
 ENTER ONE OF THE FOLLOWING

Atomservice   MQconn        Webservice
Bundle        PARTItionset
CONnection    PARTNer
CORbaserver   PIpeline
DB2Conn       PROCesstype
DB2Entry      PROFile
DB2Tran       PROGram
DJar          Requestmodel
DOctemplate   Sessions
Enqmodel      TCpipservice
File          TDqueue
Ipconn        TErminal
JOurnalmodel  TRANClass
JVmserver     TRANSaction
LIbrary       TSmodel
LSrpool       TYpeterm
MApset        Urimap


                                          SYSID=CICS APPLID=CICSTS42


PF 1 HELP       3 END           6 CRSR          9 MSG          12 CNCL
```

Now type MAPSET and you will see this screen:

```
DEF mapset
 OVERTYPE TO MODIFY                            CICS RELEASE = 0670
  CEDA  DEFine MApset(          )
   MApset        ==>
   Group         ==>
   DEScription   ==>
   REsident      ==> No                No | Yes
   USAge         ==> Normal            Normal | Transient
   USElpacopy    ==> No                No | Yes
   Status        ==> Enabled           Enabled | Disabled
   RSl           : 00                  0-24 | Public
  DEFINITION SIGNATURE
   DEFinetime    :
   CHANGETime    :
   CHANGEUsrid   :
   CHANGEAGEnt   :                     CSDApi | CSDBatch
   CHANGEAGRel   :




  MESSAGES: 2 SEVERE
                                          SYSID=CICS APPLID=CICSTS42

PF 1 HELP 2 COM 3 END            6 CRSR 7 SBH 8 SFH 9 MSG 10 SB 11 SF 12 CNCL
```

534

You will need to enter the mapset name, the CICS group and a description for the transaction. This information is listed in the design, except for the group name. In this textbook we are specifying USER01 as our group name. In your case, you'll be assigned a group name by your system administrator (it might be your user id by default). Let's enter our info as follows:

```
DEF mapset
 OVERTYPE TO MODIFY                                     CICS RELEASE = 0670
  CEDA  DEFine MApset(          )
   MApset        ==> EMPMMNU
   Group         ==> USER01
   DEScription   ==> EMPLOYEE SUPPORT MAIN MENU
   REsident      ==> No                   No | Yes
   USAge         ==> Normal               Normal | Transient
   USElpacopy    ==> No                   No | Yes
   Status        ==> Enabled              Enabled | Disabled
   RSl           :   00                   0-24 | Public
  DEFINITION SIGNATURE
   DEFinetime    :
   CHANGETime    :
   CHANGEUsrid   :
   CHANGEAGEnt   :                        CSDApi | CSDBatch
   CHANGEAGRel   :


                                          SYSID=CICS APPLID=CICSTS42

PF 1 HELP 2 COM 3 END           6 CRSR 7 SBH 8 SFH 9 MSG 10 SB 11 SF 12 CNCL
```

When you press ENTER, you will see this showing the define is successful:

```
 OVERTYPE TO MODIFY                                     CICS RELEASE = 0670
  CEDA  DEFine MApset( EMPMMNU  )
   MApset        :  EMPMMNU
   Group         :  USER01
   DEScription   ==> EMPLOYEE SUPPORT MAIN MENU
   REsident      ==> No                   No | Yes
   USAge         ==> Normal               Normal | Transient
   USElpacopy    ==> No                   No | Yes
   Status        ==> Enabled              Enabled | Disabled
   RSl           :   00                   0-24 | Public
  DEFINITION SIGNATURE
   DEFinctime    :  11/27/18 04:09:56
   CHANGETime    :  11/27/18 04:09:56
   CHANGEUsrid   :  USER01
   CHANGEAGEnt   :  CSDApi               CSDApi | CSDBatch
   CHANGEAGRel   :  0670


                                          SYSID=CICS APPLID=CICSTS42
   DEFINE SUCCESSFUL                      TIME: 04.09.56  DATE: 11/27/18
 PF 1 HELP 2 COM 3 END           6 CRSR 7 SBH 8 SFH 9 MSG 10 SB 11 SF 12 CNCL
```

Installing a Mapset

After defining your mapset, you must install it. You can follow the menu prompts again, or you can simply enter the full command as follows:

```
INSTALL MAPSET(EMPMMNU) GROUP(USER01)
 OVERTYPE TO MODIFY
  CEDA  Install
   ATomservice  ==>
   Bundle       ==>
   CONnection   ==>
   CORbaserver  ==>
   DB2Conn      ==>
   DB2Entry     ==>
   DB2Tran      ==>
   DJar         ==>
   DOctemplate  ==>
   Enqmodel     ==>
   File         ==>
   Ipconn       ==>
   JOurnalmodel ==>
   JVmserver    ==>
   LIBrary      ==>
   LSrpool      ==>
+  MApset       ==> EMPMMNU

                                      SYSID=CICS APPLID=CICSTS42
  INSTALL SUCCESSFUL               TIME: 04.12.08  DATE: 11/27/18
 PF 1 HELP       3 END       6 CRSR 7 SBH 8 SFH 9 MSG 10 SB 11 SF 12 CNCL
```

Displaying a Map

Now you can check the format of the screen by displaying the map in CICS. You cannot use the transaction name itself to display the map because we have yet not defined the transaction or program. However you can use the CECI utility as follows.

```
CECI SEND MAP(EMPMNU) MAPSET(EMPMMNU)
```

Now you'll see this screen. Press ENTER.

```
SEND MAP(EMPMNU) MAPSET(EMPMMNU)
 STATUS:  ABOUT TO EXECUTE COMMAND                          NAME=
  EXEC CICS  SENd Map( 'EMPMNU ' )
   << FROm() > < LEngth() > < DAtaonly > | MAPOnly >
   < MAPSet( 'EMPMMNU' ) >
   < FMhparm() >
   < Reqid() >
   < LDc() | < ACTpartn() > < Outpartn() > >
   < MSr() >
   < Cursor() >
   < Set() < MAPPingdev() > | PAging | Terminal < Wait > < LAst > >
   < PRint >
   < FREekb >
   < ALArm >
   < L40 | L64 | L80 | Honeom >
   < NLeom >
   < ERASE < DEfault | ALTernate > | ERASEAup >
   < ACCum >
   < FRSet >
+  < NOflush >


PF 1 HELP 2 HEX 3 END 4 EIB 5 VAR 6 USER 7 SBH 8 SFH 9 MSG 10 SB 11 SF
```

Now you'll see the formatted map:

```
EMPMMNU                      EMPLOYEE SUPPORT MENU                      EMNU

          ENTER THE NUMBER OF YOUR SELECTION,   THEN PRESS ENTER.


                     _   1. EMPLOYEE INQUIRY

                         2. EMPLOYEE ADD

                         3. EMPLOYEE CHANGE

                         4. EMPLOYEE DELETE




     F3 EXIT
```

Congratulations, you've just created, deployed and displayed a CICS component!

Compiling and Viewing the Rest of the Screens

We could move on to the programming part of our project by starting the program EMPPGMNU. Instead, I suggest defining, installing and displaying all your screens first. Doing so means you can perform your programming tasks without getting sidetracked by an improperly formatted screen. Let's go ahead and do the remaining four screen definitions.

Note: on the data entry screens, we initialize or pre-fill the enterable fields with X's or with the format required (e.g., YYYY-MM-DD for dates) until the user fills in data or the program fills it in with data from a query. An example appears below.

```
EMPMINQ                     EMPLOYEE INQUIRY                        EMIN


      EMPLOYEE ->              ENTER EMPLOYEE ID, THEN PRESS ENTER

      EMPLOYEE ID    XXXX

      EMP LAST NAME  XXXX

      EMP FIRST NAME XXXX

      EMP SOCIAL SEC XXXXXXXXX

      EMP YEARS SRVC 00

      EMP LAST PROM  YYYY-MM-DD




   F2 INQ   F3 EXIT   F4 ADD   F5 CHG   F6 DEL
```

This technique is not required, but I recommend doing the field pre-fill during development because it helps to see the data format. For production you can remove the pre-fill characters unless your user base finds it worthwhile. Some do, some don't.

EMPMINQ

Here is the BMS code for the employee inquiry screen EMPMINQ.   Be sure to use tag names that make sense. The maintenance programmers that follow you will appreciate that.

```
EMPMINQ    DFHMSD TYPE=&SYSPARM,MODE=INOUT,CTRL=(FREEKB,FRSET),          X
               LANG=COBOL,TIOAPFX=YES,                                   X
               DSATTS=(COLOR,HILIGHT),                                   X
               MAPATTS=(COLOR,HILIGHT),                                  X
               STORAGE=AUTO
EMPINQ     DFHMDI SIZE=(24,80)
           DFHMDF POS=(01,1),LENGTH=07,COLOR=BLUE,                       X
               INITIAL='EMPMINQ'
           DFHMDF POS=(01,28),LENGTH=16,COLOR=BLUE,                      X
               INITIAL='EMPLOYEE INQUIRY'
TRANID     DFHMDF POS=(01,76),LENGTH=04,INITIAL='EMIN',COLOR=BLUE
           DFHMDF POS=(04,08),LENGTH=11,COLOR=BLUE,                      X
               INITIAL='EMPLOYEE ->'
EMPIN      DFHMDF POS=(04,23),LENGTH=04,ATTRB=(IC,UNPROT),               X
               COLOR=GREEN,HILIGHT=UNDERLINE
           DFHMDF POS=(04,28),LENGTH=01,ATTRB=ASKIP
           DFHMDF POS=(04,34),LENGTH=35,COLOR=BLUE,                      X
               INITIAL='ENTER EMPLOYEE ID, THEN PRESS ENTER'
           DFHMDF POS=(06,08),LENGTH=14,COLOR=BLUE,                      X
               INITIAL='EMPLOYEE ID   '
EMPNO      DFHMDF POS=(06,23),LENGTH=04,COLOR=GREEN,                     X
               INITIAL='XXXX'
           DFHMDF POS=(08,08),LENGTH=14,COLOR=BLUE,                      X
               INITIAL='EMP LAST NAME '
LNAME      DFHMDF POS=(08,23),LENGTH=30,COLOR=GREEN,                     X
               INITIAL='XXXX'
           DFHMDF POS=(10,08),LENGTH=14,COLOR=BLUE,                      X
               INITIAL='EMP FIRST NAME'
FNAME      DFHMDF POS=(10,23),LENGTH=20,COLOR=GREEN,                     X
               INITIAL='XXXX'
           DFHMDF POS=(12,08),LENGTH=14,COLOR=BLUE,                      X
               INITIAL='EMP SOCIAL SEC'
SOCSEC     DFHMDF POS=(12,23),LENGTH=09,COLOR=GREEN,                     X
               INITIAL='XXXXXXXXX'
           DFHMDF POS=(14,08),LENGTH=14,COLOR=BLUE,                      X
               INITIAL='EMP YEARS SRVC'
YRSSVC     DFHMDF POS=(14,23),LENGTH=02,COLOR=GREEN,                     X
               INITIAL='00'
           DFHMDF POS=(16,08),LENGTH=14,COLOR=BLUE,                      X
               INITIAL='EMP LAST PROM '
LSTPRM     DFHMDF POS=(16,23),LENGTH=10,COLOR=GREEN,                     X
               INITIAL='YYYY-MM-DD'
MESSAGE    DFHMDF POS=(23,02),LENGTH=67,COLOR=YELLOW
           DFHMDF POS=(24,02),LENGTH=43,ATTRB=PROT,COLOR=BLUE,           X
               INITIAL='F2 INQ   F3 EXIT   F4 ADD   F5 CHG   F6 DEL'
           DFHMSD TYPE=FINAL
           END
```

Go ahead and compile the mapset. Then define, install and display it. The CICS commands are as follows:

```
def mapset
 OVERTYPE TO MODIFY                                        CICS RELEASE = 0670
  CEDA  DEFine MApset(            )
   MApset       ==> EMPMINQ
   Group        ==> USER01
   DEScription  ==> EMPLOYEE INQUIRY
   REsident     ==> No              No | Yes
   USAge        ==> Normal          Normal | Transient
   USElpacopy   ==> No              No | Yes
   Status       ==> Enabled         Enabled | Disabled
   RSl          : 00                0-24 | Public
  DEFINITION SIGNATURE
   DEFinetime   :
   CHANGETime   :
   CHANGEUsrid  :
   CHANGEAGEnt  :                   CSDApi | CSDBatch
   CHANGEAGRel  :




                                           SYSID=CICS APPLID=CICSTS42

 PF 1 HELP 2 COM 3 END           6 CRSR 7 SBH 8 SFH 9 MSG 10 SB 11 SF 12 CNCL
```

```
INSTALL MAPSET(EMPMINQ) GROUP(USER01)
 OVERTYPE TO MODIFY
  CEDA  Install
   ATomservice  ==>
   Bundle       ==>
   CONnection   ==>
   CORbaserver  ==>
   DB2Conn      ==>
   DB2Entry     ==>
   DB2Tran      ==>
   DJar         ==>
   DOctemplate  ==>
   Enqmodel     ==>
   File         ==>
   Ipconn       ==>
   JOurnalmodel ==>
   JVmserver    ==>
   LIBrary      ==>
   LSrpool      ==>
+  MApset       ==> EMPMINQ

                                           SYSID=CICS APPLID=CICSTS42
  INSTALL SUCCESSFUL                       TIME: 04.36.21  DATE: 11/27/18
 PF 1 HELP       3 END           6 CRSR 7 SBH 8 SFH 9 MSG 10 SB 11 SF 12 CNCL
```

Now let's display the `EMPINQ` map using CECI:

```
CECI SEND MAP(EMPINQ) MAPSET(EMPMINQ)

SEND MAP(EMPINQ) MAPSET(EMPMINQ)
 STATUS:  ABOUT TO EXECUTE COMMAND                        NAME=
  EXEC CICS  SENd Map( 'EMPINQ ' )
   << FROm() > < LEngth() > < DAtaonly > | MAPOnly >
   < MAPSet( 'EMPMINQ' ) >
   < FMhparm() >
   < Reqid() >
   < LDc() | < ACTpartn() > < Outpartn() > >
   < MSr() >
   < Cursor() >
   < Set() < MAPPingdev() > | PAging | Terminal < Wait > < LAst > >
   < PRint >
   < FREekb >
   < ALArm >
   < L40 | L64 | L80 | Honeom >
   < NLeom >
   < ERASE < DEfault | ALTernate > | ERASEAup >
   < ACCum >
   < FRSet >
+  < NOflush >



PF 1 HELP 2 HEX 3 END 4 EIB 5 VAR 6 USER 7 SBH 8 SFH 9 MSG 10 SB 11 SF
```

When you press ENTER you will see the Employee Inquiry screen map as follows.

```
----+----1----+----2----+----3----+----4----+----5----+----6----+----7----+----8
EMPMINQ                    EMPLOYEE INQUIRY                             EMIN


        EMPLOYEE ->                ENTER EMPLOYEE ID, THEN PRESS ENTER

        EMPLOYEE ID    XXXX

        EMP LAST NAME  XXXX

        EMP FIRST NAME XXXX

        EMP SOCIAL SEC XXXXXXXXX

        EMP YEARS SRVC 00

        EMP LAST PROM  YYYY-MM-DD






  F2 INQ   F3 EXIT   F4 ADD   F5 CHG   F6 DEL
```

Now that you've got the hang of it, go ahead and compile, define, install and display the three other maps and mapsets. Makes sure they are as you want them to be displayed. I recommend that you try this on your own. You have the basics that you need, so give it a shot.

When you finish, turn the page and I'll give you my version of the three other mapsets. If you have any problems with compiling or displaying the maps, check to make sure your offsets are correct and that you don't have any overlapping fields.

EMPMADD

Here's the BMS code for the EMPMADD mapset.

```
EMPMADD   DFHMSD TYPE=&SYSPARM,MODE=INOUT,CTRL=(FREEKB,FRSET),          X
                 LANG=COBOL,TIOAPFX=YES,                                X
                 DSATTS=(COLOR,HILIGHT),                                X
                 MAPATTS=(COLOR,HILIGHT),                               X
                 STORAGE=AUTO
EMPADD    DFHMDI SIZE=(24,80)
          DFHMDF POS=(01,1),LENGTH=07,COLOR=BLUE,                       X
                 INITIAL='EMPMADD'
          DFHMDF POS=(01,28),LENGTH=12,COLOR=BLUE,                      X
                 INITIAL='EMPLOYEE ADD'
TRANID    DFHMDF POS=(01,76),LENGTH=04,INITIAL='EMAD',COLOR=BLUE
          DFHMDF POS=(04,23),LENGTH=35,COLOR=BLUE,                      X
                 INITIAL='ENTER EMPLOYEE INFO, THEN PRESS PF4'
          DFHMDF POS=(06,08),LENGTH=14,COLOR=BLUE,                      X
                 INITIAL='EMPLOYEE ID   '
EMPNO     DFHMDF POS=(06,23),LENGTH=04,COLOR=GREEN,                     X
                 HILIGHT=UNDERLINE,ATTRB=(IC,UNPROT)
          DFHMDF POS=(06,28),LENGTH=01,ATTRB=ASKIP
          DFHMDF POS=(08,08),LENGTH=14,COLOR=BLUE,                      X
                 INITIAL='EMP LAST NAME '
LNAME     DFHMDF POS=(08,23),LENGTH=30,COLOR=GREEN,ATTRB=(UNPROT),      X
                 HILIGHT=UNDERLINE
          DFHMDF POS=(08,54),LENGTH=01,ATTRB=ASKIP
          DFHMDF POS=(10,08),LENGTH=14,COLOR=BLUE,                      X
                 INITIAL='EMP FIRST NAME'
FNAME     DFHMDF POS=(10,23),LENGTH=20,COLOR=GREEN,ATTRB=(UNPROT),      X
                 HILIGHT=UNDERLINE
          DFHMDF POS=(10,44),LENGTH=01,ATTRB=ASKIP
          DFHMDF POS=(12,08),LENGTH=14,COLOR=BLUE,                      X
                 INITIAL='EMP SOCIAL SEC'
SOCSEC    DFHMDF POS=(12,23),LENGTH=09,COLOR=GREEN,ATTRB=(UNPROT),      X
                 HILIGHT=UNDERLINE
          DFHMDF POS=(12,33),LENGTH=01,ATTRB=ASKIP
          DFHMDF POS=(14,08),LENGTH=14,COLOR=BLUE,                      X
                 INITIAL='EMP YEARS SRVC'
YRSSVC    DFHMDF POS=(14,23),LENGTH=02,COLOR=GREEN,ATTRB=(UNPROT),      X
                 HILIGHT=UNDERLINE
          DFHMDF POS=(14,26),LENGTH=01,ATTRB=ASKIP
          DFHMDF POS=(16,08),LENGTH=14,COLOR=BLUE,                      X
                 INITIAL='EMP LAST PROM '
LSTPRM    DFHMDF POS=(16,23),LENGTH=10,COLOR=GREEN,ATTRB=(UNPROT),      X
```

```
            HILIGHT=UNDERLINE  '
         DFHMDF POS=(16,34),LENGTH=01,ATTRB=ASKIP
MESSAGE  DFHMDF POS=(23,02),LENGTH=67,COLOR=YELLOW
         DFHMDF POS=(24,02),LENGTH=43,ATTRB=PROT,COLOR=BLUE,            X
            INITIAL='F2 INQ   F3 EXIT   F4 ADD   F5 CHG   F6 DEL'
         DFHMSD TYPE=FINAL
         END
```

After you compile and install it, the add screen should display as follows using CECI:

```
----+----1----+----2----+----3----+----4----+----5----+----6----+----7----+----8
EMPMADD                      EMPLOYEE ADD                               EMAD


        EMPLOYEE ->    ENTER EMPLOYEE INFO, THEN PRESS PF4PRESS ENTER

        EMPLOYEE ID

        EMP LAST NAME

        EMP FIRST NAME

        EMP SOCIAL SEC

        EMP YEARS SRVC

        EMP LAST PROM




   F2 INQ   F3 EXIT   F4 ADD   F5 CHG   F6 DEL
```

EMPMCHG

Now here's the BMS code for the change screen:

```
EMPMCHG  DFHMSD TYPE=&SYSPARM,MODE=INOUT,CTRL=(FREEKB,FRSET),          X
            LANG=COBOL,TIOAPFX=YES,                                    X
            DSATTS=(COLOR,HILIGHT),                                    X
            MAPATTS=(COLOR,HILIGHT),                                   X
            STORAGE=AUTO
EMPCHG   DFHMDI SIZE=(24,80)
         DFHMDF POS=(01,1),LENGTH=07,COLOR=BLUE,                       X
            INITIAL='EMPMCHG'
         DFHMDF POS=(01,28),LENGTH=15,COLOR=BLUE,                      X
            INITIAL='EMPLOYEE CHANGE'
TRANID   DFHMDF POS=(01,76),LENGTH=04,INITIAL='EMCH',COLOR=BLUE
         DFHMDF POS=(04,08),LENGTH=11,COLOR=BLUE,                      X
            INITIAL='EMPLOYEE ->'
EMPIN    DFHMDF POS=(04,23),LENGTH=04,ATTRB=(IC,UNPROT),COLOR=GREEN,   X
            HILIGHT=UNDERLINE
         DFHMDF POS=(04,28),LENGTH=01,ATTRB=ASKIP
```

543

```
            DFHMDF POS=(04,34),LENGTH=35,COLOR=BLUE,                    X
                   INITIAL='ENTER EMPLOYEE ID, THEN PRESS ENTER'
            DFHMDF POS=(06,08),LENGTH=14,COLOR=BLUE,                    X
                   INITIAL='EMPLOYEE ID  '
EMPNO       DFHMDF POS=(06,23),LENGTH=04,COLOR=GREEN,                   X
                   INITIAL='XXXX'
            DFHMDF POS=(08,08),LENGTH=14,COLOR=BLUE,                    X
                   INITIAL='EMP LAST NAME '
LNAME       DFHMDF POS=(08,23),LENGTH=30,COLOR=GREEN,ATTRB=UNPROT,      X
                   INITIAL='XXXX'
            DFHMDF POS=(10,08),LENGTH=14,COLOR=BLUE,                    X
                   INITIAL='EMP FIRST NAME'
FNAME       DFHMDF POS=(10,23),LENGTH=20,COLOR=GREEN,ATTRB=UNPROT,      X
                   INITIAL='XXXX'
            DFHMDF POS=(12,08),LENGTH=14,COLOR=BLUE,                    X
                   INITIAL='EMP SOCIAL SEC'
SOCSEC      DFHMDF POS=(12,23),LENGTH=09,COLOR=GREEN,ATTRB=UNPROT,      X
                   INITIAL='XXXXXXXXX'
            DFHMDF POS=(14,08),LENGTH=14,COLOR=BLUE,                    X
                   INITIAL='EMP YEARS SRVC'
YRSSVC      DFHMDF POS=(14,23),LENGTH=02,COLOR=GREEN,ATTRB=UNPROT,      X
                   INITIAL='00'
            DFHMDF POS=(16,08),LENGTH=14,COLOR=BLUE,                    X
                   INITIAL='EMP LAST PROM '
LSTPRM      DFHMDF POS=(16,23),LENGTH=10,COLOR=GREEN,ATTRB=UNPROT,      X
                   INITIAL='YYYY-MM-DD'
MESSAGE     DFHMDF POS=(23,02),LENGTH=67,COLOR=YELLOW
            DFHMDF POS=(24,02),LENGTH=45,ATTRB=PROT,COLOR=BLUE,         X
                   INITIAL='F2 INQ   F3 EXIT   F4 ADD   F5 CHG   F6 DEL'
            DFHMSD TYPE=FINAL
            END
```

The change screen should display as follows:

```
----+----1----+----2----+----3----+----4----+----5----+----6----+----7----+----8
EMPMCHG                      EMPLOYEE CHANGE                          EMCH


        EMPLOYEE ->                ENTER EMPLOYEE ID, THEN PRESS ENTER

        EMPLOYEE ID   XXXX

        EMP LAST NAME  ____

        EMP FIRST NAME ____

        EMP SOCIAL SEC _____

        EMP YEARS SRVC __

        EMP LAST PROM  _____




  F2 INQ   F3 EXIT   F4 ADD   F5 CHG   F6 DEL
```

544

EMPMDEL

Now let's look at the BMS code for the delete screen.

```
EMPMDEL
EMPMDEL  DFHMSD TYPE=&SYSPARM,MODE=INOUT,CTRL=(FREEKB,FRSET),        X
                LANG=COBOL,TIOAPFX=YES,                              X
                DSATTS=(COLOR,HILIGHT),                              X
                MAPATTS=(COLOR,HILIGHT),                             X
                STORAGE=AUTO
EMPDEL   DFHMDI SIZE=(24,80)
         DFHMDF POS=(01,1),LENGTH=07,COLOR=BLUE,                    X
                INITIAL='EMPMDEL'
         DFHMDF POS=(01,28),LENGTH=15,COLOR=BLUE,                   X
                INITIAL='EMPLOYEE DELETE'
TRANID   DFHMDF POS=(01,76),LENGTH=04,INITIAL='EMDE',COLOR=BLUE
         DFHMDF POS=(04,08),LENGTH=11,COLOR=BLUE,                   X
                INITIAL='EMPLOYEE ->'
EMPIN    DFHMDF POS=(04,23),LENGTH=04,ATTRB=(IC,UNPROT),COLOR=GREEN, X
                HILIGHT=UNDERLINE
         DFHMDF POS=(04,28),LENGTH=01,ATTRB=ASKIP
         DFHMDF POS=(04,34),LENGTH=35,COLOR=BLUE,                   X
                INITIAL='ENTER EMPLOYEE ID, THEN PRESS ENTER'
         DFHMDF POS=(06,08),LENGTH=14,COLOR=BLUE,                   X
                INITIAL='EMPLOYEE ID   '
EMPNO    DFHMDF POS=(06,23),LENGTH=04,COLOR=GREEN,                  X
                INITIAL='XXXX'
         DFHMDF POS=(08,08),LENGTH=14,COLOR=BLUE,                   X
                INITIAL='EMP LAST NAME '
LNAME    DFHMDF POS=(08,23),LENGTH=30,COLOR=GREEN,ATTRB=UNPROT,     X
                INITIAL='XXXX'
         DFHMDF POS=(10,08),LENGTH=14,COLOR=BLUE,                   X
                INITIAL='EMP FIRST NAME'
FNAME    DFHMDF POS=(10,23),LENGTH=20,COLOR=GREEN,ATTRB=UNPROT,     X
                INITIAL='XXXX'
         DFHMDF POS=(12,08),LENGTH=14,COLOR=BLUE,                   X
                INITIAL='EMP SOCIAL SEC'
SOCSEC   DFHMDF POS=(12,23),LENGTH=09,COLOR=GREEN,ATTRB=UNPROT,     X
                INITIAL='XXXXXXXXX'
         DFHMDF POS=(14,08),LENGTH=14,COLOR=BLUE,                   X
                INITIAL='EMP YEARS SRVC'
YRSSVC   DFHMDF POS=(14,23),LENGTH=02,COLOR=GREEN,ATTRB=UNPROT,     X
                INITIAL='00'
         DFHMDF POS=(16,08),LENGTH=14,COLOR=BLUE,                   X
                INITIAL='EMP LAST PROM '
LSTPRM   DFHMDF POS=(16,23),LENGTH=10,COLOR=GREEN,ATTRB=UNPROT,     X
                INITIAL='YYYY-MM-DD'
MESSAGE  DFHMDF POS=(23,02),LENGTH=67,COLOR=YELLOW
         DFHMDF POS=(24,02),LENGTH=45,ATTRB=PROT,COLOR=BLUE,        X
                INITIAL='F2 INQ   F3 EXIT   F4 ADD   F5 CHG   F6 DEL'
         DFHMSD TYPE=FINAL
         END
```

The delete screen should display as follows:

```
----+----1----+----2----+----3----+----4----+----5----+----6----+----7----+----8
EMPMDEL                      EMPLOYEE DELETE                          EMDE


         EMPLOYEE ->              ENTER EMPLOYEE ID, THEN PRESS ENTER

         EMPLOYEE ID

         EMP LAST NAME

         EMP FIRST NAME

         EMP SOCIAL SEC

         EMP YEARS SRVC

         EMP LAST PROM




  F2 INQ   F3 EXIT   F4 ADD   F5 CHG   F6 DEL
```

Final Notes on Mapsets

When you make a modification to a mapset that has already been defined and installed into production, you must refresh the copy that resides in the online executable library. The reason is that a mapset is an executable program, and the load library that you compile into is not the same as the live load library that CICS uses to load programs. To refresh a modified program in CICS, you use the CEMT utility with the SET PROGRAM command and NEWCOPY parameter. For example, to refresh the menu program you would logon to CICS and enter the following:

```
    CEMT SET PROGRAM(EMPMMNU) NEWCOPY
```

The above is also true for the application programs we'll soon be writing. Anytime you change the program, you must reload it into CICS using the CEMT utility.

Program Construction and Testing

Now we are ready to write the program code according to some basic specifications. For each program we will specify the name, purpose, data access requirements, SQL (if any), and some test cases. The test cases especially will help us understand how to code the program. Finally, we'll do unit testing of each program.

Employee Menu Program

The following are rudimentary specifications for our menu program.

```
Name:         EMPPGMNU
Purpose:      Employee Support Main Menu
Data Access:  None
SQL:          None (non-DB2)
```

EMPPGMNU Test Cases:

Note: four of the test cases below are shaded to indicate we cannot test them yet because they transfer to programs that we haven't written yet. We'll come back to those in integration testing.

Case	Condition	Expected Result	Actual Result
1	Initial screen display	All literals displayed, empty action field	
2	Invalid selection entered	Error message returned: ENTER A VALID ACTION: 1, 2, 3 OR 4	
3	Invalid key pressed	Error message: INVALID KEY PRESSED	
4	Option 1 entered	Transfer to EMIN transaction	
5	Option 2 entered	Transfer to EMAD transaction	
6	Option 3 entered	Transfer to EMCH transaction	
7	Option 4 entered	Transfer to EMDE transaction	

EMPPGMNU Program Code

We've already provided this code earlier as a demonstrator, but here it is again.

```
       IDENTIFICATION DIVISION.
       PROGRAM-ID. EMPPGMNU.
      ****************************************************
      *   MENU PROGRAM FOR EMPLOYEE APPLICATION          *
      *                                                  *
      *   AUTHOR      : ROBERT WINGATE                   *
      *   DATE-WRITTEN : 2018-07-26                      *
      ****************************************************
```

```cobol
ENVIRONMENT DIVISION.

DATA DIVISION.

WORKING-STORAGE SECTION.

01 WS-FLAGS.
   05 SW-VALID-SELECTION      PIC X(1) VALUE 'N'.
      88  VALID-SELECTION              VALUE 'Y'.
      88  NOT-VALID-SELECTION          VALUE 'N'.

01 WS-VARS.
   05 COMM-AREA               PIC X(20) VALUE SPACE.
   05 PROGRAM-NAME            PIC X(08) VALUE SPACES.
   05 INVALID-ACTION-MSG      PIC X(34)
      VALUE 'ENTER A VALID ACTION: 1, 2, 3 OR 4'.

   COPY EMPMMNU.
   COPY DFHAID.
   COPY DFHBMSCA.

LINKAGE SECTION.

01 DFHCOMMAREA          PIC X(20).

PROCEDURE DIVISION.

   IF EIBCALEN > ZERO
     MOVE DFHCOMMAREA  TO COMM-AREA
   END-IF.

   EVALUATE TRUE

     WHEN EIBCALEN = ZERO
       MOVE LOW-VALUES   TO EMPMNUO
       PERFORM SEND-MAP

     WHEN EIBAID = DFHCLEAR
       MOVE LOW-VALUES   TO EMPMNUO
       PERFORM SEND-MAP

     WHEN EIBAID = DFHPA1 OR DFHPA2 OR DFHPA3
       CONTINUE

     WHEN EIBAID = DFHPF3
       MOVE LOW-VALUES TO EMPMNUO
       MOVE "BYE, PRESS CLEAR KEY TO ENTER A TRANSACTION ID"
            TO MESSAGEO
       PERFORM SEND-MAP-DATAONLY

       EXEC CICS
```

```
              RETURN
         END-EXEC

      WHEN EIBAID = DFHENTER
         PERFORM MAIN-PROCESS-PARA

      WHEN OTHER
         MOVE LOW-VALUES TO EMPMNUO
         MOVE "INVALID KEY PRESSED" TO MESSAGEO
         PERFORM SEND-MAP-DATAONLY

   END-EVALUATE.

   EXEC CICS
      RETURN TRANSID('EMNU')
      COMMAREA (COMM-AREA)
   END-EXEC.

MAIN-PROCESS-PARA.

   PERFORM RECEIVE-MAP.

   IF ACTIONI NOT = '1' AND '2' AND '3' AND '4'
      MOVE DFHREVRS TO ACTIONH
      MOVE INVALID-ACTION-MSG TO MESSAGEO
      SET NOT-VALID-SELECTION TO TRUE
   ELSE
      SET VALID-SELECTION TO TRUE
   END-IF.

   IF VALID-SELECTION
      EVALUATE ACTIONI
         WHEN '1'
            MOVE 'EMPPGINQ' TO PROGRAM-NAME
         WHEN '2'
            MOVE 'EMPPGADD' TO PROGRAM-NAME
         WHEN '3'
            MOVE 'EMPPGCHG' TO PROGRAM-NAME
         WHEN '4'
            MOVE 'EMPPGDEL' TO PROGRAM-NAME
      END-EVALUATE

      PERFORM BRANCH-TO-PROGRAM

   END-IF.

   PERFORM SEND-MAP-DATAONLY.

BRANCH-TO-PROGRAM.

   EXEC CICS
      XCTL PROGRAM(PROGRAM-NAME)
```

```
                     END-EXEC

                     MOVE 'PROGRAM NOT AVAILABLE' TO MESSAGEO.

             SEND-MAP.
                 EXEC CICS SEND
                     MAP    ('EMPMNU')
                     MAPSET ('EMPMMNU')
                     FROM   (EMPMNUO)
                     ERASE
                 END-EXEC.

             SEND-MAP-DATAONLY.
                 EXEC CICS SEND
                     MAP    ('EMPMNU')
                     MAPSET ('EMPMMNU')
                     FROM   (EMPMNUO)
                     DATAONLY
                 END-EXEC.


             RECEIVE-MAP.
                  EXEC CICS RECEIVE
                     MAP    ('EMPMNU')
                     MAPSET ('EMPMMNU')
                     INTO   (EMPMNUI)
                  END-EXEC.
```

Compiling a Program

To compile the COBOL program, you'll need to locate the appropriate compile JCL used in your shop. Check with your supervisor or technical leader for the needed JCL. When you complete the compile process, always check the output for errors. Once there are no errors, you can proceed to define and install the program.

Here's the compile JCL I use:

```
//USER01D JOB MSGLEVEL=(1,1),NOTIFY=&SYSUID
//*  COBOL + DB2 + CICS COMPILE JCL
//*
//DBONL    EXEC DB2CICSC,
//             COPYLIB=USER01.COPYLIB,            <= COPYBOOK LIBRARY
//             DCLGLIB=USER01.DCLGEN.COBOL,       <= DCLGEN LIBRARY
//             DBRMLIB=USER01.DBRMLIB,            <= DBRM LIBRARY
//             SRCLIB=USER01.CICS.SRCLIB,         <= SOURCE LIBRARY
//             MEMBER=EMPPGADD                    <= SOURCE MEMBER
```

Defining a Program

Type DEF or DEFINE and press enter. You will see this screen.

```
DEF
 ENTER ONE OF THE FOLLOWING

Atomservice  MQconn        Webservice
Bundle       PARTItionset
CONnection   PARTNer
CORbaserver  PIpeline
DB2Conn      PROCesstype
DB2Entry     PROFile
DB2Tran      PROGram
DJar         Requestmodel
DOctemplate  Sessions
Enqmodel     TCpipservice
File         TDqueue
Ipconn       TErminal
JOurnalmodel TRANClass
JVmserver    TRANSaction
LIbrary      TSmodel
LSrpool      TYpeterm
MApset       Urimap


                                          SYSID=CICS APPLID=CICSTS42


PF 1 HELP      3 END          6 CRSR        9 MSG          12 CNCL
```

Now type PROGRAM and you will see this screen:

```
DEF PROGRAM
 OVERTYPE TO MODIFY                           CICS RELEASE = 0670
  CEDA  DEFine PROGram(        )
   PROGram      ==>
   Group        ==>
   DEScription  ==>
   Language     ==>               CObol | Assembler | Le370 | C | Pli
   RELoad       ==> No            No | Yes
   RESident     ==> No            No | Yes
   USAge        ==> Normal        Normal | Transient
   USElpacopy   ==> No            No | Yes
   Status       ==> Enabled       Enabled | Disabled
   RSl          : 00              0-24 | Public
   CEdf         ==> Yes           Yes | No
   DAtalocation ==> Below         Below | Any
   EXECKey      ==> User          User | Cics
   COncurrency  ==> Quasirent     Quasirent | Threadsafe | Required
   Api          ==> Cicsapi       Cicsapi | Openapi
  REMOTE ATTRIBUTES
+  DYnamic      ==> No            No | Yes
  MESSAGES: 2 SEVERE
                                          SYSID=CICS APPLID=CICSTS42

PF 1 HELP 2 COM 3 END        6 CRSR 7 SBH 8 SFH 9 MSG 10 SB 11 SF 12 CNCL
```

551

You will need to enter the mapset name, the CICS group and a description for the transaction. This information is in the design. So we enter it as follows:

```
DEF PROGRAM
 OVERTYPE TO MODIFY                                    CICS RELEASE = 0670
  CEDA  DEFine PROGram(          )
   PROGram       ==> EMPPGMNU
   Group         ==> USER01
   DEScription   ==> EMPLOYEE SUPPORT MAIN MENU
   Language      ==> COBOL           CObol | Assembler | Le370 | C | Pli
   RELoad        ==> No              No | Yes
   RESident      ==> No              No | Yes
   USAge         ==> Normal          Normal | Transient
   USElpacopy    ==> No              No | Yes
   Status        ==> Enabled         Enabled | Disabled
   RSl           : 00                0-24 | Public
   CEdf          ==> Yes             Yes | No
   DAtalocation  ==> Below           Below | Any
   EXECKey       ==> User            User | Cics
   COncurrency   ==> Quasirent       Quasirent | Threadsafe | Required
   Api           ==> Cicsapi         Cicsapi | Openapi
  REMOTE ATTRIBUTES
+  DYnamic       ==> No              No | Yes
  MESSAGES: 2 SEVERE
                                            SYSID=CICS APPLID=CICSTS42

PF 1 HELP 2 COM 3 END          6 CRSR 7 SBH 8 SFH 9 MSG 10 SB 11 SF 12 CNCL
```

When you press ENTER, you will see this screen, and always check for the DEFINE SUCCESS-FUL message at the bottom:

```
 OVERTYPE TO MODIFY                                    CICS RELEASE = 0670
  CEDA  DEFine PROGram( EMPPGMNU )
   PROGram           : EMPPGMNU
   Group             : USER01
   DEScription   ==> EMPLOYEE SUPPORT MAIN MENU
   Language      ==> COBOL           CObol | Assembler | Le370 | C | Pli
   RELoad        ==> No              No | Yes
   RESident      ==> No              No | Yes
   USAge         ==> Normal          Normal | Transient
   USElpacopy    ==> No              No | Yes
   Status        ==> Enabled         Enabled | Disabled
   RSl           : 00                0-24 | Public
   CEdf          ==> Yes             Yes | No
   DAtalocation  ==> Below           Below | Any
   EXECKey       ==> User            User | Cics
   COncurrency   ==> Quasirent       Quasirent | Threadsafe | Required
   Api           ==> Cicsapi         Cicsapi | Openapi
  REMOTE ATTRIBUTES
+  DYnamic       ==> No              No | Yes

                                            SYSID=CICS APPLID=CICSTS42
  DEFINE SUCCESSFUL                   TIME: 06.27.01  DATE: 11/27/18
PF 1 HELP 2 COM 3 END          6 CRSR 7 SBH 8 SFH 9 MSG 10 SB 11 SF 12 CNCL
```

Installing a Program

Now you must install the program, which is very similar to installing the mapset except you specify program. You can follow the menu prompts again, or you can simply enter the full command as follows:

```
INSTALL PROGRAM(EMPPGMNU) GROUP(USER01)
 OVERTYPE TO MODIFY
  CEDA  Install
   ATomservice  ==>
   Bundle       ==>
   CONnection   ==>
   CORbaserver  ==>
   DB2Conn      ==>
   DB2Entry     ==>
   DB2Tran      ==>
   DJar         ==>
   DOctemplate  ==>
   Enqmodel     ==>
   File         ==>
   Ipconn       ==>
   JOurnalmodel ==>
   JVmserver    ==>
   LIBrary      ==>
   LSrpool      ==>
 + MApset       ==>


                                   SYSID=CICS APPLID=CICSTS42
   INSTALL SUCCESSFUL              TIME: 06.30.42  DATE: 11/27/18
 PF 1 HELP        3 END      6 CRSR 7 SBH 8 SFH 9 MSG 10 SB 11 SF 12 CNCL
```

Defining a Transaction

Next you must define and install the corresponding transaction. You do this in a similar way, but there is no source code. Go through the menu system or simply type DEF TRANS.

```
DEF TRANSACTION
 OVERTYPE TO MODIFY                                    CICS RELEASE = 0670
  CEDA  DEFine TRANSaction(        )
   TRANSaction  ==> EMNU
   Group        ==> USER01
   DEScription  ==> EMPLOYEE SUPPORT MAIN MENU
   PROGram      ==> EMPPGMNU
   TWasize      ==> 00000            0-32767
   PROFile      ==> DFHCICST
   PArtitionset ==>
   STAtus       ==> Enabled          Enabled | Disabled
   PRIMedsize    : 00000             0-65520
   TASKDATALoc  ==> Below            Below | Any
   TASKDATAKey  ==> User             User | Cics
   STOrageclear ==> No               No | Yes
   RUnaway      ==> System           System | 0 | 500-2700000
   SHutdown     ==> Disabled         Disabled | Enabled
   ISolate      ==> Yes              Yes | No
   Brexit       ==>
 + REMOTE ATTRIBUTES


                                      SYSID=CICS APPLID=CICSTS42


 PF 1 HELP 2 COM 3 END          6 CRSR 7 SBH 8 SFH 9 MSG 10 SB 11 SF 12 CNCL
```

```
 OVERTYPE TO MODIFY                                    CICS RELEASE = 0670
   CEDA  DEFine TRANSaction( EMNU )
    TRANSaction   : EMNU
    Group         : USER01
    DEScription  ==> EMPLOYEE SUPPORT MAIN MENU
    PROGram      ==> EMPPGMNU
    TWasize      ==> 00000            0-32767
    PROFile      ==> DFHCICST
    PArtitionset ==>
    STAtus       ==> Enabled          Enabled | Disabled
    PRIMedsize    : 00000             0-65520
    TASKDATALoc  ==> Below            Below | Any
    TASKDATAKey  ==> User             User | Cics
    STOrageclear ==> No               No | Yes
    RUnaway      ==> System           System | 0 | 500-2700000
    SHutdown     ==> Disabled         Disabled | Enabled
    ISolate      ==> Yes              Yes | No
    Brexit       ==>
 + REMOTE ATTRIBUTES

                                      SYSID=CICS APPLID=CICSTS42
   DEFINE SUCCESSFUL                  TIME: 06.38.01  DATE: 11/27/18
 PF 1 HELP 2 COM 3 END          6 CRSR 7 SBH 8 SFH 9 MSG 10 SB 11 SF 12 CNCL
```

Installing a Transaction

Now you must install the transaction, just like with the mapset and program. Here is the command and the result:

```
INSTALL TRANSACTION(EMNU) GROUP(USER01)
 OVERTYPE TO MODIFY
  CEDA  Install
   ATomservice   ==>
   Bundle        ==>
   CONnection    ==>
   CORbaserver   ==>
   DB2Conn       ==>
   DB2Entry      ==>
   DB2Tran       ==>
   DJar          ==>
   DOctemplate   ==>
   Enqmodel      ==>
   File          ==>
   Ipconn        ==>
   JOurnalmodel  ==>
   JVmserver     ==>
   LIBrary       ==>
   LSrpool       ==>
+  MApset        ==>

                                          SYSID=CICS APPLID=CICSTS42
   INSTALL SUCCESSFUL                 TIME: 06.40.01  DATE: 11/27/18
 PF 1 HELP        3 END        6 CRSR 7 SBH 8 SFH 9 MSG 10 SB 11 SF 12 CNCL
```

Testing a Transaction

Now you can test the transaction, program and mapset by simply invoking the transaction. Let's try it. With a clear screen, enter transaction EMNU. It should display correctly on your screen.

```
EMPMMNU                    EMPLOYEE SUPPORT MENU                      EMNU

          ENTER THE NUMBER OF YOUR SELECTION,   THEN PRESS ENTER.


                       1. EMPLOYEE INQUIRY

                       2. EMPLOYEE ADD

                       3. EMPLOYEE CHANGE

                       4. EMPLOYEE DELETE




   F3 EXIT
```

Great, it displays! However, we cannot test it further until the other programs are built and developed (so that the menu options can actually transfer you to those programs). So for now we'll place our menu program on the shelf as we construct the four DB2-CICS programs to access and operate on the employee data.

CICS DB2 Programs

Employee Inquiry Program

The following are basic specifications for our inquiry program. Pair this with the screen specifications and we can code this program.

```
Name:        EMPPGINQ
Purpose:     Display Employee Information
Data Access: Read-Only on EMPLOYEE
SQL:
             SELECT EMP_ID,
                    EMP_LAST_NAME,
                    EMP_FIRST_NAME,
                    EMP_SSN,
                    EMP_SERVICE_YEARS,
                    EMP_PROMOTION_DATE
               INTO
                    :EMP-ID,
                    :EMP-LAST-NAME,
                    :EMP-FIRST-NAME,
                    :EMP-SSN,
                    :EMP-SERVICE-YEARS,
                    :EMP-PROMOTION-DATE
             FROM USER01.EMPLOYEE
             WHERE EMP_ID = :EMP-ID
```

EMPPGINQ Test Cases:

Case	Condition	Expected Result	Actual Result
1	Initial screen display	All literals displayed, empty employee id	
2	Valid employee number entered	Detail returned for employee	
3	Invalid employee number entered	Error message that employee does not exist	
4	Enter pressed without changing anything	No change	
5	PF2 pressed	Refresh screen display	

6	PF3 pressed	Message to clear screen and enter a transaction id	
7	Other PF Keys pressed (PF1, PF7, PF8)	Error message - invalid key	
8	Employee id passed from other screen	Detail returned for employee	
9	PF4 pressed	Transfer to EMAD transaction	
10	PF5 pressed	Transfer to EMCH transaction	
11	PF6 Pressed	Transfer to EMDE transaction	

Note: the shaded test cases cannot be tested yet because the programs do not exist. We will bring these back and apply them in integration testing.

EMPPGINQ Program Code

The following is the program code for EMPPGINQ. Take a good look at it, and then we'll discuss how/why we coded it this way.

```
      IDENTIFICATION DIVISION.
      PROGRAM-ID. EMPPGINQ.
     ****************************************************
     *  COBOL/CICS/DB2 PROGRAM TO DISPLAY AN EMPLOYEE *
     *                                                *
     *  AUTHOR        : ROBERT WINGATE                *
     *  DATE-WRITTEN  : 2018-07-19                    *
     ****************************************************
      ENVIRONMENT DIVISION.
      DATA DIVISION.
      WORKING-STORAGE SECTION.
      01 WS-EMPNO          PIC 9(4).
      01 WS-EMP-SRV-YRS    PIC 9(2).
      01 WS-SQLCODE        PIC S9(08).
      01 WS-COMMAREA.
         05 WS-EMP-PASS    PIC 9(04) VALUE ZERO.
         05 WS-PGM-PASS    PIC X(08) VALUE SPACES.
         05 FILLER         PIC X(08).

      01 PROGRAM-NAME      PIC X(08) VALUE SPACES.

      01 SW-PASSED-DATA-SWITCH   PIC X(1) VALUE 'N'.
         88  SW-PASSED-DATA            VALUE 'Y'.
         88  SW-NO-PASSED-DATA         VALUE 'N'.

         COPY EMPMINQ.
         COPY DFHAID.
         COPY DFHBMSCA.

         EXEC SQL
```

557

```
        INCLUDE SQLCA
     END-EXEC.

     EXEC SQL
        INCLUDE EMPLOYEE
     END-EXEC.

LINKAGE SECTION.
01 DFHCOMMAREA          PIC X(20).

PROCEDURE DIVISION.

     SET SW-NO-PASSED-DATA TO TRUE

     IF EIBCALEN > ZERO
        MOVE DFHCOMMAREA  TO WS-COMMAREA
     END-IF.

     EVALUATE TRUE

        WHEN EIBCALEN = ZERO
           MOVE LOW-VALUES   TO  EMPINQO
           PERFORM SEND-MAP

        WHEN EIBAID = DFHCLEAR
           MOVE LOW-VALUES   TO  EMPINQO
           PERFORM SEND-MAP

        WHEN EIBAID = DFHPA1 OR DFHPA2 OR DFHPA3
           CONTINUE

        WHEN EIBAID = DFHPF2
           IF WS-PGM-PASS NOT EQUAL "EMPPGINQ"
              SET SW-PASSED-DATA TO TRUE
           ELSE
              SET SW-NO-PASSED-DATA TO TRUE
           END-IF
           PERFORM PROCESS-PARA

        WHEN EIBAID = DFHPF3
           MOVE LOW-VALUES TO  EMPINQO
           MOVE "BYE, PRESS CLEAR KEY TO ENTER A TRANSACTION ID"
                TO MESSAGEO
           PERFORM SEND-MAP-DATA

           EXEC CICS
              RETURN
           END-EXEC

        WHEN EIBAID = DFHPF4
           MOVE 'EMPPGADD' TO PROGRAM-NAME
           PERFORM BRANCH-TO-PROGRAM
```

```
            EXEC CICS
               RETURN
            END-EXEC

         WHEN EIBAID = DFHPF5
            MOVE 'EMPPGCHG' TO PROGRAM-NAME
            PERFORM BRANCH-TO-PROGRAM

            EXEC CICS
               RETURN
            END-EXEC

         WHEN EIBAID = DFHPF6
            MOVE 'EMPPGDEL' TO PROGRAM-NAME
            PERFORM BRANCH-TO-PROGRAM

            EXEC CICS
               RETURN
            END-EXEC

         WHEN EIBAID = DFHENTER
            PERFORM PROCESS-PARA

         WHEN OTHER
            MOVE LOW-VALUES TO EMPINQO
            MOVE "INVALID KEY PRESSED" TO MESSAGEO
            PERFORM SEND-MAP-DATA

      END-EVALUATE.

      EXEC CICS
         RETURN TRANSID('EMIN')
         COMMAREA (WS-COMMAREA)
         LENGTH(20)
      END-EXEC.

  PROCESS-PARA.

      PERFORM RECEIVE-MAP.
      INITIALIZE DCLEMPLOYEE MESSAGEO

      IF SW-PASSED-DATA
         MOVE WS-EMP-PASS TO EMP-ID
      ELSE
         MOVE EMPINI    TO WS-EMPNO WS-EMP-PASS
         MOVE WS-EMPNO  TO EMP-ID
      END-IF

      EXEC SQL
         SELECT EMP_ID,
                EMP_LAST_NAME,
```

559

```
              EMP_FIRST_NAME,
              EMP_SSN,
              EMP_SERVICE_YEARS,
              EMP_PROMOTION_DATE
          INTO
              :EMP-ID,
              :EMP-LAST-NAME,
              :EMP-FIRST-NAME,
              :EMP-SSN,
              :EMP-SERVICE-YEARS,
              :EMP-PROMOTION-DATE
          FROM USER01.EMPLOYEE
          WHERE EMP_ID = :EMP-ID
END-EXEC.

MOVE SQLCODE   TO  WS-SQLCODE.

EVALUATE SQLCODE
   WHEN 0
      MOVE EMP-ID                TO   WS-EMPNO
      MOVE WS-EMPNO              TO   EMPNOO EMPINO
      MOVE EMP-LAST-NAME-TEXT  TO LNAMEO
      MOVE EMP-FIRST-NAME-TEXT TO FNAMEO
      MOVE EMP-SSN               TO SOCSECO
      MOVE EMP-SERVICE-YEARS    TO WS-EMP-SRV-YRS
      MOVE WS-EMP-SRV-YRS       TO YRSSVCO
      MOVE EMP-PROMOTION-DATE   TO LSTPRMO

   WHEN 100
      STRING "EMPLOYEE ID " DELIMITED BY SIZE
      WS-EMPNO DELIMITED BY SPACE
      " NOT FOUND" DELIMITED BY SIZE INTO MESSAGEO
      MOVE WS-EMPNO       TO EMPINO
      MOVE SPACES        TO EMPNOO
      MOVE SPACES        TO LNAMEO
      MOVE SPACES        TO FNAMEO
      MOVE SPACES        TO SOCSECO
      MOVE SPACES        TO YRSSVCO
      MOVE SPACES        TO LSTPRMO

   WHEN OTHER
      STRING "ERROR - SQL CODE: " DELIMITED BY SIZE
            WS-SQLCODE    DELIMITED BY SIZE
         INTO MESSAGEO
END-EVALUATE.

MOVE DFHBMFSE   TO EMPINF
MOVE -1 TO EMPINL
MOVE "EMPPGINQ" TO WS-PGM-PASS

IF SW-PASSED-DATA
   PERFORM SEND-MAP
```

560

```
        ELSE
            PERFORM SEND-MAP-DATA
        END-IF.

    BRANCH-TO-PROGRAM.

        EXEC CICS
            XCTL PROGRAM(PROGRAM-NAME)
            COMMAREA (WS-COMMAREA)
            LENGTH(20)
        END-EXEC

        MOVE 'PROGRAM NOT AVAILABLE' TO MESSAGEO.

    SEND-MAP.
        EXEC CICS SEND
            MAP    ('EMPINQ')
            MAPSET ('EMPMINQ')
            FROM   (EMPINQO)
            ERASE
        END-EXEC.

    SEND-MAP-DATA.
        EXEC CICS SEND
            MAP    ('EMPINQ')
            MAPSET ('EMPMINQ')
            FROM   (EMPINQO)
            DATAONLY
        END-EXEC.

    RECEIVE-MAP.
        EXEC CICS RECEIVE
            MAP    ('EMPINQ')
            MAPSET ('EMPMINQ')
            INTO   (EMPINQI)
        END-EXEC.
```

Ok, first let's take a look at the working storage section below. We have a work variable for employee number to convert it to numeric for our DB2 query. Similarly we have a work variable for employee service years. We've added a picture field for the DB2 SQLCODE.

Then we have a communication area for transferring data between programs. Note that the communication area WS-COMMAREA has two fields: the passed employee id and the name of the program that passed the data. This is important because if another program passes an employee id, then we will first retrieve the data and load the symbolic map. Then we'll display the initial screen which will already be populated with data for the employee number that was passed to the program. We'll set a switch that will let us know there is passed data – this will help us to determine certain logic branches later in our processing of the user request.

Finally notice we include two DB2 copybooks. These are for the SQLCA and for the EMPLOYEE DB2 table.

```
WORKING-STORAGE SECTION.
01 WS-EMPNO          PIC 9(4).
01 WS-EMP-SRV-YRS    PIC 9(2).
01 WS-SQLCODE        PIC S9(08).
01 WS-COMMAREA.
   05 WS-EMP-PASS    PIC 9(04) VALUE ZERO.
   05 WS-PGM-PASS    PIC X(08) VALUE SPACES.
   05 FILLER         PIC X(08).
01 PROGRAM-NAME      PIC X(08) VALUE SPACES.
01 SW-PASSED-DATA-SWITCH   PIC X(1) VALUE 'N'.
   88  SW-PASSED-DATA              VALUE 'Y'.
   88  SW-NO-PASSED-DATA           VALUE 'N'.

    COPY EMPMINQ.
    COPY DFHAID.
    COPY DFHBMSCA.

    EXEC SQL
      INCLUDE SQLCA
    END-EXEC.

    EXEC SQL
      INCLUDE EMPLOYEE
    END-EXEC.

LINKAGE SECTION.

01 DFHCOMMAREA        PIC X(20).
```

Now let's take a look at the procedures to send and receive data. These look just like the ones in the menu program, except for the names of the map and mapset. The initial screen will be sent using SEND-MAP. Subsequent screens will be sent using SEND-MAP-DATA. The user's screen input will be received into the program and loaded to the symbolic map using RECEIVE-MAP.

```
SEND-MAP.
    EXEC CICS SEND
        MAP    ('EMPINQ')
        MAPSET ('EMPMINQ')
        FROM   (EMPINQO)
        ERASE
    END-EXEC.

SEND-MAP-DATA.
    EXEC CICS SEND
        MAP    ('EMPINQ')
        MAPSET ('EMPMINQ')
        FROM   (EMPINQO)
        DATAONLY
```

```
          END-EXEC.


     RECEIVE-MAP.
          EXEC CICS RECEIVE
            MAP    ('EMPINQ')
            MAPSET ('EMPMINQ')
            INTO   (EMPINQI)
          END-EXEC.
```

Now let's look briefly at the BRANCH-TO-PROGRAM paragraph. This one is slightly different. Besides the program name to transfer to, we've added two parameters: COMMAREA and LENGTH. These are somewhat self-explanatory (the named communication area we defined in working storage, and it's length), and they are absolutely necessary to transfer data from one program to another. We're going to code for that but we won't be able to test that part of the program until the other programs are completed.

```
     BRANCH-TO-PROGRAM.

          EXEC CICS
            XCTL PROGRAM(PROGRAM-NAME)
            COMMAREA (WS-COMMAREA)
            LENGTH(20)
          END-EXEC

          MOVE 'PROGRAM NOT AVAILABLE' TO MESSAGEO.
```

Now let's look at our program mainline and see what's different from the menu program. For our inquiry program, if the pressed key is PF2 and the name of the program passing data is not this program's name (EMPPGINQ), then we know this program was arrived at by transfer from another program. So we turn on the SW-PASSED-DATA switch. We'll use that later. In either case, since the PF2 key was pressed we call the processing paragraph. We'll look at that momentarily.

Meanwhile, notice that if the PF4, PF5 or PF6 key is pressed, the program will perform a transfer to the add, change or delete program, respectively. Since we haven't programmed those yet, you might want to comment out that code (or just remember not to press those keys in testing until the programs are written – otherwise your session will freeze or possibly abend).

```
          SET SW-NO-PASSED-DATA TO TRUE

       IF EIBCALEN > ZERO
         MOVE DFHCOMMAREA  TO WS-COMMAREA
       END-IF.
```

```
EVALUATE TRUE

   WHEN EIBCALEN = ZERO
     MOVE LOW-VALUES    TO  EMPINQO
     PERFORM SEND-MAP

   WHEN EIBAID = DFHCLEAR
     MOVE LOW-VALUES    TO  EMPINQO
     PERFORM SEND-MAP

   WHEN EIBAID = DFHPA1 OR DFHPA2 OR DFHPA3
     CONTINUE

   WHEN EIBAID = DFHPF2
     IF WS-PGM-PASS NOT EQUAL "EMPPGINQ"
        SET SW-PASSED-DATA TO TRUE
     ELSE
        SET SW-NO-PASSED-DATA TO TRUE
     END-IF
     PERFORM PROCESS-PARA

   WHEN EIBAID = DFHPF3
     MOVE LOW-VALUES TO  EMPINQO
     MOVE "BYE, PRESS CLEAR KEY TO ENTER A TRANSACTION ID"
          TO MESSAGEO
     PERFORM SEND-MAP-DATA

     EXEC CICS
        RETURN
     END-EXEC

   WHEN EIBAID = DFHPF4
     MOVE 'EMPPGADD' TO PROGRAM-NAME
     PERFORM BRANCH-TO-PROGRAM

     EXEC CICS
        RETURN
     END-EXEC

   WHEN EIBAID = DFHPF5
     MOVE 'EMPPGCHG' TO PROGRAM-NAME
     PERFORM BRANCH-TO-PROGRAM

     EXEC CICS
        RETURN
     END-EXEC

   WHEN EIBAID = DFHPF6
     MOVE 'EMPPGDEL' TO PROGRAM-NAME
     PERFORM BRANCH-TO-PROGRAM

     EXEC CICS
        RETURN
     END-EXEC

   WHEN EIBAID = DFHENTER
     PERFORM PROCESS-PARA
```

564

```
        WHEN OTHER
          MOVE LOW-VALUES TO EMPINQO
          MOVE "INVALID KEY PRESSED" TO MESSAGEO
          PERFORM SEND-MAP-DATA

     END-EVALUATE.

     EXEC CICS
        RETURN TRANSID('EMIN')
        COMMAREA (WS-COMMAREA)
        LENGTH(20)
     END-EXEC.
```

Now let's look at the main processing paragraph. Here we will load the employee id to be retrieved, do the DB2 processing, and then either load the screen with employee data (if successful) or report an error. Then we'll perform a send map action.

```
     PROCESS-PARA.

        PERFORM RECEIVE-MAP.
        INITIALIZE DCLEMPLOYEE MESSAGEO

        IF SW-PASSED-DATA
           MOVE WS-EMP-PASS TO EMP-ID
        ELSE
           MOVE EMPINI    TO WS-EMPNO WS-EMP-PASS
           MOVE WS-EMPNO  TO EMP-ID
        END-IF

        EXEC SQL
           SELECT EMP_ID,
                  EMP_LAST_NAME,
                  EMP_FIRST_NAME,
                  EMP_SSN,
                  EMP_SERVICE_YEARS,
                  EMP_PROMOTION_DATE
              INTO
                  :EMP-ID,
                  :EMP-LAST-NAME,
                  :EMP-FIRST-NAME,
                  :EMP-SSN,
                  :EMP-SERVICE-YEARS,
                  :EMP-PROMOTION-DATE
              FROM USER01.EMPLOYEE
              WHERE EMP_ID = :EMP-ID
        END-EXEC.

        MOVE SQLCODE   TO   WS-SQLCODE.

        EVALUATE SQLCODE
           WHEN 0
```

```
         MOVE EMP-ID                TO  WS-EMPNO
         MOVE WS-EMPNO              TO  EMPNOO EMPINO
         MOVE EMP-LAST-NAME-TEXT   TO LNAMEO
         MOVE EMP-FIRST-NAME-TEXT  TO FNAMEO
         MOVE EMP-SSN               TO SOCSECO
         MOVE EMP-SERVICE-YEARS    TO WS-EMP-SRV-YRS
         MOVE WS-EMP-SRV-YRS       TO YRSSVCO
         MOVE EMP-PROMOTION-DATE   TO LSTPRMO

      WHEN 100
         STRING "EMPLOYEE ID " DELIMITED BY SIZE
         WS-EMPNO DELIMITED BY SPACE
         " NOT FOUND" DELIMITED BY SIZE INTO MESSAGEO
         MOVE WS-EMPNO        TO EMPINO
         MOVE SPACES          TO EMPNOO
         MOVE SPACES          TO LNAMEO
         MOVE SPACES          TO FNAMEO
         MOVE SPACES          TO SOCSECO
         MOVE SPACES          TO YRSSVCO
         MOVE SPACES          TO LSTPRMO


      WHEN OTHER
         STRING "ERROR - SQL CODE: " DELIMITED BY SIZE
                WS-SQLCODE   DELIMITED BY SIZE
            INTO MESSAGEO
   END-EVALUATE.

   MOVE DFHBMFSE    TO EMPINF
   MOVE -1 TO EMPINL
   MOVE "EMPPGINQ" TO WS-PGM-PASS

   IF SW-PASSED-DATA
      PERFORM SEND-MAP
   ELSE
      PERFORM SEND-MAP-DATA
   END-IF.
```

Notice that we check SW-PASSED-DATA to see if data was passed to the program, and if it was, we use the passed employee id value for the query. Otherwise we will use the employee id the user entered on the screen which resides in the input symbolic map.

Also we load the inquiry program name into WS-PGM-PASS so on the next time through the program we know to use whatever employee id is on the screen (the user may have changed it) instead of the previously passed employee id.

Finally, we check to see if data was passed and if so we will use the send map routine that clears the screen and writes the entire map (including literals) rather than the data-only version that is used when the user has entered something on an already presented employee inquiry screen.

Now we can compile the program, define and install it in CICS, and test it as well. Again you will need to check with your supervisor or a fellow programmer for the correct JCL to compile a CICS-DB2 program. Mine looks like this:

```
//USER01D JOB MSGLEVEL=(1,1),NOTIFY=&SYSUID
//*
//*  COBOL + DB2 + CICS COMPILE JCL
//*
//DBONL    EXEC DB2CICSC,
//              COPYLIB=USER01.COPYLIB,        <= COPYBOOK LIBRARY
//              DCLGLIB=USER01.DCLGEN.COBOL,   <= DCLGEN LIBRARY
//              DBRMLIB=USER01.DBRMLIB,        <= DBRM LIBRARY
//              SRCLIB=USER01.CICS.SRCLIB,     <= SOURCE LIBRARY
//              MEMBER=EMPPGCHG                <= SOURCE MEMBER
```

You will also need to bind the program of course. I use this DB2 command to do it:

```
BIND  MEMBER     (EMPPGINQ) -
PLAN      (MRWP01) -
ACTION    (REPLACE)  -
ISOLATION (CS)        -
EXPLAIN   (YES)       -
VALIDATE  (BIND)      -
RELEASE   (COMMIT)    -
OWNER     (USER01)    -
QUALIFIER (USER01)    -
ENCODING  (1047)
```

Take particular note of the plan name because you'll need it when defining the DB2ENTRY CICS entity that is required for each DB2 program. Note also that my plan name is rather arbitrary. You will need to use a plan name that is consistent with your installation's standards. Sometimes a DBA will assign the plan name for you to use.

Ok, next you can define the program and transaction exactly the way you defined the menu program:

```
OVERTYPE TO MODIFY                                      CICS RELEASE = 0670
  CEDA  DEFine PROGram( EMPPGINQ )
  PROGram       : EMPPGINQ
  Group         : USER01
  DEScription  ==> EMPLOYEE INQUIRY
  Language     ==> CObol           CObol | Assembler | Le370 | C | Pli
  RELoad       ==> No              No | Yes
  RESident     ==> No              No | Yes
  USAge        ==> Normal          Normal | Transient
  USElpacopy   ==> No              No | Yes
  Status       ==> Enabled         Enabled | Disabled
  RSl           : 00               0-24 | Public
  CEdf         ==> Yes             Yes | No
  DAtalocation ==> Below           Below | Any
  EXECKey      ==> User            User | Cics
  COncurrency  ==> Quasirent       Quasirent | Threadsafe | Required
  Api          ==> Cicsapi         Cicsapi | Openapi
  REMOTE ATTRIBUTES
+ DYnamic      ==> No              No | Yes


                                          SYSID=CICS APPLID=CICSTS42
  DEFINE SUCCESSFUL                       TIME: 07.26.57  DATE: 11/27/18
PF 1 HELP 2 COM 3 END        6 CRSR 7 SBH 8 SFH 9 MSG 10 SB 11 SF 12 CNCL




OVERTYPE TO MODIFY                                      CICS RELEASE = 0670
  CEDA  DEFine TRANSaction( EMIN )
  TRANSaction   : EMIN
  Group         : USER01
  DEScription  ==> EMPLOYEE INQUIRY
  PROGram      ==> EMPPGINQ
  TWasize      ==> 00000           0-32767
  PROFile      ==> DFHCICST
  PArtitionset ==>
  STAtus       ==> Enabled         Enabled | Disabled
  PRIMedsize    : 00000            0-65520
  TASKDATALoc  ==> Below           Below | Any
  TASKDATAKey  ==> User            User | Cics
  STOrageclear ==> No              No | Yes
  RUnaway      ==> System          System | 0 | 500-2700000
  SHutdown     ==> Disabled        Disabled | Enabled
  ISolate      ==> Yes             Yes | No
  Brexit       ==>
+ REMOTE ATTRIBUTES

                                          SYSID=CICS APPLID=CICSTS42
  DEFINE SUCCESSFUL                       TIME: 07.32.26  DATE: 11/27/18
PF 1 HELP 2 COM 3 END        6 CRSR 7 SBH 8 SFH 9 MSG 10 SB 11 SF 12 CNCL
```

And use these commands to install the program and transaction.

```
CEDA INSTALL PROGRAM(EMPPGINQ) GROUP(USER01)

CEDA INSTALL TRANS(EMIN) GROUP(USER01)
```

Finally, we need a `DB2ENTRY` to provide DB2 connectivity to the transaction. Here's how to define it – make sure to enter your DB2 plan name in the `PLAN` field and your transaction id in the transid field. Otherwise your program will fail when you run it.

```
DEF DB2ENTRY
OVERTYPE TO MODIFY                                    CICS RELEASE = 0670
 CEDA  DEFine DB2Entry(            )
  DB2Entry     ==> EMPPGINQ
  Group        ==> USER01
  DEScription  ==> EMPLOYEE INQUIRY
 THREAD SELECTION ATTRIBUTES
  TRansid      ==> EMIN
 THREAD OPERATION ATTRIBUTES
  ACcountrec   ==> None           None | TXid | TAsk | Uow
  AUTHId       ==>
  AUTHType     ==>                Userid | Opid | Group | Sign | TErm
                                  | TX
  DRollback    ==> Yes            Yes | No
  PLAN         ==> MRWP01
  PLANExitname ==>
  PRIority     ==> High           High | Equal | Low
  PROtectnum   ==> 0000           0-2000
  THREADLimit  ==>                0-2000
+ THREADWait   ==> Pool           Pool | Yes | No

                                           SYSID=CICS APPLID=CICSTS42


PF 1 HELP 2 COM 3 END            6 CRSR 7 SBH 8 SFH 9 MSG 10 SB 11 SF 12 CNCL
```

```
OVERTYPE TO MODIFY                                    CICS RELEASE = 0670
 CEDA  DEFine DB2Entry( EMPPGINQ )
  DB2Entry     : EMPPGINQ
  Group        : USER01
  DEScription  ==> EMPLOYEE INQUIRY
 THREAD SELECTION ATTRIBUTES
  TRansid      ==> EMIN
 THREAD OPERATION ATTRIBUTES
  ACcountrec   ==> None           None | TXid | TAsk | Uow
  AUTHId       ==>
  AUTHType     ==> Userid         Userid | Opid | Group | Sign | TErm
                                  | TX
  DRollback    ==> Yes            Yes | No
  PLAN         ==> MRWP01
  PLANExitname ==>
  PRIority     ==> High           High | Equal | Low
  PROtectnum   ==> 0000           0-2000
  THREADLimit  ==> 0000           0-2000
+ THREADWait   ==> Pool           Pool | Yes | No

                                           SYSID=CICS APPLID=CICSTS42
                                    TIME: 07.42.25  DATE: 11/27/18
  DEFINE SUCCESSFUL
PF 1 HELP 2 COM 3 END            6 CRSR 7 SBH 8 SFH 9 MSG 10 SB 11 SF 12 CNCL
```

569

Now you can install the DB2ENTRY as follows:

> **CEDA INSTALL DB2ENTRY(EMPPGINQ) GROUP(USER01)**

And now we are ready to test our inquiry program! Enter the EMIN transaction in CICS and you should see this screen.

```
EMPMINQ                  EMPLOYEE INQUIRY                        EMIN


     EMPLOYEE ->                 ENTER EMPLOYEE ID, THEN PRESS ENTER

     EMPLOYEE ID   XXXX

     EMP LAST NAME  XXXX

     EMP FIRST NAME XXXX

     EMP SOCIAL SEC XXXXXXXXX

     EMP YEARS SRVC 00

     EMP LAST PROM  YYYY-MM-DD




 F2 INQ   F3 EXIT   F4 ADD   F5 CHG   F6 DEL
```

Now key employee id 3217 into the EMPLOYEE field and press ENTER. Your result should be similar to this:

```
EMPMINQ                  EMPLOYEE INQUIRY                        EMIN


     EMPLOYEE ->    3217        ENTER EMPLOYEE ID, THEN PRESS ENTER

     EMPLOYEE ID    3217

     EMP LAST NAME  JOHNSON

     EMP FIRST NAME EDWARD

     EMP SOCIAL SEC 397342007

     EMP YEARS SRVC 07

     EMP LAST PROM  2017-01-01




 F2 INQ   F3 EXIT   F4 ADD   F5 CHG   F6 DEL
```

There you go! You've completed a CICS DB2 screen, program, transaction and DB2Entry. Now go ahead and execute all the test cases except the ones that transfer to other programs (PF4, PF5 and PF6), or that process data that is received from these programs. Make sure all functions work as intended before going on.

Now we have three more programs to write. These will be very similar to the inquiry program except of course we will be adding, updating or deleting data. That means we'll need field edits and the use of attributes to place the cursor, and a few other new features. Let's proceed with the add program.

Employee Add Program

The following are specifications for our add program.

```
Name:          EMPPGADD
Purpose:       Add a new employee to the EMPLOYEE table.
Data Access:   INSERT on EMPLOYEE

SQL Statements:

INSERT INTO USER01.EMPLOYEE
(EMP_ID,
 EMP_LAST_NAME,
 EMP_FIRST_NAME,
 EMP_SERVICE_YEARS,
 EMP_PROMOTION_DATE,
 EMP_SSN)

VALUES
(:EMP-ID,
 :EMP-LAST-NAME,
 :EMP-FIRST-NAME,
 :EMP-SERVICE-YEARS,
 :EMP-PROMOTION-DATE,
 :EMP-SSN)
```

EMPPGADD Test Cases:

Case	Condition	Expected Result	Actual Result
1	Initial screen display	Blank with enterable fields	
2	Enter pressed without changing anything	No change	
3	Employee number is blank	Error – employee number is required	
4	Employee number not numeric	Error – employee number must be numeric	

571

5	Last name is blank	Error – Last Name is required	
6	First name is blank	Error – Fast Name is required	
7	Social Security number is blank	Error – Social Security Number is required	
8	Social Security number is not numeric	Error – Social Security Number must be numeric	
9	Years of Service is blank	Error – Years of Service is required	
10	Years of Service not numeric	Error – Years of Service must be numeric	
11	Last Promotion Date is blank	Error – Last Promotion Date is required	
12	PF2 pressed	Transfer to EMIN transaction	
13	PF3 pressed	Message to clear screen and enter a transaction id	
14	PF4 pressed	Add record if no errors, message that record was added successfully	
15	PF5 pressed	Transfer to EMCH transaction	
16	PF6 Pressed	Transfer to EMDE transaction	
17	Other PF Keys pressed (PF1, PF7, PF8)	Error message – invalid key	

Employee Add Program Code

Here is the code for the add program. Much of the program looks like the inquiry program. But please review this add program carefully to see how we handle certain requirements such as field edits and errors. Also notice that the "action" key for this transaction is PF4 (not EN-TER).

```
       IDENTIFICATION DIVISION.
       PROGRAM-ID. EMPPGADD.
      **************************************************
      *   COBOL/CICS/DB2 PROGRAM TO ADD AN EMPLOYEE    *
      *                                                *
      *   AUTHOR      : ROBERT WINGATE                 *
      *   DATE-WRITTEN : 2018-07-21                    *
      **************************************************
       ENVIRONMENT DIVISION.
       DATA DIVISION.

       WORKING-STORAGE SECTION.
       01 WS-EMPNO         PIC 9(04).
       01 WS-EMP-SRV-YRS   PIC 9(02).
       01 WS-SQLCODE       PIC 9(08).
       01 WS-COMMAREA.
          05 WS-EMP-PASS   PIC 9(04).
          05 WS-PGM-PASS   PIC X(08).
```

```
    05 FILLER          PIC X(08).

01 PROGRAM-NAME        PIC X(08) VALUE SPACES.
01 POS-CTR             PIC S9(9) USAGE COMP VALUE +0.

01 SW-SPACE-FOUND-SWITCH   PIC X(1) VALUE 'N'.
   88  SW-SPACE-FOUND                 VALUE 'Y'.
   88  SW-SPACE-NOT-FOUND             VALUE 'N'.

   COPY EMPMADD.
   COPY DFHAID.
   COPY DFHBMSCA.

   EXEC SQL
     INCLUDE SQLCA
   END-EXEC.

   EXEC SQL
     INCLUDE EMPLOYEE
   END-EXEC.

LINKAGE SECTION.
01 DFHCOMMAREA          PIC X(20).

PROCEDURE DIVISION.

   IF EIBCALEN > ZERO
     MOVE DFHCOMMAREA  TO WS-COMMAREA
   END-IF.

   EVALUATE TRUE

     WHEN EIBCALEN = ZERO
       MOVE LOW-VALUES   TO  EMPADDO
       MOVE -1 TO EMPNOL
       PERFORM SEND-MAP

     WHEN EIBAID = DFHCLEAR
       MOVE LOW-VALUES   TO  EMPADDO
       MOVE -1 TO EMPNOL
       PERFORM SEND-MAP

     WHEN EIBAID = DFHPA1 OR DFHPA2 OR DFHPA3
       CONTINUE

     WHEN EIBAID = DFHPF2
       MOVE 'EMPPGINQ' TO PROGRAM-NAME
       PERFORM BRANCH-TO-PROGRAM

       EXEC CICS
         RETURN
       END-EXEC
```

```
               WHEN EIBAID = DFHPF3
                 MOVE LOW-VALUES TO  EMPADDO
                 MOVE -1 TO EMPNOL
                 MOVE "BYE, PRESS CLEAR KEY TO ENTER A TRANSACTION ID"
                     TO MESSAGEO
                 PERFORM SEND-MAP-DATA

                 EXEC CICS
                   RETURN
                 END-EXEC

               WHEN EIBAID = DFHPF4
 *                 PERFORM THE EDITS AND VALIDATIONS
 *                 IF NO ERRORS THEN INSERT THE RECORDS

                 IF WS-PGM-PASS NOT EQUAL "EMPPGADD"
                     MOVE LOW-VALUES TO EMPADDO
                     MOVE -1 TO EMPNOL
                     MOVE DFHBMFSE TO EMPNOF
                     MOVE
                     "ENTER DATA FOR NEW EMPLOYEE, THEN PRESS PF4 TO ADD"
                         TO MESSAGEO
                     PERFORM SEND-MAP
                 ELSE
                     PERFORM VALIDATE-DATA
                 END-IF

               WHEN EIBAID = DFHPF5
                 MOVE 'EMPPGCHG' TO PROGRAM-NAME
                 PERFORM BRANCH-TO-PROGRAM

               WHEN EIBAID = DFHPF6
                 MOVE 'EMPPGDEL' TO PROGRAM-NAME
                 PERFORM BRANCH-TO-PROGRAM

               WHEN EIBAID = DFHENTER
                 PERFORM PROCESS-PARA

               WHEN OTHER
                 MOVE LOW-VALUES TO EMPADDO
                 MOVE -1 TO EMPNOL
                 MOVE "INVALID KEY PRESSED" TO MESSAGEO
                 PERFORM SEND-MAP-DATA

           END-EVALUATE.

           EXEC CICS
               RETURN TRANSID('EMAD')
               COMMAREA (WS-COMMAREA)
           END-EXEC.
```

```
PROCESS-PARA.

    PERFORM RECEIVE-MAP.
    INITIALIZE DCLEMPLOYEE MESSAGEO

    MOVE DFHBMFSE TO EMPNOF
    MOVE DFHBMFSE TO LNAMEF
    MOVE DFHBMFSE TO FNAMEF
    MOVE DFHBMFSE TO SOCSECF
    MOVE DFHBMFSE TO YRSSVCF
    MOVE DFHBMFSE TO LSTPRMF

    MOVE -1 TO EMPNOL.
    MOVE "ENTER DATA FOR NEW EMPLOYEE, THEN PRESS PF4 TO ADD"
        TO MESSAGEO

    MOVE "EMPPGADD" TO WS-PGM-PASS
    PERFORM SEND-MAP-ALL.

VALIDATE-DATA.

    PERFORM RECEIVE-MAP
    INITIALIZE DCLEMPLOYEE MESSAGEO

    MOVE DFHBMFSE TO EMPNOF
    MOVE DFHBMFSE TO LNAMEF
    MOVE DFHBMFSE TO FNAMEF
    MOVE DFHBMFSE TO SOCSECF
    MOVE DFHBMFSE TO YRSSVCF
    MOVE DFHBMFSE TO LSTPRMF

    EVALUATE TRUE

        WHEN EMPNOI EQUAL SPACES OR EMPNOL EQUAL ZERO
            MOVE "EMPLOYEE NUMBER IS REQUIRED" TO MESSAGEO
            MOVE -1 TO EMPNOL

        WHEN EMPNOI IS NOT NUMERIC
            MOVE "EMPLOYEE NUMBER MUST BE NUMERIC" TO MESSAGEO
            MOVE -1 TO EMPNOL

        WHEN LNAMEI EQUAL SPACES OR LNAMEL EQUAL ZERO
            MOVE "EMPLOYEE LAST NAME IS REQUIRED" TO MESSAGEO
            MOVE -1 TO LNAMEL

        WHEN FNAMEI EQUAL SPACES OR FNAMEL EQUAL ZERO
            MOVE "EMPLOYEE FIRST NAME IS REQUIRED" TO MESSAGEO
            MOVE -1 TO FNAMEL

        WHEN SOCSECI EQUAL SPACES OR SOCSECL EQUAL ZERO
            MOVE "SOCIAL SECURITY NUMBER IS REQUIRED" TO MESSAGEO
            MOVE -1 TO SOCSECL
```

```
        WHEN SOCSECI IS NOT NUMERIC
            MOVE "SOCIAL SECURITY MUST BE NUMERIC" TO MESSAGEO
            MOVE -1 TO SOCSECL

        WHEN YRSSVCI EQUAL SPACES OR YRSSVCL EQUAL ZERO
            MOVE "YEARS OF SERVICE IS REQUIRED" TO MESSAGEO
            MOVE -1 TO YRSSVCL

        WHEN YRSSVCI IS NOT NUMERIC
            MOVE "YEARS OF SERVICE MUST BE NUMERIC" TO MESSAGEO
            MOVE -1 TO YRSSVCL

        WHEN LSTPRMI EQUAL SPACES OR LSTPRML EQUAL ZERO
            MOVE "LAST PROMOTION DATE IS REQUIRED" TO MESSAGEO
            MOVE -1 TO LSTPRML

        WHEN OTHER
            PERFORM ADD-RECORD
            MOVE -1 TO EMPNOL

    END-EVALUATE.

    PERFORM SEND-MAP-DATA.


ADD-RECORD.

*  MAP INPUT FIELDS TO DB2 RECORD

    MOVE EMPNOI              TO WS-EMPNO
    MOVE WS-EMPNO            TO EMP-ID
    MOVE LNAMEI             TO EMP-LAST-NAME-TEXT

    SET SW-SPACE-NOT-FOUND TO TRUE

    PERFORM VARYING POS-CTR FROM +1 BY +1
       UNTIL (POS-CTR = 30) OR SW-SPACE-FOUND
          IF EMP-LAST-NAME-TEXT(POS-CTR:1) = SPACE
             SET SW-SPACE-FOUND TO TRUE
          END-IF
    END-PERFORM

    IF POS-CTR EQUAL +30
       MOVE +30 TO EMP-LAST-NAME-LEN
    ELSE
       SUBTRACT +1 FROM POS-CTR
       MOVE POS-CTR TO EMP-LAST-NAME-LEN
    END-IF

    MOVE FNAMEI              TO EMP-FIRST-NAME-TEXT
```

```
        SET SW-SPACE-NOT-FOUND TO TRUE

        PERFORM VARYING POS-CTR FROM +1 BY +1
           UNTIL (POS-CTR = 20) OR SW-SPACE-FOUND
              IF EMP-FIRST-NAME-TEXT(POS-CTR:1) = SPACE
                 SET SW-SPACE-FOUND TO TRUE
              END-IF
        END-PERFORM

        IF POS-CTR EQUAL +20
           MOVE +20 TO EMP-FIRST-NAME-LEN
        ELSE
           SUBTRACT +1 FROM POS-CTR
           MOVE POS-CTR TO EMP-FIRST-NAME-LEN
        END-IF

        MOVE SOCSECI           TO EMP-SSN
        MOVE YRSSVCI           TO WS-EMP-SRV-YRS
        MOVE WS-EMP-SRV-YRS    TO EMP-SERVICE-YEARS
        MOVE LSTPRMI           TO EMP-PROMOTION-DATE

*   INSERT THE RECORD

        EXEC SQL
           INSERT INTO USER01.EMPLOYEE
           (EMP_ID,
            EMP_LAST_NAME,
            EMP_FIRST_NAME,
            EMP_SERVICE_YEARS,
            EMP_PROMOTION_DATE,
            EMP_SSN)

           VALUES
           (:EMP-ID,
            :EMP-LAST-NAME,
            :EMP-FIRST-NAME,
            :EMP-SERVICE-YEARS,
            :EMP-PROMOTION-DATE,
            :EMP-SSN)
        END-EXEC

*  HANDLE THE SQLCODE AND RETURN

        MOVE SQLCODE TO WS-SQLCODE

        EVALUATE SQLCODE
           WHEN 0
              MOVE "EMPLOYEE ADDED SUCCESSFULLY" TO MESSAGEO
              MOVE -1 TO EMPNOL
              MOVE WS-EMPNO TO WS-EMP-PASS
           WHEN -803
              MOVE "ERROR - RECORD ALREADY EXISTS" TO MESSAGEO
```

```
                MOVE -1 TO EMPNOL
                MOVE WS-EMPNO TO WS-EMP-PASS
            WHEN OTHER
                STRING "ERROR - DB2 SQLCODE IS " DELIMITED BY SIZE
                   WS-SQLCODE DELIMITED BY SIZE
                      INTO MESSAGEO
                MOVE -1 TO EMPNOL

        END-EVALUATE.

 BRANCH-TO-PROGRAM.

        EXEC CICS
            XCTL PROGRAM(PROGRAM-NAME)
            COMMAREA (WS-COMMAREA)
            LENGTH(10)
        END-EXEC

        MOVE 'PROGRAM NOT AVAILABLE' TO MESSAGEO.

 SEND-MAP.
        EXEC CICS SEND
            MAP    ('EMPADD')
            MAPSET ('EMPMADD')
            FROM   (EMPADDO)
            CURSOR
            ERASE
        END-EXEC.

 SEND-MAP-DATA.
        EXEC CICS SEND
            MAP    ('EMPADD')
            MAPSET ('EMPMADD')
            FROM   (EMPADDO)
            CURSOR
            DATAONLY
        END-EXEC.

 SEND-MAP-ALL.
        EXEC CICS SEND
            MAP    ('EMPADD')
            MAPSET ('EMPMADD')
            FROM   (EMPADDO)
            CURSOR
        END-EXEC.

 RECEIVE-MAP.
        EXEC CICS RECEIVE
            MAP    ('EMPADD')
            MAPSET ('EMPMADD')
            INTO   (EMPADDI)
        END-EXEC.
```

Notice in the error handling that if a data value is missing or invalid, we set the length field for that data element to -1. This has the effect of placing the cursor on that screen field. For example:

```
WHEN LNAMEI EQUAL SPACES OR LNAMEL EQUAL ZERO
    MOVE "EMPLOYEE LAST NAME IS REQUIRED" TO MESSAGEO
    MOVE -1 TO LNAMEL
```

Also notice the following code in the data validation routine.

```
MOVE DFHBMFSE TO EMPNOF
MOVE DFHBMFSE TO LNAMEF
MOVE DFHBMFSE TO FNAMEF
MOVE DFHBMFSE TO SOCSECF
MOVE DFHBMFSE TO YRSSVCF
MOVE DFHBMFSE TO LSTPRMF
```

The code turns on the data-changed attribute for each field to ensure that the data is actually sent in the send map routine. Ordinarily you do not need the above code – once you type data into the screen field, the appropriate F variable (such as LNAMEF) will be set by your 3270 emulator program to DFHBMFSE automatically to indicate the field data changed.

I included this code in my program because one of the 3270 emulation products I use has a bug and the data-changed attribute was not getting set. This caused the program to pass a zero length data area which caused an abend. So I forced the data-changed attribute to be set in the program code for each entry field (until I get an updated version of my 3270 emulator). If you ever have this problem, this is one workaround. Hopefully it won't be an issue for you.

Ok, let's get back now and compile, bind and test the program. Make sure you go through all the test cases. My test results are as follows:

```
EMPMADD                    EMPLOYEE ADD                              EMAD


                    ENTER EMPLOYEE INFO, THEN PRESS PF4

        EMPLOYEE ID

        EMP LAST NAME

        EMP FIRST NAME

        EMP SOCIAL SEC

        EMP YEARS SRVC

        EMP LAST PROM
```

EMPLOYEE NUMBER IS REQUIRED
```
F2 INQ   F3 EXIT   F4 ADD   F5 CHG   F6 DEL
```

```
EMPMADD                    EMPLOYEE ADD


                    ENTER EMPLOYEE INFO, THEN PRESS PF4

        EMPLOYEE ID    666N

        EMP LAST NAME

        EMP FIRST NAME

        EMP SOCIAL SEC

        EMP YEARS SRVC

        EMP LAST PROM
```

EMPLOYEE NUMBER MUST BE NUMERIC
```
F2 INQ   F3 EXIT   F4 ADD   F5 CHG   F6 DEL
```

```
EMPMADD                    EMPLOYEE ADD                         EMAD


                   ENTER EMPLOYEE INFO, THEN PRESS PF4

          EMPLOYEE ID    6666

          EMP LAST NAME

          EMP FIRST NAME

          EMP SOCIAL SEC

          EMP YEARS SRVC

          EMP LAST PROM
```

EMPLOYEE LAST NAME IS REQUIRED
```
F2 INQ    F3 EXIT    F4 ADD    F5 CHG    F6 DEL
```


```
EMPMADD                    EMPLOYEE ADD                         EMAD


                   ENTER EMPLOYEE INFO, THEN PRESS PF4

          EMPLOYEE ID    6666

          EMP LAST NAME   EDEN

          EMP FIRST NAME

          EMP SOCIAL SEC

          EMP YEARS SRVC

          EMP LAST PROM
```

EMPLOYEE FIRST NAME IS REQUIRED
```
F2 INQ    F3 EXIT    F4 ADD    F5 CHG    F6 DEL
```

```
EMPMADD                    EMPLOYEE ADD                        EMAD


                  ENTER EMPLOYEE INFO, THEN PRESS PF4

          EMPLOYEE ID    6666

          EMP LAST NAME  EDEN

          EMP FIRST NAME HELEN

          EMP SOCIAL SEC

          EMP YEARS SRVC

          EMP LAST PROM
```

SOCIAL SECURITY NUMBER IS REQUIRED
```
F2 INQ   F3 EXIT   F4 ADD   F5 CHG   F6 DEL
```

```
EMPMADD                    EMPLOYEE ADD                        EMAD


                  ENTER EMPLOYEE INFO, THEN PRESS PF4

          EMPLOYEE ID    6666

          EMP LAST NAME  EDEN

          EMP FIRST NAME HELEN

          EMP SOCIAL SEC 45239866C

          EMP YEARS SRVC

          EMP LAST PROM
```

SOCIAL SECURITY MUST BE NUMERIC
```
F2 INQ   F3 EXIT   F4 ADD   F5 CHG   F6 DEL
```

```
EMPMADD                     EMPLOYEE ADD                              EMAD


                    ENTER EMPLOYEE INFO, THEN PRESS PF4

           EMPLOYEE ID    6666

           EMP LAST NAME  EDEN

           EMP FIRST NAME HELEN

           EMP SOCIAL SEC 452398667

           EMP YEARS SRVC

           EMP LAST PROM
```

YEARS OF SERVICE IS REQUIRED
```
F2 INQ    F3 EXIT    F4 ADD    F5 CHG    F6 DEL
```


```
EMPMADD                     EMPLOYEE ADD                              EMAD


                    ENTER EMPLOYEE INFO, THEN PRESS PF4

           EMPLOYEE ID    6666

           EMP LAST NAME  EDEN

           EMP FIRST NAME HELEN

           EMP SOCIAL SEC 452398667

           EMP YEARS SRVC D7

           EMP LAST PROM
```

YEARS OF SERVICE MUST BE NUMERIC
```
F2 INQ    F3 EXIT    F4 ADD    F5 CHG    F6 DEL
```

```
EMPMADD                    EMPLOYEE ADD                           EMAD


                    ENTER EMPLOYEE INFO, THEN PRESS PF4

        EMPLOYEE ID    6666

        EMP LAST NAME  EDEN

        EMP FIRST NAME HELEN

        EMP SOCIAL SEC 452398667

        EMP YEARS SRVC 17

        EMP LAST PROM
```

LAST PROMOTION DATE IS REQUIRED
```
F2 INQ   F3 EXIT   F4 ADD   F5 CHG   F6 DEL
```

```
EMPMADD                    EMPLOYEE ADD                           EMAD


                    ENTER EMPLOYEE INFO, THEN PRESS PF4

        EMPLOYEE ID    6666

        EMP LAST NAME  EDEN

        EMP FIRST NAME HELEN

        EMP SOCIAL SEC 452398667

        EMP YEARS SRVC 17

        EMP LAST PROM  2018-02-30
```

ERROR - DB2 SQLCODE IS 00000181
```
F2 INQ   F3 EXIT   F4 ADD   F5 CHG   F6 DEL
```

Note: the above error means a bad date. You could have just coded the message, but I wanted to demonstrate an example of formatting an SQL code. For user friendliness, you can change the message to "Last promotion date must be valid" or something like that.

```
EMPMADD                    EMPLOYEE ADD                         EMAD

                  ENTER EMPLOYEE INFO, THEN PRESS PF4

        EMPLOYEE ID    6666

        EMP LAST NAME  EDEN

        EMP FIRST NAME HELEN

        EMP SOCIAL SEC 452398667

        EMP YEARS SRVC 17

        EMP LAST PROM  2018-01-01




  EMPLOYEE ADDED SUCCESSFULLY
  F2 INQ    F3 EXIT    F4 ADD    F5 CHG    F6 DEL
```

Ok, we successfully added the record. Now try to add the record again by pressing PF4. You should get an error message that the record already exists.

```
EMPMADD                    EMPLOYEE ADD                         EMAD

                  ENTER EMPLOYEE INFO, THEN PRESS PF4

        EMPLOYEE ID    6666

        EMP LAST NAME  EDEN

        EMP FIRST NAME HELEN

        EMP SOCIAL SEC 452398667

        EMP YEARS SRVC 17

        EMP LAST PROM  2018-01-01




  ERROR - RECORD ALREADY EXISTS
  F2 INQ    F3 EXIT    F4 ADD    F5 CHG    F6 DEL
```

Also try an invalid key such as PF7:

```
EMPMADD                        EMPLOYEE ADD                              EMAD


                      ENTER EMPLOYEE INFO, THEN PRESS PF4

           EMPLOYEE ID    1234

           EMP LAST NAME  Test

           EMP FIRST NAME Test

           EMP SOCIAL SEC 999999999

           EMP YEARS SRVC 22

           EMP LAST PROM  2018-01-01




      INVALID KEY PRESSED
      F2 INQ   F3 EXIT   F4 ADD   F5 CHG   F6 DEL
```

Last, press PF3 to end the transaction, then the clear key to clear the screen.

```
EMPMADD                        EMPLOYEE ADD                              EMAD


                      ENTER EMPLOYEE INFO, THEN PRESS PF4

           EMPLOYEE ID    6666

           EMP LAST NAME  EDEN

           EMP FIRST NAME HELEN

           EMP SOCIAL SEC 452398667

           EMP YEARS SRVC 17

           EMP LAST PROM  2018-01-01




      BYE, PRESS CLEAR KEY TO ENTER A TRANSACTION ID
      F2 INQ   F3 EXIT   F4 ADD   F5 CHG   F6 DEL
```

Ok that's it for the add program. Let's move on to the change program, EMPPGCHG.

Employee Change Program

The following are specifications for our change program. Note that there are two SQL statements, one to retrieve the record and one to update it.

```
Name:        EMPPGCHG
Purpose:     Change an employee in the EMPLOYEE table.
Data Access: Read and Update on EMPLOYEE

SQL Statements:

SELECT EMP_ID,
       EMP_LAST_NAME,
       EMP_FIRST_NAME,
       EMP_SSN,
       EMP_SERVICE_YEARS,
       EMP_PROMOTION_DATE
  INTO
       :EMP-ID,
       :EMP-LAST-NAME,
       :EMP-FIRST-NAME,
       :EMP-SSN,
       :EMP-SERVICE-YEARS,
       :EMP-PROMOTION-DATE
  FROM USER01.EMPLOYEE
  WHERE EMP_ID = :EMP-ID


UPDATE USER01.EMPLOYEE
SET EMP_LAST_NAME      = :EMP-LAST-NAME,
    EMP_FIRST_NAME     = :EMP-FIRST-NAME,
    EMP_SERVICE_YEARS  = :EMP-SERVICE-YEARS,
    EMP_PROMOTION_DATE = :EMP-PROMOTION-DATE,
    EMP_SSN            = :EMP-SSN
WHERE EMP_ID = :EMP-ID
```

EMPPGCHG Test Cases:

Case	Condition	Expected Result	Actual Result
1	Initial screen display	Literals displayed, all enterable fields unprotected.	
2	Valid employee number	Display detail for employee	
3	Invalid employee number	Error message that emp number is invalid	
4	Enter pressed without changing anything	No change	
5	Employee number is blank	Error – employee number is required	
6	Employee number not numeric	Error – employee number must be numeric	
7	Last name is blank	Error – Last Name is required	

8	First name is blank	Error - Fast Name is required	
9	Social Security number is blank	Error - Social Security Number is required	
10	Social Security number is not numeric	Error - Social Security Number must be numeric	
11	Years of Service is blank	Error - Years of Service is required	
12	Years of Service not numeric	Error - Years of Service must be numeric	
13	Last Promotion Date is blank	Error - Last Promotion Date is required	
14	PF2 pressed	Transfer to EMIN transaction	
15	PF3 pressed	Message to clear screen and enter a transaction id	
16	PF4 pressed	Transfer to EMAD transaction	
17	PF5 pressed	Change the record if no errors, message that record was changed	
18	PF6 Pressed	Transfer to EMDE transaction	
19	Other PF Keys pressed (PF1, PF7, PF8)	Error message - invalid key	

Employee Change Program Code

Here is the code for the change program. Much of the program looks like the add program, except we also do an initial display of the record before we edit and update it. Please review this change program carefully to see how we handle field edits and errors. Also notice that the "action" key for this transaction is PF5 (not ENTER, although you use ENTER if you are keying in a new employee id for display).

```
        IDENTIFICATION DIVISION.
        PROGRAM-ID. EMPPGCHG.
       *****************************************************
       *   COBOL/CICS/DB2 PROGRAM TO CHANGE AN EMPLOYEE *
       *                                                 *
       *   AUTHOR       : ROBERT WINGATE                 *
       *   DATE-WRITTEN : 2018-07-23                     *
       *****************************************************
        ENVIRONMENT DIVISION.
        DATA DIVISION.

        WORKING-STORAGE SECTION.
        01 WS-EMPNO        PIC 9(4).
        01 WS-EMP-SRV-YRS  PIC 9(02).
        01 WS-SQLCODE      PIC 9(08).
        01 WS-COMMAREA.
           05 WS-EMP-PASS  PIC 9(04).
           05 WS-PGM-PASS  PIC X(08).
```

```cobol
    05 FILLER          PIC X(08).

01 PROGRAM-NAME       PIC X(08) VALUE SPACES.
01 POS-CTR            PIC S9(9) USAGE COMP VALUE +0.

01 SW-SPACE-FOUND-SWITCH   PIC X(1) VALUE 'N'.
   88   SW-SPACE-FOUND                VALUE 'Y'.
   88   SW-SPACE-NOT-FOUND            VALUE 'N'.

01 SW-PASSED-DATA-SWITCH   PIC X(1) VALUE 'N'.
   88   SW-PASSED-DATA                VALUE 'Y'.
   88   SW-NO-PASSED-DATA             VALUE 'N'.

   COPY EMPMCHG.
   COPY DFHAID.
   COPY DFHBMSCA.

   EXEC SQL
     INCLUDE SQLCA
   END-EXEC.

   EXEC SQL
     INCLUDE EMPLOYEE
   END-EXEC.

LINKAGE SECTION.
01 DFHCOMMAREA        PIC X(20).

PROCEDURE DIVISION.

   SET SW-NO-PASSED-DATA TO TRUE

   IF EIBCALEN > ZERO
     MOVE DFHCOMMAREA  TO WS-COMMAREA
   END-IF.

   EVALUATE TRUE

     WHEN EIBCALEN = ZERO
       MOVE LOW-VALUES   TO  EMPCHGO
       MOVE -1 TO EMPINL
       PERFORM SEND-MAP

     WHEN EIBAID = DFHCLEAR
       MOVE LOW-VALUES   TO  EMPCHGO
       MOVE -1 TO EMPINL
       PERFORM SEND-MAP

     WHEN EIBAID = DFHPA1 OR DFHPA2 OR DFHPA3
       CONTINUE

     WHEN EIBAID = DFHPF2
```

```
                MOVE 'EMPPGINQ' TO PROGRAM-NAME
                PERFORM BRANCH-TO-PROGRAM

                EXEC CICS
                  RETURN
                END-EXEC

            WHEN EIBAID = DFHPF3
                MOVE LOW-VALUES TO  EMPCHGO
                MOVE -1 TO EMPINL
                MOVE "BYE, PRESS CLEAR KEY TO ENTER A TRANSACTION ID"
                    TO MESSAGEO
                PERFORM SEND-MAP-DATA

                EXEC CICS
                  RETURN
                END-EXEC

            WHEN EIBAID = DFHPF4
                MOVE 'EMPPGADD' TO PROGRAM-NAME
                PERFORM BRANCH-TO-PROGRAM

                EXEC CICS
                  RETURN
                END-EXEC

            WHEN EIBAID = DFHPF5
    *           PERFORM THE EDITS AND VALIDATIONS
    *           IF NO ERRORS THEN MODIFY THE RECORD

                IF WS-PGM-PASS NOT EQUAL "EMPPGCHG"
                   SET SW-PASSED-DATA TO TRUE
                   PERFORM PROCESS-PARA
                ELSE
                   PERFORM VALIDATE-DATA
                END-IF

            WHEN EIBAID = DFHPF6
                MOVE 'EMPPGDEL' TO PROGRAM-NAME
                PERFORM BRANCH-TO-PROGRAM

                EXEC CICS
                  RETURN
                END-EXEC

            WHEN EIBAID = DFHENTER
                PERFORM PROCESS-PARA

            WHEN OTHER
                MOVE LOW-VALUES TO EMPCHGO
                MOVE -1 TO EMPINL
                MOVE "INVALID KEY PRESSED" TO MESSAGEO
```

```
        PERFORM SEND-MAP-DATA

    END-EVALUATE.

    EXEC CICS
        RETURN TRANSID('EMCH')
        COMMAREA (WS-COMMAREA)
        LENGTH(20)
    END-EXEC.

PROCESS-PARA.

    PERFORM RECEIVE-MAP.
    INITIALIZE DCLEMPLOYEE MESSAGEO

    IF SW-PASSED-DATA
        MOVE WS-EMP-PASS TO EMP-ID
    ELSE
        MOVE EMPINI     TO WS-EMPNO WS-EMP-PASS
        MOVE WS-EMPNO   TO EMP-ID
    END-IF

    MOVE DFHBMFSE TO EMPINF
    MOVE DFHBMFSE TO EMPNOF
    MOVE DFHBMFSE TO LNAMEF
    MOVE DFHBMFSE TO FNAMEF
    MOVE DFHBMFSE TO SOCSECF
    MOVE DFHBMFSE TO YRSSVCF
    MOVE DFHBMFSE TO LSTPRMF

    EXEC SQL
        SELECT EMP_ID,
               EMP_LAST_NAME,
               EMP_FIRST_NAME,
               EMP_SSN,
               EMP_SERVICE_YEARS,
               EMP_PROMOTION_DATE
          INTO
               :EMP-ID,
               :EMP-LAST-NAME,
               :EMP-FIRST-NAME,
               :EMP-SSN,
               :EMP-SERVICE-YEARS,
               :EMP-PROMOTION-DATE
          FROM USER01.EMPLOYEE
          WHERE EMP_ID = :EMP-ID
    END-EXEC.

    MOVE SQLCODE        TO  WS-SQLCODE.
    DISPLAY "SQLCODE: " WS-SQLCODE.

    EVALUATE SQLCODE
```

```
        WHEN 0
           MOVE EMP-ID              TO WS-EMPNO WS-EMP-PASS
           MOVE WS-EMPNO            TO EMPNOO EMPINO
           MOVE EMP-LAST-NAME-TEXT  TO LNAMEO
           MOVE EMP-FIRST-NAME-TEXT TO FNAMEO
           MOVE EMP-SSN             TO SOCSECO
           MOVE EMP-SERVICE-YEARS   TO WS-EMP-SRV-YRS
           MOVE WS-EMP-SRV-YRS      TO YRSSVCO
           MOVE EMP-PROMOTION-DATE  TO LSTPRMO

           MOVE "MAKE CHANGES AND THEN PRESS PF5" TO MESSAGEO

        WHEN 100
           STRING "EMPLOYEE ID " DELIMITED BY SIZE
           WS-EMPNO DELIMITED BY SPACE
           " NOT FOUND" DELIMITED BY SIZE INTO MESSAGEO
           MOVE WS-EMPNO      TO EMPINO
           MOVE SPACES        TO EMPNOO
           MOVE SPACES        TO LNAMEO
           MOVE SPACES        TO FNAMEO
           MOVE SPACES        TO SOCSECO
           MOVE SPACES        TO YRSSVCO
           MOVE SPACES        TO LSTPRMO

        WHEN OTHER
           STRING "SQL CODE: " DELIMITED BY SIZE
                  WS-SQLCODE   DELIMITED BY SIZE
              INTO MESSAGEO

    END-EVALUATE.

    MOVE DFHBMFSE    TO EMPINF
    MOVE -1 TO EMPINL.
    MOVE "EMPPGCHG" TO WS-PGM-PASS

    IF SW-PASSED-DATA
       PERFORM SEND-MAP
    ELSE
       PERFORM SEND-MAP-DATA
    END-IF.

VALIDATE-DATA.

    PERFORM RECEIVE-MAP
    INITIALIZE DCLEMPLOYEE MESSAGEO

    EVALUATE TRUE

       WHEN EMPNOI EQUAL SPACES OR EMPNOL EQUAL ZERO
          MOVE "EMPLOYEE NUMBER IS REQUIRED" TO MESSAGEO
          MOVE -1 TO EMPNOL
```

```
            WHEN EMPNOI IS NOT NUMERIC
                MOVE "EMPLOYEE NUMBER MUST BE NUMERIC" TO MESSAGEO
                MOVE -1 TO EMPNOL

            WHEN LNAMEI EQUAL SPACES OR LNAMEL EQUAL ZERO
                MOVE "EMPLOYEE LAST NAME IS REQUIRED" TO MESSAGEO
                MOVE -1 TO LNAMEL

            WHEN FNAMEI EQUAL SPACES OR FNAMEL EQUAL ZERO
                MOVE "EMPLOYEE FIRST NAME IS REQUIRED" TO MESSAGEO
                MOVE -1 TO FNAMEL

            WHEN SOCSECI EQUAL SPACES OR SOCSECL EQUAL ZERO
                MOVE "SOCIAL SECURITY NUMBER IS REQUIRED" TO MESSAGEO
                MOVE -1 TO SOCSECL

            WHEN SOCSECI IS NOT NUMERIC
                MOVE "SOCIAL SECURITY MUST BE NUMERIC" TO MESSAGEO
                MOVE -1 TO SOCSECL

            WHEN YRSSVCI EQUAL SPACES OR YRSSVCL EQUAL ZERO
                MOVE "YEARS OF SERVICE IS REQUIRED" TO MESSAGEO
                MOVE -1 TO YRSSVCL

            WHEN YRSSVCI IS NOT NUMERIC
                MOVE "YEARS OF SERVICE MUST BE NUMERIC" TO MESSAGEO
                MOVE -1 TO YRSSVCL

            WHEN LSTPRMI EQUAL SPACES OR LSTPRML EQUAL ZERO
                MOVE "LAST PROMOTION DATE IS REQUIRED" TO MESSAGEO
                MOVE -1 TO LSTPRML

            WHEN OTHER
                PERFORM CHANGE-RECORD
                MOVE -1 TO EMPINL

        END-EVALUATE.

        MOVE DFHBMFSE TO EMPINF
        MOVE DFHBMFSE TO EMPNOF
        MOVE DFHBMFSE TO LNAMEF
        MOVE DFHBMFSE TO FNAMEF
        MOVE DFHBMFSE TO SOCSECF
        MOVE DFHBMFSE TO YRSSVCF
        MOVE DFHBMFSE TO LSTPRMF

        MOVE "EMPPGCHG" TO WS-PGM-PASS
        PERFORM SEND-MAP-DATA.

    CHANGE-RECORD.

    *   MAP INPUT FIELDS TO DB2 RECORD
```

```
      MOVE EMPNOI              TO WS-EMPNO
      MOVE WS-EMPNO            TO EMP-ID
      MOVE LNAMEI              TO EMP-LAST-NAME-TEXT

      SET SW-SPACE-NOT-FOUND TO TRUE

      PERFORM VARYING POS-CTR FROM +1 BY +1
         UNTIL (POS-CTR = 30) OR SW-SPACE-FOUND
            IF EMP-LAST-NAME-TEXT(POS-CTR:1) = SPACE
               SET SW-SPACE-FOUND TO TRUE
            END-IF
      END-PERFORM

      IF POS-CTR EQUAL +30
         MOVE +30 TO EMP-LAST-NAME-LEN
      ELSE
         SUBTRACT +1 FROM POS-CTR
         MOVE POS-CTR TO EMP-LAST-NAME-LEN
      END-IF

      MOVE FNAMEI              TO EMP-FIRST-NAME-TEXT

      SET SW-SPACE-NOT-FOUND TO TRUE

      PERFORM VARYING POS-CTR FROM +1 BY +1
         UNTIL (POS-CTR = 20) OR SW-SPACE-FOUND
            IF EMP-FIRST-NAME-TEXT(POS-CTR:1) = SPACE
               SET SW-SPACE-FOUND TO TRUE
            END-IF
      END-PERFORM

      IF POS-CTR EQUAL +20
         MOVE +20 TO EMP-FIRST-NAME-LEN
      ELSE
         SUBTRACT +1 FROM POS-CTR
         MOVE POS-CTR TO EMP-FIRST-NAME-LEN
      END-IF

      MOVE SOCSECI             TO EMP-SSN
      MOVE YRSSVCI             TO WS-EMP-SRV-YRS
      MOVE WS-EMP-SRV-YRS      TO EMP-SERVICE-YEARS
      MOVE LSTPRMI             TO EMP-PROMOTION-DATE

*  UPDATE THE RECORD

      EXEC SQL
         UPDATE USER01.EMPLOYEE
         SET EMP_LAST_NAME       = :EMP-LAST-NAME,
             EMP_FIRST_NAME      = :EMP-FIRST-NAME,
             EMP_SERVICE_YEARS   = :EMP-SERVICE-YEARS,
             EMP_PROMOTION_DATE  = :EMP-PROMOTION-DATE,
```

594

```
            EMP_SSN              = :EMP-SSN
        WHERE EMP_ID = :EMP-ID

    END-EXEC

* HANDLE THE SQLCODE AND RETURN

    MOVE SQLCODE TO WS-SQLCODE

    EVALUATE SQLCODE
        WHEN 0
            MOVE "EMPLOYEE MODIFIED SUCCESSFULLY" TO MESSAGEO
            MOVE -1 TO EMPNOL
        WHEN +100
            MOVE "ERROR - RECORD NOT FOUND" TO MESSAGEO
            MOVE -1 TO EMPNOL
        WHEN OTHER
            STRING "ERROR - DB2 SQLCODE IS " DELIMITED BY SIZE
               WS-SQLCODE DELIMITED BY SIZE
                  INTO MESSAGEO
            MOVE -1 TO EMPNOL

    END-EVALUATE.

 BRANCH-TO-PROGRAM.

    EXEC CICS
        XCTL PROGRAM(PROGRAM-NAME)
        COMMAREA (WS-COMMAREA)
        LENGTH(20)
    END-EXEC

    MOVE 'PROGRAM NOT AVAILABLE' TO MESSAGEO.

 SEND-MAP.
    EXEC CICS SEND
        MAP    ('EMPCHG')
        MAPSET ('EMPMCHG')
        FROM   (EMPCHGO)
        CURSOR
        ERASE
    END-EXEC.

 SEND-MAP-DATA.
    EXEC CICS SEND
        MAP    ('EMPCHG')
        MAPSET ('EMPMCHG')
        FROM   (EMPCHGO)
        DATAONLY
        CURSOR
    END-EXEC.
```

```
     SEND-MAP-ALL.
         EXEC CICS SEND
             MAP     ('EMPCHG')
             MAPSET  ('EMPMCHG')
             FROM    (EMPCHGO)
             CURSOR
         END-EXEC.

     RECEIVE-MAP.
          EXEC CICS RECEIVE
             MAP     ('EMPCHG')
             MAPSET  ('EMPMCHG')
             INTO    (EMPCHGI)
          END-EXEC.
```

The edits and validations are the same as for the add program. Also for error handling if a data value is missing or invalid, we set the length field for that data element to -1. This has the effect of placing the cursor on that screen field. For example:

```
     WHEN LNAMEI EQUAL SPACES OR LNAMEL EQUAL ZERO
         MOVE "EMPLOYEE LAST NAME IS REQUIRED" TO MESSAGEO
         MOVE -1 TO LNAMEL
```

Also notice that we check to see whether another program has passed data to this one, and if it has we use that passed employee id to populate the initial screen display with data. Otherwise we simply display an empty screen.

Now you can go ahead and compile, link and bind the program. Next, define and install the program in CICS. Remember to include a DB2ENTRY resource definition or your program won't work. Finally, run the test cases to make sure all functions work as intended.

When you've finished testing the Change program, let's move on to the Delete program.

Employee Delete Program
The following are specifications for our delete program. Again there are two SQL statements, one to retrieve the record for display, and the other to delete it.

```
Name:         EMPPGDEL
Purpose:      Delete an employee from the EMPLOYEE table.
Data Access: Read and Delete on EMPLOYEE
SQL Statements:

SELECT EMP_ID,
       EMP_LAST_NAME,
       EMP_FIRST_NAME,
```

```
        EMP_SSN,
        EMP_SERVICE_YEARS,
        EMP_PROMOTION_DATE
    INTO
        :EMP-ID,
        :EMP-LAST-NAME,
        :EMP-FIRST-NAME,
        :EMP-SSN,
        :EMP-SERVICE-YEARS,
        :EMP-PROMOTION-DATE
    FROM USER01.EMPLOYEE
    WHERE EMP_ID = :EMP-ID

DELETE FROM USER01.EMPLOYEE
    WHERE EMP_ID = :EMP-ID
```

EMPPGDEL Test Cases:

Case	Condition	Expected Result	Actual Result
1	Initial screen display	Literals displayed, all enterable fields unprotected.	
2	Valid employee number entered	Display detail for employee	
3	Invalid employee number entered	Error message that emp number is invalid	
4	Enter pressed without changing anything	No change	
5	PF2 pressed	Transfer to EMIN transaction	
6	PF3 pressed	Message to clear screen and enter a transaction id	
7	PF4 pressed	Transfer to EMAD transaction	
8	PF5 pressed	Transfer to EMCH transaction	
9	PF6 Pressed	Delete the record if no errors, message that record was deleted	
10	Other PF Keys pressed	Error message - invalid key	

Employee Delete Program Code

Here is the code for the delete program. This one is very much like the change program where we do an initial display of the record before we delete it. Please review this program carefully. Also notice that the "action" key for this transaction is PF6 (not ENTER).

```
        IDENTIFICATION DIVISION.
        PROGRAM-ID. EMPPGDEL.
        *************************************************
        *  COBOL/CICS/DB2 PROGRAM TO DELETE AN EMPLOYEE *
        *                                               *
```

```
*   AUTHOR       : ROBERT WINGATE                  *
*   DATE-WRITTEN : 2018-07-25                       *
****************************************************
 ENVIRONMENT DIVISION.
 DATA DIVISION.
 WORKING-STORAGE SECTION.
 01 WS-EMPNO         PIC 9(4).
 01 WS-EMP-SRV-YRS   PIC 9(2).
 01 WS-SQLCODE       PIC -9(08).
 01 WS-COMMAREA.
    05 WS-EMP-PASS   PIC 9(04).
    05 WS-PGM-PASS   PIC X(08).
    05 FILLER        PIC X(08).

 01 PROGRAM-NAME     PIC X(08) VALUE SPACES.

 01 SW-PASSED-DATA-SWITCH   PIC X(1) VALUE 'N'.
    88  SW-PASSED-DATA                VALUE 'Y'.
    88  SW-NO-PASSED-DATA             VALUE 'N'.

    COPY EMPMDEL.
    COPY DFHAID.
    COPY DFHBMSCA.

    EXEC SQL
      INCLUDE SQLCA
    END-EXEC.

    EXEC SQL
      INCLUDE EMPLOYEE
    END-EXEC.

 LINKAGE SECTION.
 01 DFHCOMMAREA          PIC X(20).

 PROCEDURE DIVISION.

    IF EIBCALEN > ZERO
      MOVE DFHCOMMAREA  TO WS-COMMAREA
    END-IF.

    EVALUATE TRUE

      WHEN EIBCALEN = ZERO
        MOVE LOW-VALUES   TO  EMPDELO
        PERFORM SEND-MAP

      WHEN EIBAID = DFHCLEAR
        MOVE LOW-VALUES   TO  EMPDELO
        PERFORM SEND-MAP

      WHEN EIBAID = DFHPA1 OR DFHPA2 OR DFHPA3
```

```
          CONTINUE

       WHEN EIBAID = DFHPF2
         MOVE 'EMPPGINQ' TO PROGRAM-NAME
         PERFORM BRANCH-TO-PROGRAM

       WHEN EIBAID = DFHPF3
         MOVE LOW-VALUES TO  EMPDELO
         MOVE "BYE, PRESS CLEAR KEY TO ENTER A TRANSACTION ID"
              TO MESSAGEO
         PERFORM SEND-MAP-DATA

         EXEC CICS
           RETURN
         END-EXEC

       WHEN EIBAID = DFHPF4
         MOVE 'EMPPGADD' TO PROGRAM-NAME
         PERFORM BRANCH-TO-PROGRAM

       WHEN EIBAID = DFHPF5
         MOVE 'EMPPGCHG' TO PROGRAM-NAME
         PERFORM BRANCH-TO-PROGRAM

       WHEN EIBAID = DFHPF6

         IF WS-PGM-PASS NOT EQUAL "EMPPGDEL"
            SET SW-PASSED-DATA TO TRUE
            PERFORM PROCESS-PARA
         ELSE
            SET SW-NO-PASSED-DATA TO TRUE
            PERFORM DELETE-RECORD
         END-IF

       WHEN EIBAID = DFHENTER
         PERFORM PROCESS-PARA

       WHEN OTHER
         MOVE LOW-VALUES TO EMPDELO
         MOVE "INVALID KEY PRESSED" TO MESSAGEO
         PERFORM SEND-MAP-DATA

    END-EVALUATE.

    EXEC CICS
       RETURN TRANSID('EMDE')
       COMMAREA (WS-COMMAREA)
    END-EXEC.

PROCESS-PARA.

    PERFORM RECEIVE-MAP.
```

```
            INITIALIZE DCLEMPLOYEE MESSAGEO

IF SW-PASSED-DATA
   MOVE WS-EMP-PASS TO EMP-ID
ELSE
   MOVE EMPINI     TO WS-EMPNO WS-EMP-PASS
   MOVE WS-EMPNO   TO EMP-ID
END-IF

EXEC SQL
   SELECT EMP_ID,
          EMP_LAST_NAME,
          EMP_FIRST_NAME,
          EMP_SSN,
          EMP_SERVICE_YEARS,
          EMP_PROMOTION_DATE
     INTO
          :EMP-ID,
          :EMP-LAST-NAME,
          :EMP-FIRST-NAME,
          :EMP-SSN,
          :EMP-SERVICE-YEARS,
          :EMP-PROMOTION-DATE
     FROM USER01.EMPLOYEE
     WHERE EMP_ID = :EMP-ID
END-EXEC.

MOVE SQLCODE        TO  WS-SQLCODE.

EVALUATE SQLCODE
  WHEN 0
    MOVE EMP-ID               TO  WS-EMPNO
    MOVE WS-EMPNO             TO  EMPNOO EMPINO
    MOVE EMP-LAST-NAME-TEXT   TO LNAMEO
    MOVE EMP-FIRST-NAME-TEXT  TO FNAMEO
    MOVE EMP-SSN              TO SOCSECO
    MOVE EMP-SERVICE-YEARS    TO WS-EMP-SRV-YRS
    MOVE WS-EMP-SRV-YRS       TO YRSSVCO
    MOVE EMP-PROMOTION-DATE   TO LSTPRMO
    MOVE "EMPPGDEL"           TO WS-PGM-PASS

    MOVE "PRESS PF6 TO DELETE EMPLOYEE" TO MESSAGEO

  WHEN 100
    STRING "EMPLOYEE ID " DELIMITED BY SIZE
    WS-EMPNO DELIMITED BY SPACE
    " NOT FOUND" DELIMITED BY SIZE INTO MESSAGEO
    MOVE WS-EMPNO        TO EMPINO
    MOVE SPACES          TO EMPNOO
    MOVE SPACES          TO LNAMEO
    MOVE SPACES          TO FNAMEO
    MOVE SPACES          TO SOCSECO
```

600

```
                MOVE SPACES         TO YRSSVCO
                MOVE SPACES         TO LSTPRMO

            WHEN OTHER
                STRING "SQL CODE: " DELIMITED BY SIZE
                       WS-SQLCODE   DELIMITED BY SIZE
                    INTO MESSAGEO
        END-EVALUATE.

        MOVE DFHBMFSE TO EMPINF
        MOVE -1 TO EMPINL.
        MOVE "EMPPGDEL" TO WS-PGM-PASS

        IF SW-PASSED-DATA
            PERFORM SEND-MAP
        ELSE
            PERFORM SEND-MAP-DATA
        END-IF.

    DELETE-RECORD.

        PERFORM RECEIVE-MAP.
        INITIALIZE DCLEMPLOYEE MESSAGEO

*   MAP INPUT FIELDS TO DB2 RECORD

        MOVE EMPINI          TO WS-EMPNO
        MOVE WS-EMPNO         TO EMP-ID

*   DELETE THE RECORD

        EXEC SQL
            DELETE FROM USER01.EMPLOYEE
                WHERE EMP_ID = :EMP-ID
        END-EXEC

*   HANDLE THE SQLCODE AND RETURN

        MOVE SQLCODE TO WS-SQLCODE

        EVALUATE SQLCODE
            WHEN 0
                MOVE "EMPLOYEE DELETED SUCCESSFULLY" TO MESSAGEO
                MOVE -1 TO EMPNOL
                MOVE WS-EMPNO TO WS-EMP-PASS
            WHEN +100
                MOVE "ERROR - RECORD DOES NOT EXIST" TO MESSAGEO
                MOVE -1 TO EMPNOL
                MOVE WS-EMPNO TO WS-EMP-PASS
            WHEN OTHER
                STRING "ERROR - DB2 SQLCODE IS " DELIMITED BY SPACE
                       WS-SQLCODE DELIMITED BY SPACE
```

```
                    INTO MESSAGEO
              MOVE -1 TO EMPNOL


       END-EVALUATE.

       MOVE DFHBMFSE TO EMPINF
       MOVE -1 TO EMPINL.
       MOVE "EMPPGDEL" TO WS-PGM-PASS
       PERFORM SEND-MAP-DATA.

   BRANCH-TO-PROGRAM.

       EXEC CICS
          XCTL PROGRAM(PROGRAM-NAME)
          COMMAREA (WS-COMMAREA)
          LENGTH(10)
       END-EXEC

       MOVE 'PROGRAM NOT AVAILABLE' TO MESSAGEO.

   SEND-MAP.
       EXEC CICS SEND
          MAP    ('EMPDEL')
          MAPSET ('EMPMDEL')
          FROM   (EMPDELO)
          ERASE
       END-EXEC.

   SEND-MAP-DATA.
       EXEC CICS SEND
          MAP    ('EMPDEL')
          MAPSET ('EMPMDEL')
          FROM   (EMPDELO)
          DATAONLY
       END-EXEC.

   SEND-MAP-ALL.
       EXEC CICS SEND
          MAP    ('EMPDEL')
          MAPSET ('EMPMDEL')
          FROM   (EMPDELO)
       END-EXEC.

   RECEIVE-MAP.
        EXEC CICS RECEIVE
          MAP    ('EMPDEL')
          MAPSET ('EMPMDEL')
          INTO   (EMPDELI)
        END-EXEC.
```

Now you can go ahead and compile, link and bind the program. Next, define and install the program in CICS. Remember to include a DB2ENTRY resource definition or your program won't work. Finally, run the test cases to make sure all functions work as intended.

When you've finished testing the Delete program, let's circle back to the Menu program.

Employee Support Menu

Now we're back to our menu program which should work now because the other programs exist and we can transfer to them. Go ahead and compile, link, bind and test the menu program.

```
Program:      EMPPGMNU
Purpose:      Driver for employee support functions
Data Access:  None
```

EMPPGMNU Test Cases

Here are the test cases for the menu program. Since the four function programs have been created we can now test the menu program.

Case	Condition	Expected Result	Actual Result
1	Initial screen display	All literals displayed, empty action field	
2	Invalid selection entered	Error message returned: ENTER A VALID ACTION: 1, 2, 3 OR 4	
3	Invalid key pressed	Error message: INVALID KEY PRESSED	
4	PF3 pressed	Message to clear screen and enter a trans id	
5	Option 1 entered	Transfer to EMIN transaction	
6	Option 2 entered	Transfer to EMAD transaction	
7	Option 3 entered	Transfer to EMCH transaction	
8	Option 4 entered	Transfer to EMDE transaction	

Employee Support Menu Program Code

We've already provided the code for the menu program earlier, but to keep everything together, we'll repeat it here.

```
IDENTIFICATION DIVISION.
PROGRAM-ID. EMPPGMNU.
*****************************************************
*   COBOL/CICS EMPLOYEE SUPPORT MENU PROGRAM       *
*                                                  *
```

```cobol
*   AUTHOR        : ROBERT WINGATE              *
*   DATE-WRITTEN  : 2018-07-26                  *
**************************************************
 ENVIRONMENT DIVISION.
 DATA DIVISION.
 WORKING-STORAGE SECTION.

 01 WS-FLAGS.
     05 SW-VALID-SELECTION     PIC X(1) VALUE 'N'.
        88   VALID-SELECTION            VALUE 'Y'.
        88   NOT-VALID-SELECTION        VALUE 'N'.

     05 SW-SEND-FLAGS          PIC X(1).
        88   SEND-ERASE                 VALUE '1'.
        88   SEND-DATAONLY              VALUE '2'.
        88   SEND-DATAONLY-ALARM        VALUE '3'.

 01 WS-VARS.
     05 COMM-AREA              PIC X(20) VALUE SPACE.
     05 WS-SQLCODE             PIC -9(08).
     05 PROGRAM-NAME           PIC X(08) VALUE SPACES.
     05 RESPONSE-CODE          PIC S9(08) COMP.
     05 INVALID-ACTION-MSG     PIC X(31)
        VALUE 'ENTER A VALID ACTION: 1, 2 OR 3'.
     05 END-SESSION-MSG        PIC X(23)
        VALUE 'THIS SESSION HAS ENDED.'.

     COPY EMPMMNU.
     COPY DFHAID.
     COPY DFHBMSCA.

 LINKAGE SECTION.
 01 DFHCOMMAREA          PIC X(10).

 PROCEDURE DIVISION.

     IF EIBCALEN > ZERO
       MOVE DFHCOMMAREA  TO COMM-AREA
     END-IF.

     EVALUATE TRUE

       WHEN EIBCALEN = ZERO
         MOVE LOW-VALUES   TO EMPMNUO
         SET SEND-ERASE TO TRUE
         PERFORM SEND-MAP

       WHEN EIBAID = DFHCLEAR
         MOVE LOW-VALUES   TO EMPMNUO
         SET SEND-ERASE TO TRUE
         PERFORM SEND-MAP

       WHEN EIBAID = DFHPA1 OR DFHPA2 OR DFHPA3
         CONTINUE

       WHEN EIBAID = DFHPF3
         MOVE LOW-VALUES TO  EMPMNUO
         MOVE "BYE, PRESS CLEAR KEY TO ENTER A TRANSACTION ID"
```

```
                  TO MESSAGEO
         PERFORM SEND-MAP-DATAONLY

         EXEC CICS
            RETURN
         END-EXEC

     WHEN EIBAID = DFHENTER
        PERFORM MAIN-MAIN-PROCESS-PARA

     WHEN OTHER
        MOVE LOW-VALUES TO EMPMNUO
        MOVE "INVALID KEY PRESSED" TO MESSAGEO
        PERFORM SEND-MAP-DATAONLY

     END-EVALUATE.

     EXEC CICS
        RETURN TRANSID('EMNU')
        COMMAREA (COMM-AREA)
     END-EXEC.

MAIN-MAIN-PROCESS-PARA.

     PERFORM RECEIVE-MAP.

     IF ACTIONI NOT = '1' AND '2' AND '3' AND '4'
        MOVE DFHREVRS TO ACTIONH
        MOVE INVALID-ACTION-MSG TO MESSAGEO
        SET NOT-VALID-SELECTION TO TRUE
     ELSE
        SET VALID-SELECTION TO TRUE
     END-IF.

     IF VALID-SELECTION
        EVALUATE ACTIONI
           WHEN '1'
              MOVE 'EMPPGINQ' TO PROGRAM-NAME
           WHEN '2'
              MOVE 'EMPPGADD' TO PROGRAM-NAME
           WHEN '3'
              MOVE 'EMPPGCHG' TO PROGRAM-NAME
           WHEN '4'
              MOVE 'EMPPGDEL' TO PROGRAM-NAME
        END-EVALUATE

        PERFORM BRANCH-TO-PROGRAM

     END-IF.

     SET SEND-DATAONLY-ALARM TO TRUE.
     PERFORM SEND-MAP-DATAONLY-ALARM.

BRANCH-TO-PROGRAM.

     EXEC CICS
        XCTL PROGRAM(PROGRAM-NAME)
     END-EXEC
```

605

```
        MOVE 'PROGRAM NOT AVAILABLE' TO MESSAGEO.

    SEND-MAP.
        EXEC CICS SEND
            MAP    ('EMPMNU')
            MAPSET ('EMPMMNU')
            FROM   (EMPMNUO)
            ERASE
        END-EXEC.

    SEND-MAP-DATAONLY.
        EXEC CICS SEND
            MAP    ('EMPMNU')
            MAPSET ('EMPMMNU')
            FROM   (EMPMNUO)
            DATAONLY
        END-EXEC.

    SEND-MAP-DATAONLY-ALARM.
        EXEC CICS SEND
            MAP    ('EMPMNU')
            MAPSET ('EMPMMNU')
            FROM   (EMPMNUO)
            DATAONLY
            ALARM
        END-EXEC.

    RECEIVE-MAP.
        EXEC CICS RECEIVE
            MAP    ('EMPMNU')
            MAPSET ('EMPMMNU')
            INTO   (EMPMNUI)
        END-EXEC.
```

Once you have a good unit test, let's proceed to integration testing.

Integration Testing

Now we must test to make sure transfers from one program to the other work correctly, and that the information that needs to be passed in the communication area does in fact get passed. So execute all the program transfers, and then also retest the individual programs to make sure they work correctly when simply entering data on the screen (without data being passed from another program).

My integration test results follow.

EMPPGMNU

Ok, begin with the menu program EMPPGMNU. We'll simply verify that we can navigate to all the screens and that they set up properly. We'll choose the menu options for inquiry, add, change and delete. First inquiry.

```
EMPMMNU                      EMPLOYEE SUPPORT MENU                      EMNU

           ENTER THE NUMBER OF YOUR SELECTION,   THEN PRESS ENTER.


                 1    1. EMPLOYEE INQUIRY

                      2. EMPLOYEE ADD

                      3. EMPLOYEE CHANGE

                      4. EMPLOYEE DELETE










F3 EXIT
```

```
EMPMINQ                      EMPLOYEE INQUIRY                           EMIN


        EMPLOYEE ->              ENTER EMPLOYEE ID, THEN PRESS ENTER

        EMPLOYEE ID    XXXX

        EMP LAST NAME  XXXX

        EMP FIRST NAME XXXX

        EMP SOCIAL SEC XXXXXXXX

        EMP YEARS SRVC 00

        EMP LAST PROM  YYYY-MM-DD






F2 INQ   F3 EXIT   F4 ADD   F5 CHG   F6 DEL
```

Next go back to the main menu and choose the add option.[21]

```
EMPMMNU                    EMPLOYEE SUPPORT MENU                      EMNU

              ENTER THE NUMBER OF YOUR SELECTION,   THEN PRESS ENTER.


                   2   1. EMPLOYEE INQUIRY

                       2. EMPLOYEE ADD

                       3. EMPLOYEE CHANGE

                       4. EMPLOYEE DELETE









F3 EXIT


EMPMADD                       EMPLOYEE ADD                            EMAD

                     ENTER EMPLOYEE INFO, THEN PRESS PF4

          EMPLOYEE ID

          EMP LAST NAME

          EMP FIRST NAME

          EMP SOCIAL SEC

          EMP YEARS SRVC

          EMP LAST PROM




F2 INQ   F3 EXIT   F4 ADD   F5 CHG   F6 DEL
```

---

21 By now we can see it would be helpful to have an exit from each of the detail programs back to the main menu. I leave this as an enhancement for you so you can go through a maintenance cycle.

Now go back to the menu and choose the change option.

```
EMPMMNU                    EMPLOYEE SUPPORT MENU                    EMNU

             ENTER THE NUMBER OF YOUR SELECTION,   THEN PRESS ENTER.


                      3    1. EMPLOYEE INQUIRY

                           2. EMPLOYEE ADD

                           3. EMPLOYEE CHANGE

                           4. EMPLOYEE DELETE






     F3 EXIT
```

```
EMPMCHG                      EMPLOYEE CHANGE                        EMCH

        EMPLOYEE ->      ____        ENTER EMPLOYEE ID, THEN PRESS ENTER

        EMPLOYEE ID    XXXX

        EMP LAST NAME  XXXX

        EMP FIRST NAME XXXX

        EMP SOCIAL SEC XXXXXXXXX

        EMP YEARS SRVC 00

        EMP LAST PROM  YYYY-MM-DD




     F2 INQ   F3 EXIT   F4 ADD   F5 CHG   F6 DEL
```

Finally choose the delete option.

```
EMPMMNU                    EMPLOYEE SUPPORT MENU                      EMNU

             ENTER THE NUMBER OF YOUR SELECTION,   THEN PRESS ENTER.


                        4   1. EMPLOYEE INQUIRY

                            2. EMPLOYEE ADD

                            3. EMPLOYEE CHANGE

                            4. EMPLOYEE DELETE








     F3 EXIT

EMPMDEL                     EMPLOYEE DELETE                           EMDE


        EMPLOYEE ->              ENTER EMPLOYEE ID, THEN PRESS ENTER

        EMPLOYEE ID   XXXX

        EMP LAST NAME  XXXX

        EMP FIRST NAME XXXX

        EMP SOCIAL SEC XXXXXXXX

        EMP YEARS SRVC 00

        EMP LAST PROM  YYYY-MM-DD




     F2 INQ   F3 EXIT   F4 ADD   F5 CHG   F6 DEL
```

All looks good. Now on to the data operations.

EMPPGINQ

Ok let's start out with the primary display and entry after choosing from the main menu. Let's use 7777 as the employee id. Enter 7777 and press ENTER.

```
EMPMINQ                        EMPLOYEE INQUIRY                        EMIN


       EMPLOYEE ->    7777      ENTER EMPLOYEE ID, THEN PRESS ENTER

       EMPLOYEE ID    XXXX

       EMP LAST NAME   XXXX

       EMP FIRST NAME XXXX

       EMP SOCIAL SEC XXXXXXXXX

       EMP YEARS SRVC 00

       EMP LAST PROM  YYYY-MM-DD




    F2 INQ   F3 EXIT   F4 ADD   F5 CHG   F6 DEL


EMPMINQ                        EMPLOYEE INQUIRY                        EMIN


       EMPLOYEE ->    7777      ENTER EMPLOYEE ID, THEN PRESS ENTER

       EMPLOYEE ID    7777

       EMP LAST NAME   JACKSON

       EMP FIRST NAME JOSEPH

       EMP SOCIAL SEC 382746236

       EMP YEARS SRVC 17

       EMP LAST PROM  2017-01-01




    F2 INQ   F3 EXIT   F4 ADD   F5 CHG   F6 DEL
```

This looks good. Now let's try passing the same employee number from another program, such as the change program. Bring up employee 7777 on the change screen, and then press PF2 to transfer to the inquiry program.

```
EMPMCHG                   EMPLOYEE CHANGE                         EMCH


        EMPLOYEE ->    7777      ENTER EMPLOYEE ID, THEN PRESS ENTER

        EMPLOYEE ID    7777

        EMP LAST NAME  JACKSON

        EMP FIRST NAME JOSEPH

        EMP SOCIAL SEC 382746236

        EMP YEARS SRVC 17

        EMP LAST PROM  2017-01-01




MAKE CHANGES AND THEN PRESS PF5
F2 INQ   F3 EXIT   F4 ADD   F5 CHG   F6 DEL
```

```
EMPMINQ                   EMPLOYEE INQUIRY                        EMIN

        EMPLOYEE ->    7777      ENTER EMPLOYEE ID, THEN PRESS ENTER

        EMPLOYEE ID    7777

        EMP LAST NAME  JACKSON

        EMP FIRST NAME JOSEPH

        EMP SOCIAL SEC 382746236

        EMP YEARS SRVC 17

        EMP LAST PROM  2017-01-01




F2 INQ   F3 EXIT   F4 ADD   F5 CHG   F6 DEL
```

Finally we should test the transfers from the inquiry screen to the add, change and delete screens. Start with inquiry and press PF4 to go to the add screen. Note that the employee id won't appear on the add screen because the add screen assumes we are going to enter a new number.

```
EMPMINQ                   EMPLOYEE INQUIRY                      EMIN


      EMPLOYEE ->    7777      ENTER EMPLOYEE ID, THEN PRESS ENTER

      EMPLOYEE ID    7777

      EMP LAST NAME  JACKSON

      EMP FIRST NAME JOSEPH

      EMP SOCIAL SEC 382746236

      EMP YEARS SRVC 17

      EMP LAST PROM  2017-01-01




 F2 INQ   F3 EXIT   F4 ADD   F5 CHG   F6 DEL
EMPMADD                   EMPLOYEE ADD                          EMAD

                  ENTER EMPLOYEE INFO, THEN PRESS PF4

      EMPLOYEE ID

      EMP LAST NAME

      EMP FIRST NAME

      EMP SOCIAL SEC

      EMP YEARS SRVC

      EMP LAST PROM




 ENTER DATA FOR NEW EMPLOYEE, THEN PRESS PF4 TO ADD
 F2 INQ   F3 EXIT   F4 ADD   F5 CHG   F6 DEL
```

Also go ahead and add an employee to ensure all attribute setting, switches and variables are correct.

```
EMPMADD                    EMPLOYEE ADD                          EMAD


                    ENTER EMPLOYEE INFO, THEN PRESS PF4

         EMPLOYEE ID    1111

         EMP LAST NAME  stone

         EMP FIRST NAME steven

         EMP SOCIAL SEC 385610088

         EMP YEARS SRVC 12

         EMP LAST PROM  2016-01-01




  ENTER DATA FOR NEW EMPLOYEE, THEN PRESS PF4 TO ADD
  F2 INQ    F3 EXIT    F4 ADD    F5 CHG    F6 DEL
```

```
EMPMADD                    EMPLOYEE ADD                          EMAD


                    ENTER EMPLOYEE INFO, THEN PRESS PF4

         EMPLOYEE ID    1111

         EMP LAST NAME  STONE

         EMP FIRST NAME STEVEN

         EMP SOCIAL SEC 385610088

         EMP YEARS SRVC 12

         EMP LAST PROM  2016-01-01




  EMPLOYEE ADDED SUCCESSFULLY
  F2 INQ    F3 EXIT    F4 ADD    F5 CHG    F6 DEL
```

All looks well, so let's move on to the add screen.

EMPPGADD

We just added a record, so let's try transferring to the inquiry screen, the change screen and then the delete screen. Press PF2.

```
EMPMADD                         EMPLOYEE ADD                              EMAD


                      ENTER EMPLOYEE INFO, THEN PRESS PF4

        EMPLOYEE ID    1111

        EMP LAST NAME  STONE

        EMP FIRST NAME STEVEN

        EMP SOCIAL SEC 385610088

        EMP YEARS SRVC 12

        EMP LAST PROM  2016-01-01




EMPLOYEE ADDED SUCCESSFULLY
F2 INQ   F3 EXIT   F4 ADD    F5 CHG    F6 DEL


EMPMINQ                         EMPLOYEE INQUIRY                          EMIN


        EMPLOYEE ->              ENTER EMPLOYEE ID, THEN PRESS ENTER

        EMPLOYEE ID    1111

        EMP LAST NAME  STONE

        EMP FIRST NAME STEVEN

        EMP SOCIAL SEC 385610088

        EMP YEARS SRVC 12

        EMP LAST PROM  2016-01-01




  F2 INQ   F3 EXIT   F4 ADD    F5 CHG    F6 DEL
```

Next let's add two more records, and in the process transfer to the change and delete screens, respectively.

```
EMPMADD                     EMPLOYEE ADD


                    ENTER EMPLOYEE INFO, THEN PRESS PF4

         EMPLOYEE ID    1212

         EMP LAST NAME  SAMPLE

         EMP FIRST NAME RECORD

         EMP SOCIAL SEC 373737373

         EMP YEARS SRVC 04

         EMP LAST PROM  2017-01-01




  EMPLOYEE ADDED SUCCESSFULLY
  F2 INQ   F3 EXIT   F4 ADD   F5 CHG   F6 DEL
```

Now press PF5.

```
EMPMCHG                   EMPLOYEE CHANGE                      EMCH

       EMPLOYEE ->              ENTER EMPLOYEE ID, THEN PRESS ENTER

       EMPLOYEE ID    1212

       EMP LAST NAME  SAMPLE

       EMP FIRST NAME RECORD

       EMP SOCIAL SEC 373737373

       EMP YEARS SRVC 04

       EMP LAST PROM  2017-01-01




  MAKE CHANGES AND THEN PRESS PF5
  F2 INQ   F3 EXIT   F4 ADD   F5 CHG   F6 DEL
```

Transfer back to the add screen and add the second record.

```
EMPMADD                      EMPLOYEE ADD                         EMAD


                   ENTER EMPLOYEE INFO, THEN PRESS PF4

        EMPLOYEE ID    2424

        EMP LAST NAME   TWO

        EMP FIRST NAME SAMPLE

        EMP SOCIAL SEC 747474747

        EMP YEARS SRVC 13

        EMP LAST PROM  2016-01-01




   EMPLOYEE ADDED SUCCESSFULLY
   F2 INQ    F3 EXIT   F4 ADD   F5 CHG   F6 DEL
```

Now press PF6.

```
EMPMDEL                      EMPLOYEE DELETE                      EMDE


        EMPLOYEE ->              ENTER EMPLOYEE ID, THEN PRESS ENTER

        EMPLOYEE ID    2424

        EMP LAST NAME   TWO

        EMP FIRST NAME SAMPLE

        EMP SOCIAL SEC 747474747

        EMP YEARS SRVC 13

        EMP LAST PROM  2016-01-01




   PRESS PF6 TO DELETE THIS RECORD
   F2 INQ    F3 EXIT   F4 ADD   F5 CHG   F6 DEL
```

All looks good with transferring from the add screen. Let's move on to the change screen.

EMPPGCHG

Let's bring up the 2424 record that we just created on the change screen. Now transfer to the inquiry screen, the delete screen, and finally the add screen (the latter will not process any transferred data except the program name).

```
EMPMCHG                     EMPLOYEE CHANGE                         EMCH


        EMPLOYEE ->    2424       ENTER EMPLOYEE ID, THEN PRESS ENTER

        EMPLOYEE ID    2424

        EMP LAST NAME  TWO

        EMP FIRST NAME SAMPLE

        EMP SOCIAL SEC 747474747

        EMP YEARS SRVC 13

        EMP LAST PROM  2016-01-01




    MAKE CHANGES AND THEN PRESS PF5
    F2 INQ   F3 EXIT   F4 ADD   F5 CHG   F6 DEL
```

Press PF2.

```
EMPMINQ                     EMPLOYEE INQUIRY                        EMIN


        EMPLOYEE ->    2424       ENTER EMPLOYEE ID, THEN PRESS ENTER

        EMPLOYEE ID    2424

        EMP LAST NAME  TWO

        EMP FIRST NAME SAMPLE

        EMP SOCIAL SEC 747474747

        EMP YEARS SRVC 13

        EMP LAST PROM  2016-01-01




    F2 INQ   F3 EXIT   F4 ADD   F5 CHG   F6 DEL
```

Transfer back to the change screen.

```
EMPMCHG                    EMPLOYEE CHANGE                          EMCH


        EMPLOYEE ->    2424      ENTER EMPLOYEE ID, THEN PRESS ENTER

        EMPLOYEE ID    2424

        EMP LAST NAME  TWO

        EMP FIRST NAME SAMPLE

        EMP SOCIAL SEC 747474747

        EMP YEARS SRVC 13

        EMP LAST PROM  2016-01-01




 MAKE CHANGES AND THEN PRESS PF5
 F2 INQ   F3 EXIT   F4 ADD   F5 CHG   F6 DEL
```

Now press PF6.

```
EMPMDEL                    EMPLOYEE DELETE                          EMDE


        EMPLOYEE ->    2424      ENTER EMPLOYEE ID, THEN PRESS ENTER

        EMPLOYEE ID    2424

        EMP LAST NAME  TWO

        EMP FIRST NAME SAMPLE

        EMP SOCIAL SEC 747474747

        EMP YEARS SRVC 13

        EMP LAST PROM  2016-01-01




 PRESS PF6 TO DELETE THIS RECORD
 F2 INQ   F3 EXIT   F4 ADD   F5 CHG   F6 DEL
```

Now go back to the change screen.

```
EMPMCHG                    EMPLOYEE CHANGE                        EMCH

        EMPLOYEE ->    2424      ENTER EMPLOYEE ID, THEN PRESS ENTER

        EMPLOYEE ID    2424

        EMP LAST NAME  TWO

        EMP FIRST NAME SAMPLE

        EMP SOCIAL SEC 747474747

        EMP YEARS SRVC 13

        EMP LAST PROM  2016-01-01




   MAKE CHANGES AND THEN PRESS PF5
   F2 INQ   F3 EXIT   F4 ADD   F5 CHG   F6 DEL
```

Press PF4.

```
EMPMADD                    EMPLOYEE ADD                           EMAD

                  ENTER EMPLOYEE INFO, THEN PRESS PF4

        EMPLOYEE ID

        EMP LAST NAME

        EMP FIRST NAME

        EMP SOCIAL SEC

        EMP YEARS SRVC

        EMP LAST PROM




   ENTER DATA FOR NEW EMPLOYEE, THEN PRESS PF4 TO ADD
   F2 INQ   F3 EXIT   F4 ADD   F5 CHG   F6 DEL
```

Again, the above is correct because the add screen should not process any passed employee number.

EMPPGDEL

Finally, let's test the delete screen. We can use employee 3333. Let's bring it up on the delete screen.

```
EMPMDEL                    EMPLOYEE DELETE                          EMDE


       EMPLOYEE ->    3333       ENTER EMPLOYEE ID, THEN PRESS ENTER

       EMPLOYEE ID    3333

       EMP LAST NAME  RADISSON

       EMP FIRST NAME BENTLEY

       EMP SOCIAL SEC 777777777

       EMP YEARS SRVC 46

       EMP LAST PROM  2015-07-01




 PRESS PF6 TO DELETE THIS RECORD
 F2 INQ   F3 EXIT   F4 ADD   F5 CHG   F6 DEL
```

Now transfer to the inquiry screen with PF2.

```
EMPMINQ                    EMPLOYEE INQUIRY                         EMIN


       EMPLOYEE ->    3333       ENTER EMPLOYEE ID, THEN PRESS ENTER

       EMPLOYEE ID    3333

       EMP LAST NAME  RADISSON

       EMP FIRST NAME BENTLEY

       EMP SOCIAL SEC 777777777

       EMP YEARS SRVC 46

       EMP LAST PROM  2015-07-01




 F2 INQ   F3 EXIT   F4 ADD   F5 CHG   F6 DEL
```

Now transfer back to the delete program

```
EMPMDEL                     EMPLOYEE DELETE                          EMDE


       EMPLOYEE ->    3333       ENTER EMPLOYEE ID, THEN PRESS ENTER

       EMPLOYEE ID    3333

       EMP LAST NAME  RADISSON

       EMP FIRST NAME BENTLEY

       EMP SOCIAL SEC 777777777

       EMP YEARS SRVC 46

       EMP LAST PROM  2015-07-01




   PRESS PF6 TO DELETE THIS RECORD
   F2 INQ   F3 EXIT   F4 ADD   F5 CHG   F6 DEL
```

And then transfer to the change program using PF5.

```
EMPMCHG                     EMPLOYEE CHANGE                          EMCH


       EMPLOYEE ->    3333       ENTER EMPLOYEE ID, THEN PRESS ENTER

       EMPLOYEE ID    3333

       EMP LAST NAME  RADISSON

       EMP FIRST NAME BENTLEY

       EMP SOCIAL SEC 777777777

       EMP YEARS SRVC 46

       EMP LAST PROM  2015-07-01




   MAKE CHANGES AND THEN PRESS PF5
   F2 INQ   F3 EXIT   F4 ADD   F5 CHG   F6 DEL
```

Now transfer back to the delete program

```
EMPMDEL                   EMPLOYEE DELETE                        EMDE


       EMPLOYEE ->    3333      ENTER EMPLOYEE ID, THEN PRESS ENTER

       EMPLOYEE ID    3333

       EMP LAST NAME  RADISSON

       EMP FIRST NAME BENTLEY

       EMP SOCIAL SEC 777777777

       EMP YEARS SRVC 46

       EMP LAST PROM  2015-07-01




 PRESS PF6 TO DELETE THIS RECORD
 F2 INQ   F3 EXIT   F4 ADD   F5 CHG   F6 DEL
```

Now transfer to the add program. Notice we do not pass the employee number to the add program since there is no need to add an already existing employee.

```
EMPMADD                   EMPLOYEE ADD                           EMAD

                     ENTER EMPLOYEE INFO, THEN PRESS PF4

       EMPLOYEE ID

       EMP LAST NAME

       EMP FIRST NAME

       EMP SOCIAL SEC

       EMP YEARS SRVC

       EMP LAST PROM




 ENTER DATA FOR NEW EMPLOYEE, THEN PRESS PF4 TO ADD
 F2 INQ   F3 EXIT   F4 ADD   F5 CHG   F6 DEL
```

That's it, looks like everything works. Our integration test is finished. This also completes our work on CICS with DB2. Now let's move on to CICS with VSAM.

CICS With VSAM

VSAM Quick Review
Introduction
Virtual Storage Access Method (VSAM) is an IBM DASD (direct access storage device) file storage access method. It has been used for many years, including with the Multiple Virtual Storage (MVS) architecture and now in z/OS. VSAM offers four data set organizations:

1. Key Sequenced Data Set (KSDS)

2. Entry Sequenced Data Set (ESDS)

3. Relative Record Data Set (RRDS)

4. Linear Data Set (LDS)

The KSDS, RRDS and ESDS organizations all are record-based. The LDS organization uses a sequence of pages without a predefined record structure.

VSAM records are either fixed or variable length. Records are organized in fixed-size blocks called Control Intervals (CIs). The CI's are organized into larger structures called Control Areas (CAs). Control Interval sizes are measured in bytes — for example 4 kilobytes — while Control Area sizes are measured in disk tracks or cylinders. When a VSAM file is read, a complete Control Interval will be transferred to memory.

The Access Method Services utility program `IDCAMS` is used to define and delete VSAM data sets. In addition, you can write custom programs in COBOL, PLI and Assembler to access VSAM datasets using Data Definition (DD) statements in Job Control Language (JCL), or via dynamic allocation or in online regions such as in Customer Information Control System (CICS).

Types of VSAM Files

Key Sequence Data Set (KSDS)
This organization type is the most commonly used. Each record has one or more key fields and a record can be retrieved (or inserted) by key value. This provides random access to data. Records are of variable length.

Entry Sequence Data Set (ESDS)

This organization keeps records in the order in which they were entered. Records must be accessed sequentially. This organization is used by IMS for overflow datasets.

Relative Record Data Set (RRDS)

This organization is based on record retrieval number; record 1, record 2, etc. This provides random access but the program must have a way of knowing which record number it is looking for.

Linear Data Set (LDS)

This organization is a byte-stream data set. It is rarely used by application programs.

We'll focus on KSDS files because they are the most commonly used and the most useful. Also this is the kind of storage and retrieval method we need for our Employee Support application.

A KSDS cluster consists of following two components:

Index – the index component of the KSDS cluster is comprised of the list of key values for the records in the cluster with pointers to the corresponding records in the data component. The index component relates the key of each record to the record's relative location in the data set. When a record is added or deleted, this index is updated.

Data – the data component of the KSDS cluster contains the actual data. Each record in the data component of a KSDS cluster contains a key field with same number of characters and occurs in the same relative position in each record.

This is just a brief introduction to VSAM. If you need more background, check out **Quick Start Training for IBM z/OS Application Developers, Volume 2.** The first chapter gives you a more complete introduction to VSAM.

Creating the VSAM File

You create VSAM files using the IDCAMS utility. Here is sample JCL to create our employee file. We'll use 80 byte records which will be enough to accommodate all our field data. Also we'll define the key as the first 4 bytes of each record which is the employee id.

```
//USER01D JOB MSGLEVEL=(1,1),NOTIFY=&SYSUID
//*
//************************************************
//* DEFINE VSAM KSDS CLUSTER
//************************************************
//JS010    EXEC PGM=IDCAMS
//SYSUDUMP DD SYSOUT=*
//SYSPRINT DD SYSOUT=*
//SYSOUT   DD SYSOUT=*
//SYSIN    DD  *
  DEFINE CLUSTER(NAME(USER01.EMPLOYEE)   -
  RECSZ(80 80)    -
  TRK(2,1)        -
  FREESPACE(5,10) -
  KEYS(4,0)       -
  CISZ(4096)      -
  VOLUMES(DEVHD1) -
  INDEXED)        -
  INDEX(NAME(USER01.EMPLOYEE.INDEX)) -
  DATA(NAME(USER01.EMPLOYEE.DATA))
/*
//SYSPRINT DD SYSOUT=*
//SYSOUT   DD SYSOUT=*
//SYSUDUMP DD SYSOUT=*
//*
```

This creates a catalog entry with two datasets, one for data and one for the index.

```
DSLIST - Data Sets Matching USER01.EMPLOYEE                Row 1 of 11
Command ===>                                          Scroll ===> CSR

Command - Enter "/" to select action          Message        Volume
--------------------------------------------------------------------------
        USER01.EMPLOYEE                                       *VSAM*
        USER01.EMPLOYEE.DATA                                  DEVHD1
        USER01.EMPLOYEE.INDEX                                 DEVHD1
***************************** End of Data Set list ****************************
```

Loading and Unloading VSAM Files

You can add data to a VSAM KSDS in several ways:

- Copying data from a flat file with the IDCAMS utility.

- Using IBM's File Manager product.

- Using an application program.

We'll show examples of loading data using a flat file with the `IDCAMS` utility, and also entering data with the File Manager utility. Here are the columns and data types for our table which we will name `EMPLOYEE`.

Field Name	Type
EMP_ID	Numeric 4 bytes
EMP_LAST_NAME	Character(30)
EMP_FIRST_NAME	Character(20)
EMP_SERVICE_YEARS	Numeric 2 bytes
EMP_PROMOTION_DATE	Date in format YYYY-MM-DD

Now let's say we have created a text data file in this format. We can browse it:

```
BROWSE     USER01.EMPLOYEE.LOAD                    Line 00000000 Col 001 080
 Command ===>                                              Scroll ===> CSR
----+----1----+----2----+----3----+----4----+----5----+----6----+----7----+----8
3217JOHNSON                    EDWARD               042017-01-01
7459STEWART                    BETTY                072016-07-31
9134FRANKLIN                   BRIANNA              032016-10-01
4720SCHULTZ                    TIM                  092017-01-01
6288WILLARD                    JOE                  062016-01-01
1122JENKINS                    DEBORAH              052016-09-01
```

Note however that before we load the VSAM file, we need to sort the input records in key sequence. Otherwise `IDCAMS` will give us an error. You can simply edit the file in ISPF and issue a `SORT 1 4` command (space between 1 and 4) to sort the records. Here is the resulting sorted data. Now we are ready.

```
BROWSE     USER01.EMPLOYEE.LOAD                    Line 00000000 Col 001 080
 Command ===>                                              Scroll ===> CSR
----+----1----+----2----+----3----+----4----+----5----+----6----+----7----+----8
1122JENKINS                    DEBORAH              052016-09-01
3217JOHNSON                    EDWARD               042017-01-01
4720SCHULTZ                    TIM                  092017-01-01
6288WILLARD                    JOE                  062016-01-01
7459STEWART                    BETTY                072016-07-31
9134FRANKLIN                   BRIANNA              032016-10-01
```

VSAM Batch Updates with IDCAMS

We can use the following `IDCAMS` JCL to load the VSAM file.

```
//USER01D JOB 'WINGATE',MSGLEVEL=(1,1),NOTIFY=&SYSUID
//*
//* REPRO/COPY DATA FROM PS TO VSAM KSDS
//*
//JS010   EXEC PGM=IDCAMS
//SYSPRINT DD SYSOUT=*
```

```
//SYSOUT   DD SYSOUT=*
//SYSUDUMP DD SYSOUT=*
//SYSIN    DD *
  REPRO - INDATASET (USER01.EMPLOYEE.LOAD) -
  OUTDATASET(USER01.EMPLOYEE)
/*
//
```

Once loaded, we can view the data using the Browse function.

```
Browse              USER01.EMPLOYEE.DATA                    Top of 6
Command ===>                                               Scroll PAGE
                        Type DATA    RBA                    Format CHAR
                                           Col 1
----+----10---+----2----+----3----+----4----+----5----+----6----+----7----+----
1122JENKINS                    DEBORAH              052016-09-01
3217JOHNSON                    EDWARD               042017-01-01
4720SCHULTZ                    TIM                  092017-01-01
6288WILLARD                    JOE                  062016-01-01
7459STEWART                    BETTY                072016-07-31
9134FRANKLIN                   BRIANNA              032016-10-01
```

To edit the data you will need to use a tool such as File Manager. Let's do this next.

VSAM Updates with File Manager

You can perform adds, changes and deletes to data records in File Manager. First, it will be useful if we create a file layout to assist us with viewing and updating data. Let's create a COBOL layout as follows.

```
BROWSE     USER01.COPYLIB(EMPLOYEE) - 01.00      Line 00000000 Col 001 080
  Command ===>                                          Scroll ===> CSR
  ***********************************************************************
  * COBOL DECLARATION FOR VSAM FILE EMPLOYEE                          *
  ***********************************************************************
  01   EMPLOYEE.
       05 EMP-ID             PIC 9(04).
       05 EMP-LAST-NAME      PIC X(30).
       05 EMP-FIRST-NAME     PIC X(20).
       05 EMP-SERVICE-YEARS  PIC 9(02).
       05 EMP-PROMOTION-DATE PIC X(10).
       05 EMP-SSN            PIC X(09).
       05 FILLER             PIC X(05).
```

Now let's go to File Manager. Below is the main FM menu. Select the EDIT option (2).

```
File Manager                 Primary Option Menu
Command ===>

0   Settings      Set processing options          User ID . : USER01
1   View          View data                       System ID : MATE
2   Edit          Edit data                       Appl ID . : FMN
3   Utilities     Perform utility functions       Version . : 11.1.0
4   Tapes         Tape specific functions         Terminal. : 3278
5   Disk/VSAM     Disk track and VSAM CI functions   Screen. . : 2
6   OAM           Work with OAM objects           Date. . . : 2018/03/07
7   Templates     Template and copybook utilities  Time. . . : 02:41
8   HFS           Access Hierarchical File System
9   WebSphere MQ  List, view and edit MQ data
X   Exit          Terminate File Manager
```

Enter your file name, copybook file name, and select the processing option 1.

```
File Manager                 Edit Entry Panel
Command ===>

Input Partitioned, Sequential or VSAM Data Set, or HFS file:
   Data set/path name 'USER01.EMPLOYEE'                          +
   Member . . . . . .              (Blank or pattern for member list)
   Volume serial  . .              (If not cataloged)
   Start position . .                             +
   Record limit . . .              Record sampling
   Inplace edit . . .              (Prevent inserts and deletes)
Copybook or Template:
   Data set name  . . 'USER01.COPYLIB(EMPLOYEE)'
   Member . . . . . .              (Blank or pattern for member list)
Processing Options:
 Copybook/template    Start position type   Enter "/" to select option
 1  1. Above          1. Key                Edit template    Type (1,2,S)
    2. Previous       2. RBA                Include only selected records
    3. None           3. Record number      Binary mode, reclen 80
    4. Create dynamic 4. Formatted key      Create audit trail
```

629

Now you will see this screen. Notice the format is TABL which shows the data in list format. If you want to change it to show one record at a time, type over the TABL with SNGL (which means single record).

```
Edit                   USER01.EMPLOYEE                        Top of 6
Command ===>                                                  Scroll PAGE
      Key                        Type KSDS    RBA             Format TABL
         EMP-ID EMP-LAST-NAME                 EMP-FIRST-NAME  EMP-SERVICE-
            #2 #3                              #4                       #5
         ZD 1:4 AN 5:30                        AN 35:20             ZD 55:2
            <---> <---+----1----+----2----+----> <---+----1----+---->   <->
   ****** ****   Top of data   ****
   000001   1122 JENKINS                       DEBORAH              5
   000002   3217 JOHNSON                       EDWARD               4
   000003   4720 SCHULTZ                        TIM                 9
   000004   6288 WILLARD                        JOE                 6
   000005   7459 STEWART                       BETTY                7
   000006   9134 FRANKLIN                      BRIANNA              3
****** ****    End of data   ****
```

Now you can edit each field on the record except the key. You cannot change the key, although you can specify a different key to bring up a different record. Let's bring up employee 6288.

```
Edit                   USER01.EMPLOYEE                        Rec 1 of 6
Command ===>                                                  Scroll PAGE
Key 1122                     Type KSDS    RBA 0               Format SNGL
                                              Top Line is 1    of 6
Current 01: EMPLOYEE                                          Length 80
Field                  Data
EMP-ID                  1122
EMP-LAST-NAME          JENKINS
EMP-FIRST-NAME         DEBORAH
EMP-SERVICE-YEARS        5
EMP-PROMOTION-DATE     2016-09-01
FILLER
***   End of record   ***
```

Now we can change this record. Let's change the years of service to 8.

```
Edit                   USER01.EMPLOYEE                        Rec 4 of 6
Command ===>                                                  Scroll PAGE
Key 6288                     Type KSDS    RBA 240             Format SNGL
                                              Top Line is 1    of 6
Current 01: EMPLOYEE                                          Length 80
Field                  Data
EMP-ID                  6288
EMP-LAST-NAME          WILLARD
EMP-FIRST-NAME         JOE
EMP-SERVICE-YEARS        6
EMP-PROMOTION-DATE     2016-01-01
FILLER
```

Now you can type SAVE on the command line or simply PF3 to exit from the record.

```
File Manager                      Edit Entry Panel              1 record(s) updated
Command ===>

Input Partitioned, Sequential or VSAM Data Set, or HFS file:
   Data set/path name 'USER01.EMPLOYEE'                                    +
   Member . . . . . .              (Blank or pattern for member list)
   Volume serial  . .              (If not cataloged)
   Start position . .                               +
   Record limit . . .           Record sampling
   Inplace edit . . .              (Prevent inserts and deletes)
Copybook or Template:
   Data set name  . . 'USER01.COPYLIB(EMPLOYEE)'
   Member . . . . . .              (Blank or pattern for member list)
Processing Options:
 Copybook/template    Start position type     Enter "/" to select option
 1  1. Above            1. Key                   Edit template    Type (1,2,S)
    2. Previous         2. RBA                   Include only selected records
    3. None             3. Record number         Binary mode, reclen 80
    4. Create dynamic   4. Formatted key         Create audit trail
```

Now let's see how we can insert and delete records. Actually it is pretty simple. If you are in table mode, you just use the I line command to insert a record, or the D line command to delete one. Let's add a record for employee 1111 who is Sandra Smith with 9 years of service and a promotion date of 01/01/2017. To do this, type I on the first line of detail.

```
Edit              USER01.EMPLOYEE                    Rec 1 of 6
Command ===>                                        Scroll PAGE
     Key 1122              Type KSDS     RBA 0       Format TABL
       EMP-ID EMP-LAST-NAME              EMP-FIRST-NAME    EMP-SERVICE-
          #2 #3                          #4                       #5
       ZD 1:4 AN 5:30                    AN 35:20          ZD 55:2
       <--->  <---+----1----+----2----+----> <---+----1----+----->    <->
I00001  1122 JENKINS                     DEBORAH                  5
000002  3217 JOHNSON                     EDWARD                   4
000003  4720 SCHULTZ                     TIM                      9
000004  6288 WILLARD                     JOE                      8
000005  7459 STEWART                     BETTY                    7
000006  9134 FRANKLIN                    BRIANNA                  3
****** ****  End of data   ****
```

631

Now you can enter the data. You will need to scroll to the right to add the correct years of service and promotion date. When you finish press ENTER and you'll see this screen.

```
Edit              USER01.EMPLOYEE                        Rec 1 of 7
Command ===>                                             Scroll PAGE
      Key 1122               Type KSDS     RBA 0         Format TABL
            EMP-ID EMP-LAST-NAME                EMP-FIRST-NAME    EMP-SERVICE-
              #2 #3                             #4                        #5
            ZD 1:4 AN 5:30                      AN 35:20              ZD 55:2
               <---> <---+----1----+----2----+----> <---+----1----+---->    <->
000001   1122 JENKINS                   DEBORAH                         5
000002   1111 SMITH                     SANDRA                          0
000003   3217 JOHNSON                   EDWARD                          4
000004   4720 SCHULTZ                   TIM                             9
000005   6288 WILLARD                   JOE                             8
000006   7459 STEWART                   BETTY                           7
000007   9134 FRANKLIN                  BRIANNA                         3
****** ****   End of data   ****
```

You could also switch to SNGL mode to make it easier to enter the data on one page. Let's do this.

```
Edit              USER01.EMPLOYEE                        Rec 1 of 7
Command ===>                                             Scroll PAGE
Key 1111                     Type KSDS     RBA 0         Format SNGL
                                               Top Line is 1    of 6
Current 01: EMPLOYEE                                     Length 80
Field              Data
EMP-ID                1111
EMP-LAST-NAME      SMITH
EMP-FIRST-NAME     SANDRA
EMP-SERVICE-YEARS      9
EMP-PROMOTION-DATE 2017-01-01
FILLER
***   End of record   ***
```

Now type SAVE on the command line.

```
Edit              USER01.EMPLOYEE                        Rec 1 of 7
Command ===>       SAVE                                  Scroll PAGE
Key 1111                     Type KSDS     RBA 0         Format SNGL
                                               Top Line is 1    of 6
Current 01: EMPLOYEE                                     Length 80
Field              Data
EMP-ID                1111
EMP-LAST-NAME      SMITH
EMP-FIRST-NAME     SANDRA
EMP-SERVICE-YEARS      9
EMP-PROMOTION-DATE 2017-01-01
FILLER
***   End of record   ***
```

When you press Enter you can verify the record was saved.

```
Edit              USER01.EMPLOYEE                         1 record(s) updated
Command ===>                                                   Scroll PAGE
Key 1111                   Type KSDS     RBA 0               Format SNGL
                                                      Top Line is 1    of 6
Current 01: EMPLOYEE                                        Length 80
Field                  Data
EMP-ID                  1111
EMP-LAST-NAME          SMITH
EMP-FIRST-NAME         SANDRA
EMP-SERVICE-YEARS        9
EMP-PROMOTION-DATE    2017-01-01
FILLER
***  End of record  ***
```

Finally, to delete a record, just go to TABL mode, find the record you want to delete, and use a D line action. Let's delete the record we just added.

```
Edit              USER01.EMPLOYEE                       Rec 1 of 7
Command ===>                                              Scroll PAGE
    Key 1111               Type KSDS     RBA 0           Format TABL
        EMP-ID EMP-LAST-NAME              EMP-FIRST-NAME    EMP-SERVICE-
           #2 #3                          #4                         #5
        ZD 1:4 AN 5:30                    AN 35:20            ZD 55:2
        <--->  <---+----1----+----2----+----> <---+----1----+----->   <->
D00001    1111 SMITH                     SANDRA                      9
000002    1122 JENKINS                   DEBORAH                     5
000003    3217 JOHNSON                   EDWARD                      4
000004    4720 SCHULTZ                   TIM                         9
000005    6288 WILLARD                   JOE                         8
000006    7459 STEWART                   BETTY                       7
000007    9134 FRANKLIN                  BRIANNA                     3
****** ****   End of data   ****
```

The record has disappeared from the display.

```
Edit              USER01.EMPLOYEE                       Rec 1 of 6
  Command ===>                                            Scroll PAGE
      Key 1122               Type KSDS     RBA 80         Format TABL
        EMP-ID EMP-LAST-NAME              EMP-FIRST-NAME    EMP-SERVICE-
           #2 #3                          #4                         #5
        ZD 1:4 AN 5:30                    AN 35:20            ZD 55:2
        <--->  <---+----1----+----2----+----> <---+----1----+----->   <->
000001    1122 JENKINS                   DEBORAH                     5
000002    3217 JOHNSON                   EDWARD                      4
000003    4720 SCHULTZ                   TIM                         9
000004    6288 WILLARD                   JOE                         8
000005    7459 STEWART                   BETTY                       7
000006    9134 FRANKLIN                  BRIANNA                     3
****** ****   End of data   ****
```

You can either type SAVE on the command line, or simply exit the file and the delete action will be saved.

```
 Edit                USER01.EMPLOYEE                          1 record(s) updated
 Command ===>                                                         Scroll PAGE
      Key 1122                   Type KSDS     RBA 80               Format TABL
          EMP-ID EMP-LAST-NAME                EMP-FIRST-NAME        EMP-SERVICE-
              #2 #3                           #4                             #5
          ZD 1:4 AN 5:30                      AN 35:20              ZD 55:2
             <---> <---+----1----+----2----+----> <---+----1----+---->     <->
 000001    1122 JENKINS                       DEBORAH                           5
 000002    3217 JOHNSON                       EDWARD                            4
 000003    4720 SCHULTZ                       TIM                               9
 000004    6288 WILLARD                       JOE                               8
 000005    7459 STEWART                       BETTY                             7
 000006    9134 FRANKLIN                      BRIANNA                           3
 ****** ****   End of data   ****
```

Now that we have some data, let's do some programming.

CICS Application Programming with VSAM

As we introduce VSAM as our data store, our application design does not change. All the screens will look exactly the same. The program structure and most of the logic is the same. The only thing that will change is the data store and the commands we use to access and update the data.

Record Structure

We introduced our COBOL copybook earlier. To make the fewest changes to our programs, we must make sure that our field names match what's already in the program. We'll only change one thing. We used fixed length records in our VSAM file, so we don't need the length and text fields for the last name and first name.

Here's the layout.

```
 01  DCLEMPLOYEE.
       05 EMP-ID               PIC 9(04).
       05 EMP-LAST-NAME        PIC X(30).
       05 EMP-FIRST-NAME       PIC X(20).
       05 EMP-SERVICE-YEARS    PIC 9(02).
       05 EMP-PROMOTION-DATE   PIC X(10).
       05 EMP-SSN              PIC X(09).
       05 FILLER               PIC X(05).
```

How to Define the VSAM File to CICS

Now we need to define our VSAM file to CICS. Invoke the CEDA transaction and type DEF
FILE. We'll give the VSAM file a CICS name of EMPLOYEE and we must also specify the actual
z/OS file name.

```
CEDA DEF FILE
 OVERTYPE TO MODIFY                                  CICS RELEASE = 0670
  CEDA  DEFine File(          )
   File         ==> EMPLOYEE
   Group        ==> USER01
   DEScription  ==> EMPLOYEE SUPPORT FILE
  VSAM PARAMETERS
   DSNAme       ==> USER01.EMPLOYEE
   Password     ==>                      PASSWORD NOT SPECIFIED
   RLsaccess    ==> No                   Yes | No
   LSRPOOLId    : 1                      1-8 | None
   LSRPOOLNum   ==> 001                  1-255 | None
   READInteg    ==> Uncommitted          Uncommitted | Consistent | Repeatable
   DSNSharing   ==> Allreqs              Allreqs | Modifyreqs
   STRings      ==> 001                  1-255
   Nsrgroup     ==>
  REMOTE ATTRIBUTES
   REMOTESystem ==>
   REMOTEName   ==>
+ REMOTE AND CFDATATABLE PARAMETERS
                                              SYSID=CICS APPLID=CICSTS42

PF 1 HELP 2 COM 3 END          6 CRSR 7 SBH 8 SFH 9 MSG 10 SB 11 SF 12 CNCL
```

Press Enter and you should get a Define Successful message at the bottom.

```
 OVERTYPE TO MODIFY                                  CICS RELEASE = 0670
  CEDA  DEFine File( EMPLOYEE )
   File         : EMPLOYEE
   Group        : USER01
   DEScription  ==> EMPLOYEE SUPPORT FILE
  VSAM PARAMETERS
   DSNAme       ==> USER01.EMPLOYEE
   Password     ==>                      PASSWORD NOT SPECIFIED
   RLsaccess    ==> No                   Yes | No
   LSRPOOLId    : 1                      1-8 | None
   LSRPOOLNum   --> 001                  1-255 | None
   READInteg    ==> Uncommitted          Uncommitted | Consistent | Repeatable
   DSNSharing   ==> Allreqs              Allreqs | Modifyreqs
   STRings      ==> 001                  1-255
   Nsrgroup     ==>
  REMOTE ATTRIBUTES
   REMOTESystem ==>
   REMOTEName   ==>
+ REMOTE AND CFDATATABLE PARAMETERS

                                              SYSID=CICS APPLID=CICSTS42
  DEFINE SUCCESSFUL               TIME: 02.52.08  DATE: 12/05/18
PF 1 HELP 2 COM 3 END          6 CRSR 7 SBH 8 SFH 9 MSG 10 SB 11 SF 12 CNCL
```

Next, scroll forward a couple of pages until you see this screen. I suggest you specify YES on each operation. This means you can add, update, delete, read and browse the file.

```
OVERTYPE TO MODIFY OR PRESS ENTER TO EXECUTE            CICS RELEASE = 0670
  CEDA  DEFine File( EMPLOYEE )
+ DATA FORMAT
  RECORDFormat ==> V                      V | F
  OPERATIONS
  Add              ==> yes                No | Yes
  BRowse           ==> yes                No | Yes
  DELete           ==> yes                No | Yes
  READ             ==> yes                Yes | No
  UPDATE           ==> yes                No | Yes
  AUTO JOURNALLING
  JOurnal          ==> No                 No | 1-99
  JNLRead          ==> None               None | Updateonly | Readonly | All
  JNLSYNCRead      ==> No                 No | Yes
  JNLUpdate        ==> No                 No | Yes
  JNLAdd           ==> None               None | Before | AFter | ALl
  JNLSYNCWrite     ==> Yes                Yes | No
  RECOVERY PARAMETERS
+ RECOVery         ==> None               None | Backoutonly | All

                                          SYSID=CICS APPLID=CICSTS42

PF 1 HELP 2 COM 3 END        6 CRSR 7 SBH 8 SFH 9 MSG 10 SB 11 SF 12 CNCL
```

Next, we must install the file with this command:

```
CEDA INSTALL FILE(EMPLOYEE) GROUP(USER01)
```

Finally, you must open the file using the following CEMT command:

```
CEMT SET FILE(EMPLOYEE) OPEN
```

CICS Commands to Access the VSAM File
Now let's look at the CICS commands that are necessary to carry out file operations on our VSAM file.

Reading a Record
To perform direct access on the EMPLOYEE file, we issue the following CICS command:

```
EXEC CICS
READ
FILE('EMPLOYEE')
INTO (DCLEMPLOYEE)
RIDFLD(EMP-ID)
EQUAL
RESP(RESPONSE-CODE)
END-EXEC.
```

636

We will also need to check the status of the read. We can do this using the RESPONSE-CODE variable that we'll add to the program. This will replace our check of the SQLCODE that we used in the DB2 version of the program.

```
IF DFHRESP(NORMAL)
    <Do normal processing>
ELSE
    <Do error processing>
END-IF
```

If the return code is CICS-NORMAL then it means we got a good return from the action.

Adding a Record
You add a record by loading the employee record structure and then do a WRITE.

```
EXEC CICS
WRITE
FILE('EMPLOYEE')
FROM (DCLEMPLOYEE)
RESP(RESPONSE-CODE)
END-EXEC.
```

Updating a Record
Updating a record in CICS is two step. It involves reading the record with a lock on it, and then performing a rewrite command. To lock the record we include the keyword UPDATE in the READ command. Here are both the READ and REWRITE commands:

```
EXEC CICS
READ
FILE('EMPLOYEE')
INTO (DCLEMPLOYEE)
RIDFLD(EMP-ID)
UPDATE
EQUAL
RESP(RESPONSE-CODE)
END-EXEC.

EXEC CICS
REWRITE
FILE('EMPLOYEE')
FROM (DCLEMPLOYEE)
RESP(RESPONSE-CODE)
END-EXEC.
```

Deleting a Record

Deleting a record is fairly straightforward. You do not need to first retrieve the record. Simply load the key value into the record structure and issue the delete as follows:

```
EXEC CICS READ
DELETE
FILE('EMPLOYEE')
RIDFLD(EMP-ID)
RESP(RESPONSE-CODE)
END-EXEC.
```

Revised Employee Support Programs

EMPPGINQ

We will make the following revisions to our inquiry program:

1. Remove DB2 include commands.
2. Add the new EMPLOYEE record structure.
3. Remove the –TEXT suffixes from EMP-LAST-NAME and EMP-FIRST-NAME fields.
4. Remove the DB2 query and replace with CICS READ.
5. Check for success using the CICS variables for good read (and missing record).

Now we'll make revisions to our employee support programs to use VSAM instead of DB2. Here is our revised code for the Employee Inquiry program:

```
IDENTIFICATION DIVISION.
PROGRAM-ID. EMPPGINQ.
****************************************************
*   COBOL/CICS/DB2 PROGRAM TO DISPLAY AN EMPLOYEE *
*                                                 *
*   AUTHOR         : ROBERT WINGATE               *
*   DATE-WRITTEN   : 2018-07-19                   *
****************************************************
ENVIRONMENT DIVISION.
DATA DIVISION.
WORKING-STORAGE SECTION.
01 WS-EMPNO         PIC 9(4).
01 WS-EMP-SRV-YRS   PIC 9(2).
01 WS-COMMAREA.
   05 WS-EMP-PASS   PIC 9(04) VALUE ZERO.
   05 WS-PGM-PASS   PIC X(08) VALUE SPACES.
   05 FILLER        PIC X(08).

01 RESPONSE-CODE    PIC S9(08) COMP.
01 RESPONSE-DISPLAY PIC S9(08) USAGE DISPLAY.
```

```cobol
01 PROGRAM-NAME      PIC X(08) VALUE SPACES.

01  DCLEMPLOYEE.
    05 EMP-ID               PIC 9(04).
    05 EMP-LAST-NAME        PIC X(30).
    05 EMP-FIRST-NAME       PIC X(20).
    05 EMP-SERVICE-YEARS    PIC 9(02).
    05 EMP-PROMOTION-DATE   PIC X(10).
    05 EMP-SSN              PIC X(09).
    05 FILLER               PIC X(05).

01 SW-PASSED-DATA-SWITCH  PIC X(1) VALUE 'N'.
   88  SW-PASSED-DATA               VALUE 'Y'.
   88  SW-NO-PASSED-DATA            VALUE 'N'.

    COPY EMPMINQ.
    COPY DFHAID.
    COPY DFHBMSCA.

LINKAGE SECTION.
01 DFHCOMMAREA        PIC X(20).

PROCEDURE DIVISION.

    SET SW-NO-PASSED-DATA TO TRUE
    IF EIBCALEN > ZERO
      MOVE DFHCOMMAREA  TO WS-COMMAREA
    END-IF.

    EVALUATE TRUE

      WHEN EIBCALEN = ZERO
        MOVE LOW-VALUES   TO  EMPINQO
        PERFORM SEND-MAP

      WHEN EIBAID = DFHCLEAR
        MOVE LOW-VALUES   TO  EMPINQO
        PERFORM SEND-MAP

      WHEN EIBAID = DFHPA1 OR DFHPA2 OR DFHPA3
        CONTINUE

      WHEN EIBAID = DFHPF2
        IF WS-PGM-PASS NOT EQUAL "EMPPGINQ"
           SET SW-PASSED-DATA TO TRUE
        END-IF
        PERFORM PROCESS-PARA

      WHEN EIBAID = DFHPF3
        MOVE LOW-VALUES TO  EMPINQO
        MOVE "BYE, PRESS CLEAR KEY TO ENTER A TRANSACTION ID"
```

```
                  TO MESSAGEO
              PERFORM SEND-MAP-DATA

              EXEC CICS
                RETURN
              END-EXEC

          WHEN EIBAID = DFHPF4
            MOVE 'EMPPGADD' TO PROGRAM-NAME
            PERFORM BRANCH-TO-PROGRAM

              EXEC CICS
                RETURN
              END-EXEC

          WHEN EIBAID = DFHPF5
            MOVE 'EMPPGCHG' TO PROGRAM-NAME
            PERFORM BRANCH-TO-PROGRAM

              EXEC CICS
                RETURN
              END-EXEC

          WHEN EIBAID = DFHPF6
            MOVE 'EMPPGDEL' TO PROGRAM-NAME
            PERFORM BRANCH-TO-PROGRAM

              EXEC CICS
                RETURN
              END-EXEC

          WHEN EIBAID = DFHENTER
            PERFORM PROCESS-PARA

          WHEN OTHER
            MOVE LOW-VALUES TO EMPINQO
            MOVE "INVALID KEY PRESSED" TO MESSAGEO
            PERFORM SEND-MAP-DATA

      END-EVALUATE.

      EXEC CICS
          RETURN TRANSID('EMIN')
          COMMAREA (WS-COMMAREA)
          LENGTH(20)
      END-EXEC.

  PROCESS-PARA.

      PERFORM RECEIVE-MAP.
      INITIALIZE DCLEMPLOYEE MESSAGEO
```

```
IF SW-PASSED-DATA
    MOVE WS-EMP-PASS TO EMP-ID
ELSE
    MOVE EMPINI     TO WS-EMPNO
    MOVE WS-EMPNO   TO EMP-ID
END-IF

EXEC CICS
    READ
    FILE('EMPLOYEE')
    INTO(DCLEMPLOYEE)
    RIDFLD(EMP-ID)
    EQUAL
    RESP(RESPONSE-CODE)
END-EXEC.

EVALUATE RESPONSE-CODE
    WHEN DFHRESP(NORMAL)
        MOVE EMP-ID               TO   WS-EMPNO WS-EMP-PASS
        MOVE WS-EMPNO             TO   EMPNOO
        MOVE EMP-LAST-NAME        TO LNAMEO
        MOVE EMP-FIRST-NAME       TO FNAMEO
        MOVE EMP-SSN              TO SOCSECO
        MOVE EMP-SERVICE-YEARS    TO WS-EMP-SRV-YRS
        MOVE WS-EMP-SRV-YRS       TO YRSSVCO
        MOVE EMP-PROMOTION-DATE   TO LSTPRMO

    WHEN DFHRESP(NOTFND)
        STRING "EMPLOYEE ID " DELIMITED BY SIZE
        WS-EMPNO DELIMITED BY SPACE
        " NOT FOUND" DELIMITED BY SIZE INTO MESSAGEO
        MOVE SPACES      TO EMPNOO
        MOVE SPACES      TO LNAMEO
        MOVE SPACES      TO FNAMEO
        MOVE SPACES      TO SOCSECO
        MOVE SPACES      TO YRSSVCO
        MOVE SPACES      TO LSTPRMO

    WHEN OTHER
        MOVE RESPONSE-CODE TO RESPONSE-DISPLAY
        STRING "UNKNOWN ERROR CODE: " DELIMITED BY SIZE
            RESPONSE-DISPLAY DELIMITED BY SIZE
          INTO MESSAGEO

END-EVALUATE.

MOVE DFHBMFSE TO EMPINF
MOVE -1 TO EMPINL
MOVE "EMPPGINQ" TO WS-PGM-PASS

IF SW-PASSED-DATA
    PERFORM SEND-MAP
```

```
            ELSE
                PERFORM SEND-MAP-DATA
            END-IF.

        BRANCH-TO-PROGRAM.

            EXEC CICS
                XCTL PROGRAM(PROGRAM-NAME)
                COMMAREA (WS-COMMAREA)
                LENGTH(20)
            END-EXEC

            MOVE 'PROGRAM NOT AVAILABLE' TO MESSAGEO.
            MOVE DFHBMFSE TO EMPINF
            MOVE -1 TO EMPINL
            PERFORM SEND-MAP-DATA.

        SEND-MAP.
            EXEC CICS SEND
                MAP    ('EMPINQ')
                MAPSET ('EMPMINQ')
                FROM   (EMPINQO)
                ERASE
            END-EXEC.

        SEND-MAP-DATA.
            EXEC CICS SEND
                MAP    ('EMPINQ')
                MAPSET ('EMPMINQ')
                FROM   (EMPINQO)
                DATAONLY
            END-EXEC.

        RECEIVE-MAP.
            EXEC CICS RECEIVE
                MAP    ('EMPINQ')
                MAPSET ('EMPMINQ')
                INTO   (EMPINQI)
            END-EXEC.
```

Now we need to do a straight CICS compile and link (no DB2 bind). Here is the JCL I use for that.

```
//USER01D JOB MSGLEVEL=(1,1),NOTIFY=&SYSUID
//*
//*  COMPILE A COBOL + CICS PROGRAM
//*
//CICSCOB  EXEC CICSCOBC,
//              COPYLIB=USER01.COPYLIB,         <= COPYBOOK LIBRARY
//              SRCLIB=USER01.CICS.SRCLIB,      <= SOURCE LIBRARY
//              MEMBER=EMPPGINQ                 <= SOURCE MEMBER
```

Next we need to go into CICS and refresh the program load module using CEMT:

```
CEMT SET PROGRAM(EMPPGINQ) NEWCOPY
```

Finally we are ready to test. Let's bring up the EMIN screen and then display data for employee 3217:

```
EMPMINQ                    EMPLOYEE INQUIRY                      EMIN


        EMPLOYEE ->     ____        ENTER EMPLOYEE ID, THEN PRESS ENTER

        EMPLOYEE ID    XXXX

        EMP LAST NAME  XXXX

        EMP FIRST NAME XXXX

        EMP SOCIAL SEC XXXXXXXXX

        EMP YEARS SRVC 00

        EMP LAST PROM  YYYY-MM-DD




     F2 INQ   F3 EXIT   F4 ADD   F5 CHG   F6 DEL
```

Enter 3217, press ENTER and here is the result:

```
EMPMINQ                    EMPLOYEE INQUIRY                      EMIN


        EMPLOYEE ->     3217        ENTER EMPLOYEE ID, THEN PRESS ENTER

        EMPLOYEE ID    3217

        EMP LAST NAME  JOHNSON

        EMP FIRST NAME EDWARD

        EMP SOCIAL SEC 493082938

        EMP YEARS SRVC 04

        EMP LAST PROM  2017-01-01




     F2 INQ   F3 EXIT   F4 ADD   F5 CHG   F6 DEL
```

Great, it works! If your version does not work, please make sure to check each step and correct if necessary.

Now, to finish the DB2-to-VSAM conversion we'll need to do regression testing. So run all the unit test cases to make sure they still work. I've done mine and didn't find any errors.

Now move on to the add program!

EMPPGADD

For the add program we'll do these same conversions as for the inquiry except of course we will be writing a record instead of reading it. Make these changes:

1. Remove DB2 include commands
2. Add the new EMPLOYEE record structure
3. Remove the –TEXT suffixes from EMP-LAST-NAME and EMP-FIRST-NAME fields
4. Remove any code that attempts to calculate the length of the employee last name or first name fields
5. Remove the DB2 INSERT query and replace with CICS WRITE
6. Check for success using the CICS variables for successful transaction as well as missing record.

Give this a try, then take a break and come back and we'll compare code.
…..

Ok, I'm back. Here's my code for the add program:

```
        IDENTIFICATION DIVISION.
        PROGRAM-ID. EMPPGADD.
        ************************************************
        *   COBOL/CICS/VSAM PROGRAM TO ADD AN EMPLOYEE   *
        *                                                *
        *   AUTHOR      : ROBERT WINGATE                 *
        *   DATE-WRITTEN : 2018-07-21                    *
        ************************************************
        ENVIRONMENT DIVISION.
        DATA DIVISION.
        WORKING-STORAGE SECTION.
        01 WS-EMPNO          PIC 9(04).
        01 WS-EMP-SRV-YRS    PIC 9(02).
        01 WS-COMMAREA.
           05 WS-EMP-PASS    PIC 9(04).
           05 WS-PGM-PASS    PIC X(08).
           05 FILLER         PIC X(08).
```

```
01 PROGRAM-NAME     PIC X(08) VALUE SPACES.

01 RESPONSE-CODE    PIC S9(08) VALUE 0.
01 RESPONSE-DISPLAY PIC  9(08) VALUE 0.

01 RESPONSE-CODE2   PIC S9(08) VALUE 0.
01 RESPONSE-DISPLA2 PIC  9(08) VALUE 0.

01  DCLEMPLOYEE.
    05 EMP-ID               PIC 9(04).
    05 EMP-LAST-NAME        PIC X(30).
    05 EMP-FIRST-NAME       PIC X(20).
    05 EMP-SERVICE-YEARS    PIC 9(02).
    05 EMP-PROMOTION-DATE   PIC X(10).
    05 EMP-SSN              PIC X(09).
    05 FILLER              PIC X(05).

    COPY EMPMADD.
    COPY DFHAID.
    COPY DFHBMSCA.

LINKAGE SECTION.
01 DFHCOMMAREA          PIC X(20).

PROCEDURE DIVISION.

    IF EIBCALEN > ZERO
      MOVE DFHCOMMAREA  TO WS-COMMAREA
    END-IF.

    EVALUATE TRUE

      WHEN EIBCALEN = ZERO
        MOVE LOW-VALUES    TO   EMPADDO
        MOVE -1 TO EMPNOL
        PERFORM SEND-MAP

      WHEN EIBAID = DFHCLEAR
        MOVE LOW-VALUES    TO   EMPADDO
        MOVE -1 TO EMPNOL
        PERFORM SEND-MAP

      WHEN EIBAID = DFHPA1 OR DFHPA2 OR DFHPA3
        CONTINUE

      WHEN EIBAID = DFHPF2
        MOVE 'EMPPGINQ' TO PROGRAM-NAME
        PERFORM BRANCH-TO-PROGRAM

        EXEC CICS
          RETURN
```

```
              END-EXEC

          WHEN EIBAID = DFHPF3
            MOVE LOW-VALUES TO  EMPADDO
            MOVE -1 TO EMPNOL
            MOVE "BYE, PRESS CLEAR KEY TO ENTER A TRANSACTION ID"
                TO MESSAGEO
            PERFORM SEND-MAP-DATA

            EXEC CICS
              RETURN
            END-EXEC

          WHEN EIBAID = DFHPF4
*           PERFORM THE EDITS AND VALIDATIONS
*           IF NO ERRORS THEN INSERT THE RECORDS

            IF WS-PGM-PASS NOT EQUAL "EMPPGADD"
               MOVE LOW-VALUES   TO  EMPADDO
               MOVE -1 TO EMPNOL
               MOVE
               "ENTER DATA FOR NEW EMPLOYEE, THEN PRESS PF4 TO ADD"
               TO MESSAGEO
               MOVE "EMPPGADD" TO WS-PGM-PASS
               PERFORM SEND-MAP
            ELSE
               PERFORM VALIDATE-DATA
            END-IF

          WHEN EIBAID = DFHPF5
            MOVE 'EMPPGCHG' TO PROGRAM-NAME
            PERFORM BRANCH-TO-PROGRAM

          WHEN EIBAID = DFHPF6
            MOVE 'EMPPGDEL' TO PROGRAM-NAME
            PERFORM BRANCH-TO-PROGRAM

          WHEN OTHER
            MOVE LOW-VALUES TO EMPADDO
            MOVE -1 TO EMPNOL
            MOVE "INVALID KEY PRESSED" TO MESSAGEO
            PERFORM SEND-MAP-DATA

        END-EVALUATE.

        EXEC CICS
           RETURN TRANSID('EMAD')
           COMMAREA (WS-COMMAREA)
        END-EXEC.

   PROCESS-PARA.
```

```
        PERFORM RECEIVE-MAP.
        INITIALIZE DCLEMPLOYEE MESSAGEO

        MOVE EMPNOI     TO WS-EMPNO
        MOVE WS-EMPNO   TO EMP-ID

        MOVE DFHBMFSE TO EMPNOF
        MOVE DFHBMFSE TO LNAMEF
        MOVE DFHBMFSE TO FNAMEF
        MOVE DFHBMFSE TO SOCSECF
        MOVE DFHBMFSE TO YRSSVCF
        MOVE DFHBMFSE TO LSTPRMF

        MOVE -1 TO EMPNOL.
        MOVE "ENTER DATA FOR NEW EMPLOYEE, THEN PRESS PF4 TO ADD"
            TO MESSAGEO

        MOVE "EMPPGADD" TO WS-PGM-PASS
        PERFORM SEND-MAP-ALL.

    VALIDATE-DATA.

        PERFORM RECEIVE-MAP
        INITIALIZE DCLEMPLOYEE MESSAGEO

        MOVE DFHBMFSE TO EMPNOF
        MOVE DFHBMFSE TO LNAMEF
        MOVE DFHBMFSE TO FNAMEF
        MOVE DFHBMFSE TO SOCSECF
        MOVE DFHBMFSE TO YRSSVCF
        MOVE DFHBMFSE TO LSTPRMF

        EVALUATE TRUE

            WHEN EMPNOI EQUAL SPACES OR EMPNOL EQUAL ZERO
                MOVE "EMPLOYEE NUMBER IS REQUIRED" TO MESSAGEO
                MOVE -1 TO EMPNOL

            WHEN EMPNOI IS NOT NUMERIC
                MOVE "EMPLOYEE NUMBER MUST BE NUMERIC" TO MESSAGEO
                MOVE -1 TO EMPNOL

            WHEN LNAMEI EQUAL SPACES OR LNAMEL EQUAL ZERO
                MOVE "EMPLOYEE LAST NAME IS REQUIRED" TO MESSAGEO
                MOVE -1 TO LNAMEL

            WHEN FNAMEI EQUAL SPACES OR FNAMEL EQUAL ZERO
                MOVE "EMPLOYEE FIRST NAME IS REQUIRED" TO MESSAGEO
                MOVE -1 TO FNAMEL

            WHEN SOCSECI EQUAL SPACES OR SOCSECL EQUAL ZERO
                MOVE "SOCIAL SECURITY NUMBER IS REQUIRED" TO MESSAGEO
```

```
                    MOVE -1 TO SOCSECL

              WHEN SOCSECI IS NOT NUMERIC
                    MOVE "SOCIAL SECURITY MUST BE NUMERIC" TO MESSAGEO
                    MOVE -1 TO SOCSECL

              WHEN YRSSVCI EQUAL SPACES OR YRSSVCL EQUAL ZERO
                    MOVE "YEARS OF SERVICE IS REQUIRED" TO MESSAGEO
                    MOVE -1 TO YRSSVCL

              WHEN YRSSVCI IS NOT NUMERIC
                    MOVE "YEARS OF SERVICE MUST BE NUMERIC" TO MESSAGEO
                    MOVE -1 TO YRSSVCL

              WHEN LSTPRMI EQUAL SPACES OR LSTPRML EQUAL ZERO
                    MOVE "LAST PROMOTION DATE IS REQUIRED" TO MESSAGEO
                    MOVE -1 TO LSTPRML

              WHEN OTHER
                    PERFORM ADD-RECORD
                    MOVE -1 TO EMPNOL

         END-EVALUATE.

         PERFORM SEND-MAP-DATA.

     ADD-RECORD.

*    MAP INPUT FIELDS TO VSAM RECORD STRUCTURE

         MOVE EMPNOI             TO WS-EMPNO
         MOVE WS-EMPNO           TO EMP-ID
         MOVE LNAMEI             TO EMP-LAST-NAME
         MOVE FNAMEI             TO EMP-FIRST-NAME
         MOVE SOCSECI            TO EMP-SSN
         MOVE YRSSVCI            TO WS-EMP-SRV-YRS
         MOVE WS-EMP-SRV-YRS     TO EMP-SERVICE-YEARS
         MOVE LSTPRMI            TO EMP-PROMOTION-DATE

*    INSERT THE RECORD

         EXEC CICS
            WRITE
            FILE('EMPLOYEE')
            FROM(DCLEMPLOYEE)
            RIDFLD(EMP-ID)
            RESP(RESPONSE-CODE)
            RESP2(RESPONSE-CODE2)
         END-EXEC.

         EVALUATE RESPONSE-CODE
```

```
        WHEN DFHRESP(NORMAL)
            MOVE "EMPLOYEE ADDED SUCCESSFULLY" TO MESSAGEO
            MOVE -1 TO EMPNOL
            MOVE WS-EMPNO TO WS-EMP-PASS

        WHEN DFHRESP(DUPREC)
            MOVE "ERROR - RECORD ALREADY EXISTS" TO MESSAGEO
            MOVE -1 TO EMPNOL
            MOVE WS-EMPNO TO WS-EMP-PASS

        WHEN OTHER
            MOVE RESPONSE-CODE  TO RESPONSE-DISPLAY
            MOVE RESPONSE-CODE2 TO RESPONSE-DISPLA2
            STRING "UNKNOWN ERROR CODE: " DELIMITED BY SIZE
                RESPONSE-DISPLAY DELIMITED BY SIZE
                " "              DELIMITED BY SIZE
                RESPONSE-DISPLA2 DELIMITED BY SIZE
              INTO MESSAGEO


    END-EVALUATE.

BRANCH-TO-PROGRAM.

    EXEC CICS
        XCTL PROGRAM(PROGRAM-NAME)
        COMMAREA (WS-COMMAREA)
        LENGTH(10)
    END-EXEC

    MOVE 'PROGRAM NOT AVAILABLE' TO MESSAGEO.

SEND-MAP.
    EXEC CICS SEND
        MAP    ('EMPADD')
        MAPSET ('EMPMADD')
        FROM   (EMPADDO)
        CURSOR
        ERASE
    END-EXEC.

SEND-MAP-DATA.
    EXEC CICS SEND
        MAP    ('EMPADD')
        MAPSET ('EMPMADD')
        FROM   (EMPADDO)
        CURSOR
        DATAONLY
    END-EXEC.

SEND-MAP-ALL.
    EXEC CICS SEND
        MAP    ('EMPADD')
```

```
              MAPSET ('EMPMADD')
              FROM   (EMPADDO)
              CURSOR
         END-EXEC.

    RECEIVE-MAP.
         EXEC CICS RECEIVE
              MAP    ('EMPADD')
              MAPSET ('EMPMADD')
              INTO   (EMPADDI)
         END-EXEC.
```

Ok, let's test the program. We'll add a record.

```
EMPMADD                    EMPLOYEE ADD                              EMAD


               ENTER EMPLOYEE INFO, THEN PRESS PF4

        EMPLOYEE ID    9461

        EMP LAST NAME  berry

        EMP FIRST NAME julie

        EMP SOCIAL SEC 947294888

        EMP YEARS SRVC 34

        EMP LAST PROM  2018-01-01




   F2 INQ   F3 EXIT   F4 ADD   F5 CHG   F6 DEL
```

Press **ENTER** and this is the result, a successful add.

```
EMPMADD                         EMPLOYEE ADD                              EMAD

                        ENTER EMPLOYEE INFO, THEN PRESS PF4

           EMPLOYEE ID    9461

           EMP LAST NAME  BERRY

           EMP FIRST NAME JULIE

           EMP SOCIAL SEC 947294888

           EMP YEARS SRVC 34

           EMP LAST PROM  2018-01-01




   EMPLOYEE ADDED SUCCESSFULLY
   F2 INQ   F3 EXIT   F4 ADD   F5 CHG   F6 DEL
```

Now we'll try adding the same record to test the duplicate record logic.

```
EMPMADD                         EMPLOYEE ADD                              EMAD

                        ENTER EMPLOYEE INFO, THEN PRESS PF4

           EMPLOYEE ID    9461

           EMP LAST NAME  BERRY

           EMP FIRST NAME JULIE

           EMP SOCIAL SEC 947294888

           EMP YEARS SRVC 34

           EMP LAST PROM  2018-01-01




   ERROR - RECORD ALREADY EXISTS
   F2 INQ   F3 EXIT   F4 ADD   F5 CHG   F6 DEL
```

Great, now go ahead and do your full unit test, then we'll do the change program.

EMPPGCHG

At this point we have everything we need to modify the change program. Simply copy the appropriate code from the inquiry and add programs (but remember to specify the UPDATE option on the read command – this is necessary to lock the record until you've finished updating it).

Go ahead and do these changes, and then we'll compare code.

......

Ok, back again. Did you run into any difficulties? I bet not, but just for comparison, here's my code.

```
        IDENTIFICATION DIVISION.
        PROGRAM-ID. EMPPGCHG.
       ****************************************************
       *   COBOL/CICS/DB2 PROGRAM TO CHANGE AN EMPLOYEE  *
       *                                                 *
       *   AUTHOR       : ROBERT WINGATE                 *
       *   DATE-WRITTEN : 2018-07-23                     *
       ****************************************************
        ENVIRONMENT DIVISION.
        DATA DIVISION.
        WORKING-STORAGE SECTION.
        01 WS-EMPNO        PIC 9(4).
        01 WS-EMP-SRV-YRS  PIC 9(02).
        01 WS-SQLCODE      PIC 9(08).
        01 WS-COMMAREA.
           05 WS-EMP-PASS  PIC 9(04).
           05 WS-PGM-PASS  PIC X(08).
           05 FILLER       PIC X(08).

        01 RESPONSE-CODE    PIC S9(08) COMP.
        01 RESPONSE-DISPLAY PIC S9(08) USAGE DISPLAY.

        01 RESPONSE-CODE2   PIC S9(08) VALUE 0.
        01 RESPONSE-DISPLA2 PIC  9(08) VALUE 0.

        01 PROGRAM-NAME     PIC X(08) VALUE SPACES.

        01 SW-PASSED-DATA-SWITCH   PIC X(1) VALUE 'N'.
           88  SW-PASSED-DATA               VALUE 'Y'.
           88  SW-NO-PASSED-DATA            VALUE 'N'.

          COPY EMPMCHG.
          COPY DFHAID.
          COPY DFHBMSCA.
```

```
01 DCLEMPLOYEE.
    05 EMP-ID                 PIC 9(04).
    05 EMP-LAST-NAME          PIC X(30).
    05 EMP-FIRST-NAME         PIC X(20).
    05 EMP-SERVICE-YEARS      PIC 9(02).
    05 EMP-PROMOTION-DATE     PIC X(10).
    05 EMP-SSN                PIC X(09).
    05 FILLER                 PIC X(05).

LINKAGE SECTION.
01 DFHCOMMAREA            PIC X(20).

PROCEDURE DIVISION.

    SET SW-NO-PASSED-DATA TO TRUE
    IF EIBCALEN > ZERO
      MOVE DFHCOMMAREA  TO WS-COMMAREA
    END-IF.

    EVALUATE TRUE

      WHEN EIBCALEN = ZERO
        MOVE LOW-VALUES    TO  EMPCHGO
        MOVE -1 TO EMPINL
        PERFORM SEND-MAP

      WHEN EIBAID = DFHCLEAR
        MOVE LOW-VALUES    TO  EMPCHGO
        MOVE -1 TO EMPINL
        PERFORM SEND-MAP

      WHEN EIBAID = DFHPA1 OR DFHPA2 OR DFHPA3
        CONTINUE

      WHEN EIBAID = DFHPF2
        MOVE 'EMPPGINQ' TO PROGRAM-NAME
        PERFORM BRANCH-TO-PROGRAM

        EXEC CICS
          RETURN
        END-EXEC

      WHEN EIBAID = DFHPF3
        MOVE LOW-VALUES TO  EMPCHGO
        MOVE -1 TO EMPINL
        MOVE "BYE, PRESS CLEAR KEY TO ENTER A TRANSACTION ID"
             TO MESSAGEO
        PERFORM SEND-MAP-DATA

        EXEC CICS
          RETURN
```

```
                END-EXEC

            WHEN EIBAID = DFHPF4
              MOVE 'EMPPGADD' TO PROGRAM-NAME
              PERFORM BRANCH-TO-PROGRAM

              EXEC CICS
                RETURN
              END-EXEC

            WHEN EIBAID = DFHPF5
*              PERFORM THE EDITS AND VALIDATIONS
*               IF NO ERRORS THEN MODIFY THE RECORD

              IF WS-PGM-PASS NOT EQUAL "EMPPGCHG"
                 SET SW-PASSED-DATA TO TRUE
                 PERFORM PROCESS-PARA
              ELSE
                 PERFORM VALIDATE-DATA
              END-IF

            WHEN EIBAID = DFHPF6
              MOVE 'EMPPGDEL' TO PROGRAM-NAME
              PERFORM BRANCH-TO-PROGRAM

              EXEC CICS
                RETURN
              END-EXEC

            WHEN EIBAID = DFHENTER
              PERFORM PROCESS-PARA

            WHEN OTHER
              MOVE LOW-VALUES TO EMPCHGO
              MOVE -1 TO EMPINL
              MOVE "INVALID KEY PRESSED" TO MESSAGEO
              PERFORM SEND-MAP-DATA

       END-EVALUATE.

       EXEC CICS
          RETURN TRANSID('EMCH')
          COMMAREA (WS-COMMAREA)
          LENGTH(20)
       END-EXEC.

    PROCESS-PARA.

       PERFORM RECEIVE-MAP.
       INITIALIZE DCLEMPLOYEE MESSAGEO

       IF SW-PASSED-DATA
```

654

```
            MOVE WS-EMP-PASS TO EMP-ID
ELSE
    MOVE EMPINI     TO WS-EMPNO
    MOVE WS-EMPNO   TO EMP-ID
END-IF

MOVE DFHBMFSE TO EMPINF
MOVE DFHBMFSE TO EMPNOF
MOVE DFHBMFSE TO LNAMEF
MOVE DFHBMFSE TO FNAMEF
MOVE DFHBMFSE TO SOCSECF
MOVE DFHBMFSE TO YRSSVCF
MOVE DFHBMFSE TO LSTPRMF

EXEC CICS
    READ
    FILE('EMPLOYEE')
    INTO(DCLEMPLOYEE)
    UPDATE
    RIDFLD(EMP-ID)
    EQUAL
    RESP(RESPONSE-CODE)
END-EXEC.

EVALUATE RESPONSE-CODE
    WHEN DFHRESP(NORMAL)
        MOVE EMP-ID              TO  WS-EMPNO WS-EMP-PASS
        MOVE WS-EMPNO            TO  EMPNOO
        MOVE EMP-LAST-NAME     TO LNAMEO
        MOVE EMP-FIRST-NAME    TO FNAMEO
        MOVE EMP-SSN           TO SOCSECO
        MOVE EMP-SERVICE-YEARS TO WS-EMP-SRV-YRS
        MOVE WS-EMP-SRV-YRS    TO YRSSVCO
        MOVE EMP-PROMOTION-DATE TO LSTPRMO

        MOVE "MAKE CHANGES AND THEN PRESS PF5" TO MESSAGEO

    WHEN DFHRESP(NOTFND)
        STRING "EMPLOYEE ID " DELIMITED BY SIZE
        WS-EMPNO DELIMITED BY SPACE
        " NOT FOUND" DELIMITED BY SIZE INTO MESSAGEO
        MOVE SPACES        TO EMPNOO
        MOVE SPACES        TO LNAMEO
        MOVE SPACES        TO FNAMEO
        MOVE SPACES        TO SOCSECO
        MOVE SPACES        TO YRSSVCO
        MOVE SPACES        TO LSTPRMO

    WHEN OTHER
        MOVE RESPONSE-CODE TO RESPONSE-DISPLAY
        STRING "UNKNOWN ERROR CODE: " DELIMITED BY SIZE
            RESPONSE-DISPLAY DELIMITED BY SIZE
```

```
                INTO MESSAGEO

        END-EVALUATE.

        MOVE -1 TO EMPINL.

        MOVE "EMPPGCHG" TO WS-PGM-PASS

        IF SW-PASSED-DATA
           PERFORM SEND-MAP
        ELSE
           PERFORM SEND-MAP-DATA
        END-IF.

    VALIDATE-DATA.

        PERFORM RECEIVE-MAP
        INITIALIZE DCLEMPLOYEE MESSAGEO

        EVALUATE TRUE

            WHEN EMPNOI EQUAL SPACES OR EMPNOL EQUAL ZERO
                MOVE "EMPLOYEE NUMBER IS REQUIRED" TO MESSAGEO
                MOVE -1 TO EMPNOL

            WHEN EMPNOI IS NOT NUMERIC
                MOVE "EMPLOYEE NUMBER MUST BE NUMERIC" TO MESSAGEO
                MOVE -1 TO EMPNOL

            WHEN LNAMEI EQUAL SPACES OR LNAMEL EQUAL ZERO
                MOVE "EMPLOYEE LAST NAME IS REQUIRED" TO MESSAGEO
                MOVE -1 TO LNAMEL

            WHEN FNAMEI EQUAL SPACES OR FNAMEL EQUAL ZERO
                MOVE "EMPLOYEE FIRST NAME IS REQUIRED" TO MESSAGEO
                MOVE -1 TO FNAMEL

            WHEN SOCSECI EQUAL SPACES OR SOCSECL EQUAL ZERO
                MOVE "SOCIAL SECURITY NUMBER IS REQUIRED" TO MESSAGEO
                MOVE -1 TO SOCSECL

            WHEN SOCSECI IS NOT NUMERIC
                MOVE "SOCIAL SECURITY MUST BE NUMERIC" TO MESSAGEO
                MOVE -1 TO SOCSECL

            WHEN YRSSVCI EQUAL SPACES OR YRSSVCL EQUAL ZERO
                MOVE "YEARS OF SERVICE IS REQUIRED" TO MESSAGEO
                MOVE -1 TO YRSSVCL

            WHEN YRSSVCI IS NOT NUMERIC
                MOVE "YEARS OF SERVICE MUST BE NUMERIC" TO MESSAGEO
                MOVE -1 TO YRSSVCL
```

```
                WHEN LSTPRMI EQUAL SPACES OR LSTPRML EQUAL ZERO
                    MOVE "LAST PROMOTION DATE IS REQUIRED" TO MESSAGEO
                    MOVE -1 TO LSTPRML

                WHEN OTHER
                    PERFORM CHANGE-RECORD
                    MOVE -1 TO EMPINL

            END-EVALUATE.

            MOVE DFHBMFSE TO EMPINF
            MOVE DFHBMFSE TO EMPNOF
            MOVE DFHBMFSE TO LNAMEF
            MOVE DFHBMFSE TO FNAMEF
            MOVE DFHBMFSE TO SOCSECF
            MOVE DFHBMFSE TO YRSSVCF
            MOVE DFHBMFSE TO LSTPRMF

            MOVE "EMPPGCHG" TO WS-PGM-PASS
            PERFORM SEND-MAP-DATA.

    CHANGE-RECORD.

*   FIRST REGET THE RECORD FOR UPDATE

            MOVE EMPINI    TO WS-EMPNO
            MOVE WS-EMPNO  TO EMP-ID

        EXEC CICS
            READ
            FILE('EMPLOYEE')
            INTO(DCLEMPLOYEE)
            UPDATE
            RIDFLD(EMP-ID)
            EQUAL
            RESP(RESPONSE-CODE)
        END-EXEC.

            IF RESPONSE-CODE EQUAL DFHRESP(NORMAL)

*           DO THE UPDATE

            MOVE EMPNOI            TO WS-EMPNO
            MOVE WS-EMPNO          TO EMP-ID
            MOVE LNAMEI            TO EMP-LAST-NAME
            MOVE FNAMEI            TO EMP-FIRST-NAME
            MOVE SOCSECI           TO EMP-SSN
            MOVE YRSSVCI           TO WS-EMP-SRV-YRS
            MOVE WS-EMP-SRV-YRS    TO EMP-SERVICE-YEARS
            MOVE LSTPRMI           TO EMP-PROMOTION-DATE
```

```
        EXEC CICS
            REWRITE
            FILE('EMPLOYEE')
            FROM(DCLEMPLOYEE)
            RESP(RESPONSE-CODE)
            RESP2(RESPONSE-CODE2)
        END-EXEC

        EVALUATE RESPONSE-CODE

            WHEN DFHRESP(NORMAL)
                MOVE "EMPLOYEE MODIFED SUCCESSFULLY" TO MESSAGEO
                MOVE -1 TO EMPNOL
                MOVE WS-EMPNO TO WS-EMP-PASS

            WHEN DFHRESP(NOTFND)
                MOVE -1 TO EMPNOL
                MOVE WS-EMPNO TO WS-EMP-PASS
                STRING "EMPLOYEE ID " DELIMITED BY SIZE
                WS-EMPNO DELIMITED BY SPACE
                " NOT FOUND" DELIMITED BY SIZE INTO MESSAGEO
                MOVE SPACES         TO EMPNOO
                MOVE SPACES         TO LNAMEO
                MOVE SPACES         TO FNAMEO
                MOVE SPACES         TO SOCSECO
                MOVE SPACES         TO YRSSVCO
                MOVE SPACES         TO LSTPRMO
                MOVE -1 TO EMPNOL

            WHEN OTHER
                MOVE RESPONSE-CODE  TO RESPONSE-DISPLAY
                MOVE RESPONSE-CODE2 TO RESPONSE-DISPLA2
                STRING "UNKNOWN ERROR CODE: " DELIMITED BY SIZE
                    RESPONSE-DISPLAY DELIMITED BY SIZE
                    " "                DELIMITED BY SIZE
                    RESPONSE-DISPLA2 DELIMITED BY SIZE
                  INTO MESSAGEO


        END-EVALUATE

    ELSE

        MOVE RESPONSE-CODE TO RESPONSE-DISPLAY
        STRING "UNKNOWN ERROR CODE: " DELIMITED BY SIZE
            RESPONSE-DISPLAY DELIMITED BY SIZE
          INTO MESSAGEO

    END-IF.

BRANCH-TO-PROGRAM.

    EXEC CICS
```

```
            XCTL PROGRAM(PROGRAM-NAME)
            COMMAREA (WS-COMMAREA)
            LENGTH(20)
        END-EXEC

        MOVE 'PROGRAM NOT AVAILABLE' TO MESSAGEO.

    SEND-MAP.
        EXEC CICS SEND
            MAP    ('EMPCHG')
            MAPSET ('EMPMCHG')
            FROM   (EMPCHGO)
            CURSOR
            ERASE
        END-EXEC.

    SEND-MAP-DATA.
        EXEC CICS SEND
            MAP    ('EMPCHG')
            MAPSET ('EMPMCHG')
            FROM   (EMPCHGO)
            DATAONLY
            CURSOR
        END-EXEC.

    SEND-MAP-ALL.
        EXEC CICS SEND
            MAP    ('EMPCHG')
            MAPSET ('EMPMCHG')
            FROM   (EMPCHGO)
            CURSOR
        END-EXEC.

    RECEIVE-MAP.
        EXEC CICS RECEIVE
            MAP    ('EMPCHG')
            MAPSET ('EMPMCHG')
            INTO   (EMPCHGI)
        END-EXEC.
```

Now let's change a record and see if our changes took. Let's pick employee 3217, and we'll change the years of service to 7 and the last promotion date to 2018-01-01. First, bring up the change screen for employee 3217:

```
EMPMCHG                    EMPLOYEE CHANGE                        EMCH


        EMPLOYEE ->    3217      ENTER EMPLOYEE ID, THEN PRESS ENTER

        EMPLOYEE ID    3217

        EMP LAST NAME  JOHNSON

        EMP FIRST NAME EDWARD

        EMP SOCIAL SEC 493082938

        EMP YEARS SRVC 04

        EMP LAST PROM  2017-01-01




    MAKE CHANGES AND THEN PRESS PF5
    F2 INQ   F3 EXIT   F4 ADD   F5 CHG   F6 DEL
```

Now make the changes and press PF5 to do the update.

```
EMPMCHG                    EMPLOYEE CHANGE                        EMCH


        EMPLOYEE ->    3217      ENTER EMPLOYEE ID, THEN PRESS ENTER

        EMPLOYEE ID    3217

        EMP LAST NAME  JOHNSON

        EMP FIRST NAME EDWARD

        EMP SOCIAL SEC 493082938

        EMP YEARS SRVC 07

        EMP LAST PROM  2018-01-01




    EMPLOYEE MODIFED SUCCESSFULLY
    F2 INQ   F3 EXIT   F4 ADD   F5 CHG   F6 DEL
```

Also, let's check to make sure the error logic still works if we enter a non-existent employee id.

```
EMPMCHG                    EMPLOYEE CHANGE                        EMCH


      EMPLOYEE ->    4321      ENTER EMPLOYEE ID, THEN PRESS ENTER

      EMPLOYEE ID

      EMP LAST NAME

      EMP FIRST NAME

      EMP SOCIAL SEC

      EMP YEARS SRVC

      EMP LAST PROM




EMPLOYEE ID 4321 NOT FOUND
F2 INQ   F3 EXIT   F4 ADD   F5 CHG   F6 DEL
```

Excellent, now go ahead and do the full unit test. Then we'll finish with the delete program.

EMPPGDEL

The delete program is easy to change. We just need to change the initial inquiry for display, and then code the delete after the user presses PF6. At this point you have everything you need to do the coding. Go ahead and give it a try. Then take a good coffee/tea/soda break and come back and compare code.

......

Ok, time to compare code. Here is mine:

```
      IDENTIFICATION DIVISION.
      PROGRAM-ID. EMPPGDEL.
     **************************************************
     *   COBOL/CICS/DB2 PROGRAM TO DELETE AN EMPLOYEE *
     *                                                *
     *   AUTHOR       : ROBERT WINGATE                *
     *   DATE-WRITTEN : 2018-07-25                    *
     **************************************************
      ENVIRONMENT DIVISION.
      DATA DIVISION.
      WORKING-STORAGE SECTION.
      01 WS-EMPNO        PIC 9(4).
      01 WS-EMP-SRV-YRS  PIC 9(2).
```

```
01 WS-COMMAREA.
   05 WS-EMP-PASS    PIC 9(04).
   05 WS-PGM-PASS    PIC X(08).
   05 FILLER         PIC X(08).

01 RESPONSE-CODE     PIC S9(08) VALUE 0.
01 RESPONSE-DISPLAY PIC  9(08) VALUE 0.

01 RESPONSE-CODE2    PIC S9(08) VALUE 0.
01 RESPONSE-DISPLA2 PIC  9(08) VALUE 0.

01 PROGRAM-NAME      PIC X(08) VALUE SPACES.

01 SW-PASSED-DATA-SWITCH   PIC X(1) VALUE 'N'.
   88  SW-PASSED-DATA                VALUE 'Y'.
   88  SW-NO-PASSED-DATA             VALUE 'N'.

   COPY EMPMDEL.
   COPY DFHAID.
   COPY DFHBMSCA.

01 DCLEMPLOYEE.
   05 EMP-ID             PIC 9(04).
   05 EMP-LAST-NAME      PIC X(30).
   05 EMP-FIRST-NAME     PIC X(20).
   05 EMP-SERVICE-YEARS  PIC 9(02).
   05 EMP-PROMOTION-DATE PIC X(10).
   05 EMP-SSN            PIC X(09).
   05 FILLER             PIC X(05).

LINKAGE SECTION.
01 DFHCOMMAREA          PIC X(20).

PROCEDURE DIVISION.

   IF EIBCALEN > ZERO
     MOVE DFHCOMMAREA  TO WS-COMMAREA
   END-IF.

   EVALUATE TRUE

     WHEN EIBCALEN = ZERO
       MOVE LOW-VALUES  TO  EMPDELO
       PERFORM SEND-MAP

     WHEN EIBAID = DFHCLEAR
       MOVE LOW-VALUES  TO  EMPDELO
       PERFORM SEND-MAP

     WHEN EIBAID = DFHPA1 OR DFHPA2 OR DFHPA3
       CONTINUE
```

```
      WHEN EIBAID = DFHPF2
        MOVE 'EMPPGINQ' TO PROGRAM-NAME
        PERFORM BRANCH-TO-PROGRAM

      WHEN EIBAID = DFHPF3
        MOVE LOW-VALUES TO  EMPDELO
        MOVE "BYE, PRESS CLEAR KEY TO ENTER A TRANSACTION ID"
            TO MESSAGEO
        PERFORM SEND-MAP-DATA

        EXEC CICS
          RETURN
        END-EXEC

      WHEN EIBAID = DFHPF4
        MOVE 'EMPPGADD' TO PROGRAM-NAME
        PERFORM BRANCH-TO-PROGRAM

      WHEN EIBAID = DFHPF5
        MOVE 'EMPPGCHG' TO PROGRAM-NAME
        PERFORM BRANCH-TO-PROGRAM

      WHEN EIBAID = DFHPF6

        IF WS-PGM-PASS NOT EQUAL "EMPPGDEL"
           SET SW-PASSED-DATA TO TRUE
           PERFORM PROCESS-PARA
        ELSE
           SET SW-NO-PASSED-DATA TO TRUE
           PERFORM DELETE-RECORD
        END-IF

      WHEN EIBAID = DFHENTER
        PERFORM PROCESS-PARA

      WHEN OTHER
        MOVE LOW-VALUES TO EMPDELO
        MOVE "INVALID KEY PRESSED" TO MESSAGEO
        PERFORM SEND-MAP-DATA

   END-EVALUATE.

   EXEC CICS
      RETURN TRANSID('EMDE')
      COMMAREA (WS-COMMAREA)
   END-EXEC.

PROCESS-PARA.

   PERFORM RECEIVE-MAP.
   INITIALIZE DCLEMPLOYEE MESSAGEO
```

```
IF SW-PASSED-DATA
    MOVE WS-EMP-PASS TO WS-EMPNO
ELSE
    MOVE EMPINI      TO WS-EMPNO
END-IF

MOVE WS-EMPNO  TO EMP-ID

EXEC CICS
    READ
    FILE('EMPLOYEE')
    INTO(DCLEMPLOYEE)
    RIDFLD(EMP-ID)
    EQUAL
    RESP(RESPONSE-CODE)
END-EXEC.

EVALUATE RESPONSE-CODE

    WHEN DFHRESP(NORMAL)
        MOVE EMP-ID              TO  WS-EMPNO WS-EMP-PASS
        MOVE WS-EMPNO            TO  EMPNOO
        MOVE EMP-LAST-NAME       TO LNAMEO
        MOVE EMP-FIRST-NAME      TO FNAMEO
        MOVE EMP-SSN             TO SOCSECO
        MOVE EMP-SERVICE-YEARS   TO WS-EMP-SRV-YRS
        MOVE WS-EMP-SRV-YRS      TO YRSSVCO
        MOVE EMP-PROMOTION-DATE  TO LSTPRMO
        MOVE "PRESS PF6 TO DELETE THIS RECORD" TO MESSAGEO

    WHEN DFHRESP(NOTFND)
        STRING "EMPLOYEE ID " DELIMITED BY SIZE
        WS-EMPNO DELIMITED BY SPACE
        " NOT FOUND" DELIMITED BY SIZE INTO MESSAGEO
        MOVE SPACES      TO EMPNOO
        MOVE SPACES      TO LNAMEO
        MOVE SPACES      TO FNAMEO
        MOVE SPACES      TO SOCSECO
        MOVE SPACES      TO YRSSVCO
        MOVE SPACES      TO LSTPRMO

    WHEN OTHER
        MOVE RESPONSE-CODE TO RESPONSE-DISPLAY
        STRING "UNKNOWN ERROR CODE: " DELIMITED BY SIZE
            RESPONSE-DISPLAY DELIMITED BY SIZE
          INTO MESSAGEO

END-EVALUATE.

MOVE DFHBMFSE TO EMPINF
MOVE -1 TO EMPINL.
```

```
           MOVE EMP-ID      TO WS-EMP-PASS
           MOVE "EMPPGDEL" TO WS-PGM-PASS

           IF SW-PASSED-DATA
              PERFORM SEND-MAP
           ELSE
              PERFORM SEND-MAP-DATA
           END-IF.

       DELETE-RECORD.

           PERFORM RECEIVE-MAP
           INITIALIZE DCLEMPLOYEE MESSAGEO

   *   MAP INPUT FIELDS TO DB2 RECORD

           MOVE EMPINI              TO WS-EMPNO
           MOVE WS-EMPNO            TO EMP-ID

   *   DELETE THE RECORD

           EXEC CICS
              DELETE
              FILE('EMPLOYEE')
              RIDFLD(EMP-ID)
              RESP(RESPONSE-CODE)
              RESP2(RESPONSE-CODE2)
           END-EXEC

           EVALUATE RESPONSE-CODE

              WHEN DFHRESP(NORMAL)
                 MOVE "EMPLOYEE DELETED SUCCESSFULLY" TO MESSAGEO
                 MOVE -1 TO EMPNOL
                 MOVE WS-EMPNO TO WS-EMP-PASS

              WHEN DFHRESP(NOTFND)
                 MOVE -1 TO EMPNOL
                 MOVE WS-EMPNO TO WS-EMP-PASS
                 STRING "EMPLOYEE ID " DELIMITED BY SIZE
                 WS-EMPNO DELIMITED BY SPACE
                 " NOT FOUND" DELIMITED BY SIZE INTO MESSAGEO
                 MOVE SPACES          TO EMPNOO
                 MOVE SPACES          TO LNAMEO
                 MOVE SPACES          TO FNAMEO
                 MOVE SPACES          TO SOCSECO
                 MOVE SPACES          TO YRSSVCO
                 MOVE SPACES          TO LSTPRMO
                 MOVE -1 TO EMPNOL

              WHEN OTHER
                 MOVE RESPONSE-CODE  TO RESPONSE-DISPLAY
```

```
                MOVE RESPONSE-CODE2 TO RESPONSE-DISPLA2
                STRING "UNKNOWN ERROR CODE: " DELIMITED BY SIZE
                    RESPONSE-DISPLAY DELIMITED BY SIZE
                    " "                 DELIMITED BY SIZE
                    RESPONSE-DISPLA2 DELIMITED BY SIZE
                INTO MESSAGEO

        END-EVALUATE

        MOVE DFHBMFSE TO EMPINF
        MOVE -1 TO EMPINL.
        MOVE "EMPPGDEL" TO WS-PGM-PASS
        PERFORM SEND-MAP-DATA.

BRANCH-TO-PROGRAM.

        EXEC CICS
            XCTL PROGRAM(PROGRAM-NAME)
            COMMAREA (WS-COMMAREA)
            LENGTH(10)
        END-EXEC

        MOVE 'PROGRAM NOT AVAILABLE' TO MESSAGEO.

SEND-MAP.
        EXEC CICS SEND
            MAP    ('EMPDEL')
            MAPSET ('EMPMDEL')
            FROM   (EMPDELO)
            ERASE
        END-EXEC.

SEND-MAP-DATA.
        EXEC CICS SEND
            MAP    ('EMPDEL')
            MAPSET ('EMPMDEL')
            FROM   (EMPDELO)
            DATAONLY
        END-EXEC.

SEND-MAP-ALL.
        EXEC CICS SEND
            MAP    ('EMPDEL')
            MAPSET ('EMPMDEL')
            FROM   (EMPDELO)
        END-EXEC.

RECEIVE-MAP.
        EXEC CICS RECEIVE
            MAP    ('EMPDEL')
            MAPSET ('EMPMDEL')
            INTO   (EMPDELI)
```

```
          END-EXEC.
```

Compile and link the program, remember to refresh it in CICS with CEMT. Now let's test it.
Bring up employee number 9999 on the delete screen.

```
EMPMDEL                    EMPLOYEE DELETE                          EMDE


       EMPLOYEE ->    9999      ENTER EMPLOYEE ID, THEN PRESS ENTER

       EMPLOYEE ID    9999

       EMP LAST NAME  WINGATE

       EMP FIRST NAME ROBERT

       EMP SOCIAL SEC 999999999

       EMP YEARS SRVC 22

       EMP LAST PROM  2018-01-01




 PRESS PF6 TO DELETE THIS RECORD
 F2 INQ   F3 EXIT   F4 ADD   F5 CHG   F6 DEL
```

Press PF6 to get the result:

```
EMPMDEL                    EMPLOYEE DELETE                          EMDE


       EMPLOYEE ->    9999      ENTER EMPLOYEE ID, THEN PRESS ENTER

       EMPLOYEE ID    9999

       EMP LAST NAME  WINGATE

       EMP FIRST NAME ROBERT

       EMP SOCIAL SEC 999999999

       EMP YEARS SRVC 22

       EMP LAST PROM  2018-01-01




 EMPLOYEE DELETED SUCCESSFULLY
 F2 INQ   F3 EXIT   F4 ADD   F5 CHG   F6 DEL
```

Finally, go ahead and complete the full unit test for EMPPGDEL. Then we'll complete our project with some integration testing.

Integration Testing

EMPPGMU

Ok, let's repeat our integration test, beginning with the menu program **EMPPGMNU**. We'll simply verify that we can still navigate to all the screens and that they display properly. We'll choose the menu options for inquiry, add, change and delete. First inquiry.

```
EMPMMNU                      EMPLOYEE SUPPORT MENU                    EMNU

            ENTER THE NUMBER OF YOUR SELECTION,   THEN PRESS ENTER.


                    1    1. EMPLOYEE INQUIRY

                         2. EMPLOYEE ADD

                         3. EMPLOYEE CHANGE

                         4. EMPLOYEE DELETE




F3 EXIT
```

```
EMPMINQ                      EMPLOYEE INQUIRY                         EMIN


        EMPLOYEE ->             ENTER EMPLOYEE ID, THEN PRESS ENTER

        EMPLOYEE ID    XXXX

        EMP LAST NAME  XXXX

        EMP FIRST NAME XXXX

        EMP SOCIAL SEC XXXXXXXX

        EMP YEARS SRVC 00

        EMP LAST PROM  YYYY-MM-DD



F2 INQ   F3 EXIT   F4 ADD   F5 CHG   F6 DEL
```

Next choose the add option.

```
EMPMMNU                    EMPLOYEE SUPPORT MENU                    EMNU

           ENTER THE NUMBER OF YOUR SELECTION,   THEN PRESS ENTER.


                    2   1. EMPLOYEE INQUIRY

                        2. EMPLOYEE ADD

                        3. EMPLOYEE CHANGE

                        4. EMPLOYEE DELETE









F3 EXIT
```

```
EMPMADD                    EMPLOYEE ADD                             EMAD

                   ENTER EMPLOYEE INFO, THEN PRESS PF4

        EMPLOYEE ID

        EMP LAST NAME

        EMP FIRST NAME

        EMP SOCIAL SEC

        EMP YEARS SRVC

        EMP LAST PROM




F2 INQ   F3 EXIT   F4 ADD   F5 CHG   F6 DEL
```

Now return to the main menu and choose the change option.

```
EMPMMNU                    EMPLOYEE SUPPORT MENU                    EMNU

            ENTER THE NUMBER OF YOUR SELECTION,   THEN PRESS ENTER.


                    3    1. EMPLOYEE INQUIRY

                         2. EMPLOYEE ADD

                         3. EMPLOYEE CHANGE

                         4. EMPLOYEE DELETE








      F3 EXIT
```

```
EMPMCHG                       EMPLOYEE CHANGE                       EMCH


         EMPLOYEE ->      ____       ENTER EMPLOYEE ID, THEN PRESS ENTER

         EMPLOYEE ID    XXXX

         EMP LAST NAME  XXXX

         EMP FIRST NAME XXXX

         EMP SOCIAL SEC XXXXXXXXX

         EMP YEARS SRVC 00

         EMP LAST PROM  YYYY-MM-DD




      F2 INQ   F3 EXIT   F4 ADD   F5 CHG   F6 DEL
```

Finally, return to the main menu and choose the delete option.

```
EMPMMNU                      EMPLOYEE SUPPORT MENU                      EMNU

           ENTER THE NUMBER OF YOUR SELECTION,   THEN PRESS ENTER.


                       4   1. EMPLOYEE INQUIRY

                           2. EMPLOYEE ADD

                           3. EMPLOYEE CHANGE

                           4. EMPLOYEE DELETE








    F3 EXIT
```

```
EMPMDEL                       EMPLOYEE DELETE                           EMDE


        EMPLOYEE ->                ENTER EMPLOYEE ID, THEN PRESS ENTER

        EMPLOYEE ID    XXXX

        EMP LAST NAME  XXXX

        EMP FIRST NAME XXXX

        EMP SOCIAL SEC XXXXXXXXX

        EMP YEARS SRVC 00

        EMP LAST PROM  YYYY-MM-DD




    F2 INQ   F3 EXIT   F4 ADD   F5 CHG   F6 DEL
```

All looks good. Now on to the data operations.

EMPPGINQ

Ok let's start out with the primary display and entry after choosing from the main menu. Let's use 7777 as the employee id.

```
EMPMINQ                    EMPLOYEE INQUIRY                        EMIN

        EMPLOYEE ->    7777      ENTER EMPLOYEE ID, THEN PRESS ENTER

        EMPLOYEE ID   XXXX

        EMP LAST NAME  XXXX

        EMP FIRST NAME XXXX

        EMP SOCIAL SEC XXXXXXXXX

        EMP YEARS SRVC 00

        EMP LAST PROM  YYYY-MM-DD




    F2 INQ   F3 EXIT   F4 ADD   F5 CHG   F6 DEL


EMPMINQ                    EMPLOYEE INQUIRY                        EMIN

        EMPLOYEE ->    7777      ENTER EMPLOYEE ID, THEN PRESS ENTER

        EMPLOYEE ID   7777

        EMP LAST NAME  JACKSON

        EMP FIRST NAME JOSEPH

        EMP SOCIAL SEC 382746236

        EMP YEARS SRVC 17

        EMP LAST PROM  2017-01-01




    F2 INQ   F3 EXIT   F4 ADD   F5 CHG   F6 DEL
```

This looks good. Now let's try passing the same employee number from another program, such as the change program. Bring up employee 7777 again, and then press PF2 to transfer to the inquiry program.

```
EMPMCHG                   EMPLOYEE CHANGE                        EMCH


        EMPLOYEE ->    7777      ENTER EMPLOYEE ID, THEN PRESS ENTER

        EMPLOYEE ID   7777

        EMP LAST NAME  JACKSON

        EMP FIRST NAME JOSEPH

        EMP SOCIAL SEC 382746236

        EMP YEARS SRVC 17

        EMP LAST PROM  2017-01-01




 MAKE CHANGES AND THEN PRESS PF5
 F2 INQ   F3 EXIT   F4 ADD   F5 CHG   F6 DEL



EMPMINQ                   EMPLOYEE INQUIRY                       EMIN


        EMPLOYEE ->    7777      ENTER EMPLOYEE ID, THEN PRESS ENTER

        EMPLOYEE ID   7777

        EMP LAST NAME  JACKSON

        EMP FIRST NAME JOSEPH

        EMP SOCIAL SEC 382746236

        EMP YEARS SRVC 17

        EMP LAST PROM  2017-01-01




 F2 INQ   F3 EXIT   F4 ADD   F5 CHG   F6 DEL
```

Finally we should test the transfers from the inquiry screen to the add, change and delete screens. Let's do that to make sure the program interfaces are working correctly. Start with inquiry to add.

```
EMPMINQ                    EMPLOYEE INQUIRY                         EMIN

        EMPLOYEE ->     7777      ENTER EMPLOYEE ID, THEN PRESS ENTER

        EMPLOYEE ID     7777

        EMP LAST NAME   JACKSON

        EMP FIRST NAME  JOSEPH

        EMP SOCIAL SEC  382746236

        EMP YEARS SRVC  17

        EMP LAST PROM   2017-01-01




 F2 INQ   F3 EXIT   F4 ADD   F5 CHG   F6 DEL

EMPMADD                     EMPLOYEE ADD                           EMAD

                    ENTER EMPLOYEE INFO, THEN PRESS PF4

        EMPLOYEE ID

        EMP LAST NAME

        EMP FIRST NAME

        EMP SOCIAL SEC

        EMP YEARS SRVC

        EMP LAST PROM




 ENTER DATA FOR NEW EMPLOYEE, THEN PRESS PF4 TO ADD
 F2 INQ   F3 EXIT   F4 ADD   F5 CHG   F6 DEL
```

Also go ahead and add an employee to ensure all attribute setting, switches and variables are correct.

```
EMPMADD                     EMPLOYEE ADD                              EMAD

                    ENTER EMPLOYEE INFO, THEN PRESS PF4

        EMPLOYEE ID    1111

        EMP LAST NAME  stone

        EMP FIRST NAME steven

        EMP SOCIAL SEC 385610088

        EMP YEARS SRVC 12

        EMP LAST PROM  2016-01-01




 ENTER DATA FOR NEW EMPLOYEE, THEN PRESS PF4 TO ADD
 F2 INQ   F3 EXIT   F4 ADD   F5 CHG   F6 DEL
```

```
EMPMADD                     EMPLOYEE ADD                              EMAD

                    ENTER EMPLOYEE INFO, THEN PRESS PF4

        EMPLOYEE ID    1111

        EMP LAST NAME  STONE

        EMP FIRST NAME STEVEN

        EMP SOCIAL SEC 385610088

        EMP YEARS SRVC 12

        EMP LAST PROM  2016-01-01




 EMPLOYEE ADDED SUCCESSFULLY
 F2 INQ   F3 EXIT   F4 ADD   F5 CHG   F6 DEL
```

All looks well, so let's move on to the add screen.

EMPPGADD

We just added a record, so let's try transferring to the inquiry screen, the change screen and then the delete screen. Press PF2.

```
EMPMINQ                    EMPLOYEE INQUIRY                         EMIN


        EMPLOYEE ->              ENTER EMPLOYEE ID, THEN PRESS ENTER

        EMPLOYEE ID    1111

        EMP LAST NAME  STONE

        EMP FIRST NAME STEVEN

        EMP SOCIAL SEC 385610088

        EMP YEARS SRVC 12

        EMP LAST PROM  2016-01-01




    F2 INQ   F3 EXIT   F4 ADD   F5 CHG   F6 DEL


EMPMINQ                    EMPLOYEE INQUIRY                         EMIN


        EMPLOYEE ->              ENTER EMPLOYEE ID, THEN PRESS ENTER

        EMPLOYEE ID    1111

        EMP LAST NAME  STONE

        EMP FIRST NAME STEVEN

        EMP SOCIAL SEC 385610088

        EMP YEARS SRVC 12

        EMP LAST PROM  2016-01-01




    F2 INQ   F3 EXIT   F4 ADD   F5 CHG   F6 DEL
```

Next let's add two more records, and transfer to the change and delete screens, respectively.

```
EMPMADD                     EMPLOYEE ADD

                     ENTER EMPLOYEE INFO, THEN PRESS PF4

        EMPLOYEE ID    1212

        EMP LAST NAME  SAMPLE

        EMP FIRST NAME RECORD

        EMP SOCIAL SEC 373737373

        EMP YEARS SRVC 04

        EMP LAST PROM  2017-01-01




  EMPLOYEE ADDED SUCCESSFULLY
  F2 INQ   F3 EXIT   F4 ADD   F5 CHG   F6 DEL



EMPMCHG                   EMPLOYEE CHANGE                    EMCH

        EMPLOYEE ->             ENTER EMPLOYEE ID, THEN PRESS ENTER

        EMPLOYEE ID    1212

        EMP LAST NAME  SAMPLE

        EMP FIRST NAME RECORD

        EMP SOCIAL SEC 373737373

        EMP YEARS SRVC 04

        EMP LAST PROM  2017-01-01




  MAKE CHANGES AND THEN PRESS PF5
  F2 INQ   F3 EXIT   F4 ADD   F5 CHG   F6 DEL
```

```
EMPMADD                    EMPLOYEE ADD                         EMAD


                 ENTER EMPLOYEE INFO, THEN PRESS PF4

         EMPLOYEE ID    2424

         EMP LAST NAME  TWO

         EMP FIRST NAME SAMPLE

         EMP SOCIAL SEC 747474747

         EMP YEARS SRVC 13

         EMP LAST PROM  2016-01-01




   EMPLOYEE ADDED SUCCESSFULLY
   F2 INQ   F3 EXIT   F4 ADD   F5 CHG   F6 DEL



EMPMDEL                    EMPLOYEE DELETE                      EMDE


         EMPLOYEE ->              ENTER EMPLOYEE ID, THEN PRESS ENTER

         EMPLOYEE ID    2424

         EMP LAST NAME  TWO

         EMP FIRST NAME SAMPLE

         EMP SOCIAL SEC 747474747

         EMP YEARS SRVC 13

         EMP LAST PROM  2016-01-01




   PRESS PF6 TO DELETE THIS RECORD
   F2 INQ   F3 EXIT   F4 ADD   F5 CHG   F6 DEL
```

All looks good with transferring from the add screen. Let's move on to the change screen.

EMPPGCHG

Let's bring up the 2424 record that we just created on the change screen. Now transfer to the inquiry screen, the delete screen, and finally the add screen (the latter will not process any transferred data except the program name).

```
EMPMCHG                      EMPLOYEE CHANGE                        EMCH


        EMPLOYEE ->    2424       ENTER EMPLOYEE ID, THEN PRESS ENTER

        EMPLOYEE ID    2424

        EMP LAST NAME  TWO

        EMP FIRST NAME SAMPLE

        EMP SOCIAL SEC 747474747

        EMP YEARS SRVC 13

        EMP LAST PROM  2016-01-01




  MAKE CHANGES AND THEN PRESS PF5
  F2 INQ   F3 EXIT   F4 ADD   F5 CHG   F6 DEL


EMPMINQ                      EMPLOYEE INQUIRY                       EMIN


        EMPLOYEE ->    2424       ENTER EMPLOYEE ID, THEN PRESS ENTER

        EMPLOYEE ID    2424

        EMP LAST NAME  TWO

        EMP FIRST NAME SAMPLE

        EMP SOCIAL SEC 747474747

        EMP YEARS SRVC 13

        EMP LAST PROM  2016-01-01




  F2 INQ   F3 EXIT   F4 ADD   F5 CHG   F6 DEL
```

```
EMPMCHG                    EMPLOYEE CHANGE                          EMCH


        EMPLOYEE ->   2424      ENTER EMPLOYEE ID, THEN PRESS ENTER

        EMPLOYEE ID   2424

        EMP LAST NAME   TWO

        EMP FIRST NAME SAMPLE

        EMP SOCIAL SEC 747474747

        EMP YEARS SRVC 13

        EMP LAST PROM  2016-01-01




 MAKE CHANGES AND THEN PRESS PF5
 F2 INQ   F3 EXIT   F4 ADD   F5 CHG   F6 DEL


 EMPMDEL                    EMPLOYEE DELETE                          EMDE


        EMPLOYEE ->   2424      ENTER EMPLOYEE ID, THEN PRESS ENTER

        EMPLOYEE ID   2424

        EMP LAST NAME   TWO

        EMP FIRST NAME SAMPLE

        EMP SOCIAL SEC 747474747

        EMP YEARS SRVC 13

        EMP LAST PROM  2016-01-01




 PRESS PF6 TO DELETE THIS RECORD
 F2 INQ   F3 EXIT   F4 ADD   F5 CHG   F6 DEL
```

EMPPGDEL

Finally, let's test the delete screen. We'll delete transfer to the three other programs. We can use employee 3333.

```
EMPMDEL                      EMPLOYEE DELETE                          EMDE


        EMPLOYEE ->     3333        ENTER EMPLOYEE ID, THEN PRESS ENTER

        EMPLOYEE ID     3333

        EMP LAST NAME   RADISSON

        EMP FIRST NAME  BENTLEY

        EMP SOCIAL SEC  777777777

        EMP YEARS SRVC  46

        EMP LAST PROM   2015-07-01




   PRESS PF6 TO DELETE THIS RECORD
   F2 INQ   F3 EXIT   F4 ADD   F5 CHG   F6 DEL
```

Now transfer to the inquiry screen with PF2.

```
EMPMINQ                      EMPLOYEE INQUIRY                         EMIN


        EMPLOYEE ->     3333        ENTER EMPLOYEE ID, THEN PRESS ENTER

        EMPLOYEE ID     3333

        EMP LAST NAME   RADISSON

        EMP FIRST NAME  BENTLEY

        EMP SOCIAL SEC  777777777

        EMP YEARS SRVC  46

        EMP LAST PROM   2015-07-01




   F2 INQ   F3 EXIT   F4 ADD   F5 CHG   F6 DEL
```

Now transfer back to the delete program

```
EMPMDEL                    EMPLOYEE DELETE                        EMDE

        EMPLOYEE ->    3333       ENTER EMPLOYEE ID, THEN PRESS ENTER

        EMPLOYEE ID    3333

        EMP LAST NAME  RADISSON

        EMP FIRST NAME BENTLEY

        EMP SOCIAL SEC 777777777

        EMP YEARS SRVC 46

        EMP LAST PROM  2015-07-01




    PRESS PF6 TO DELETE THIS RECORD
    F2 INQ   F3 EXIT   F4 ADD   F5 CHG   F6 DEL
```

And then transfer to the change program using PF5.

```
EMPMCHG                    EMPLOYEE CHANGE                        EMCH

        EMPLOYEE ->    3333       ENTER EMPLOYEE ID, THEN PRESS ENTER

        EMPLOYEE ID    3333

        EMP LAST NAME  RADISSON

        EMP FIRST NAME BENTLEY

        EMP SOCIAL SEC 777777777

        EMP YEARS SRVC 46

        EMP LAST PROM  2015-07-01




    MAKE CHANGES AND THEN PRESS PF5
    F2 INQ   F3 EXIT   F4 ADD   F5 CHG   F6 DEL
```

Now transfer back to the delete program

```
EMPMDEL                    EMPLOYEE DELETE                        EMDE


        EMPLOYEE ->    3333      ENTER EMPLOYEE ID, THEN PRESS ENTER

        EMPLOYEE ID    3333

        EMP LAST NAME  RADISSON

        EMP FIRST NAME BENTLEY

        EMP SOCIAL SEC 777777777

        EMP YEARS SRVC 46

        EMP LAST PROM  2015-07-01




 PRESS PF6 TO DELETE THIS RECORD
 F2 INQ   F3 EXIT   F4 ADD   F5 CHG   F6 DEL
```

And then transfer to the add program. Notice we do not pass the employee number to the add program since there is no need to add an already existing employee.

```
EMPMADD                    EMPLOYEE ADD                           EMAD

                    ENTER EMPLOYEE INFO, THEN PRESS PF4

        EMPLOYEE ID

        EMP LAST NAME

        EMP FIRST NAME

        EMP SOCIAL SEC

        EMP YEARS SRVC

        EMP LAST PROM




 ENTER DATA FOR NEW EMPLOYEE, THEN PRESS PF4 TO ADD
 F2 INQ   F3 EXIT   F4 ADD   F5 CHG   F6 DEL
```

That's it, looks like everything works. Our integration test is finished.

We've come to the end, so let me close with a sincere "best of luck". It's been a pleasure walking you through all these IBM developer concepts, methods, languages and tools. I truly hope you do exceptionally well as an IBM application developer, and that you have every success!

Best regards,

Robert Wingate

IBM Certified Application Developer – DB2 11 for z/OS

Chapter 7 Review Questions

1. What does the CICS acronym stand for?

2. In which programming languages can you develop CICS programs?

3. What is BMS?

4. In a COBOL program, how are CICS commands enclosed?

5. What are the CICS commands for handling screen interactions?

6. What does the symbolic map include?

7. What is the difference between a physical BMS mapset and a BMS symbolic mapset?

8. What are the three macros used in building a mapset?

9. What attributes are used to make a field normal intensity on the screen and protected (so that the field cannot change)?

10. What attribute will cause the cursor to be placed on a field when the screen displays?

11. Which EIB field contains the length of the data that was passed to the program?

12. The CICS commands for processing records from VSAM files are:

13. What typically causes a MAPFAIL condition when processing a RECEIVE MAP command?

14. The communication or COMMAREA must be included in the Linkage section of the program. What is the required name for this field in the Linkage section?

16. What is the syntax of the RECEIVE MAP command?

17. When do you need to use the NEWCOPY keyword?

18. What is the syntax of the XCTL command?

19. On a SEND command, what happens if you don't specify MAPONLY or DATAONLY?

20. What is the syntax of the READ command when you intend to update the record later?

21. If an application READS a VSAM KSDS file with UPDATE, and decides not to update the record, what command can be issued to release exclusive control on the record?

22. What exceptional condition will be raised if you try to insert a record into a file and a record with the same key is already there?

Appendices

Chapter Questions and Answers

Chapter 1 Review Questions

1. What statement do you use to delimit an instream procedure?

   Use the PEND statement to mark the end of an instream procedure in a JCL.

2. What does COND=(0,NE) mean?

   Literally it means if zero is not equal to the highest condition code encountered in the job, then do not execute this step. Another way of saying it is the condition code is not equal to zero.

3. What keyword on a SPACE allocation statement returns unused space to the system when the dataset is closed?

   The RLSE keyword frees any allocated but unused space when the dataset is closed.

4. How do you reference the latest version of a GDG-based file?

   You refer to it as generation zero. For example, the most recent generation of a dataset named XXX.FILE1 would be referenced in JCL as:

   ```
   //DD1 DD DSN=XXX.FILE1(0),DISP=SHR
   ```

5. Explain JOBLIB and STEPLIB in JCL.

   JOBLIB specifies the library where the load modules can be found before searching the linklist (i.e., linklist includes the default the system libraries). JOBLIB is coded after the JOB statement and applies to the entire job or JCL.

STEPLIB specifies a load module library but applies only to the step for which it is coded. If both JOBLIB and STEPLIB are coded, the STEPLIB takes precedence over the JOBLIB.

6. How is a temporary dataset coded?

    A temporary dataset is coded with the symbols && in front of a dataset name. Since it is a temporary dataset, it will not be cataloged. An example is:

```
//DD1   DD   DSN=&&TEMPFILE,DISP=(NEW,PASS),
//           UNIT=SYSDA,SPACE=(TRK,(1,1),RLSE)
//           RECFM=FB,LRECL=80,BLKSIZE=0
```

7. How do you concatenate datasets in JCL?

    The first dataset is written as normal, i.e., DDNAME DD DSN=. Any subsequent datasets to be concatenated are added on the following line just like the first the line except you omit the DDNAME. Here's an example:

```
//DDNAME DD DSN=FILE1,DISP=SHR
//          DD DSN=FILE2,DISP=SHR
//          DD DSN=FILE3,DISP=SHR
```

8. How do you designate a comment in JCL?

    The comment statement is //* followed by the comments. For example:

    //* This is a comment and will not be processed

9. What does a disposition of (NEW,CATLG,DELETE) mean?

    This disposition means to allocate a new dataset (NEW). If the job step is successful it will catalog the new dataset (CATLG), and if the job step is not successful it will delete (DELETE) the dataset.

10. What are the required attributes for an output DD statement?

Unless already allocated earlier, the statement must have a DSN, DISP, UNIT, SPACE and DCB (or the sub-parameters of the DCB: RECFM, LRECL, BLK-SIZE).

An example is:

```
//DD1   DD   DSN=XXX.FILE1,
//           DISP=(NEW,CATLG,DELETE),
//           UNIT=SYSDA,
//           SPACE=(TRK,(1,1),RLSE)
//           DCB=(RECFM=FB,
//           LRECL=80,
//           BLKSIZE=0)
```

11. What does COND=EVEN mean in JCL?

It means to execute this step even if any of the previous steps abended.

12. What does the following statement mean?

```
     SYSIN   DD   *
```

Instream data follows, and is terminated when a line containing /* in columns 1 and 2 is encountered.

Chapter 2 Review Questions

1. What is the purpose of the IEFBR14 utility?

   IEFBR14 has a variety of uses. Most often it is used to catalog or delete a file. Also, by using (MOD,DELETE,DELETE) as the disposition, IEFBR14 first creates the catalog entry for the file if it does not already exist. Why do it this way? Because it is flexible, allowing for the possibility that the files to be deleted do not actually exist every time the job runs. If the files do exist, they get deleted. If they do not exist, you prevent a runtime error (trying to delete a non-existent file) by specifying MOD as the first sub-parameter of the disposition.

2. What IDCAMS keyword do you specify to copy a file?

   REPRO is used to copy a file. A coding example is:

```
//**********************************************
//* IDCAMS TO COPY A DATA SET
//**********************************************
//*
//STEP1     EXEC PGM=IDCAMS
//SYSPRINT  DD SYSOUT=*
//SYSUDUMP  DD SYSOUT=*
//FILEIN    DD DSN=DSNAME.TEST.FILE,DISP=SHR
//FILEOUT   DD DSN=DSNAME.TEST.FILE2,
//             DISP=(NEW,CATLG,DELETE),
//             UNIT=SYSDA,
//             SPACE=(TRK,(5,5),RLSE),
//             RECFM=FB,LRECL=4096,BLKSIZE=4096
//SYSIN     DD *
 REPRO INFILE (FILEIN) OUTFILE (FILEOUT)
/*
```

3. Which IBM JCL utility is used to duplicate a partitioned dataset?

   IEBCOPY is used to copy a partitioned dataset. Here is a code example that copies PDS DSNAME.PDS.FILE to DSNAME.PDS.FILE.BACKUP:

```
//*
//STEP1      EXEC PGM=IEBCOPY
//SYSIN      DD *
  COPY INDD=SYSUT1,OUTDD=SYSUT2
/*
//SYSUT1     DD DSN=DSNAME.PDS.FILE,DISP=SHR
//SYSUT2     DD DSN=DSNAME.PDS.FILE.BACKUP,
//              DISP=(NEW,CATLG,DELETE),
//              UNIT=TAPE,RECFM=FB,LRECL=80,
//              DSORG=PO,BLKSIZE=27920
//SYSUT3     DD UNIT=SYSDA,
//              SPACE=(TRK,(100,100),RLSE)
//SYSUT4     DD UNIT=SYSDA,
//              SPACE=(TRK,(100,100),RLSE)
//SYSPRINT   DD SYSOUT=*
//SYSUDUMP   DD SYSOUT=*
```

4. How do you catalog an uncataloged dataset with a JCL?

By using the UNIT and VOL serial parameter in the dataset DD statement, and specifying disposition (OLD,CATLG,DELETE). For example:

```
//*
//****************************************************************
//* DEFINE GDG BASE AND A MODEL DATA SET
//****************************************************************
//STEP10    EXEC PGM=IEFBR14
//SYSPRINT DD SYSOUT=*
//SYSOUT   DD SYSOUT=*
//DD1      DD DSN=XXXXXX.TEST.FILE,
//            DISP=(OLD,CATLG,DELETE),
```

5. What are some capabilities of the IDCAMS utility in JCL?

IDCAMS has a number of functions. Some of these are:

- **Copy VSAM and NON VSAM Datasets (REPRO)**
- **Create GDG**
- **Delete GDG**
- **Rename a dataset**
- **Create a VSAM dataset (Define Cluster)**
- **To run a LISTCAT**

693

Chapter 3 Review Questions

1. Name some elements of the COBOL Identification division.

   The elements include Program-Id, Author, Installation, Date-Written, Date-Compiled.

   ```
   IDENTIFICATION DIVISION.
   Program-ID. PGM12345.
   Author. John Smith.
   Installation. Sunrise Programming.
   Date-Written. 06/12/2016.
   Date-Compiled. 06/14/2016.
   ```

2. Which clause do you use to define a table in a program?

   Use the OCCURS clause to define a table. For example, to create a 50 element table and have it indexed by variable VAR1:

   ```
   77 VAR1 USAGE IS INDEX.

   01 SAMPLE-TABLE
      05   SAMPLE-COLUMN1 OCCURS 50 TIMES
           INDEXED BY VAR1.
         10   SAMPLE-FIELDA     PIC X (2).
         10   SAMPLE-FIELDB     PIC X (5).
   ```

3. What does the INITIALIZE keyword do?

   INITIALIZE assigns default values for fields and is often used to initialize a structure variable with one statement instead of several. INITIALIZE moves zeros to alphanumeric fields and spaces to alphabetic fields.

4. What is the LINKAGE SECTION used for?

   The LINKAGE SECTION is used to pass data from one program to another program, or to receive data from a JCL.

5. What verb do you use to identify external files that the program will be using?

Use the SELECT verb in the FILE-CONTROL part of the INPUT-OUTPUT SECTION of the ENVIRONMENT DIVISION. For example, if you want to relate the internal file-name EMPLOYEE to it's JCL DDNAME which is EMPFILE. Here's is the syntax:

```
SELECT EMPLOYEE ASSIGN TO AS-EMPFILE.
```

6. How do you terminate an IF/ELSE statement?

Terminate an IF/ELSE statement with END-IF. For example:

```
IF NOT S-EOF
PERFORM 100-PROCESS-DATA
ELSE
PERFORM 400-PRINT-TOTALS
END-IF
```

7. What is an 88 level data element used for?

A level 88 is always associated with another variable. It is used to set up condition names based on the data. For example if you have a gender variable and you want to use the value in the field for later branching in the program, you could define it this way:

```
05   GENDER        PIC X.
     88  MALE     VALUE "M".
     88  FEMALE   VALUE "F".
```

Now you can short-cut checking the actual value of GENDER by simply coding:

```
IF MALE PERFORM XXX.
```

```
IF FEMALE PERFORM YYY.
```

8. Explain the meaning of a PIC 9v99 field.

PIC 9v99 is a three position number field which has two positions to the right of an implied decimal point.

9. If you are not certain how many entries a table should have, how would you create a variable length table?

You add a DEPENDING ON X option to the OCCURS clause, where X is a variable. For example you could make the table variable between 1 and 50 elements depending on the value of a record counter.

```
SAMPLE-TABLE
SAMPLE-COLUMN1 OCCURS 1 to 50 TIMES DEPENDING ON REC-CNT.
10   SAMPLE-FIELDA    PIC X (2)
10   SAMPLE-FIELDB    PIC X (5)
```

10. In COBOL, how do you call a program statically? How about dynamically?

For STATIC calls just used the program name in quotes. Example:

```
CALL 'PROG1' USING <arguments>
```

For a DYNAMIC call you create a program name variable and use that in the CALL statement. For example:

```
77 WS-PROGRAM PIC X(8) VALUE 'PROG2'.
```

```
CALL WS-PROGRAM USING arguments
```

11. What type of picture can be used for alphanumeric data types?

Use PIC X for alphanumerics, e.g., you define a 10 byte alphanumeric variable called TEST-VAR as follows:

```
TEST-VAR   PIC X (10).
```

12. When you open a file in I-O mode, what verb is used to update a record?

The REWRITE verb is used with files opened in I-O mode.

13. What are the different modes for opening a file in COBOL?

Files can be opened for:

- **INPUT**
- **OUTPUT**
- **I-O**
- **EXTEND**

14. Explain what an EVALUATE statement is used for?

In COBOL, the EVALUATE verb implements the case construct. It can be used in place of nexted IFs to make code less complex and more readable. An example of EVALUATE is:

```
EVALUATE GENDER
    WHEN "M"
        MOVE "MALE" TO PRINT-GENDER
    WHEN "F"
        MOVE "FEMALE" TO PRINT-GENDER
    WHEN OTHER
        MOVE "UNKNOWN" TO PRINT-GENDER
END-EVALUATE.
```

15. If you have a complex arithmetic calculation, which verb could you use to perform the calculation with a single statement?

You can use the COMPUTE statement for most arithmetic evaluations, and often you can use a single COMPUTE statement rather than multiple ADD, SUBTRACT, MULTIPLY, and DIVIDE statements. For example:

```
Compute TOTAL = a + b / c ** d - e
```

Chapter 4 Review Questions

1. What are the three types of VSAM datasets?

   Entry-sequenced datasets (ESDS), key-sequenced datasets (KSDS) and relative record dataset (RRDS).

2. How are records stored in an ESDS (entry sequenced) dataset?

   They are stored in the order in which they are inserted into the file.

3. What VSAM feature enables you to access the records in a KSDS dataset based on a key that is different than the file's primary key?

   VSAM allows creation of an alternate index which enables you to access the records in a KSDS dataset based on that alternate index rather than the primary key.

4. What is the general purpose utility program that provides services for VSAM files?

   Access Method Services is the utility program that provides services for VSAM files. Often it is referred to as IDCAMS which is the executable program in batch.

5. Which AMS function lists information about datasets?

   The LISTCAT function lists information about datasets. An example is:

   ```
   //STEP1     EXEC PGM=IDCAMS
   //SYSPRINT  DD SYSOUT=X
   //SYSIN     DD *
   LISTCAT GDG ENT('DSNAME.GDGFILE.TEST1') ALL
   ```

6. If you are mostly going to use a KSDS file for sequential access, should you define a larger or smaller control interval when creating the file?

   For sequential access a larger control interval is desirable for performance because you maximize the data brought in with each I/O.

7. What is the basic AMS command to create a VSAM file?

   DEFINE CLUSTER is the basic command to create a VSAM file.

8. To use the REWRITE command in COBOL, the VSAM file must be opened in what mode?

   To use the REWRITE command in COBOL, the VSAM file must be opened for I-O.

9. When you define an alternate index, what is the function of the RELATE parameter?

   The RELATE parameter associates your alternate index with the base cluster that you are creating the alternate index for.

10. When you define a path using DEFINE PATH, what does the PATHENTRY parameter do?

    The PATHENTRY parameter includes the name of the alternate index that you are creating the path for.

11. After you've defined an alternate index and path, what AMS command must you issue to actually populate the alternate index?

    Issue the BLXINDEX command to populate an alternate index.

12. After you've created a VSAM file, if you need to add additional DASD volumes that can be used with that file, what command would you use?

    Use an ALTER command and specify the keyword ADDVOLUMES(XXX001 YYY002) where XXX001 and YYY002 are DASD volume names.

13. If you want to set a VSAM file to read only status, what command would you use?

    Use the ALTER command with the INHIBIT keyword. For example:

    ```
    //STEP1 EXEC PGM=IDCAMS
    //SYSPRINT DD SYSOUT=*
    //SYSIN DD *
    ALTER -
    ```

```
PROD.EMPL.DATA -
INHIBIT
ALTER -
PROD.EMPL.INDEX -
INHIBIT
/*
```

To return the file to read/update, use ALTER with the UNINHIBIT keyword.

14. What are some ways you can improve the performance of a KSDS file?

Ensure that the control interval is optimally sized (smaller for random access and larger for sequential access).

Allocate additional index buffers to reduce data I/Os by keeping needed records in virtual storage.

Ensure sufficient free space in control intervals to avoid control interval splits.

15. Do primary key values in a KSDS have to be unique?

Yes the primary key has to be unique. However, alternate index values need not be unique. For example if an EMPLOYEE file uses employee number as the primary key, then the employee number must be unique. However the EMPLOYEE file could be alternately indexed on department. In this case, the department need not be unique.

16. In the COBOL SELECT statement what organization should be specified for a KSDS file?

In a COBOL SELECT statement, the organization for a KSDS file is INDEXED.

17. In the COBOL SELECT statement for a KSDS what are the three possibilities for ACCESS?

In the COBOL SELECT statement for a KSDS, ACCESS can be SEQUENTIAL, RANDOM or DYNAMIC.

18. Is there a performance penalty for using an alternate index compared to using the primary key?

Yes because if you access a record through an ALTERNATE INDEX, the alternate key must first be located and then it points to the primary key entry which is finally used to locate the actual record.

19. What file status code will you receive if an operation succeeded?

If an operation succeeded without any problem you will receive a 00 file status code.

Chapter 5 Review Questions

1. What is the name of the interface program you call from a COBOL program to perform IMS operations?

    CBLTDLI is the normal interface program for a COBOL program to access IMS.

2. Here are some IMS return codes. Explain briefly what each of them means: blank, GE, GB, II

    Blank – successful operation
    GE – segment not found
    GB – end of database
    II – duplicate key, insert failed

3. What is an SSA?

    Segment Search Argument – it is used to select segments by name and to specify search criteria for specific segments.

4. Briefly explain these entities: DBD, PSB, PCB?

    A Database Description (DBD) specifies characteristics of a database. The name, parent, and length of each segment type in the database.

    A Program Specification Block (PSB) is the program view of one or more IMS databases. The PSB includes one or more program communication blocks (PCB) for each IMS database that the program needs access to.

    A Program Communication Block (PCB) specifies the database to be accessed, the processing options such as read-only or various updating options, and the database segments that can be accessed.

5. What is the use of CMPAT parameter in PSB ?

    It is required if you are going to run your program in Batch Mode Processing (BMP), that is - in the online region. If you always run the program in DL/I mode, you do not need the CMPAT. If you are going to run BMP, you need the CMPAT=YES specified in the PSB.

6. In IMS, what is the difference between a key field and a search field?

   A key field is used to make the record unique and to order the database. A search field is a field that is needed to search the database on but does not have to be unique and does not order the database. For example, an EMPLOYEE database might be keyed on unique EMP-NUMBER. A search field might be needed on PHONE-NUMBER or ZIP-CODE. Even though the database is not ordered by these fields, they can be made search fields to query the database.

7. What does PROCOPT mean in a PCB?

   The PROCOPT parameter specifies processing options **that are allowed for this PCB when operating on a segment.**

8. The different PROCOPTs and their meaning are:

   **G - Get segment from DB
   I - Insert segment into DB
   R - Replace segment
   D - Delete segment
   A - All the above operations**

9. What are the four basic parameters of a DLI retrieval call?

   **Function
   PCB mask
   SSAs
   IO Area**

10. What are Qualified SSA and Unqualified SSA?

   A qualified SSA specifies the segment type and the specific instance (key) of the segment to be returned. An unqualified SSA simply supplies the name of the segment type that you want to operate upon. You could use the latter if you don't care which specific segment you retrieve.

11. Which PSB parameter in a PSBGEN specifies the language in which the application program is written?

   The LANG parameter specifies the language in which the application program is written. Examples:

```
LANG=COBOL
LANG=PLI
LANG=ASSEM
```

12. What does SENSEG stand for and how is it used in a PCB?

   SENSEG is known as Segment Level Sensitivity. It defines the program's access to parts of the database and it is identified at the segment level. For example, PROCOPT=G on a SENSEG means the segment is read-only by this PCB.

13. What storage mechanism/format is used for IMS index databases?

   IMS index databases must use VSAM KSDS.

14. What are the DL/I commands to add, change and remove a segment?

   The following are the DL/I commands for adding, changing and removing a segment:

   ISRT
   REPL
   DLET

15. What return code will you receive from IMS if the DL/I call was successful?

   IMS returns blanks/spaces in the PCB STATUS-CODE field when the call was successful.

16. If you want to retrieve the last occurrence of a child segment under its parent, what command code could you use?

   Use the L command code to retrieve the last child segment under its parent. Incidentally, IMS ignores the L command code at the root level.

17. When would you use a GU call?

GU is used to retrieve a segment occurrence based on SSA supplied arguments.

18. When would you use a GHU call?

GHU (Get Hold Unique) retrieves and locks the record that you intend to update or delete.

19. What is the difference between running an IMS program as DLI and BMP ?

DLI runs within its own address space. BMP runs under the IMS online control region. The practical difference concerns programs that update the database. If performing updates, DLI requires exclusive use of the database. Running BMP does not require exclusive use because it runs under control of the online region.

20. When would you use a GNP call?

The GNP call is used for Get Next within Parent. This function is used to retrieve segment occurrences in sequence subordinate to an established parent segment.

21. Which IMS call is used to restart an abended program?

The XRST IMS call is made to restart an abended IMS program. Assuming the program has taken checkpoints during the abended program execution, the XRST call is used to restart from the last checkpoint taken instead of starting the processing all over.

22. How do you establish parentage on a segment occurrence?

By issuing a successful GU or GN (or GHU or GHN) call that retrieves the segment on which the parentage is to be established. IMS normally sets parentage at the lowest level segment retrieved in a call. If you want to establish parentage at a level other than the normal level, use the P command code.

23. What is a checkpoint?

   A checkpoint is a stage where the modifications done to a database by an application program are considered complete and are committed to the database with the CKPT IMS call.

24. How do you update the primary key of an IMS segment?

   You cannot update the primary key of a segment. If the key on a record must be changed, you can DLET the existing segment and then ISRT a new segment with the new key.

25. Do you need to use a qualified SSA with REPL/DLET calls?

   No, you don't need to include an SSA with REPL/DLET calls. This is because the target segment has already been retrieved and held by a get hold call (that is the only way you can update or delete a segment).

Chapter 6 Review Questions

1.   Which of the following is NOT a valid data type for use as an identity column?

     a. INTEGER
     b. REAL
     c. DECIMAL
     d. SMALLINT

The correct answer is B. A REAL type cannot be used as an identity field because it is considered an approximation of a number rather than an exact value. Only numeric types that have an exact value can be used as an identity field. INTEGER, DECIMAL, and SMALLINT are all incorrect here because they CAN be used as identity fields.

2.   You need to store numeric integer values of up to 5,000,000,000. What data type is appropriate for this?

     a. INTEGER
     b. BIGINT
     c. LARGEINT
     d. DOUBLE

The correct answer is B. BIGINT is an integer that can hold up to 9,223,372,036,854,775,807. INTEGER is not correct because an INTEGER can only hold up to 2,147,483,647. LARGEINT is an invalid type. DOUBLE could be used but since we are dealing with integer data, the double precision is not needed.

3.   Which of the following is NOT a LOB (Large Object) data type?

     a. CLOB
     b. BLOB
     c. DBCLOB
     d. DBBLOB

The correct answer is D. There is no DBBLOB datatype in DB2. The other data types are valid. CLOB is a character large object with maximum length

2,147,483,647 bytes. A BLOB stores binary data and has a maximum size of 2,147,483,647. A DBCLOB stores double character data and has a maximum length of 1,073,741,824.

4.   If you want to add an XML column VAR1 to table TBL1, which of the following would accomplish that?

   a.   ALTER TABLE TBL1 ADD VAR1 XML
   b.   ALTER TABLE TBL1 ADD COLUMN VAR1 XML
   c.   ALTER TABLE TBL1 ADD COLUMN VAR1 (XML)
   d.   ALTER TABLE TBL1 ADD XML COLUMN VAR1

The correct answer is B. The correct syntax is:

```
ALTER TABLE TBL1
ADD COLUMN VAR1 XML;
```

The other choices would result in a syntax error.

5.   If you want rows that have similar key values to be stored physically close to each other, what keyword should you specify when you create an index?

   a.   UNIQUE
   b.   ASC
   c.   INCLUDE
   d.   CLUSTER

The correct answer is D - CLUSTER. Specifying a CLUSTER type index means that DB2 will attempt to physically store rows with similar keys close together. This is used for performance reasons when sequential type processing is needed according to the index. UNIQUE is incorrect because this keyword simply guarantees that there can be no more than one row with the same index key. ASC is incorrect because it has to do with the sort order for the index, and does not affect the physical storage of rows. INCLUDE specifies that a non-key field or fields will be stored with the index.

6. Assume a table where certain columns contain sensitive data and you don't want all users to see these columns. Some other columns in the table must be made accessible to all users. What type of object could you create to solve this problem?

    a. INDEX
    b. SEQUENCE
    c. VIEW
    d. TRIGGER

The correct answer is C. A view is a virtual table based upon a SELECT query that can include a subset of the columns in a table. So you can create multiple views against the same base table, and control access to the views based upon userid or group.

The other answers do not address the problem of limiting access to specific columns. An INDEX is an object that stores the physical location of records and is used to improve performance and enforce uniqueness. A SEQUENCE allows for the automatic generation of sequential values, and has nothing to do with limiting access to columns in a table. A TRIGGER is an object that performs some predefined action when it is activated. A trigger is only activated by an INSERT, UPDATE or DELETE of a record in a particular table.

7. To grant a privilege to all users of the database, grant the privilege to whom?

    a. ALL
    b. PUBLIC
    c. ANY
    d. DOMAIN

The correct answer is B. PUBLIC is a special "pseudo" group that means all users of the database. The other answers ALL, ANY, and DOMAIN are incorrect because they are not valid recipients of a grant statement.

8.   Tara wants to grant CONTROL of table TBL1 to Bill, and also allow Bill to grant the same privilege to other users. What clause should Tara use on the GRANT statement?

    a.   WITH CONTROL OPTION
    b.   WITH GRANT OPTION
    c.   WITH USE OPTION
    d.   WITH REVOKE OPTION

The correct answer is B. Using the WITH GRANT OPTION permits the recipient of the grant to also grant this privilege to other users. The other choices WITH CONTROL OPTION, WITH USE OPTION, and WITH REVOKE OPTION are incorrect because they are not valid clauses on a GRANT statement.

9.   Which of the following will generate DB2 SQL data structures for a table or view that can be used in a PLI or COBOL program?

    a.   DECLARE
    b.   INCLUDE
    c.   DCLGEN
    d.   None of the above.

The correct answer is C. DCLGEN is an IBM utility that generates SQL data structures (table definition and host variables) for a table or view, stores it in a PDS and then that PDS member can be included in a PL/1 or COBOL program. DECLARE is a verb used to define a temporary table or cursor. INCLUDE can be used to embed the generated structure into the program. Assuming the structure is in member MEMBER1 of the PDS, the statement EXEC SQL INCLUDE MEMBER1 will include it in the program.

10.   Assuming you are using a DB2 precompiler, which of the following orders the DB2 program preparation steps correctly?

    a.   Precompile SQL, Bind Package, Bind Plan.
    b.   Precompile SQL, Bind Plan, Bind Package.
    c.   Bind Package, Precompile SQL, Bind Plan.
    d.   Bind Plan, Precompile SQL, Bind Package.

710

The correct answer is A. The DB2 related steps for program preparation
are:

- **Precompile SQL which produces a DBRM**
- **Bind package using the DBRM**
- **Bind plan specifying the package(s)**

11.   To end a transaction without making the changes permanent, which DB2 statement
should be issued?

    a. COMMIT
    b. BACKOUT
    c. ROLLBACK
    d. NO CHANGE

The correct answer is C. Issuing a ROLLBACK statement will end a transaction
without making the changes permanent.

12.   If you want to maximize data concurrency without seeing uncommitted data, which
isolation level should you use?

    a. RR
    b. UR
    c. RS
    d. CS

The correct answer is D (Cursor Stability). CURSOR STABILITY (CS) only
locks the row where the cursor is placed, thus maximizing concurrency compared to RR or RS. REPEATABLE READ (RR) ensures that a query issued
multiple times within the same unit of work will produce the exact same results. It does this by locking ALL rows that could affect the result, and does
not permit any changes to the table that could affect the result. With READ
STABILITY(RS), all rows that are returned by the query are locked. UNCOM-
MITTED READ (UR) is incorrect because it permits reading of uncommitted
data and the question specifically disallows that.

13. To end a transaction and make the changes visible to other processes, which statement should be issued?

    a. ROLLBACK
    b. COMMIT
    c. APPLY
    d. CALL

The correct answer is B. The COMMIT statement ends a transaction and makes the changes visible to other processes.

14. Order the isolation levels, from greatest to least impact on performance.

    a. RR, RS, CS, UR
    b. UR, RR, RS, CS
    c. CS, UR, RR, RS
    d. RS, CS, UR, RR

The correct answer is A - RR, RS, CS, UR. Repeatable Read has the greatest impact on performance because it incurs the most overhead and locks the most rows. It ensures that a query issued multiple times within the same unit of work will produce the exact same results. It does this by locking all rows that could affect the result, and does not permit any adds/changes/deletes to the table that could affect the result. Next, READ STABILITY locks for the duration of the transaction those rows that are returned by a query, but it allows additional rows to be added to the table. CURSOR STABILITY only locks the row that the cursor is placed on (and any rows it has updated during the unit of work). UNCOMMITTED READ permits reading of uncommitted changes which may never be applied to the database and does not lock any rows at all unless the row(s) is updated during the unit of work.

15. Suppose you have created a test version of a production table, and you want to to use the UNLOAD utility to extract the first 1,000 rows from the production table to load to the test version. Which keyword would you use in the UNLOAD statement?

    a. WHEN
    b. SELECT
    c. SAMPLE
    d. SUBSET

The correct answer is C. You can specify SAMPLE n where n is the number of rows to unload. For example you can limit the unloaded rows to the first 5,000 by specifying:

   SAMPLE 1000

WHEN is used to specify rows that meet a criteria such as: WHEN (EMP_SALARY < 90000).

SELECT and SUBSET are invalid clauses and would cause an error.

16. Which of the following is NOT a way you could test a DB2 SQL statement?

    a. Running the statement from the DB2 command line processor.
    b. Running the statement from the SPUFI utility.
    c. Running the statement from IBM Data Studio.
    d. All of the above are valid ways to test an SQL statement.

The correct answer is D. Any of these three methods could be used to test a DB2 SQL statement.

Chapter 7 Questions with Answers

1. What does the CICS acronym stand for?

   CICS basically stands for Customer Information Control System.

2. In which programming languages can you develop CICS programs?

   COBOL, Assembler, PLI, Java and C/C++

3. What is BMS?

   BMS is Basic Map Support. It allows you to code assembler level programs to define screens.

4. In a COBOL program, how are CICS commands enclosed?

   CICS commands are coded between the EXEC CICS and END-EXEC statements. For example:

   ```
   EXEC CICS
         <command>
   END-EXEC
   ```

5. What are the CICS commands for handling screen interactions?

   `RECEIVE MAP` **– Retrieves input from the terminal.**

   `SEND MAP` **– Sends information to the terminal.**

6. What does the symbolic map include?

   It includes (in COBOL) two 01 level structures, one for input and one for output.

7. What is the difference between a physical BMS mapset and a BMS symbolic mapset?

   The physical mapset is a load module used to map the data to the screen at execution time. The symbolic map is the actual copybook member used in the program to reference the input and output fields on the screen.

8. What are the three macros used in building a mapset?

    The three macros used in building a mapset are:

    - **DFHMSD starts the mapset.**

    - **DFHMDI starts a map within the mapset.**

    - **DFHMDF defines each field within a map.**

9. What attributes are used to make a field normal intensity on the screen and protected (so that the field cannot change)?

    The attribute value for this scenario is coded as:

    ```
    ATTRB=(NORM,PROT)
    ```

10. What attribute will cause the cursor to be placed on a field when the screen displays?

    Use the IC attribute value to cause the cursor to be placed on a field when the screen displays. For example:

    ```
    ATTRB=(NORM,UNPROT,IC)
    ```

11. Which EIB field contains the length of the data that was passed to the program?

    The EIBCALEN field contains the length of the data that was passed to the program.

12. The CICS commands for processing records from VSAM files are:

    - **WRITE – Adds a record.**

    - **READ – retrieves a record.**

    - **DELETE – deletes a record.**

    - **REWRITE – updates a record.**

13. What typically causes a MAPFAIL condition when processing a RECEIVE MAP command?

When no data was sent from the screen, this raises a `MAPFAIL` **condition.**

14. The communication or COMMAREA must be included in the Linkage section of the program. What is the required name for this field in the Linkage section?

The area must be defined as the first area in the Linkage Section and must be called `DFHCOMMAREA`.

15. What is the syntax of the RECEIVE MAP command?

The syntax of the `RECEIVE MAP` **command is as follows:**

```
EXEC CICS
      RECEIVE MAP (map name)
                     MAPSET(map set name)
                     INTO (data name)

END-EXEC.
```

So for example if your mapset name is EMPMS01 and your map is named EMPM02, and your data structure name is EMPMAP1, you would code:

```
EXEC CICS
      RECEIVE MAP (EMPM02)
             MAPSET(EMPMS01)
             INTO (EMPMAP1)
END-EXEC.
```

16. When do you need to use the NEWCOPY keyword?

You use `NEWCOPY` **with** `CEMT` **to bring the latest version of the program from the loadlib into CICS.**

For example, to bring latest version of program `EMPPGM1` **into storage, issue:**

CEMT SET PROGRAM(*EMPPGM1*) NEWCOPY

17. What is the syntax of the XCTL command?

The syntax of the XCTL command is:

```
EXEC CICS
      XCTL PROGRAM (program name)
END-EXEC.
```

18. On a SEND command, what happens if you don't specify MAPONLY or DATAONLY?

Both constant data from the physical map and modifiable data from the symbolic map are sent.

19. What is the syntax of the READ command when you intend to update the record later?

The syntax of the READ command when you intend to update the record is as follows:

```
EXEC CICS
      READ FILE (file name)
      INTO (data structure name)
      RIDFLD(field name)
      UPDATE
      RESP(RESPONSE-CODE)
END-EXEC.
```

20. If an application READS a VSAM KSDS file with UPDATE, and decides not to update the record, what command can be issued to release exclusive control on the record?

Issuing an `EXEC CICS UNLOCK FILE (filename)` **command with the File or Data-set option will release control of the record. The lock will also be released if another** `READ` **is issued to move to another record.**

21. What exceptional condition will be raised if you try to insert a record into a file and a record with the same key is already there?

A `DUPREC` **condition will be raised if you try to insert a record into a file and a record with the same key is already there.**

Index

Other Titles by Robert Wingate

Interview Questions for IBM Mainframe Developers

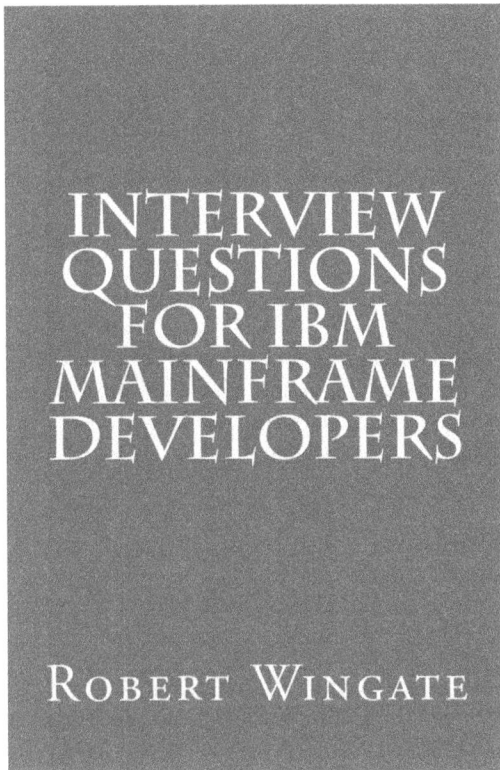

ISBN-13: 978-1539612896
This book is targeted for job seeking individuals who aspire to ace the IBM mainframe developer technical interview. Over 300 questions and answers dealing with JCL, VSAM, IMS, DB2, COBOL, PL/I and CICS. Freshen up and be prepared for your IBM mainframe developer technical interview!

CICS Basic Training for Application Developers Using DB2 and VSAM

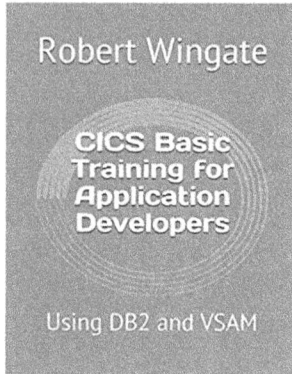

ISBN-13: 978-1794325067
This book will teach you the basic information and skills you need to develop applications with CICS on IBM mainframe computers running z/OS. The instruction, examples and sample programs in this book are a fast track to becoming productive as quickly as possible using CICS with the COBOL programming language. The content is easy to read and digest, well organized and focused on honing real job skills.

Teradata Basic Training for Application Developers

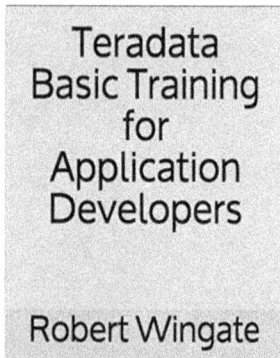

ISBN-13: 978-1082748882
This book will help you learn the basic information and skills you need to develop applications with Teradata. The instruction, examples and questions/answers in this book are a fast track to becoming productive as quickly as possible. The content is easy to read and digest, well organized and focused on honing real job skills. Programming examples are coded in both Java and C# .NET. Teradata Basic Training for Application Developers is a key step in the direction of mastering Teradata application development so you'll be ready to join a technical team.

Quick Start Training for IBM z/OS Application Developers, Volume 1

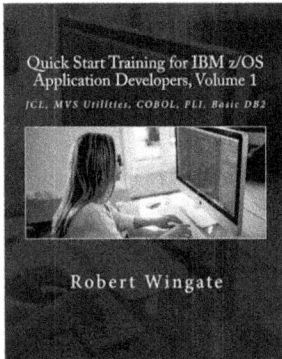

ISBN-13: 978-1986039840

This book will teach you the basic information and skills you need to develop applications on IBM mainframes running z/OS. The instruction, examples and sample programs in this book are a fast track to becoming productive as quickly as possible in JCL, MVS Utilities, COBOL, PLI and DB2. The content is easy to read and digest, well organized and focused on honing real job skills. IBM z/OS Quick Start Training for Application Developers is a key step in the direction of mastering IBM application development so you'll be ready to join a technical team.

Quick Start Training for IBM z/OS Application Developers, Volume 2

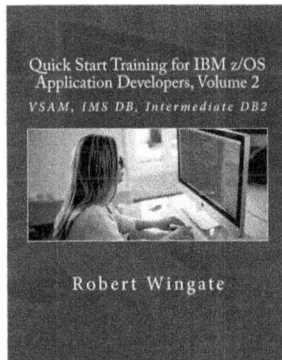

ISBN-13: 978-1717284594

This book will teach you the basic information and skills you need to develop applications on IBM mainframes running z/OS. The instruction, examples and sample programs in this book are a fast track to becoming productive as quickly as possible in VSAM, IMS and DB2. The content is easy to read and digest, well organized and focused on honing real job skills. IBM z/OS Quick Start Training for Application Developers is a key step in the direction of mastering IBM application development so you'll be ready to join a technical team.

DB2 Exam C2090-313 Preparation Guide

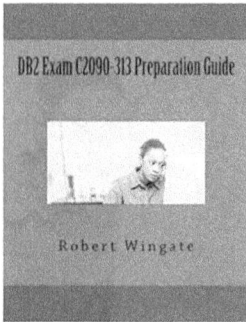

ISBN 13: 978-1548463052
This book will help you pass IBM Exam C2090-313 and become an IBM Certified Application Developer - DB2 11 for z/OS. The instruction, examples and questions/answers in the book offer you a significant advantage by helping you to gauge your readiness for the exam, to better understand the objectives being tested, and to get a broad exposure to the DB2 11 knowledge you'll be tested on.

DB2 Exam C2090-320 Preparation Guide

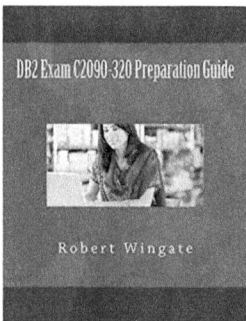

ISBN 13: 978-1544852096
This book will help you pass IBM Exam C2090-320 and become an IBM Certified Database Associate - DB2 11 Fundamentals for z/OS. The instruction, examples and questions/answers in the book offer you a significant advantage by helping you to gauge your readiness for the exam, to better understand the objectives being tested, and to get a broad exposure to the DB2 11 knowledge you'll be tested on. The book is also a fine introduction to DB2 for z/OS!

DB2 Exam C2090-313 Practice Questions

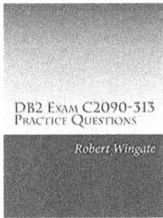

ISBN 13: 978-1534992467

This book will help you pass IBM Exam C2090-313 and become an IBM Certified Application Developer - DB2 11 for z/OS. The 180 questions and answers in the book (three full practice exams) offer you a significant advantage by helping you to gauge your readiness for the exam, to better understand the objectives being tested, and to get a broad exposure to the DB2 11 knowledge you'll be tested on.

DB2 Exam C2090-320 Practice Questions

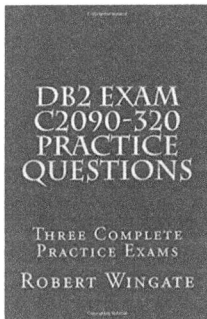

ISBN-13: 978-1539715405

This book will help you pass IBM Exam C2090-320 and become an IBM Certified Database Associate - DB2 11 Fundamentals for z/OS. The 189 questions and answers in the book (three full practice exams) offer you a significant advantage by helping you to gauge your readiness for the exam, to better understand the objectives being tested, and to get a broad exposure to the DB2 11 knowledge you'll be tested on.

IMS Basic Training for Application Developers

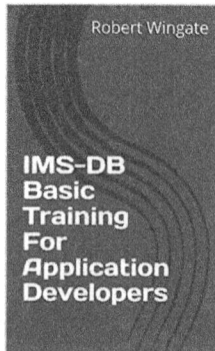

ISBN-13: 978-1793440433

This book will teach you the basic information and skills you need to develop applications with IMS on IBM mainframe computers running z/OS. The instruction, examples and sample programs in this book are a fast track to becoming productive as quickly as possible using IMS with COBOL and PLI. The content is easy to read and digest, well organized and focused on honing real job skills.

About the Author

Robert Wingate is a computer services professional with over 30 years of IBM mainframe programming experience. He holds several IBM certifications, including IBM Certified Application Developer - DB2 11 for z/OS, and IBM Certified Database Administrator for LUW. He lives in Fort Worth, Texas.

www.ingramcontent.com/pod-product-compliance
Lightning Source LLC
Chambersburg PA
CBHW081753200326

41597CB00023B/4016